GIS

Following two successful editions, the third edition of *GIS: A Computing Perspective* has been completely revised and updated, with extensive new content reflecting the significant progress that has been made in the realm of GIS within the last 20 years. Major new topics covered for the first time in this edition include: graph databases and graph query languages, ontology engineering and qualitative spatial reasoning, geosensor networks and GeoAI, decentralized computing and online algorithms, and critical GIS and data sovereignty.

Features

- Includes an entirely new chapter on AI and GIS, including ontologies and the Semantic Web, knowledge representation (KR) and spatial reasoning, machine learning and spatial analysis, and neural networks and deep learning.
- Presents new material reflecting the advances made in cloud computing, stream computing, and sensor networks, as well as extensively revised and updated content on cartography, visualization, and interaction design.
- Connects the technology to the social aspects and implications of GIS, including privacy and fair information practices, FATE (fairness, accountability, transparency, and ethics), and codes of conduct for responsible use of GIS.
- Integrates the necessary background to foundational areas, such as databases and data structures, algorithms and indexes, system architecture and AI, provided in context so readers new to those topics can still understand the concepts being discussed.
- Incorporates over 20 carefully explained spatial algorithms; over 60 inset boxes with in-depth material that enriches the central topics; and more than 300 color figures to support the reader in mastering key concepts.
- Welcomes a new coauthor, Qian (Chayn) Sun, to the third edition, who brings her expertise in topics such as web mapping, cloud computing, critical geography, and machine learning with big spatial data.

Intended for anyone interested in understanding GIS, especially students taking upper-level undergraduate and graduate courses in computer science and geography, as well as academics, researchers, practitioners, and professionals working in the field and involved in advanced GIS projects.

Matt Duckham is a professor in geospatial sciences at RMIT University, Melbourne, Australia and was awarded a PhD in GIS from the University of Glasgow, UK, in 2000. Dr. Duckham has previously held positions as Associate Dean Geospatial Sciences at RMIT University and as a professor of GIS at the University of Melbourne, Australia. He was author of the 2013 book *Decentralized Spatial Computing: Foundations of Geosensor Networks* and coauthor of more than 50 research papers in scholarly journals in the field. Together with Dr. Worboys, he is a founding editor of the *Journal of Spatial Information Science* (*JOSIS*).

Qian (Chayn) Sun is a senior lecturer in geospatial sciences at RMIT University, Melbourne, Australia and holds a PhD in GIS from Curtin University, Perth, Australia. Prior to her PhD study, she was a geospatial analyst and gained extensive experience in multi-disciplinary applied GIS research projects with both government departments and engineering consultancies in New Zealand and Australia. Dr. Sun has authored over 40 research articles in interdisciplinary journals in spatial analysis, spatial and statistical modeling, urban informatics, GeoAI, and human behavior analysis.

Michael F. Worboys is a professor of spatial informatics at the University of Greenwich, London, England and holds a PhD in mathematics from the University of Birmingham, UK. Before returning to England in 2012, he was a professor at the National Center for Geographic Information and Analysis (NCGIA) and director of the School of Computing and Information Science at the University of Maine, US. Dr. Worboys is a Distinguished Scientist of the ACM and a Life Member of the London Mathematical Society. He is also a composer and sound artist, interested in instrumental, vocal and electroacoustic music, and holds a PhD in music (composition) from the University of Durham. His instrumental and vocal music has been showcased at many venues around Britain, including the London Institute for Contemporary Arts (ICA) and the National Maritime Museum.

GIS
A Computing Perspective
Third Edition

Matt Duckham
Qian (Chayn) Sun
Michael F. Worboys

CRC Press
Taylor & Francis Group
Boca Raton London New York

CRC Press is an imprint of the
Taylor & Francis Group, an **informa** business

Designed cover image: Lou Bloomer

Third edition published 2024
by CRC Press
6000 Broken Sound Parkway NW, Suite 300, Boca Raton, FL 33487-2742

and by CRC Press
4 Park Square, Milton Park, Abingdon, Oxon, OX14 4RN

CRC Press is an imprint of Taylor & Francis Group, LLC

Second edition published by CRC Press 2004
First edition published by CRC Press 1995

Library of Congress Cataloging-in-Publication Data

Names: Duckham, Matt, 1972- author. | Sun, Qian (Chayn), author. |
Worboys, Michael, author.
Title: GIS : a computing perspective / Matt Duckham, Qian (Chayn) Sun,
Michael F. Worboys.
Description: Third edition. | Boca Raton, FL ; London : CRC Press, 2024. |
"Second edition published by CRC Press 2004. First edition published by
CRC Press 1995"--T.p. verso. | Includes bibliographical references and
index.
Identifiers: LCCN 2023004562 (print) | LCCN 2023004563 (ebook) | ISBN
9781466587199 (hbk) | ISBN 9781032539737 (pbk) | ISBN 9780429168093
(ebk)
Subjects: LCSH: Geographic information systems.
Classification: LCC G70.2 .W659 2024 (print) | LCC G70.2 (ebook) | DDC
910/.285--dc23/eng/20230505
LC record available at https://lccn.loc.gov/2023004562
LC ebook record available at https://lccn.loc.gov/2023004563

ISBN: 978-1-4665-8719-9 (hbk)
ISBN: 978-1-0325-3973-7 (pbk)
ISBN: 978-0-429-16809-3 (ebk)

DOI: 10.1201/9780429168093

Typeset in CormorantGaramond-Medium font
by KnowledgeWorks Global Ltd.

The Open Access version of chapters 1, 2, 3, 8, and 9 was funded by Matt Duckham.

Publisher's note: This book has been prepared from camera-ready copy provided by the authors.

In memory of Christian Freksa

26.06.1950–12.11.2020

Contents

Preface

Who should read this book

This book is intended for readers from any background interested in the issues that arise when computing technology meets spatial information—in other words, in GIS! You do not need to be a specialist computing scientist already; the text develops the necessary background in specialist areas, such as databases, system architecture, and AI, as it progresses. Nevertheless, some knowledge of, and interest in the basic components and functionality of computers is essential for understanding the importance of certain key issues in GIS. Where some aspect of general computing bears a direct relevance to our development, the background is given in the text.

This book can be used as a teaching text, taking readers through the main concepts by means of definitions, explications, and examples. However, the more advanced reader is not neglected, and the book attempts to highlight the threads and references that can be used to follow up on particular research topics.

Changes to the third edition

In a highly technology-led area, such as GIS, the pace of technological change sometimes feels like shifting sand under one's feet. Returning to write the third edition, it was again encouraging to see that the spine of the first edition—databases and spatial data, structures, algorithms, and indexes (Chapters 2, 3, 5, 6)—continues to stand the test of time.

While that core is still sound, the third edition has of course been extensively revised and updated, and complemented by significant new material, especially in those other areas where the field has moved more rapidly. Graph databases have now been added to the core as a major new topic. The material on time has been significantly reorganized and extended, reflecting a tighter integration of space and time in the field more generally. The fundamental material on models and modeling has also been significantly revised and reorganized to strengthen the clarity and messaging. Major new material reflecting important advances in the technology has been added in connection with architectures (Chapter 7, including cloud computing, stream computing, and sensor networks). The material on cartography and visualization (Chapter 8) has be extensively rewritten and redeveloped with largely new content. Finally, the introduction of an entirely new chapter on AI and GIS (Chapter 9, including ontologies, spatial reasoning, machine leaning and spatial analysis, and deep learning) is a reflection of the rapid advances in this area over the past decade.

In addition to the changes in content, we have striven to make further improvements to the format to produce a more attractive and readable book. The format for the book has been completely revised in full color, in particular with all figures either new or completely regenerated. The text has likewise been completely rewritten with the continuing aim of making the book more accessible to an ever-wider audience. Every chapter begins with a new summary, outlining the major ideas and learning objectives in that chapter. At its close, every chapter ends with some more personal reflections and perspectives on the topic from the authors.

Before sitting down to write a third edition, it was essential for us to move the book towards open access. We are very grateful to the Taylor & Francis team for working with us to enable five chapters—half the book—to be made freely available online and open access. Another important change, which may not be immediately obvious, is the conscious effort to rebalance the list of references. The cited references in the second edition were overwhelmingly to male authors—of more than 300 references in the second edition, over 86% were authored exclusively by men. In writing the third edition, we have attempted to address this stubborn bias by actively seeking out the abundance of excellent research by women GI scientists. It has been a joy to read and reread these works of women researchers, who continue to be systematically under-cited and under-recognized as they have been for the entire history of the field.[1]

[1] Tracking gender in bibliographies is not straightforward, but we estimate over 50% of cited references in this third edition are by female authors or include female co-authors. Nevertheless, the book remains a long way from gender parity (i.e., in terms of the total proportion of male authors, as opposed to the proportion of references with exclusively male authors) and still offers no visibility for the contributions of non-binary GI scientists.

Finally, following the pattern Mike Worboys began in 2004 by inviting Matt Duckham to become a co-author of the second edition, Matt and Mike welcome Qian (Chayn) Sun on to this third edition as a third author. With a background in geography and GIS, Chayn has brought her own fresh perspective and distinct expertise to the third edition, in particular, in web mapping, cloud computing, critical geography, and machine learning with GIS. We hope you will agree that this third edition has succeeded in maintaining the high standards set by *GIS: A Computing Perspective* back in 1994!

Formatting used in this book

Several formatting conventions, continued from the second edition, have been used in this book. Material that is relevant to the main themes in the text, but not essential to the reader, is clearly separated out in purple inset boxes, usually at the top of a page. Boxes typically contain interesting asides, more challenging material, or background to a topic, as well as references and links that readers may wish to pursue. With over 60 such boxes, a complete list is included in the front matter to the book for ease of reference. A list of insets is included in the book's front matter.

Throughout this book, we have used margin text to allow rapid reference to important terms. When an important term is first defined or introduced, that term will appear in the margin. A corresponding entry can be found in the index, with the page reference in bold typeface. This enables the reader to use the index as an extensive glossary of more than 1200 terms used in this book. Each index term has at most one bold typeface page reference, and a term can be rapidly located within a page by finding the corresponding margin entry. In addition to normal- and bold-typeface index entries, those index entries that appear in italics refer to terms that appear within an inset box.

Structure of this book

Figure 1 indicates the overall structure of interdependencies between chapters. Readers may find it helpful to refer to Figure 1 to tailor their use of this book to their own particular interests and prior background.

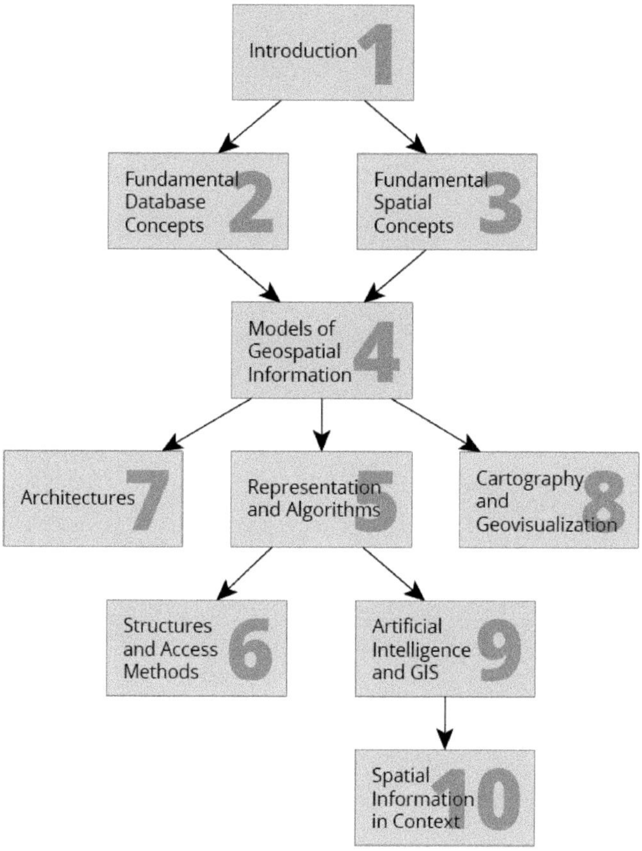

Figure 1: Relationships between chapters

Chapter 1 motivation and introduction to GIS; preparatory material on general computing; and an overview of what makes "spatial special";

Chapters 2–3 relational and graph databases; conceptual data modeling; foundations and formalisms for spatial concepts;

Chapter 4 high-level modeling of space and time; object- and field-based models; time in GIS;

Chapters 5–6 exposition of the core material; spatial algorithms and data structures; spatial indexes and access methods;

Chapter 7 GIS system architecture and distributed systems; web mapping; stream computing and sensor networks;

Chapter 8 graphic design and cartography; GIS interface and interaction design; visualization of geospatial data;

Chapter 9 ontology engineering and spatial reasoning; machine learning and deep learning; "GeoAI"; and

Chapter 10 uncertainty and imperfection in spatial information; location privacy; critical GIS.

Online resources

The website that accompanies this book can be found at:

http://gisacp.duckham.org

The resources at this site are constantly under development, but they include resources such as sample exercises, lecture slides and notes, open-source computer code, sample material, useful links, errata, and contact information. We welcome suggestions from readers as to resources that we should include on the website, or indeed any feedback or comments on the book itself. Matt can be contacted on Mastodon on @mduckham@mastodon.au or Twitter at @geospatial_md; other up-to-date contact information can be found on the website.

ACKNOWLEDGMENTS

Matt and Chayn are very grateful for the support of RMIT University friends and colleagues in Melbourne, including members of the Geospatial Sciences team, the STEM College, and the Research and Innovation teams led by Swee Mak and Calum Drummond.

Special thanks are due to Sarah Rigbye and Prag Sapkota who assisted with the conversion of second edition chapters and the development of some of the figures. The original characters adapted for use in many figures are from a series by Nadia Snopek. The authors are grateful to Mark Newman for permission to use his cartogram on page 320. We would also like to thank the Taylor and Francis production and editorial team for their support, and in particular Irma Britton for her patience and persistence over more than a decade that this edition was in the making.

Finally, Matt will be forever grateful to his family in Melbourne—Laurie, Nola, Tess, Sophie, Eloise, James, Russell, Josh, Frankie, Nina, Allison, Andy, Aidan, and Anja—for their love, patience, and encouragement, and most especially to Ingrid for sharing her joyful laughter, wise counsel, and inexhaustible kindness.

Matt Duckham, Chayn Sun, and Mike Worboys, January 2023

Artwork

I am a part Indigenous Australian artist originally from Grafton, Northern NSW, identifying as Bundjalung for the land I was born and grew up on, as well as Anaiwan and Kamilaroi—the nations of my ancestors. I draw my artistic inspiration from nature and my everyday surroundings, highlighting the similarities in our life's journeys with that of the universe and everything living within it.

This artwork reflects the geographical lines and pathways of the lands we walk upon every day. What once were carved by the oceans currents and tides, are now the hills and valleys we travel through. Much like the paths we take with every decision we make, the winds and waters travel scientifically the same.

Lou Bloomer, June 2023

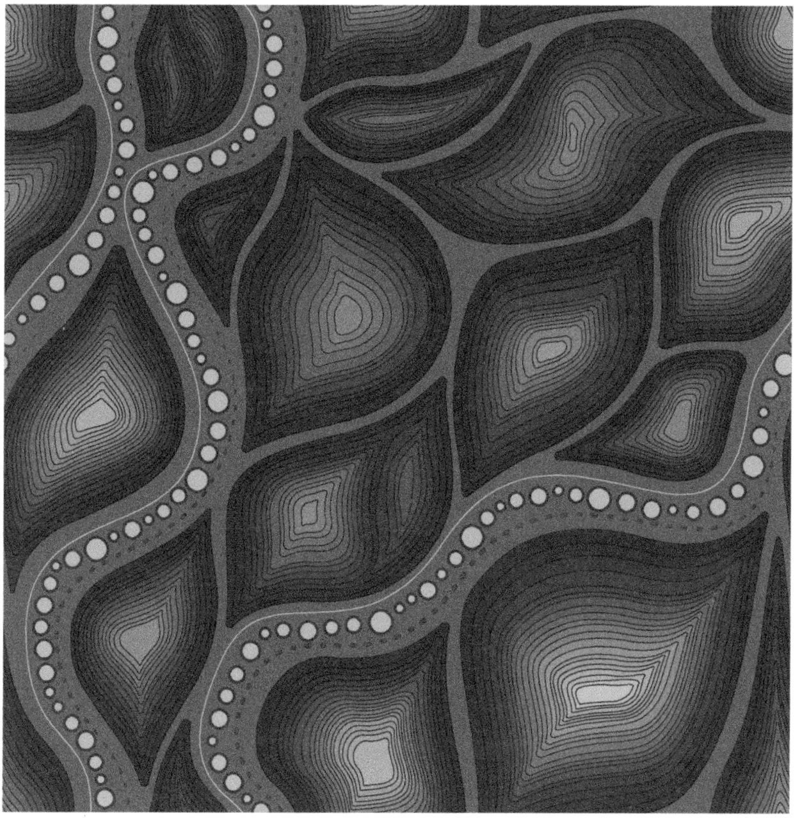

List of Inset Boxes

List of Algorithms

INTRODUCTION

1

Since their emergence in the 1960s, geographic information systems (GIS) have evolved into an array of sophisticated technologies for working with spatial data. This chapter introduces GIS: what are they and what can one use them for; what technologies and concepts are they based upon; and what makes them different and worthy of special study. Through this Introduction, you will learn to:

- define and describe the main **functions of a GIS** and discuss example **applications** of the technology;
- summarize the key features of the general **computing technologies** that underpin any GIS;
- explain the particular importance of the **database**, which lies at the heart of every GIS; and
- reflect on what makes GIS special, and more generally on why "**spatial is special.**"

SECTIONS

WHAT makes GIS special? Most people who work with geographic information systems have asked themselves this question at one time or another. This chapter starts to answer the question by describing the field of GIS against the general background of computing. First, we define the terms "information system" and "GIS," identifying what distinguishes geographic information systems from other information systems. The discussion outlines the main functions of a GIS, with a particular emphasis on those vital functions provided by the *database* (Section 1.1). Then, in Section 1.2, we look at what we can do with a GIS. Presenting some typical, example applications provides a motivation for studying GIS. Spatial data plays a key role in any application of GIS, as do the different ways we use that data to model the world around us, explored in Section 1.3. The fundamental computing technologies and concepts that enable GIS are briefly introduced in Section 1.4. Taking a step back, the chapter concludes with an analysis of what makes spatial—not just GIS—special (Section 1.5).

1.1 *What is a GIS?*

A good starting point for defining a geographic information system is to look at a general definition of an *information system*. An information system is an association of people, machines, data, and procedures working together to collect, manage, analyze, and distribute information of importance to individuals or organizations. The term "organization" is meant here in the widest sense, to encompass corporations, governments, and societies as well

information system

as diffuse groupings, such as global networks of researchers with common interests or collections of people looking at the environmental impacts of a new development. The World Wide Web (WWW) is an example of an information system. The WWW comprises data (web pages) and machines (web servers and web browsers), but also the many people across the world who use the WWW and the procedures for storing, retrieving, finding, and maintaining information on the WWW.

A GIS is a special type of information system concerned with *geographically* referenced data. Specifically:

geographic information system

> A geographic information system is a computer-based information system that enables the capture and modeling, storage and retrieval, communication and sharing, manipulation and analysis, presentation and exploration of geographically referenced data.

spatial data

Underlying the need for a GIS is the fact that *spatial data*—i.e., data about geographic spaces—requires special handling, management, technologies, and concepts, when compared with other sorts of data (such as found in a library or banking information system, for example). We sometimes also use the term "geospatial" to emphasize the "geographically referenced" aspect of GIS: that data in a GIS is not only spatial, but it is also referenced to the surface of the Earth. Several other related terms and types of information system are relevant to GIS, discussed in Box 1.1 on page 4.

geospatial

1.1.1 The "shape" of GIS

The world around us is both spatial and temporal, so we have a need for information that has spatial and temporal dimensions. Future decisions that affect us all—for example, in planning new roads or cities, formulating agricultural strategies, and locating mineral extraction sites—rely upon properly collected, managed, distributed, analyzed, and presented spatial and temporal information. A GIS may be thought of as a tool that is able to assist us with these tasks. This tool relies on several underlying functions, summarized schematically in Figure 1.1.

database

Data storage and retrieval functions At the heart of any GIS is the *database*. A database is a collection of data organized in such a way that a computer can efficiently store and retrieve the data. As we shall see, a database must also be *reliable* (continue to operate even if unexpected events occur, such as power failures); correct and *consistent* (automatically detect and protect data from errors); technology *independent* (with standardized access mechanisms that insulate access from rapidly evolving technological details); and *secure* (ensure sensitive data can be protected, with different access levels for different users). These fundamental capabilities are introduced in Chapter 2. GIS require databases with further specialized capabilities for handling spatial data, which are discussed throughout the book.

Efficient storage of data depends on not only properly structured data in the database, but also optimized indexes and algorithms for retrieval oper-

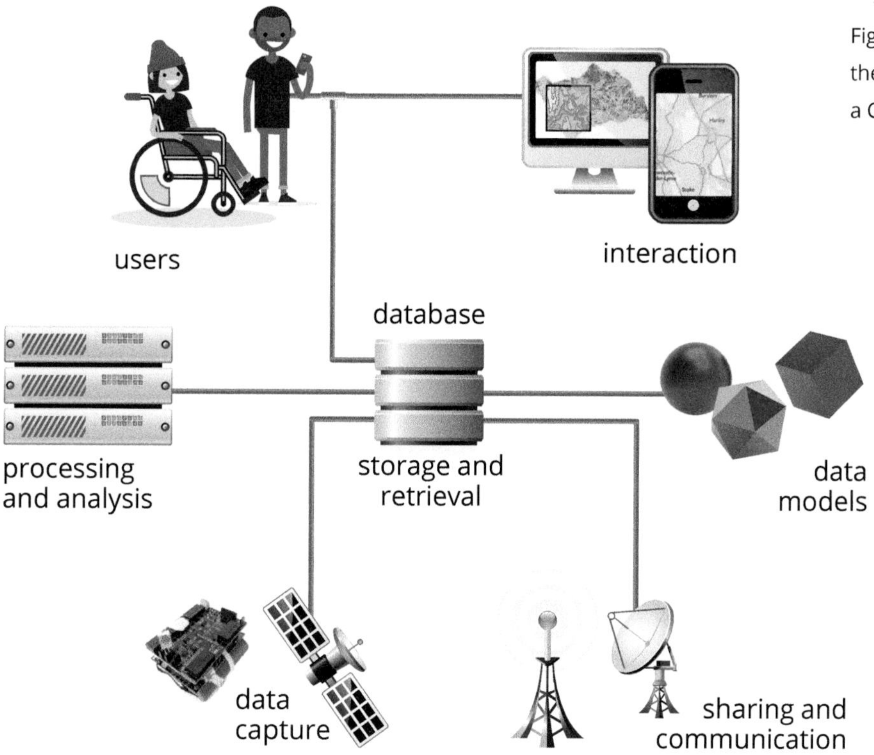

Figure 1.1: Schematic of the functional elements of a GIS

ations. Storage, retrieval, and performance raise many interesting questions for spatial data and will be a further main theme of this text (Chapter 6), also discussed further below in Section 1.4.

Data analysis and processing functions A GIS needs to have sufficiently complete functionality to provide the higher-level analysis and decision support required within an application domain. For example, a GIS that is used for managing utilities and critical infrastructure, such as electricity distribution, water and sewerage, or transportation networks, will require network processing operations (e.g., optimal routing between nodes, connectivity checks, and so forth). A GIS for urban planning will require sophisticated geometry operations and perhaps 3D modeling capabilities.

Identifying and specifying a general set of primitive operations required by a generic GIS is a major concern of the book, covered primarily in Chapter 5. However, the discussion in this book does not address whether specific spatial analyses are suitable for particular applications: those questions we leave to application domain experts.

Data capture functions The process of collecting data from observations of the physical environment is termed *data capture*. The primary source of data for a GIS is from *sensors* that measure some feature of the geographic environment. Surveyors, for example, use sensors to measure distances and

data capture

Box 1.1: GIS terminology

There are several terms in common usage that are closely related to GIS. The term *spatial information system* (SIS) is practically synonymous with GIS. Using "spatial" in place of "geographic" or "geospatial" highlights the broader connections of GIS to concepts and technologies that are not limited to geographic scales, such as room-sized or tabletop spaces. Similarly, a *spatial database* is broadly synonymous with the terms *geographic database* and *geodatabase* and provides the database functionality for a GIS. An *image database* is fundamentally different from a spatial database or geodatabase in that the images have no structural interrelationships, such as *topological* features discussed at length in this book. Image databases use quite different technology to power applications such as medical imagery and image search, although some images may also be geographically referenced (such as a database of satellite imagery). Computer-aided design (CAD) has some elements in common with GIS, and historically some GIS software packages evolved from CAD software. CAD also differs from GIS in that data need not be geographically referenced, and it can instead use coordinate systems local to that design. Closely related to CAD, *building information modeling* (BIM) likewise does not require models to be geographically referenced, and focuses more specifically on the 3D digital representation of physical buildings. The combination of geographic references and structural interrelationships between data means that GIS software is used with larger data sets and/or more complex data models than CAD, BIM, or image databases. Conventionally, and in this book, the term GIS may also be used to refer to "the field of GIS," as well as the information system itself. However, the terms *geographic information science* (GI science), *geospatial science*, or *geoinformatics* are more frequently used, and arguably more appropriate, to describe the systematic study of geographic information and geographic information systems.

angles between features located on the Earth's surface. Satellite imagery uses sensors to measure the electromagnetic radiation reflected or emitted from the Earth's surface at different wavelengths. The field of *Earth observation* (EO) more broadly is concerned with all aspects of sensing data about the Earth, and in particular using space-based remote sensing data sources.

Earth observation

Recent technological advances in MEMS (microelectromechanical systems) have resulted in an explosion in the variety of sensors available today. As you read this book, it is likely that the room you are in or the computing devices around you (perhaps even in your pocket) are bristling with sensors of different types. Miniaturized digital sensors capable of measuring temperature, light, sound, magnetic fields, pressure, and even chemicals are increasingly common in a range of applications and computing devices. For example, *sensor networks* are networks of miniaturized, sensor-enabled computing devices capable of monitoring a wide variety of environmental stimuli, explored further in Chapter 7. Naturally, sensors capable of determining location are of particular importance to GIS, also explored in more detail in Chapter 7.

sensor network

Indeed, technological advances in sensors are being matched by social advances in data capture. An increasingly important source of data is that generated by individual *people* and freely contributed and shared via online repositories, termed *user-generated content* (UGC). Much of this user-generated content is geographically referenced, sometimes also called *volunteer geographic information* (VGI). The motivations for people to capture and share such data are varied, but the impact of geographically referenced user-generated content over recent years has been immense (see Box 1.2 on page 6). As a result, GIS

user-generated content

often need to store, integrate, and process sensor-based information from many different sources.

A secondary data capture stream for GIS is from a legacy data source, such as paper maps. Maps combine the functions of data presentation and data storage; functions that GIS keep separate. Although most maps today are generated from digital spatial data, much of this data was originally captured from paper maps, and many historical maps are still in hard copy format. Converting spatial data stored in a paper map into a form that can be stored in a GIS can be difficult and costly. Automatic conversions, such as scanning a map, cannot easily capture the complex structure of the map, and the results are more similar to an image database than a GIS. Manual conversions, for example, where humans trace the features of a map using a digitizer (Figure 1.2) can capture more structure, but are time-consuming and laborious.

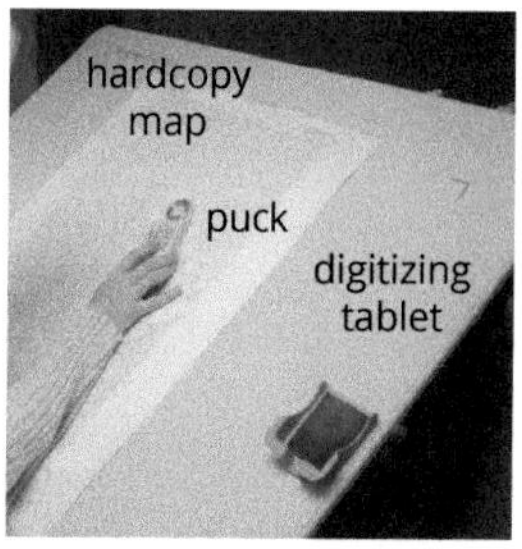

Figure 1.2: Digitizing tablet for manual map digitization

Data sharing functions GIS first began as standalone software in desktop computer labs, and the term GIS is still the label commonly used to identify the software package.[1] Today, however, most applications of GIS rely on the combination of many different software and hardware components, including spatial database servers, web browsers, smartphones and mobile devices, remote sensing satellites and sensor networks, not to mention the different people and organizations that must work together through that application.

[1] Examples of common GIS software packages today include ArcGIS, QGIS, GRASS GIS, and OpenJUMP GIS.

The different components are frequently not co-located, with data and computation *distributed* across many different physical locations. There are many natural reasons for data and systems to be distributed in this way. Data may be more appropriately associated with one site rather than another, allowing a greater degree of autonomy and easier update and maintenance. For example, details of local weather conditions may be more usefully held at a local site where local control and integrity checks may be maintained. Another advantage of a distributed database is increased reliability; failure at one site will not mean failure of the entire system. Distributed databases may also offer improved performance, sharing the computational load across multiple devices.

Because components may not be physically co-located, most GIS applications rely on a digital communications network (see Section 1.4). For example, in order to access real-time information about the best route to drive home from work, allowing for traffic jams, road works, and changing road conditions, several different distributed information system components must work together with wired and wireless communications networks. Consequently, a fundamental characteristic of any GIS is the capability to *share* data between different information systems, or between different components within a single information system. Understanding and achieving data sharing are key topics in this book, discussed under the general heading of *system architecture* (Chapter 7).

Box 1.2: User-generated content

Perhaps one of the best-known examples of user-generated geographic content is Open Street Map (OSM). OSM combines a conventional spatial database with specially designed software that enables anyone to create and edit topographic data. Founded in 2004, by 2013 OSM contained more than 100 million lines and polygons, made up of more than 1 billion points, generated by more than a million users. In 2022 it had grown to more that 1.5 TB (terabytes) of data. The importance of user generated content first achieved global prominence in January 2010. Within days of a devastating earthquake in Haiti, mappers both in Haiti and around the world had helped to generate previously unavailable detailed and up-to-date maps of the worst-affected areas (see below). These maps were used by many of the humanitarian agencies to aid in their response to the disaster.

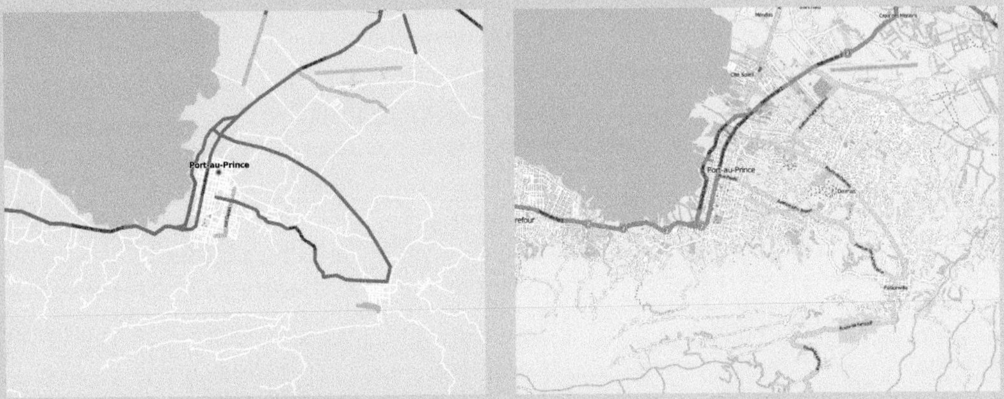

12 January 2010 14 January 2010

OSM maps of Port-au-Prince in the days following the 2010 Haiti earthquake (Source: Mikel Maron)

Modeling, interaction, and user functions Data models provide the bridges between the data and analysis functions of an information system and the users and applications of that system. Some information systems only require relatively simple data models. In a library information system, for example, data about books, users, reservations, and loans is structured in a relatively straightforward way. GIS applications, however, demand more complex data models. One of the main issues addressed in this book is the provision of facilities for handling these complex data models. Chapter 2 shows how data modeling is at the core of database design, while Chapters 3 and 4 examine in more detail what makes spatial data and spatial data models special. In preparation, Section 1.3 below introduces some fundamental data modeling concepts.

The bridges that data models provide are critical, because GIS are ultimately as useful as the decisions they support. Effective GIS require the capability to communicate data and analysis effectively to people. Many non-spatial information systems have capabilities for data presentation based around tables of data, numerical computation, and textual commentary. While these forms are also required to support decision-making using a GIS, GIS also require a whole new range of capabilities. These include *cartographic* (map-based) presentation as well as more dynamic, immersive, and exploratory forms of interaction. Interaction with data and analyses results using a GIS therefore takes us beyond the scope of many traditional software and database systems and is a further major focus of this text (Chapter 8).

1.2 *GIS applications*

As highlighted above, a GIS is an information system that has some special spatial capabilities. To demonstrate the range of capabilities of a GIS, a series of example applications are described in this section. The examples here are merely to illustrate prototypical applications, rather than in-depth explorations of application areas, however.

The applications have been chosen for a region of England, familiarly called "The Potteries" due to its dominant eponymous industry (Figure 1.3). The Potteries comprise the six pottery towns of Burslem, Fenton, Hanley, Longton, Stoke, and Tunstall, along with the neighboring town of Newcastle-under-Lyme (see Figure 1.4). The Potteries region developed rapidly during the 18th century at the vanguard of the English industrial revolution. The local communities produced ware of the highest standard (for example, from the potteries of Wedgwood and Spode, founded in the late 1700s) from conditions of poverty and cramp. The region's landscape is scarred by the extraction of coal, ironstone, and clay. In the 20th century, The Potteries declined in prosperity although the area is now a focus for regeneration given its historic and geographic centrality in the UK.

Figure 1.3: Historic bottle kilns at the famous Gladstone Pottery Museum in The Potteries today

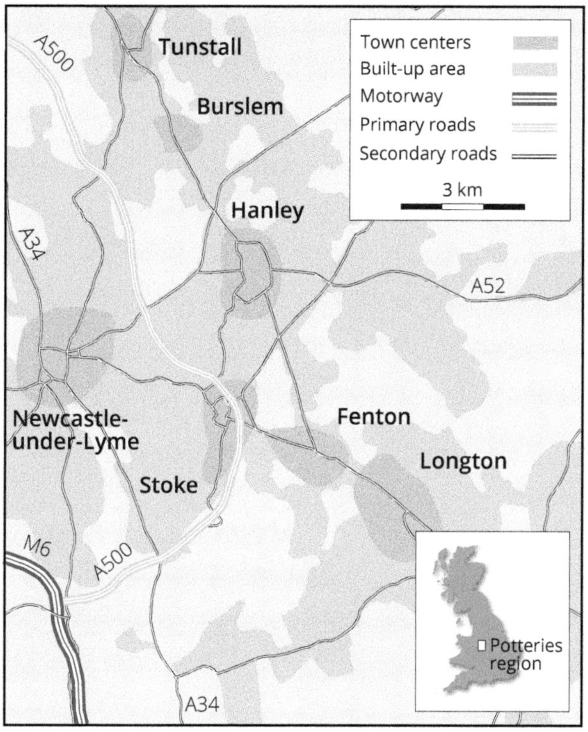

Figure 1.4: The Potteries region

1.2.1 Resources inventory: A tourist information system

resource inventory

The Potteries, because of its past, has a locally important tourist industry based upon the industrial heritage of the area. A GIS may be used to support this, by drawing together data on cultural and recreational facilities within the region, and combining this data with details of local transport infrastructure and hotel accommodation. Such an application is an example of a simple *resource inventory*. The power of almost any information system lies in its ability to relate and combine data gathered from disparate sources. This power is increased dramatically when the data is geographically referenced, provided that the different sources of spatial data are made compatible through some common spatial unit or transformation process. Figure 1.5 shows the beginnings of such a system, including some of the local tourist attractions, the major road network, and built-up areas in the region.

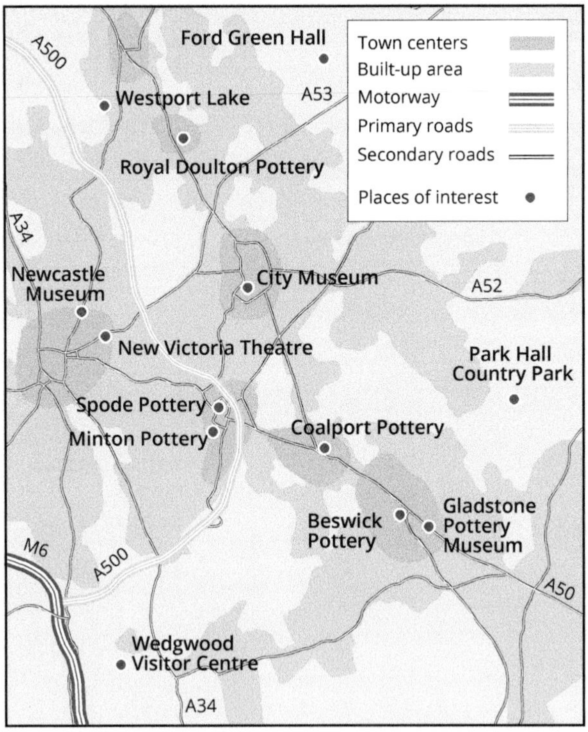

Figure 1.5: Places of interest in The Potteries region

1.2.2 Network analysis: A tour of The Potteries

network analysis

Network analysis is one of the cornerstones of GIS functionality. Applications of network analysis can be found in many areas, from transportation networks to the utilities. As simple example, the major potteries in The Potteries area are famous worldwide. Many of these potteries offer factory tours and have factory outlet stores. The problem is to provide a route using the major road network, visiting each pottery (and the City Museum) only once, while minimizing the traveling time. The data set required is a travel-time network between the potteries: an example is given in Figure 1.6.

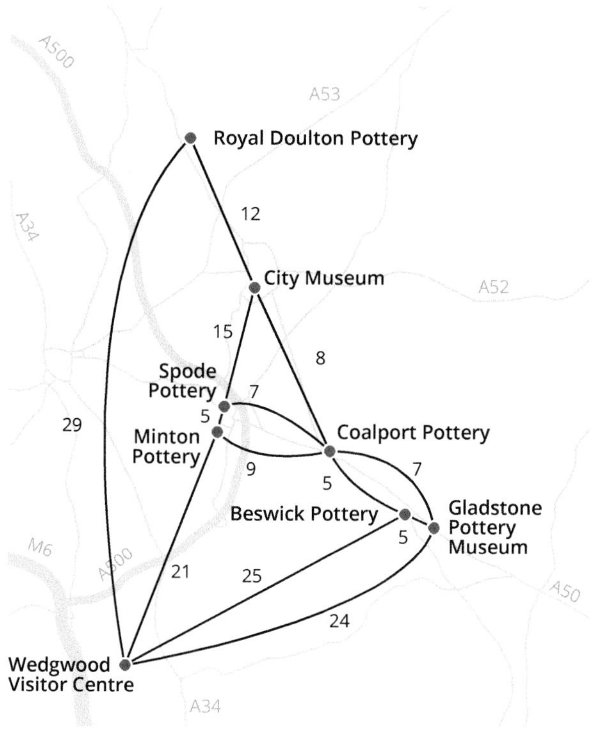

Figure 1.6: Travel network based upon travel times (in minutes)

A travel-time network such as that in Figure 1.6 can be used as a basis for generating efficient routes between the different attractions. The analysis might be dynamic, assigning weights to the edges of the network and calculating optimal routes depending upon changeable road conditions. For instance, one specific relevant analysis is the *traveling salesperson algorithm*, which constructs a minimum weight tour through a network that visits each node at least once.[2]

[2] Can you identify the quickest tour of all the potteries in Figure 1.6? The answer is not straightforward to construct and is discussed further in Chapter 5.

1.2.3 Distributed data: Navigating around The Potteries

Much of the data used in a GIS is located in different formats at physically remote locations. A GIS needs to be able to overcome these barriers to data sharing. Continuing the tourist information example, the resources inventory and network analysis above might rely on sharing and analysis of disparate spatial data about the locations of places of interest; transport infrastructure, connections, and congestion; and cultural, hospitality, and recreational resources. These resources are commonly held by different organizations at different locations. Base map data may be held in one place, such as by an online mapping service or national mapping agency. Data about transport infrastructure might be compiled by the local government or held by individual bus or train companies. The tourist information bureau will hold some data about the local amenities, although more will often reside with the individual amenities themselves, such as museums or hotels.

Figure 1.7 shows a schematic version of a distributed information system that might be used as the basis of a Potteries tourist application. Before a tourist visiting The Potteries can receive navigation directions and information about local attractions (for example, on their smartphone), data from all these different sources must be integrated, organized, and processed. Since the tourist will not usually be a GIS expert, these complex tasks will normally be done on behalf of the tourist by some tourist service provider. The service provider might gather all the information needed, either in advance or dynamically when requested, and perform the network analysis necessary to find the best route for a particular user at a particular time and location. Systems that can present data relevant to a user's current location, such as navigation instructions, are called *location-based services*, discussed in more depth in Chapter 7. Although the information actually presented to a tourist at any moment in time may be simple (such as "turn left" indicators, Figure 1.7), the task of integrating data from different sources may be complex.

Figure 1.7: Schematic view of a distributed tourist information system

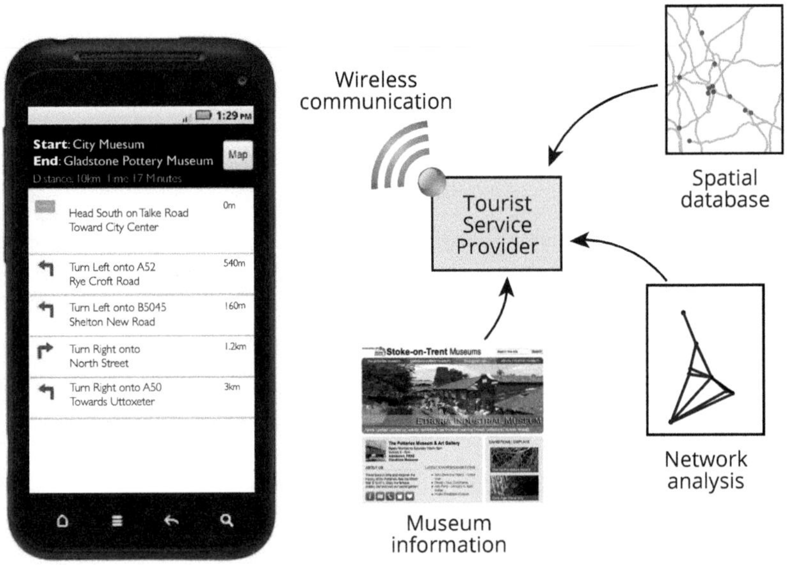

1.2.4 *Terrain analysis: Siting an opencast coal mine*

Terrain analysis operates upon data about topographic elevations across a site. Basic information about degree of slope and direction of slope (termed *aspect*) can be derived from such data sets. A more complex type of analysis, termed *visibility analysis*, concerns the visibility between locations and the generation of a *viewshed* (a map of all the points visible from some location).

aspect

visibility analysis

viewshed

The applications of terrain analysis are diverse. For our example, the search for new areas of opencast coal mining in The Potteries conurbation has in the past resulted in much interest from local communities, who might be concerned about the effects of such operations. One factor in this complex question is the visual impact of proposed opencast sites. Visibility analysis can be used to evaluate visual impact, for example, by measuring the size of the local population within a given viewshed. Sites that minimize this population may be considered more desirable.

The terrain surface of the area around Biddulph moor may be represented using a contour map, as in Figure 1.8a. Figure 1.8b shows a perspective projection of the same surface shown in Figure 1.8a. Such projections provide a powerful depiction of the terrain. Figure 1.8c shows the same surface as before, this time draped by the viewshed. The darker shaded regions give the area from which the marked point would not be visible. If we assume the point represents the location of the opencast mine, then the lighter areas provide a first approximation to the visual impact of the mine. Of course, a real case would take into account much more than just the visibility of a single point, but the principle remains.

Figure 1.8: Contour map, perspective projection, and viewshed for the marked point on terrain surface

1.2.5 *Layer-based analysis: Extraction sites for mineral ore deposits*

The Potteries area is rich in occurrences of superficial and bedrock sand and gravel, although few such sites have been worked in recent times. Determining the potential of different locations for sand and gravel extraction demands the collation and analysis of data from a variety of sources. Geological data describing the location of appropriate deposits is, of course, needed. Other important considerations are local urban structure (e.g., urban overbuilding), water table level, transportation network, land prices, and land zoning restrictions. Figure 1.9 shows a sample of the available data overlaid on a single sheet, including data on built-up areas, known sand and gravel deposits, and the major road network.

Layer-based analysis results from posing a query such as: "Find all locations that are within 0.5 km of a major road, not in a built-up area, and on a sand/-gravel deposit." Figure 1.10 illustrates the construction of an answer to this question. The shaded areas in Figure 1.10a show the region within 0.5 km of a major road (not including the motorway), termed a *buffer*. The stippled areas in Figure 1.10b indicate known sand and gravel deposits, and in Figure 1.10c the shaded areas indicate locations that are not built up. Figure 1.10d shows the overlay of the three other layers, and thus the areas that satisfy our query. The analysis here is simplistic. A more realistic exercise would take into ac-

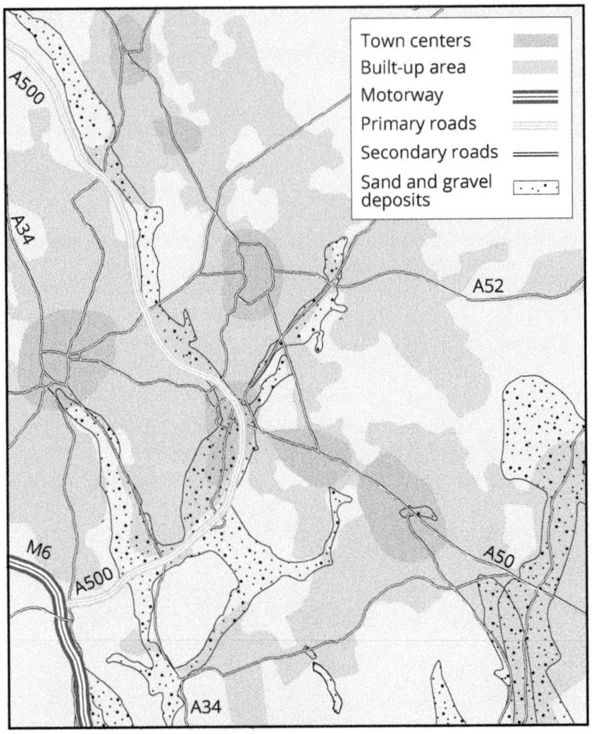

Figure 1.9: Locations of sand and gravel deposits in The Potteries region

count other factors, like the grading of the deposit, land prices, and regional legislation. However, the example does show some of the main functionality required of a GIS engaged in layer-based analysis, including:

buffering

- The formation of areas containing locations within a given range of a given set of features is termed *buffering*. Buffers are commonly circular or rectangular around points, and corridors of constant width about lines and areas.
- The combination of one or more layers into a single layer that is the union, intersection, difference, or other operations applied to the input layers is termed *overlay*.

overlay

Layer-based functionality is explored further in the context of field-based models and structures later in the book (Chapter 4).

1.2.6 Location analysis: Locating a clinic in The Potteries

Location problems have been solved in previous examples using terrain models (opencast mine example) and layer-based analysis (estimating the potential of sites for extracting sand and gravel). Our next example is the location of clinics in The Potteries area. A critical factor in the decision to use a particular clinic is the time it takes to travel to it. To assess this, we may construct the "neighborhood" of a clinic, based upon positions of nearby clinics and travel times to the clinic. With this evidence, we can then support decisions to relocate, close, or create a new clinic.

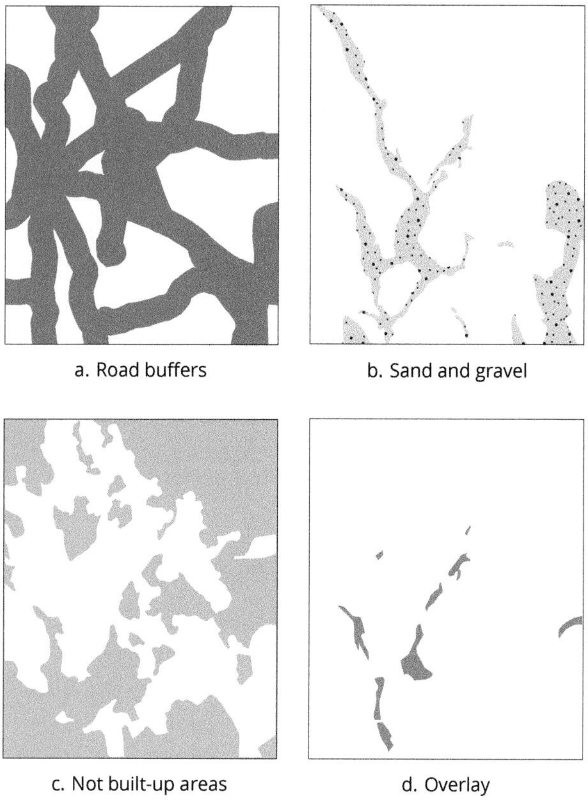

Figure 1.10: Layer-based analysis to site a mineral ore extraction facility

Figure 1.11 shows the idealized positions of clinics in The Potteries region. Assuming an "as the crow flies" travel time between points (i.e., the time is directly related to the Euclidean distance between points), Figure 1.11a shows lines connecting locations that are equally far from the clinic in terms of travel time, termed *isochrones*. It is then possible to partition the region into areas, each containing a single clinic, such that each area contains all the points that are nearest (in travel time) to its clinic, termed *proximal polygons* (Figure 1.11b). Of course, we are making a simplistic assumption about travel time. If the road network is accounted for in the travel-time analysis, then the isochrones will no longer be circular and the areal partition will no longer be polygonal. This more general situation is discussed later in the book.

isochrone

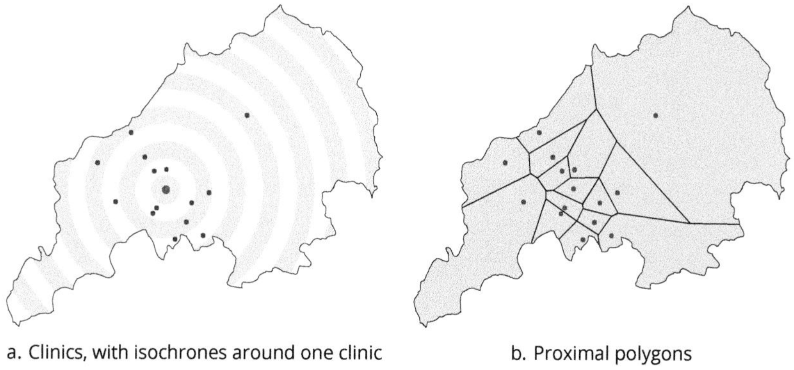

Figure 1.11: Potteries clinics and their proximal polygons

1.2.7 *Spatiotemporal information: Thirty years in The Potteries*

Spatial data sometimes becomes equated with purely static data, thus neglecting the importance of change and time. However, data about the world can always be references to three kinds of dimensions: what (attribute), where (space), *and* when (time). Our next example illustrates the spatiotemporal functionality needed by GIS.

The main period of industrial activity in The Potteries is long since past. The history of the region in the latter half of the 20th century has been one of industrial decline. Figure 1.12 shows the Cobridge area of The Potteries, recorded in snapshot at two times: 1878 and 1924. It is clear that as time has passed many changes have occurred, such as the extension of residential areas in the northwest and southeast of the map. Examples of questions that we may wish to ask of our spatiotemporal system include:

- Which streets have changed name in the period of 1878–1924?
- Which streets have been constructed between 1878 and 1924?
- In what year is the existence of the Cobridge Brick Works last recorded in the system?
- What is the overall spatial pattern of change in this region between 1878 and 1924?

a. 1878 b. 1924

Figure 1.12: History of the Cobridge area, recorded in snapshots at times 1878 and 1924 (Source: Ordnance Survey)

1.2.8 *Digital twins: Smart cities of the future*

GIS as a technology has its origins more than half a century ago. This book (with its own origins more than quarter of a century ago!) has as its focus that growing core of knowledge that lies at the stable foundations of this rapidly advancing technology. Nevertheless, the technology continues to evolve, if anything more rapidly today than ever before. One of the emerging applications of GIS likely to be increasingly important over the coming decade is *digital twins*. A digital twin is a virtual representation of part of our phys-

digital twin

ical world. Many governments around the world are building increasingly sophisticated digital twins that capture not only the appearance of precincts and cities, but are also able to support predictive simulations and fed by near real-time data from sensors, satellites, and citizens (termed *smart cities*).

smart city

As such, digital twins bring together many of the more advanced functions of a GIS in one application. A digital twin of The Potteries does not exist today, at the time of writing, but one is almost certain to be developed in the coming few years, perhaps by the time you read this. Like any digital twin, a Potteries digital twin will need not only 2D mapping capabilities, but 3D representations of the physical topography and buildings (such as in Figure 1.13). It will need to be highly interactive with the ability to support exploration by decision makers and citizens of different views and analyses of the conurbation, populated with frequently updated or even real-time data about the city. It will also need the capability run predictive models, such as of traffic or potential epidemic spread. All this must be supported by rapid storage, update, and retrieval of the spatial data that underpins these capabilities.

Figure 1.13: Digital twins are sophisticated digital representations of parts of our physical world, such as a precinct, city, or nation

1.2.9 *Summary of analysis and processing requirements*

The examples of applications reviewed in this section demonstrate some of the specialized processing functionality that a GIS needs to provide, usually termed *spatial analysis*. A GIS must handle information about phenomena embedded in the geographic world and having not only multiple spatial and even temporal dimensions, but also structural placement in multidimensional

spatial analysis

geographic models. The key analytical processing requirements that give a GIS its special flavor include:

Geometric, topological, and set-oriented analyses Most if not all geographically referenced phenomena have geometric, topological, or set-oriented properties. Set-oriented properties include membership conditions, relationships between collections of elements, and handling of hierarchies (e.g., administrative areas). Topological operations include adjacency and connectivity relationships. All these properties are key to a GIS and form a main theme of this book.

Field-based analysis Many applications involve spatial *fields*, that is, variations of attributes over a region. The terrain in Section 1.2.4 is a variation of topographic elevation over an area, for example. The gravel and sand deposits and built-up areas of Section 1.2.5 are variations of other attributes. Fields may be discrete (a location is either built up or not) or continuous (e.g., topographic elevation). Fields may be *scalar* (variations of a scalar quantity and represented as a surface) or *vector* (variations of a vector quantity such as wind velocity). Field operations include overlay (Section 1.2.5), slope and aspect analysis, path finding, flow analysis, and viewshed analysis. Fields are discussed further throughout the text.

scalar

vector

Network analysis A network is a configuration of connections between nodes. The maps of most metro systems are in network form, for example, with nodes representing stations and edges representing direct connections between stations. Networks may be directed, where edges are assigned directions (e.g., in a representation of a one-way street system) or labeled, where edges are assigned numeric or non-numeric attributes (e.g., travel-time along a rail link). Network operations include connectivity analysis, path finding (trace-out from a single node and shortest path between two nodes), flow analysis, and proximity tracing (the network equivalent of proximal polygons). Networks are another major topic in this book.

1.3 GIS models and data

Any information system relies on data. However, the data itself is often not the primary focus of the users of an information system; rather, the focus is what that data tells us about the world. This section delves into a little more detail on data, what it is, what are the special characteristics of spatial data, and how we relate data in a GIS to things in the world.

1.3.1 Models and data

model

The word "model" has been used freely in varying contexts in the preceding text. In general, a *model* can be defined as an artificial construction in which parts of one domain, termed the *source domain*, are represented in another domain, the *target domain*. The purpose of the model is to simplify and abstract away from the complexity of the source domain. Only selected aspects of the source domain are translated by the model into the target domain to be viewed and analyzed in this new target context. Insights, results, computations, or whatever has taken place in the target domain may then be interpreted back into the source domain.

A simple example of a model is a flight simulator. Objects in the real world such as an aircraft, its instrument panel, sounds, movements, views from the cockpit, and the navigation space, are simulated in an artificial environment. The pilot may manipulate the model environment, for example, by simulating a landing into Boston's Logan Airport in bad weather. This experience within the target domain may then be transferred back to experience with flying real aircraft.

The usefulness of a particular model is determined by how faithfully it can simulate the essential elements of the source domain, and how easy it is to move between the two domains. There is a mathematical concept that captures this type of structure, called a *morphism*. A morphism is a function from one domain to another that preserves some of the structure in the translation. Mapping and navigation provide a convenient example. Suppose that the geographic world is the source domain, modeled by a map (target domain). A user needing to travel from Edinburgh to London by road consults and analyzes the map, then translates the results of the analysis back in order to navigate through the UK road network. If the map is a good model of the real road network, then the user's journey may be smooth.

morphism

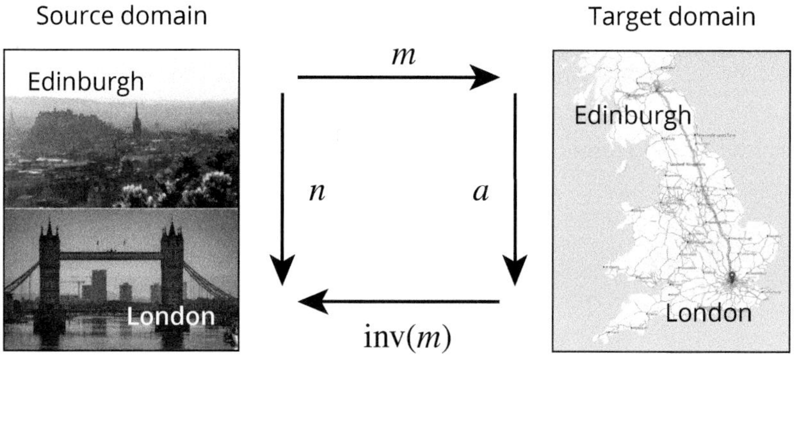

Figure 1.14: Mapping and navigation as a morphism (Map image: Open Street Map)

Figure 1.14 illustrates the concept, with the real-world Edinburgh and London in the source domain, and a mapping engine modeling those locations in the target domain. The figure also shows the modeling process more abstractly in terms of morphisms. The source domain, of real-world cities, is modeled using the morphism m. An activity n in the source domain (e.g., navigating from Edinburgh to London) is modeled by the analysis a in the target domain (e.g., the routing algorithm in the mapping platform). The result of the transformation in the target domain is then reinterpreted in source domain, using the inverse **inv**(m) of the morphism. The whole process "works" if the morphism and analysis accurately reflects the activity in the world.[3]

[3] Mathematically, this is expressed by the equation $\mathbf{inv}(m) \circ a \circ m = n$. Such structural relationships are the subject of the mathematical theory of *categories*. The satisfactory way that the functions (arrows) work together in Figure 1.14 allows us to say that the diagram *commutes*.

In fact, as we take a closer look, we will see that every GIS application encompasses multiple models at different levels of abstraction in parallel. For example, the data in a GIS may embody a model of a particular application domain, such models of transportation or of wildfire. At the same time, the GIS will need to implement certain physical computer-based models that enable efficient storage and retrieval of that data. Indeed, understanding GIS implies an understanding of the different models upon which the technology is based, how these models interact, and their in-built assumptions and limitations.

Conventionally, the different interlocking models in an information system are often decomposed into four distinct models in Figure 1.15. Moving from left to right:

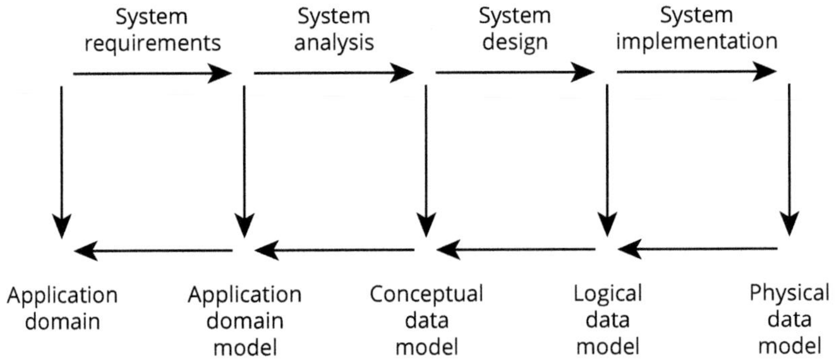

Figure 1.15: Four levels of models in information systems

application domain model

- The application domain itself is the subject of an *application domain model*. An example of an application domain might be bus transportation. In the case of bus transportation, the application domain model might describe entities such as streets and stops, buses and routes, passengers and drivers, as well as more abstract concepts, such as reliability and punctuality. In effect, the application domain model describes and delimits the scope of interest, termed the *system requirements*. Developing an application domain model often requires the assistance of domain experts, enlisted through an initial study of the application domain, sometimes called *requirements analysis*.

system requirements

requirements analysis

- In the *conceptual computational model* or *conceptual data model*, the application domain model is carefully cataloged and structured into computational elements that are more compatible with an information system. Two of the conceptual data modeling techniques we will encounter later in this book are *entity relationship modeling* and *object-oriented modeling*. For example, in the bus transportation example, the entity relationship (conceptual) model might differentiate between which domain elements are entities (such as bus routes and bus stops), which are relationships (such

Box 1.3: System development life cycle

Together, the four modeling functions in Figure 1.15—requirements, analysis, design, and implementation—make up the first four stages of the *system development life cycle*. The system development life cycle is a structured process that can be used to develop any information system, including a GIS application. Two further important stages of the system development life cycle are not modeling tasks. *System validation* follows implementation and aims to validate and refine the developed system by testing it against a range of practical scenarios in deployment. Finally, *system maintenance* is concerned with making ongoing changes and improvements to a system after deployment in response to actual system usage. The logical progression through successive stages of the system development life cycle is often termed the *waterfall model* of system development. In practice, system development is necessarily much more iterative than a steady planned progression from one stage to the next. Issues uncovered at later stages, such as implementation, may throw now light on decisions at earlier stages, such as analysis. Consequently, most system development methodologies today recognize the different stages in the waterfall model, but they allow much more flexible transitions between those stages. For example, *agile development* techniques aim to decompose the development process into multiple rapid development iterations (often called "sprints") that involve cross-functional teams of analysts, designers, developers, as well as users combining in microcosm all the development stages in each iteration.

as "stops at" relationship between bus route and bus stop), and which are attributes (such as route number and stop number). The process of constructing a conceptual model from an application domain model is usually called *system analysis*.

system analysis

- The *logical computational model* or *logical data model* develops the system analysis further by specifying how the conceptual data model can be realized within a specific system paradigm, such as a database or programming paradigm. Several different database paradigms are explored in this book, with most attention paid to relational databases and graph databases. In the case of a relational database, the logical data model will specify the structure of the specific tables (relations) that will be stored in the database, such as tables of bus stops that capture each stop's number or coordinate location. The process of developing a logical model from a conceptual model is called *system design*.

logical data model

system design

- Finally, the *physical computational model* or *physical data model* is constructed by system programmers and developers, who implement the logical computational model for a specific software system or hardware platform. Constructing a physical data model is achieved through a process of *system implementation*, which will involve decisions about exactly how to structure data such that it can be efficiently stored and retrieved.

physical data model

system implementation

The models in Figure 1.15 provide stepping stones to help smooth the transition between application domain and computer system (see also Box 1.3). Without this structure, the system development process quickly becomes unmanageable. Skipping levels tends to lead to systems that are either not useful, because they neglect the human and practical aspects of the application; or inefficient, because they neglect the computational aspects of the system.

1.3.2 *Information and data*

Having pinned down what we mean by the term "model," we should also be precise about the terms *data* and *information*. The structure and interrelationships within data, and how data is collected, processed, used, and understood within an application, form the *context* for data. An understanding of the data model and of the limitations of data are elements of that context. Data only becomes useful, taking on value as information, within this context.

context

data

For example, data about atmospheric conditions is recorded by meteorological stations across the world: there is likely to be one near you recording right now. On its own, the raw data from your nearest recording station is unlikely to be useful to you. Useful information, such as a weather forecast that helps you decide if you need to carry an umbrella today, is produced within the context of careful modeling and analysis of raw data from multiple sources (such as satellite imagery, data from other recording stations, and historical data). Accordingly, information can be defined as "data plus context":

information

$$\text{information} = \text{data} + \text{context}$$

We refine the concepts of data and information in later chapters (particularly Chapter 10), but this basic distinction between data and information is needed throughout the book. Human knowledge about the world and decisions to act in the world are (hopefully) based on information, which in turn is based on data.[4]

[4] Some texts go further and describe a continuum of levels, from data to information, and through knowledge, to wisdom. However, the dividing line between these different levels is frequently fuzzy: one person's knowledge is another person's wisdom. Rather than get tied up in such definitions, we prefer the simpler "information = data + context" in this book.

1.3.3 *Spatial data*

Spatial data is often subdivided into two great classes, *raster* and *vector*. Traditionally, systems have tended to specialize in one or another of these classes, and the division is still evident in many systems today. Figure 1.16 shows a simplified raster and vector representation of the same situation of a house, outbuilding, and pond next to a road.

Figure 1.16: Raster and vector data

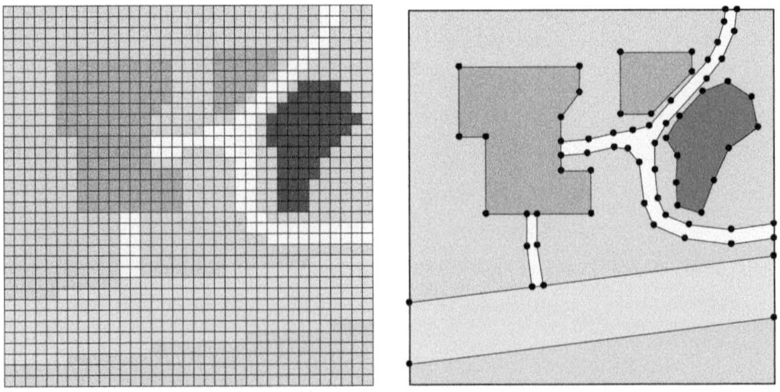

Raster data is structured as an array or grid of cells, referred to as *pixels*. Each cell in a raster is addressed by its position in the array (row and column

raster

pixels

number). The 3D equivalent of a raster is a 3D array of cubic cells, called *voxels*. Rasters are able to represent a large range of computable spatial objects. Thus, a point may be represented by a single cell, an arc by a sequence of neighboring cells and a connected area by a collection of contiguous cells. Rasters are natural structures to use in computers, because array handling and operations are widely supported by computer systems and languages. However, a raster when stored in a raw state with no compression can be inefficient and unwieldy in terms of usage of computer storage, as the number of cells in the raster increases with level of detail. We will consider efficient computational methods for raster handling in later chapters.

voxels

The other common paradigm for spatial data is the vector format. (This usage of the term "vector" is similar but not identical to "vector" in vector fields). A *vector* is a finite straight-line segment defined by its end-points. The locations of end-points are given with respect to some coordinatization of the plane or higher-dimensional space. The discretization of space into a grid of cells is not explicit as it is with the raster structure. However, it must exist implicitly in some form, because of the discrete nature of computer arithmetic. Vectors are an appropriate representation for a wide range of spatial data. Thus, a point is just given as a coordinate. An arc is discretized as a sequence of straight-line segments, each represented by a vector, and an area is defined in terms of its boundary, represented as a collection of vectors. At comparable levels of detail, the vector data representation is inherently more efficient in its use of computer storage than raster, because only points of interest need be stored. A disadvantage is the vector-representation is not so natural computational a structure to use as arrays, necessitating specialized data storage, retrieval, analysis, and presentation capabilities. We return to these issues in later chapters.

vector

1.3.4 Spatial data retrieval

As identified above, a key function of the database at the heart of a GIS is the capability to efficiently store and retrieve data. To retrieve data from the database, we may apply a filter, usually in the form of a logical expression, or *query*. For example:

query

1. Retrieve names and addresses of all potteries in Staffordshire.
2. Retrieve names and addresses of all employees of Wedgwood Pottery who earn more than half the sum earned by the managing director.
3. Retrieve the mean population of administrative districts in The Potteries area.
4. Retrieve the names of all patients at Stoke City General Hospital who are over the age of 60 and have been admitted on a previous occasion.

Assuming that the first query accesses a national database and that the county in which a mine is located is given as part of its address, then the data may be retrieved by means of a simple look-up and match. For the second query, each employee's salary would be retrieved along with the salary of the managing director and a simple numerical comparison made. For the third, populations are retrieved and then a numerical calculation is required. For the last, a more complex filter is required, including a check for multiple records for the same person.

There are spatial operators in the above queries. Thus, in the first query, whether a mine is *in* Staffordshire is a spatial question. However, it is likely that the processing for this query needs no special spatial component. In our example, there is no requirement to check whether a mine is within the spatial boundary of the county of Staffordshire, because this information is expected to be given explicitly in the stored address in the database.

In contrast, consider the following three queries (see Figure 1.17):

1. Retrieve the town that contains the lat-long coordinate location 53.0280, −2.1747 (a *point query*, answer "Hanley").
2. Retrieve names and addresses of all potteries that lie outside built-up areas (a *region query*, answer "Wedgwood Pottery Centre").
3. Retrieve the closest pottery to Minton Pottery (a *nearest neighbor query*, answer "Spode Pottery").

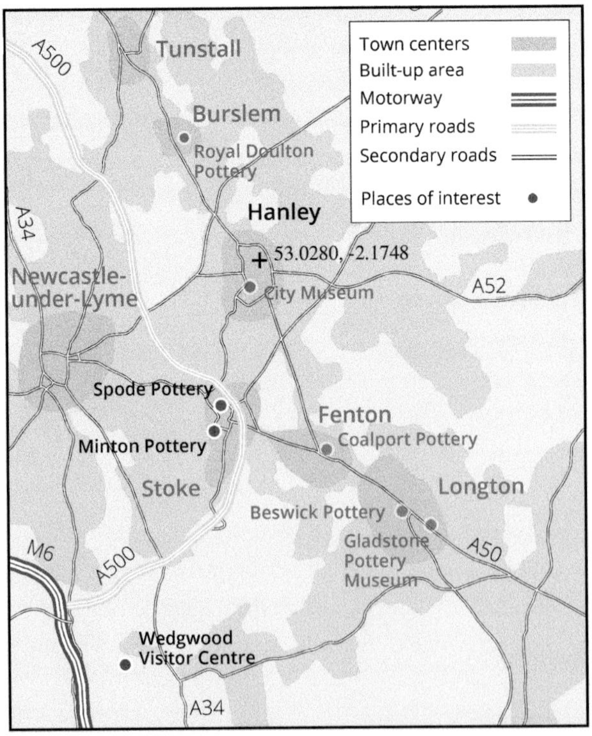

Figure 1.17: Retrieval of spatial data: point, region, and nearest neighbor queries

Responding to these three queries requires specialized spatial data storage and retrieval capabilities. It will not be possible to store the data needed to answer these queries as explicit text or numerical data. Explicitly enumerating every point inside a feature is no more possible than enumerating every region

Box 1.4: Data mining

Data mining refers to the process of discovering valuable information and meaningful patterns within big data sets. A classic example of data mining is *market basket analysis* (MBA). Retailers analyze data about millions of customer purchases to uncover items that are frequently bought together, using that information to promote additional products to customers buying one of the frequent item set. Market basket analysis is often also called *association rule learning*. Other types of data mining include clustering (identifying groups of similar data items), classification (attaching meaningful labels to data items), and anomaly detection (uncovering rare or unusual data items). An example of MBA, often quoted in university lectures, is the story of the discovery of the association between supermarket purchases of beer and diapers by data mining pioneers. The story is memorable because it resonates with deep-seated gender stereotypes about reluctant fathers (as they are presumably the ones purchasing the beer for themselves at the same time as dutifully buying diapers for their baby children, or so the narrative goes). Fortunately, the story is apocryphal, and there is no evidence for such an association occurring in actuality; beer and snacks, such as nuts or chips, is a more veridical market basket association. Indeed, the beer and diapers example serves us better as a cautionary tale about the power of appealing narratives in data mining. As GIS experts, it is advisable to cultivate a healthy skepticism when interpreting the outputs of machine intelligence, as we shall see in Chapter 9.

that contains a feature. Hence, responding to true spatial queries requires new capabilities not available in non-spatial databases.

1.3.5 Spatial data analysis

Now, consider the following two queries:

1. What locations in The Potteries satisfy the following set of conditions:
   - less than the average price for land in the area;
   - within a 15-minute drive of the motorway M6;
   - having suitable geology for gravel extraction; and
   - not subject to planning restrictions?
2. Is there any correlation between:
   - the frequency of vehicle accidents (as recorded in a hospital database); and
   - the locations of designated "accident black spots" for the area?

Satisfying such queries often requires the integration of both spatial and non-spatial information. But it also requires more than spatial data retrieval capabilities; it also requires spatial data *analysis*. Computing the average price for land in an area, the travel time isochrones, and the spatial correlations between data will require us to process spatial data in a way that moves beyond simple point, region, and nearest neighbor queries. While the first of these examples is rather more prescriptive than the second, specifying the type of relationship required, the second example requires a more open-ended, exploratory application of spatial analysis techniques. We will touch on many different types of analysis in this book (and see Box 1.4). However, our focus in this book will be on the computational aspects of spatial data analysis rather than a deeper exploration of the underpinning statistical concepts.

All of the above functionality is attractive and desirable but will be useless if not matched by commensurate performance. Performance is an even bigger issue for a spatial database than a general-purpose database. Spatial data is notoriously voluminous. In addition, spatial data is often hierarchically structured (a point being part of an arc, which in turn forms the boundary of a polygon). We shall see that these structures present problems for traditional database technology. Although databases are designed to handle multidimensional data, the dimensions are assumed independent. However, spatial data is often embedded in the Euclidean plane, requiring special storage structures and access methods. We shall return to questions of spatial data storage, access, and processing later in the book.

1.4 Computing technologies

software
hardware

The term *software* is used to refer to the instructions or programs executed by a computer system. In contrast, *hardware* is used to refer to the physical components of a computer system, such as computer chips, screens, and keyboards. The software needed for a GIS is highly specialized, and much of this book is concerned with those specialized concepts and technologies. However, most of the hardware, networking, and other computing technologies used by a GIS is broadly the same as that used within general-purpose information systems.

This section briefly summarizes the key digital computing technologies used in an information system, to the degree necessary for understanding the role that technology plays in supporting GIS. We assume readers are already somewhat familiar with the architecture of a computer, so this section contains only a brief overview of those components directly relevant to GIS.

1.4.1 The von Neumann model

von Neumann model

Despite the wide range of tasks required of computers, the overwhelming majority of computers we encounter today conform to the *von Neumann model* of computer architecture, developed by the Hungarian-born mathematician John von Neumann at Princeton University during the 1940s and 1950s. According to the von Neumann architecture, a computer system can be thought of as comprising four major subsystems, illustrated in Figure 1.18 and described below.

Processing Data processing consists of operations performed to combine and transform data. Complex data processing functions may be reduced to a small set of primitive operations.

Storage Data is held in storage so that it may be processed. This storage may range from short term (held only long enough for the processing to take place) to long term (held in case of future processing needs).

Control The storage and processing functions must be controlled by the computer, which must manage and allocate resources for the processing, storage, and movement of data.

Input/output Computers must be able to accept data input and to output the results of processing operations. Two important classes of input/output are discussed in more detail in later chapters: input/output between humans and computers and input/output between different computer systems, especially when mediated by a digital communication network.

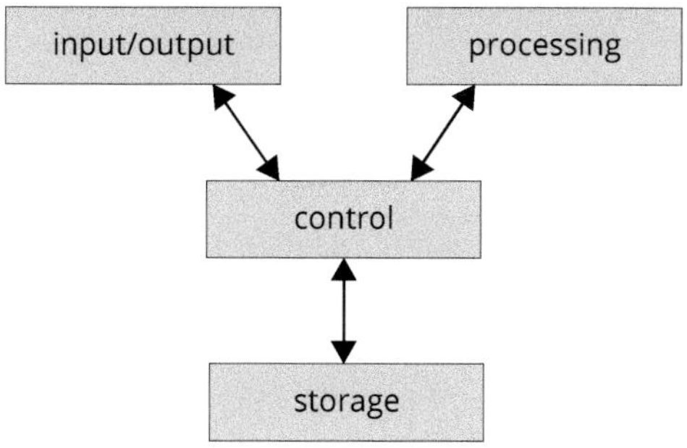

Figure 1.18: The four major functional components of a computer

In terms of components rather than functionality, the key computer system components are the CPU (responsible for processing and control), memory devices (responsible for data storage), and input/output devices (such as human input/output devices and computer networks). Each of these components is considered in turn in the following sections.

1.4.2 *Processing and control*

Processing of data in the computer hardware is handled by the *central processing unit* (CPU). The CPU's main function is to execute machine instructions, each of which performs a primitive computational operation. The CPU executes machine instructions by fetching data into special registers and then performing computer arithmetic upon them.

CPU

The CPU is itself made up of several subcomponents, most important of which are the *arithmetic/logic unit* (ALU) and the *control unit*. The control unit is responsible for the control function, managing and allocating resources. The ALU is responsible for the actual processing functions.

ALU

control unit

Operations are performed upon data sequentially, by retrieving stored data, executing the appropriate operation, and then returning the results to storage. The process of execution is known as the *instruction cycle* (also termed the *machine cycle* or the *fetch-execute cycle*). The instruction cycle involves four steps, shown in Figure 1.19.

instruction cycle

The first step is for the control unit to retrieve an instruction from storage. Next, the control unit decodes the stored instruction to determine what operation must be performed. The control unit then passes this instruction to the ALU, which completes the actual task of executing the operation. Finally,

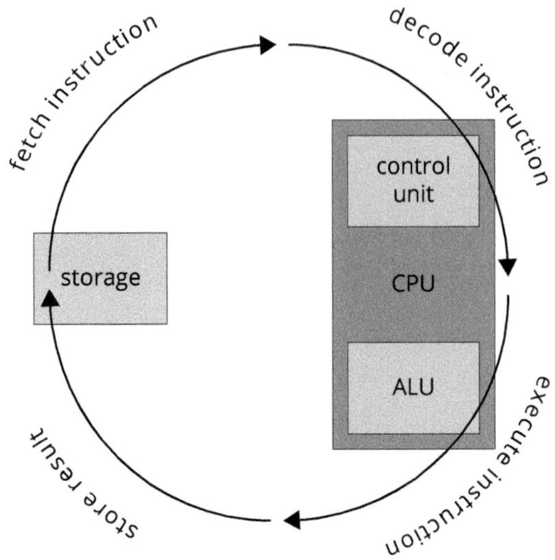

Figure 1.19: The instruction cycle

the results of the execution are returned to storage, ready to be retrieved for a subsequent instruction cycle. Connectivity between the CPU and other components in the computer is provided by dedicated communication wires, each called a *bus*. While almost all CPUs rely on the instruction cycle, computer processors differ in the types and range of instructions they implement (see Box 1.5 on the facing page).

bus

1.4.3 Storage

Digital data must be physically kept somewhere in the computer system. Storage devices differ in their capacity (how much data can be stored), performance (how quickly the data can be accessed), volatility (whether stored data persists after power to the system is turned off), and price. Storage devices can be divided into three categories:

- Storage that can be directly manipulated by the CPU is termed *primary storage*. Primary storage is typically relatively expensive per stored bit, compared to secondary storage, and is generally *volatile* (stored data is lost when the power to the storage device is turned off).
- Storage that can be accessed only indirectly by the CPU (via input/output controllers) is termed *secondary storage*. Secondary storage is relatively cheaper than primary storage and is normally *non-volatile* or *persistent* (stored data persists after the power to the storage device is turned off).
- Storage used to archive huge volumes of data that will be accessed only infrequently by computers is termed *tertiary storage*. Secondary storage is the cheapest of all storage and always non-volatile.

primary storage

volatile

secondary storage

tertiary storage

To enable efficient computation, the speed of access of primary storage must be comparable with CPU instruction cycle times. As a result, primary storage is the fastest, lowest capacity, and most expensive of all memory types.

Box 1.5: CISC and RISC

We usually think of hardware as completely separate from software: hardware is physical computing objects you can touch, software is virtual instructions and data. However, the distinction is not in truth so clear-cut. An illustration of why the distinction between hardware and software can be blurred is provided by "RISC" and "CISC" CPU architectures. The set of instructions supported by a CPU (hardware) is a determining factor in the complexity of programs (software) running on that CPU. In recent decades, most personal computer CPUs have been built using a *complex instruction set computer* (CISC) architecture, such as that found in the Intel x86 family of CPUs dating back to the 1980s. In a CISC architecture, the CPU supports complex instructions, so making software simpler and cheaper to program. However, complex instructions, each of which may require multiple instruction cycles, are generally slower and more energy-intensive to execute than simple instructions. A *reduced instruction set computer* (RISC) architecture supports only simpler instructions than a CISC architecture, each of which can be completed in a single instruction cycle. RISC has the advantage of leading to generally cheaper CPUs with lower power consumption which can as a result run at higher speeds. Many small and mobile computing devices use a low-power RISC architecture, such as the ARM family of CPUs, as well as an increasing variety of laptop, workstation, and supercomputers. However, by making the hardware simpler, RISC architectures typically need more complex software to achieve the same tasks.

Because primary storage is volatile, most information systems, including GIS, also require secondary storage: lower cost, higher volume, non-volatile data storage. Secondary storage is not directly accessible to the CPU. Instead, secondary storage is accessed indirectly, via a bus for transferring data between primary and secondary storage. Secondary storage is consequently much slower than primary storage, with data access times measured in milliseconds or microseconds rather than nanoseconds for primary storage. Therefore, efficient structuring of data files on secondary storage devices is an important factor in GIS performance. Much effort and ingenuity has been spent by computer scientists devising suitable data structures to ensure good performance with spatial data, as we shall see in Chapter 6.

Secondary storage devices can be categorized into disk drives and flash drives. The most important form of disk drive is the *hard disk drive* (HDD). In an HDD, a magnetic disk is coated with a thin layer of magnetic material. The polarization of minute regions of the disk can be accessed or changed by an electrical read/write head. A combination of disk rotation and movement of the read/write head provides access to the entire surface of the disk. *Optical disks* operate along similar principles, instead storing data as minute pits in an optical disk, read by a head that can detect differences in the laser light reflected from the pitted disk surface.

Whether magnetic or optical, disk storage relies on the physical rotation of the disk and movement of read head to store and retrieve data. *Flash storage*, such as solid state drives (SSDs), is based purely on electronic circuitry and has no moving parts. Flash storage relies on the quantum properties of certain semiconducting materials to persistently store data in an array of cells in an integrated circuit. As a result of having no moving parts, flash storage is the fastest form of secondary storage.

flash storage

Finally, tertiary storage is the lowest cost, highest volume, and slowest form of storage, used for archiving and backing up data. Tertiary storage devices include magnetic tape drives and optical jukeboxes, where libraries of magnetic tapes or optical disks are physically mounted and unmounted by robots. As a consequence, tertiary storage has access times measured in seconds. Figure 1.20 shows some common primary, secondary, and tertiary storage devices.

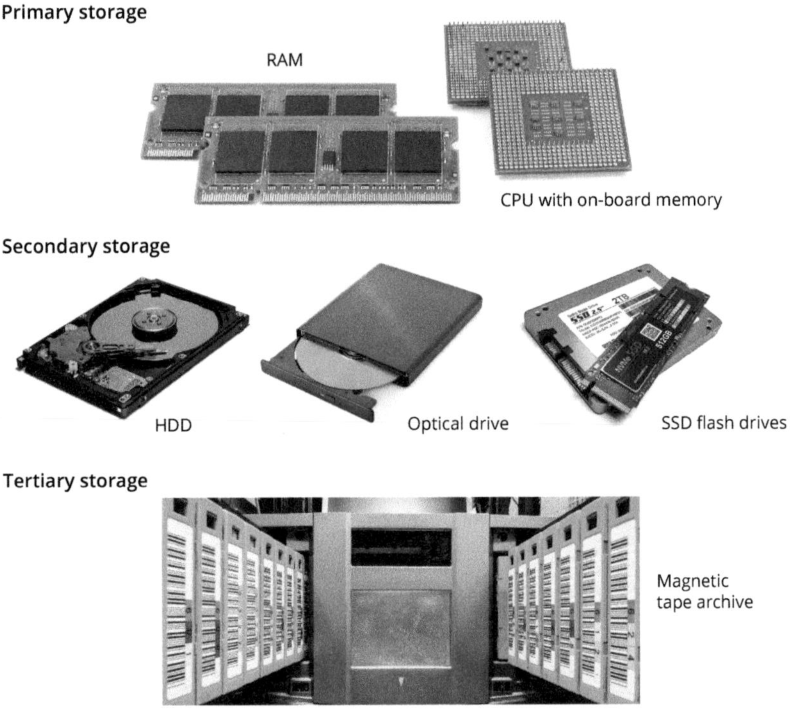

Figure 1.20: Common storage devices

1.4.4 Input/output and networks

hard copy

soft copy

Human users are often the focus of input to and output from the computer. For example, *hard copy* output, created by devices such as 2D and 3D printers and plotters, is output with physical permanence. By contrast, *soft copy* output, such as the image on a computer display screen, is transient and intangible. An array of different devices exist to enable humans to input data into the computer and to receive a computer's output, sometimes simultaneously. We return to these devices, their distinguishing characteristics, and their importance to GIS, in the context of GIS interfaces in Chapter 8.

Achieving input and output between one computer and another relies on communication networks, a fundamental component of most of today's information systems, including GIS. Most communication networks are *digital*, meaning a series of binary digits (bits) is transmitted using signal bursts corresponding to the binary values 0 and 1. Digital communication technology has superseded older technology based on *analog* signals, where the signal strength can vary continuously like a sine wave. Figure 1.21 contrasts digital (Figure 1.21a) and analog (Figure 1.21b) signals diagrammatically.

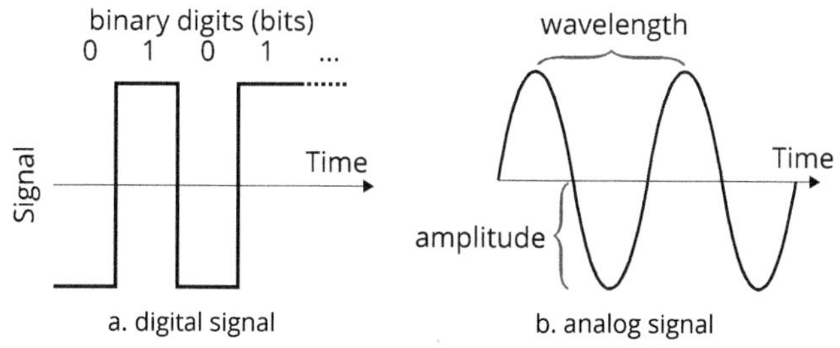

Figure 1.21: Digital and analog signals

Irrespective of whether digital or analog *signals* are being transmitted, all communication networks use electromagnetic (EM) radiation to propagate signals, termed the *carrier wave* or *carrier signal*. EM radiation can be thought of as an analog wave traveling through a medium, such as a cable or the air. The *frequency* of the carrier wave is the number of cycles a wave completes per unit time. At constant wave speed, frequency is inversely proportional to the carrier wave's *wavelength*: the length of each cycle. The *amplitude* is the intensity of the wave.

carrier signal

frequency

amplitude

EM radiation with shorter wavelengths, such as infrared and visible light, can carry more data than radiation with a longer wavelength, such as microwaves and radio waves, because the shorter wavelengths allow the signal bursts to be shorter. However, shorter wavelength EM signals degrade more quickly than longer wavelengths, and so are harder to use over longer distances. The range of wavelengths or frequencies available for data transmission is called the *bandwidth*. There is a clear relationship between the bandwidth of a signal and the amount of data that can be carried by the signal: higher bandwidth means greater data transmission capacity.

bandwidth

Encoding digital data within an analog EM carrier wave involves modulating (varying) either the amplitude or the frequency (or conversely the wavelength) of the carrier signal. A highly simplified example of *amplitude modulation* for encoding digital data within an analog signal is illustrated Figure 1.22.

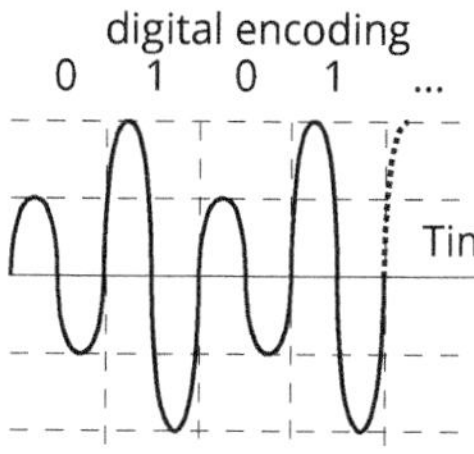

Figure 1.22: Amplitude modulation encoding digital data within an analog signal

Figure 1.23 summarizes the magnetic spectrum and its data-transmission capabilities. Radio waves, microwaves, infrared, and visible light can all be used for computer networks. High-frequency ultraviolet, X-ray, and gamma ray EM radiation are not used for data transmission, because their high energies can be hazardous to the environment and human health.

Signals carried by EM radiation can be transmitted through different media. Ethernet cables and telephone wires are typically made of metal, primarily copper. By contrast, fiber-optic cables are made of fine glass fibers. Although fiber-optic cables are more expensive than copper wire, they offer much higher bandwidth, can operate reliably over much longer distances, are are much less susceptible to interference because they rely on the transmission of visible light when compared with the radio waves transmitted through copper wires.

Data transmission can also take place unguided through the Earth's atmosphere, termed *wireless communication*. Without wires to guide signals,

wireless communication

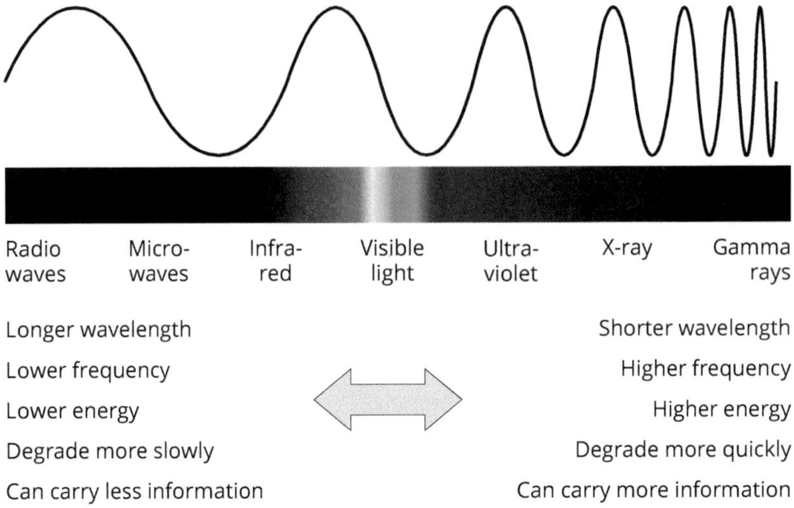

Figure 1.23: The electromagnetic spectrum

path loss

line-of-sight

wireless communication must overcome signal degradation and interference from environmental radiation (see Box 1.6 on page 32) as well as *path loss*: the reduction in signal strength as it propagates through space. Wireless communications normally use radio wave or microwave EM radiation to transmit data, which carries the advantage that the low-frequency radiation can penetrate non-metallic obstacles, such as walls and floors, while visible light and infrared cannot. As a result, networks based on visible light and infrared signals are often termed *line-of-sight* technology, and are less suitable for wireless communications.

WAN

Communication networks are also sometimes distinguished according to the size of geographic area they cover. A WAN (wide area network) operates over large-scale geographical regions, such as states and countries. The Internet is the largest WAN in existence, estimated in 2001 to connect more than 100 million hosts, and connecting more than a billion hosts at the time of writing this book.

LAN

PAN

A WAN connects groups of LANs (local area networks) together. A LAN connects groups of computers over medium- to small-scale geographical regions, for example, from city centers down to individual buildings. A (wireless) PAN (personal area network) operates over very small geographic or sub geographic areas, typically only a few meters in size. Wireless PANs operate over the shortest distances of any network. A wireless PAN is ideal for connecting multiple small computing devices together, such as smartphones, watches, speakers, personal computers, and sensor networks, without the need for wires. A familiar example of a low-power, short-range radio-wave wireless PAN technology is *Bluetooth*, named after the 10th century Danish king. Figure 1.24 summarizes the different network extents, from a few meters (PAN), through local and metropolitan area networks (LANs and metropolitan area networks, MANs) over a few hundred or thousand meters, to global WANs.

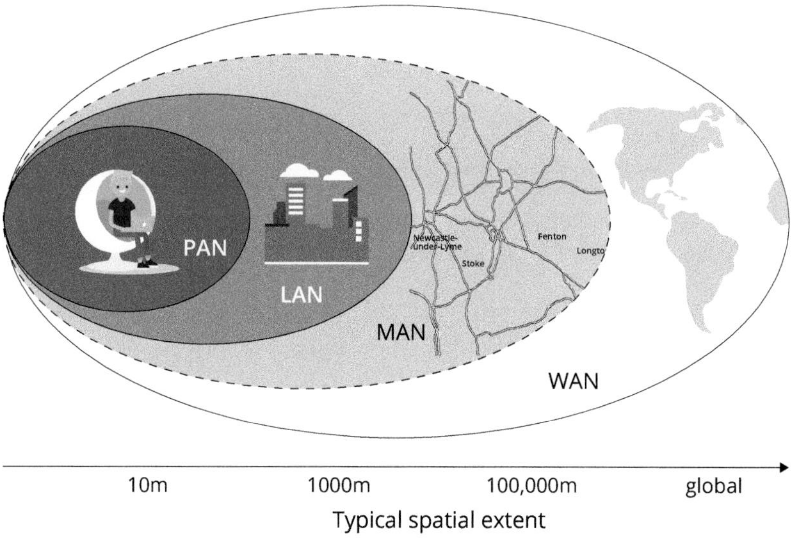

Figure 1.24: Types of network by spatial extent

From a GIS perspective, digital communications have radically altered the way people use spatial information. Digital communications promote sharing of spatial data. Many companies, mapping agencies, and government departments make a variety of spatial data available via the Internet. Digital communications are also *rapid.* The speed of digital communications makes possible certain modes of computing that would not otherwise be possible. For example, mobile GIS applications, such as the tourist navigation system described in Section 1.2.3, rely heavily on rapid and wireless digital communications. We return to the architecture of such systems in Chapter 7.

1.5 What makes spatial special?

The discussion so far has provided many clues to what makes GIS special. We will explore these themes more deeply throughout the remainder of this book. These themes include the following topics we return to later in the book.

- The need already highlighted for specialized techniques for storage and retrieval of spatial data in a database is picked up in Chapter 2.
- Chapter 3 uncovers the richness of different concepts and structures that fall under the heading of "spatial" and need to be supported by a spatial database, including geometry and topology.
- This richness and diversity in turn gives rise to the variety of different models encountered in connection with spatial data in Chapter 4.
- Armed with these concepts, Chapter 5 explores the foundations of algorithms for efficient computation with spatial data and the solutions to iconic spatial problems, such as computing shortest paths, point-in-polygon, and polygon overlay.

Box 1.6: Combating wireless interference

The problem facing all wireless network technologies is how to ensure that signals from different devices do not interfere with one another, causing data loss. There are essentially three increasingly sophisticated mechanisms for achieving this. The simplest option is to ensure that each device only uses a narrow frequency range for communication. This is rather like the way radio stations broadcast only on particular frequencies, into which you need to tune to hear the broadcast. The second option is to ensure that each device only transmits for a certain amount of time on its frequency range. This allows the same frequency to be shared by several devices, rather like the way that different radio programs occur at different times on the same radio station. The third option is to ensure that each device transmits at a range of frequencies, hopping between frequencies at different times in some sequence agreed by transmitter and receiver, termed *spread spectrum* technology. This would be rather like starting to listen to your favorite radio program on one station, and part-way through retuning your radio to another station for the remainder of the program. This "frequency hopping" makes the network more tolerant to interference from other devices and environmental background noise: if some of the signal is lost to interference, the chances are that the signal will be regained next time it hops to a new frequency. Spread spectrum technology commonly switches hundreds of times every second between dozens of randomly chosen frequencies.

- GIS technology today is highly dynamic, with individual applications integrating many different components and systems together via a digital communications network. Chapter 7 explores the major concepts and technologies behind the architecture of GIS, including distributed, decentralized, streaming, and Web GIS.
- Where Chapter 7 concerns connecting computer-based components together, Chapter 8 covers the foundations of human user interaction with GIS. GIS can draw on principles common to design more broadly, in addition to the unique knowledge and centuries-old design traditions of cartography and map-making in interacting with spatial data.
- In Chapter 9, the focus turns to the combination of artificial intelligence and GIS. So-called "GeoAI" is a burgeoning topic today, but one that draws on long-established techniques and concepts in GIS and AI.
- Finally, Chapter 10 ends with a look at some of the most enduring problems in science that also present unique challenges to GIS, such as *uncertainty*, *privacy*, *fairness*, and *equity* in information systems.

We began this chapter with the question: what makes *GIS* special? We end with the broader question: what makes *spatial* special? Many books in the field of "geographic information science" (GI science), including this one, are linked by the central idea that spatial is indeed special. But while there are almost as many possible answers to the question of "what makes spatial special" as there books in the area, it is possible to identify five key themes that recur in unique combinations at the core of the field of GI science, summarized in Figure 1.25 (Duckham, 2015, and based on a published data analysis of the GI science literature).

1.5.1 *Theme 1: Structure*

A central theme running throughout this book—and indeed underpinning any book on spatial data—is that the *structure* of spatial information is different to other types of data and requires special handling. We will encounter this theme in many different guises in this book (Figure 1.25, Theme 1): in Chapter 3 when we encounter Euclidean, topological, network, and metric spaces; in Chapter 5 when we examine the impacts of the jump from one to two spatial dimensions on computation and algorithms; and in Chapter 2 when we explore the effects of adding dependency between spatial dimensions on database indexes and spatial data retrieval. One pervasive example of the unique structure of spatial data encountered in Chapter 4 is *spatial autocorrelation*. But many other examples of the special structure of "spatial" have been proposed, such as statistical nonstationarity (Box 1.7 on the following page).

Theme 1: Structure

Theme 2: Dynamism

Theme 5: Uncertainty

Theme 3: Interfaces

Theme 4: Cognition

Figure 1.25: What makes spatial special?

1.5.2 Theme 2: Dynamism

Our geographical world is ever-changing. Nevertheless, static "snapshots" of our dynamic world are still adequate for many purposes, such as traditional cartographic maps for planning and navigation. Ignoring time, many of the foundational concepts and techniques covered in this book likewise concern storing and analyzing spatial "snapshots." However, we should not forget such snapshots are simplifications of convenience, and remain alert to the need for more sophisticated, integrated spatiotemporal concepts and techniques when it arises. Chapter 4 is the first point at which we introduce time into the equation, with inherently spatiotemporal phenomena such as movement (Figure 1.25, Theme 2). However, the importance of time, and the inherent

Box 1.7: Spatial nonstationarity

Spatial nonstationarity, sometimes also called *spatial heterogeneity*, concerns the statistical property of spatial data that no single "global" model can adequately capture its variability. In other words, there is no such thing as an "average" place. In contrast, many other data sets do exhibit stationarity, such as natural variation in human height, errors in GNSS positions, and economic processes and financial markets. Further, many other processes that are strictly speaking nonstationary, such as temperature time series, can often be transformed into stationary processes by removing trends or cyclic variations. Stationarity is important because an assumption of stationarity underpins many standard statistical tests and analyses. Because spatial data is nonstationary, such tests are often not well adapted to spatial data, and indeed the field of *geostatistics* has grown up around exactly this issue (see also Section 9.3.6). Goodchild (2004) discusses the importance of nonstationarity as a fundamental "law" governing spatial data, while any spatial analysis or geostatistics text will address the issue in more detail.

dynamism of any data about space, will recur at many points throughout the book and is a distinguishing feature of "spatial."

1.5.3 Theme 3: Interfaces

Spatial data and GIS technology pervade almost every aspect of our lives today, whether we notice it or not. From the biggest decisions of governments about how to respond to a global pandemic, to the smallest decisions about where to have lunch, spatial data is critical. The connection between data and decisions is provided by an increasingly wide variety of *interfaces*. To help people make the best possible decisions with the available data, those interfaces require careful and deliberate design. Happily, the design of interfaces to spatial data is one of the oldest and most well-understood areas of spatial expertise, embodied in maps and encompassed by cartography. The diversity of technologies that are used provides an interface to spatial data, from wristwatches to wheelchairs (Figure 1.25, Theme 3), means that modern approaches to spatial interface design must incorporate many more concepts and techniques than were required for hardcopy paper maps. Nevertheless, an understanding of the importance of interfaces, addressed in depth in Chapter 8, remains one of the central themes in "spatial" today, as it was from the earliest days of cartography.

1.5.4 Theme 4: Cognition

The contrast between humans and machines is nowhere more evident than when they must work in tandem to complete some task, for example, when flying a modern airliner. Aviation systems usually excel at reliably processing reams of data at lightning speeds; however, pilots must remain the masters of managing the unexpected, applying experience, adapting to context. Data is at its most valuable when systems help to combine these complementary strengths of humans and machines (Figure 1.25, Theme 4). Humans have evolved specialized capabilities for understanding and processing information about space. For example, like many animals, human brains possess *place cells* and *grid cells* that activate in specific ways when we move around a space, helping us innately "know where we are." Our innate abilities for navigation and wayfinding, while not as astounding as salmon or shearwater, are nonetheless impressive. Ensuring effective collaboration between humans and GIS by acknowledging their complementary strengths and weaknesses is a running theme throughout the later topics in this book, including in connection with location-based services (Chapter 7), user interfaces (Chapter 8), and spatial reasoning and artificial intelligence (Chapter 9).

1.5.5 Theme 5: Uncertainty

Just as Theme 2 acknowledges that change is a constant in the world around us, Theme 5 acknowledges that uncertainty is endemic in spatial data about that world. We can never rely on spatial data with complete certainty. For example, all spatial coordinates must necessarily be limited in the level of detail given about a location, termed *precision*. Any observation of location is likewise subject to potential errors in measurement, no matter how small. An understanding of uncertainty is at the root of many areas of science and engineering, and geographic information science is no exception. However, uncertainty in "spatial" is concerned as much with techniques for managing and working with uncertainty—even using it to our advantage—as it is to reducing or expunging uncertainty, the focus in some other areas of engineering and the measurement sciences. Although uncertainty is endemic in all spatial data, the topic is addressed directly at the end of the book in Chapter 10, providing a framework for understanding the unique characteristics of uncertainty in spatial data.

Reflections

Introductory chapters to books are the most tricky to write, because they touch on many topics but dwell on few. Like this book, most other GIS textbooks, such as the excellent Longley, Goodchild, Maguire, & Rhind (2015), begin with some definitions but aim to give just a flavor of the variety of applications of GIS as well as of what makes spatial special. Other recommended introductory texts, including Heywood, Cornelius, & Carver (2011) and Burrough, McDonnell, & Lloyd (2015), similarly give their different perspectives on these topics.

Affectionately known as the "Big Book," Longley, Goodchild, Maguire, & Rhind (2005) is a more advanced two-volume collection of chapters by many of the pioneers of GIS, each of which delves directly into more detail on one topic. Despite its age, the first edition of this book is still relevant and at the time of writing some chapters are freely available online (Maguire, Goodchild, & Rhind, 1991).

Taking a slightly different tack from many other GIS textbooks, we chose to provide a more detailed summary of the computing technologies that underpin GIS, such as storage devices, networking, and the instruction cycle in Section 1.4. This is partly because this book is, after all, "a computing perspective"; but also because we have found this provides a background and context that assists in understanding many features of information systems at the higher level of GIS (such as the relationship between power, range, and bandwidth in wireless networks, which helps when thinking about location-based services and wireless geosensor networks). Other recommended texts that provide more background on computer hardware, software, networks, and architecture include Tanenbaum (2013) and Null & Lobur (2018).

One topic that does not feature prominently in this introduction—but does in several other recommended texts such as Schuurman (2003), Chrisman (2006), Nyerges, Couclelis, & McMaster (2011), and Ballas, Clarke, Franklin, & Newing (2017)—is the importance of people and society in understanding GIS. That societal, human perspective is something we dip into throughout the book, in particular in Chapter 10. But the relative lack of coverage of this important topic is by no means a reflection of its importance, only of the focus and depth possible in a single-volume book.

One further difference between most other GIS texts and this book is the central focus on databases as the *engine* of GIS. This "database perspective" is at the heart of our "computing perspective." As a result, a solid grounding in databases is, we believe, a hallmark of the most capable GIS students and professionals, and it is the topic we turn to in the next chapter.

Fundamental Database Concepts

2

Sections

This chapter is all about databases, their structure and operations, and the role they play in a GIS. An understanding of databases is fundamental to an understanding of GIS, and indeed of many other types of information systems. Progressing through this chapter, you will learn to:

- discuss the key features of **databases** and the rationale for using them;
- contrast the characteristics and limitations of the **relational database** model and the **graph database** model, the two most important database paradigms;
- develop a database design using **entity-relationship modeling** and the standard diagrammatic language **UML**;
- build simple **queries** to a relational or graph database using a standard language, such as **SQL** or **Cypher**.

Databases are the foundation of GIS. A knowledge of the fundamental principles of databases is indispensable for understanding GIS technology. Most GIS are built upon general-purpose relational databases; certainly all GIS will connect with such systems in a distributed environment. This chapter introduces the reader to the main principles of databases. The general database approach is introduced in Section 2.1, with the principles of database development introduced in Section 2.2. The most common database model is the *relational model*, described in detail in Section 2.3. However, the graph database model is a markedly different approach to databases with growing significance for GIS, explored in Section 2.4.

2.1 *Introduction to databases*

From their earliest days, computers have always been used to convert one data set into another, accomplishing large and complex transformation processes that would be tedious for humans to complete manually. For example, we may wish to predict average annual traffic flows using a computer model of transportation. The input to this system might include data about city populations and about the road network. The transformation is encoded by the transport model, implemented in the computer as software. The output is the traffic flow prediction data, detailing expected flows on each road over time. This approach of treating the computer as a "giant calculator" is illustrated in Figure 2.1.

Figure 2.1: The "computer as a giant calculator" paradigm

Thinking of the computer as a giant calculator has several disadvantages, which are discussed in more detail below. In short, such an approach tends to lead to significant duplication of data and even of transformation processes. The alternative offered by the database approach is shown in Figure 2.2. In this case, the computer acts as a useful repository of data, allowing the deposit, storage, and retrieval of data. Data in the store can be accessed, modified, and analyzed in a standard way, ensuring that these and other basic functions are never duplicated.

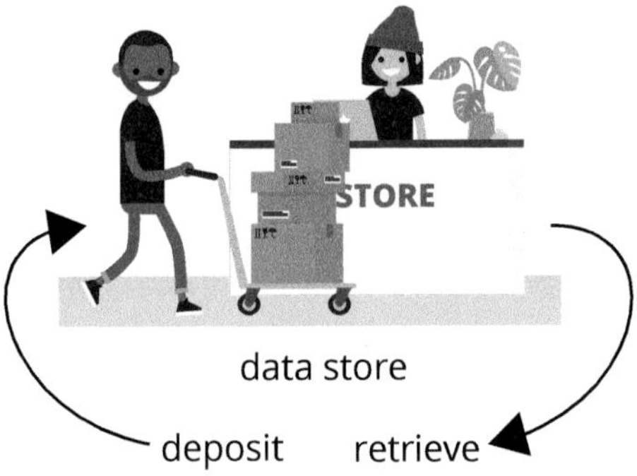

Figure 2.2: The "computer as data repository" paradigm

The remainder of this section delves into more detail on the database approach. A summary of the key features of the database approach can be found in Box 2.1 on the next page.

2.1.1 *The database approach*

The "computer as giant calculator" and "computer as data repository" are extreme positions. Most applications require a balance of calculation (processing) and a repository of data upon which the processes are to act. We illustrate this balance with a fictitious example.

"Nutty Nuggets" is a dinner kit company, established in 2018, delivering tasty vegetarian slow-food meal kits to customers to cook at home. Starting from just a handful of local customers, the business grew quickly by word-of-mouth. To help cope with their growing order book, the two owner-managers decided to apply their computer skills to help keep on top of business.

Box 2.1: Databases in a nutshell

In order to act effectively as a data store, a computer system must have the confidence of its users. A database must be dependable and continue to operate correctly even in the case of unexpected events, such as power or hardware failure (*reliability*). As far as possible, data in the database should be correct and consistent (*integrity*). A database must be able to prevent data being used in unauthorized ways (*security*), but it must offer sufficient flexibility to give different classes of users different types of access to the store (*user views*). Ideally, the database interface should be easy to use for casual or first-time users as well as offer more powerful functions for regular users (*user interface*). Most users will not be concerned with how the database works and should not be exposed to low-level database mechanisms (*data independence*). Users should be able to find out what is in the database (*self-describing*). Many users may wish to use the store, perhaps at the same time or accessing the same data (*concurrency*). Databases should be able communicate with each other in order to access remote data (*distributed database*). Finally, a database should be able to retrieve data rapidly (*performance*). All these functions are managed by a dedicated software application (*database management system*).

They started by populating a spreadsheet with details of their orders, including customer addresses, contact details, the meal type, and meal size in terms of how many people the meal kit should feed. Programs were then written to print out delivery schedules for each day, as well as meal kit contents sheets, helping to streamline the process of making up orders in the kitchen. Figure 2.3 shows on the left the order file being accessed by the two programs. On the right, Figure 2.3 shows the main constituents of order data. The order file is stored on a computer and accessed when required by the programs.

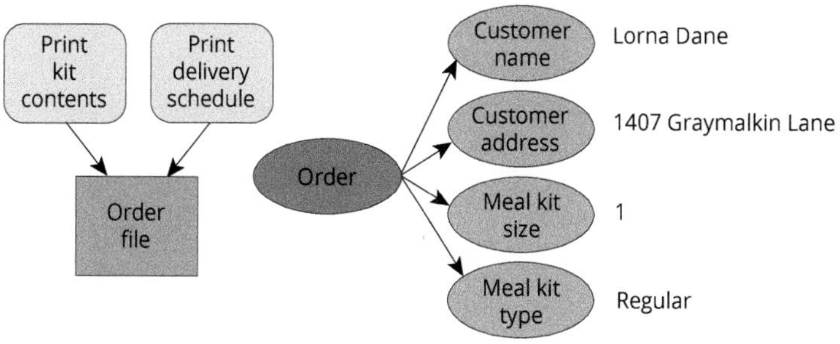

Figure 2.3: Nutty Nuggets stage one: Orders system

Time passed, the order system was successful, and the owners gained the confidence to extend the system to help with the growing issue of stock management. A stock extension was set up, consisting of a stock spreadsheet, and a program to print a daily stock report, highlighting low stock, shown in Figure 2.4. With stock and order details now together in one system, it became natural to link them. A program was written to use the data about customer orders to assist with restocking the kitchen. This program helped reduce costs and waste by helping ensure only enough perishable stock was kept to satisfy upcoming orders, shown at the bottom of Figure 2.4.

The system continued to grow, with new spreadsheets for supplier and customer details added. However, as the system became enlarged, some problems began to emerge, including:

Loss of integrity The system led to problems maintaining the structure and currency of the data. For example, when a staff member entered a new

Figure 2.4: Nutty Nuggets stage two: Orders, stock, and restocking system

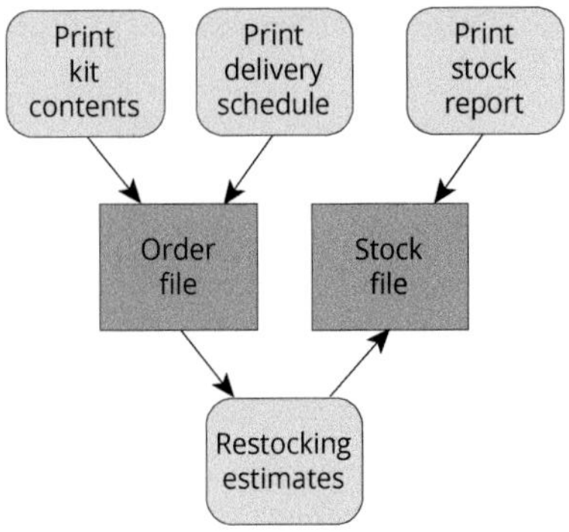

customer's address in the wrong format, the programs that accessed the data produced garbled output or crashed. Worse still, when one of the owners made a copy of the stock file to work on, the two versions rapidly became out of sync, requiring many hours' work later in the week to merge them back into a single authoritative version.

Loss of independence The programs encoded the relationships between the data items in the files. If the relationships changed, then the programs too had to be changed. This close linkage between program and data was becoming complex and costly to maintain, leading to errors. For example, when the option to pause orders was added to the order spreadsheet (popular with customers going on vacation, for example), the restocking program was not updated to incorporate the data associated with this new feature. As a result, some expensive perishable ingredients were ordered in error and went to waste.

Loss of security An unscrupulous kitchen staffmember working with the stock file also made a secret copy of the orders file, accessing customer names and personal details as the basis for an email scam.

The database philosophy is an attempt to solve these and other problems that occur in a traditional file processing system. Figure 2.5 shows a reorganization of the system, so that all data files are isolated from the rest of the system and accessible to the processes only through a controlled channel. The idea is to place as much of the structure of the information into the database as possible. For example, if there is a relationship between orders and the stock required by those orders, then this relationship should be stored with the data. Databases are *self-describing*, in the sense that they encode both data and the structure of that data. The means of expressing the structure and relationships in the data is provided by the data model.

self-describing

The data model also allows the designer to enter into the database any properties of the data that are expected always to be true, called *integrity constraints*. Integrity constraints are an aid to maintaining correctness of the data in the database, because they only allow modifications to the database that conform to these constraints. An example of an integrity constraint might be

integrity constraint

that the stored meal kit type comes from the list of available options, such as "Regular," "Lo carb," or "No gluten."

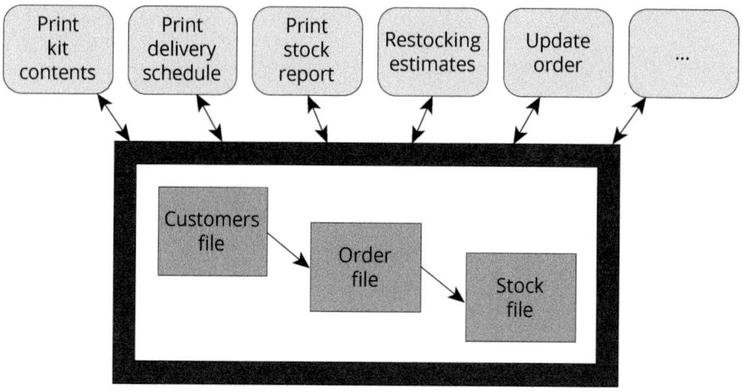

Figure 2.5: Nutty Nuggets stage three: Database approach

The data is collected in one logical, centralized location. A database is created and maintained using a general-purpose piece of software called a *database management system* (DBMS, see Box 2.2 on the following page). The DBMS manages the database by insulating the data from uncontrolled access. A DBMS allows the definition of the data model, supports the manipulation of the data, and provides controlled two-way access channels between the exterior and the database. Not all users have the same requirements of the database: each user group may require a particular window onto the data. This is the concept of a *view*. Views provide users with their own customizable data model, which is a subset of the entire data model, and the authorization to access the sectors of the database that fall within their domain.

DBMS

user view

A database also enables a designer to define the structure of the data in the database, providing levels of authorization that permit different groups of users secure access to data. Different users may access the database at the same time, termed *concurrency*. A DBMS also allows users to access the data in the database without precise knowledge of implementation details, termed *data independence*.

concurrency

data independence

A database can now be more precisely defined as a unified computer-based collection of data, shared by authorized users, with the capability for controlled definition, access, retrieval, manipulation, and presentation of data. Examples of common types of database application include:

Home/office database Home/office database systems are often relatively small and inexpensive, and they may not require concurrent multiuser access. An example is the Nutty Nuggets system.

Enterprise database Enterprise databases are widely used to manage the information of businesses, including financial and personnel data. Enterprise databases must be secure and reliable, run on many platforms, offer high performance, and allow concurrent access by different groups of users. An example might be the database used by a university to manage both staff and student information.

Box 2.2: Elements of a database management system

There are many different components in a DBMS that must work together to answer a query, such as a query to retrieve the address of a customer from the Nutty Nuggets database. The query, communicated to the DBMS using a database *query language*, must first be parsed and analyzed by the *query compiler*. Along the way, the compiler may call the *query optimizer* to optimize the code, so that performance on the retrieval is as good as possible. The *authorization controller* will also ensure the user has the correct privileges to access the requested data. Auxiliary units may be needed to handle constraint enforcement, transaction management, and concurrency control. The *stored data manager* then controls access to physical data storage devices, as well as back-up and recovery in the case of system failure. Throughout the process, stored data must be mapped to the high-level objects referred to in the query using the *metadata catalog*, also termed the *data dictionary*, which stores the information about the structure of the stored data and the data model.

Document store Document stores, also called document-oriented databases, enable storage and retrieval of a wide range of documents, such as web pages, catalogs, and social media and user-generated content. Document stores are highly flexible in the structure of stored data, and so they need specialized query capabilities to efficiently retrieve and query unstructured document contents.

Image database Image and multimedia databases allow the storage and retrieval of a wide range of media data types, such as images, audio, and video. Image and multimedia databases must offer specialized search functions, such as searching for (visually) similar images or for particular information that might appear part-way through a video.

Geodatabase Geodatabases store a combination of spatial and non-spatial data and require complex data structures and analyses. They are discussed throughout this book.

2.1.2 *Transaction management*

transaction

Databases exist to support their users. A *transaction* is the most basic unit of interaction between a user and a database. Transactions primarily involve one or more of the following operations:

- creation of data in the database,
- retrieval of data from the database,
- update of data in the database, or
- deletion of data from the database.

persistence

Data stored in such a way that it continues to exist beyond the session in which it was created—even after the DBMS and the physical storage media have been shut down—is said to be *persistent*. As we encountered in Chapter 1, data that is not persistent and that is only temporarily stored during active computing processes or sessions is termed *volatile*. Together, the four basic operations (creation, retrieval, update, and deletion) on persistent, stored data above are referred to as *CRUD*.

CRUD

A key issue that must be addressed by database transaction management is support for concurrency. Concurrent access to persistent data confronts us with this problem: if the same data item is involved in more than one concurrent transaction, the result may be a loss of database integrity.

T_1	T_2	B	X	Y
		−\$70		
$X \leftarrow B$		−\$70	−\$70	
$X \leftarrow X + \$70$		−\$70	\$0	
	$Y \leftarrow B$	−\$70	\$0	−\$70
	$Y \leftarrow Y - \$40$	−\$70	\$0	−\$110
$B \leftarrow X$		\$0	\$0	−\$110
	$B \leftarrow Y$	−\$110		−\$110

Table 2.1: Lost update for non-atomic interleaved transactions, T_1 and T_2, with variables X and Y and bank balance B

Imagine a Nutty Nuggets customer already owes \$70 on their account. Now suppose that, by chance, two transactions are in progress together. In transaction T_1, the customer is paying their outstanding bill with \$70 credit. Concurrently, a finance staff member is debiting a new invoice for \$40 to the same account in transaction T_2. Table 2.1 shows a particular sequence of the constituent operations of each transaction, termed *interleaving*. Interleaving can improve database performance, because shorter operations may be executed while more lengthy operations are still in progress. However, interleaving must be controlled carefully to avoid potential problems.

interleaving

For example, transaction T_1 begins by reading the customer's balance B from the database into a volatile program variable X. Crediting the customer's payment of \$70 then modifies X to \$0. Before the update operation can conclude by writing this volatile data to persistent storage, however, concurrent transaction T_2 begins. Transaction T_2 likewise starts by reading the same customer's balance from the database, into volatile program variable Y. Transaction T_2 then debits the new invoice from that balance, modifying Y to −\$110. T_1 then completes its update by writing to the database the new balance of $X = \$0$. When subsequently T_2 completes its update to the database, writing its new balance of $Y = -\$110$, it is as if transaction T_1 never occurred. This is a problem known as *lost update*.

lost update

To prevent lost updates and other problems that can occur with concurrent transactions, any DBMS must be able to guarantee four "ACID" properties:

ACID

Atomicity Atomicity dictates that constituent operations of a transaction must either *all* have their effect on the database or else make *no* change to the database. For example, in a transaction to transfer credit from one customer's account to another, debiting the first and crediting the second account, atomicity requires that either both or neither operations are completed, but never one without the other.

Consistency Consistency ensures that all transactions may only result in valid database states compliant with all the database constraints. For example, in a database where the customer balance is constrained to be a number, any transaction that attempts to update a customer's balance to a non-numerical value, such as "sixty dollars" or "overdrawn" should fail to be executed.

Isolation Isolation requires that concurrent transactions must leave the database in the same state as if the transactions were executed sequentially. For example, the lost update problem above is an example of a failure in isolation, where transactions T_1 and T_2 executed in sequence would result in a different outcome to the interleaved transaction in Table 2.1.

Durability Durability necessitates that transaction outcomes are persistent, even in the case of a serious system failure or power outage. One of the common protocols used by databases to support durability is discussed in more detail in Box 2.3 on the next page.

Databases are the engines of today's data-driven applications, including GIS and the WWW. Databases provide reliable, high-performance CRUD operations for authorized access to secure data, all underpinned by ACID guarantees for concurrent transactions. Most importantly, databases give users and developers confidence that data in the database will support transparent and standardized access methods to stored data, independent of the specific details of any data structures. The DBMS is the software component that is responsible for delivering these essential functions and features of any database.

2.2 Database design

A DBMS is a general-purpose information system that must be customized to meet the requirements of particular applications. In order to do this, we need to develop a precise idea of the way that information will be structured in the database, and the way data items will relate to one another. Here, we are not so much concerned with the actual data in the database, as with the *kinds* of data that we expect. For example, in a mapping application, we are not so much concerned with individual data items, "Addis Ababa," "Ethiopia," "Mount Entoto," as with data types, **city**, **country**, and **mountain**. We abstract from information system content to information structure.

As a result, this section focuses primarily on system analysis and design stages of the system development process introduced in Section 1.3.1. During these stages, the main task for the database analyst is the construction of the conceptual computational model for the database, termed a *conceptual data model*. A database designer will then tailor the conceptual data model to the particular kind of DBMS on which the system will be implemented, called a *logical data model*. We will come back to the database design when we look in more detail at the different DBMS models and structures we may need, in Sections 2.3 and 2.4.

Box 2.3: Two-phase commit (2PC) protocols

Durability and atomicity require DBMS have robust mechanism to deal with system failures. One of the most important ways that systems such as DBMS achieve durability and atomicity is the two-phase commit (2PC) protocol. The essence of 2PC is that each subsystem involved in a transaction must coordinate first in the *request* to make a change before actually *completing* any change. The 2PC begins with a transaction manager or coordinator making a request for a change to each subsystem responsible for any resources involved in that change. The subsystems then complete the transaction up to, but not including, finally committing that transaction. During the request phase, each subsystem also writes a persistent log of its own operations to aid in the recovery from any failure. Subsystems then respond directly to the transaction manager to indicate assent to the request, so completing the request phase of 2PC. If the transaction manager receives affirmative responses from all the requested resources, then a second round of messages from the transaction manager directs each subsystem to complete the pending transaction, and commit the change. However, if any subsystems did not respond correctly or in time to the request phase, the transaction manager can instead send a *rollback* or abort message, instructing subsystems to recover the state of the database immediately prior to the transaction from their persistent logs. As a result, only if all components are successful in making the update will the transaction be finally committed. A failure with any constituent operation or subsystem means that the transaction will not be committed and the database will roll back to its last valid state.

2.2.1 *Conceptual data modeling*

conceptual data model

A *conceptual data model* is a model of the proposed database system that is independent of any implementation details. A conceptual data model must express the structure of the information in the system: that is, the types of data and their interrelationships. The correctness (integrity) of the information in a system is often a critical factor in its success. Correctness is maintained as much as possible by the specification of integrity constraints that impose conditions on the static and dynamic structures of the system. A data model should allow the specification of a range of integrity constraints.

A good conceptual data model can act as an efficient means of communication between the analyst, designer, and potential users. This will aid the design and implementation of the system. Further, when the system is eventually implemented, the conceptual data model provides a basic reference for users who need to understand the structure of the data in the system (see Figure 2.6).

In summary, effective conceptual data models:

1. provide a framework that allows the expression of the structure of the system in a way that is clear and easy to communicate to non-specialists;

2. contain sufficient modeling constructs so that the complexity of the system may be captured as completely as possible; and

3. have the capability for translation into implementation-dependent models (i.e., logical and physical models), so that the system may be designed and built.

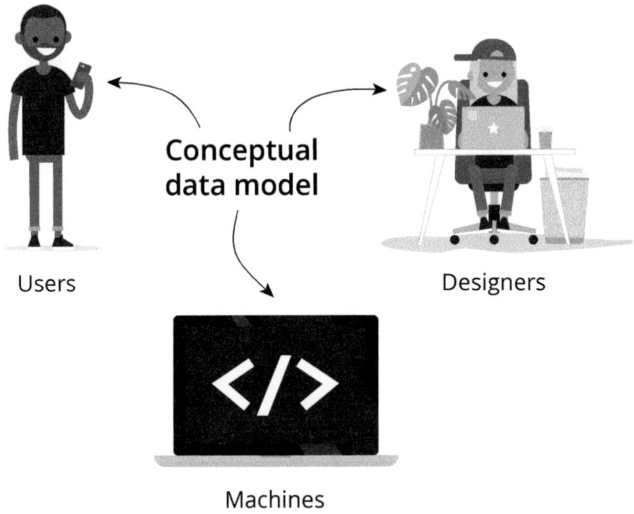

Figure 2.6: The conceptual data model as mediator between users, designers, and machines

2.2.2 *Entity-relationship models*

Imagine that we are responsible for designing the Nutty Nuggets database introduced above. This hypothetical system should contain data on customers (e.g., names, contact details), orders (e.g., meal types and sizes), delivery addresses (e.g., street addresses, zip codes); and meal kit contents (e.g., ingredients and contents). How would we make a start? Well, we have already started in that we have elicited some requirements of the system and expressed these requirements in the form of collections of entities and their relevant properties. This is the simple and powerful idea behind one of the most compelling and widely used approaches to forming a conceptual data model of an information system: the *entity-relationship model* (*E-R model*).

E-R model

entity type

An *entity type* is an abstraction that represents a collection of similar objects, about which the system is going to contain information. In our example, some of the entity types might be **customer**, **order**, and **address**. We make a distinction between the *type* of an entity and an *occurrence* or *instance* of an entity type. For example, we have entity type **address** and occurrences such as '1407 Graymalkin Lane' and '1128 Mission Street' or entity type **customer** and occurrences such as 'Lorna Dane' or 'Roberto Da Costa'. By convention for conceptual modeling, types are rendered in bold, and instance values in single quotes.

entity instance

For conciseness, we will start to use the term "entity" to refer to both "entity types" and "entity instances" as we move through the text as long as the meaning is clear from the context. For example, the caption in Figure 2.8 reads "Two entities ..." rather than "Two entity types ...". However, it is essential always to be precise in our own minds about this fundamental difference between a *type* and an *instance*, between *categories* and *individuals*.

Another aspect where precision is essential is in distinguishing between the entity instance itself and the *identifiers* of that instance. When we write about the customer Lorna Dane, the value 'Lorna Dane' is not the customer herself, of course, but a data item that serves to identify that customer. The data item 'Lorna Dane' is associated with a particular occurrence of the entity (type) **customer** as its name. Thus, entity types have properties, called *attribute types* that serve to describe them. For example, entity type **customer** has attribute types **title**, **given name**, **family name**, and telephone **contact number**. A particular occurrence of **customer** would have associated with it occurrences of these attributes, such as the value '+1-610-555-0195' assigned to the attribute (type) **contact number**.

attribute type

The attachment of attribute types to an entity type may be represented diagrammatically as an *entity-relationship diagram* (*E-R diagram*). As shown in Figure 2.7, entity and attribute types are shown in rectangular segmented boxes; entity types are labeled in the colored box header; attribute types are listed in the section below the header. The E-R diagrammatic notation presented here is common but not standard; there are many variations in use. However, this notation is based on UML (unified modeling language), a standard diagrammatic language widely used for describing many different aspects of information systems.

E-R diagram

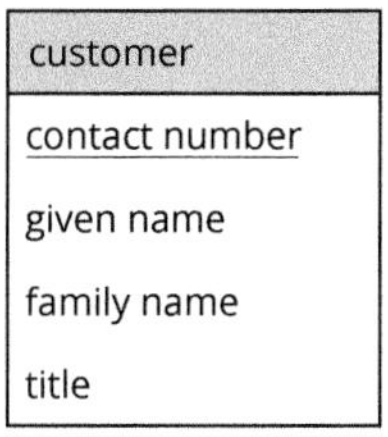

Figure 2.7: An entity type and its attribute types

An important characteristic of an entity is that any occurrence of it should be capable of unique identification. The means of unique identification is through the value of a subset of its attribute types. The name of a person, for example, is frequently insufficient to identify that person uniquely. Similarly, while there is only one place with the address '1128 Mission Street' (in South Pasadena, California), places with the address '11 Mission Street' appear in California, New York, South Carolina, and Connecticut in the US, for example, as well as in other countries around the world. An attribute type or combination of attribute types that serves to identify an entity type uniquely is termed an *identifier*. By definition, an entity type must have at least one identifier. The attributes comprising the chosen identifier are often underlined. In our example, attribute type **contact number** is assumed to be an identifier of **customer** (under the assumption that no two customers share the same telephone number).

identifier

So far, the model comprises a number of independent entity types, each with an associated set of attribute types, one or more of which serve to identify uniquely each occurrence of the type. The real power in this model comes with the next stage, which provides a means of describing connections between entity types. A question that we might want to ask of our finished system is, "Which customers placed which orders?" This question can only be answered by forming a link between customers and their orders. This connection is called a *relationship*. A *relationship type* connects one or more entity types. A *relationship occurrence* is a particular instance of a relationship type. Thus, the incidence of the customer 'Lorna Dane' with the order number '058' is an occurrence of the relationship **placed**. The relationship **placed** between entities **customer** and **order** may be captured in an E-R diagram, as shown in Figure 2.8. Relationships are shown with a labeled line connecting the two entities.

relationship

relationship type

relationship occurrence

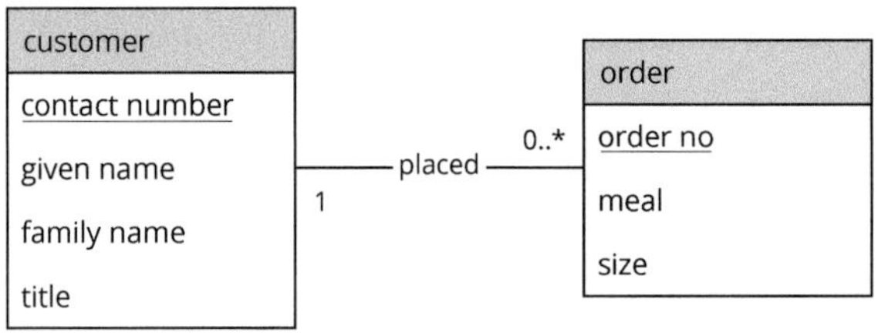

Figure 2.8: Two entities and a many-to-one relationship

Relationships may have their own attributes, which are independent of any of the attributes of the participant entities. In the above example, the relationship **placed** might have the attribute **date**, which gives the date on which the customer placed that order.

Relationship types are subdivided into many-to-one, many-to-many, and one-to-one relationships. The relationship **placed** is an example of a *many-to-one* relationship, because each customer may place several (potentially zero, one, or many) orders, but each order must be placed by exactly one customer. This constraint on the relationship is shown diagrammatically by the 0..* and 1 on each side of the relationship. Thus, E-R diagrams allow the modeler to express *cardinality conditions* upon entity occurrences entering into relationships. Cardinality conditions are a type of integrity constraint where the number of entities participating from each side of the relationship is restricted in some way.

many-to-one

cardinality conditions

Take a moment to note in particular how the location of each label in Figure 2.8 reflects the direction of the relationship. A customer can place zero or more orders so the "0..*" label appears on the right-hand side of the relationship, closest to the **order** entity type (i.e., reading the diagram as: **customer placed** 0..* **order**). An order must be placed by exactly one customer, so the "1" label appears on the left-hand side of the relationship, closest to the **customer** entity type (i.e., read: **order placed** by 1 **customer**). If these symbols were reversed, the diagram would have a very different interpretation (i.e., a rather strange business in which a single order might involve any number of customers, even none at all; but each customer would be required to place one and only one order).

Not all relationships are many-to-one. For example, consider the relationship **contained** between types **order** and **stock** shown as an E-R diagram in Figure 2.9 (top). Each stock item may appear in many different orders, and an order will naturally contain many different stock items (i.e., ingredients).

Such a relationship is called a *many-to-many* relationship. These cardinality conditions are again shown diagrammatically by the 0..* and 1..* on each side of the relationship. Note that these conditions also implicitly specify whether participation in the relationship is *optional* (i.e., 0..*, each ingredient may appear one, many or no orders currently in the system) or *mandatory* (i.e., 1..* each order may have many different ingredients but must have at least one).

many-to-many

The third and final relationship category is exemplified by the relationship **issued** between **staff member** and **staff card.** In this relationship, each staff member must have been issued a staff card, and each staff card is issued to that specific staff member. This is a *one-to-one* relationship. The form of the corresponding E-R diagram is shown in Figure 2.8. The positioning of the two 1s indicates the nature of the relationship.

one-to-one

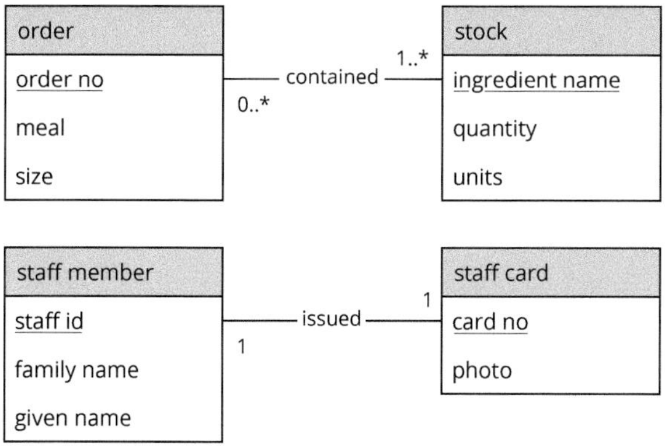

Figure 2.9: Many-to-many (top) and one-to-one (bottom) relationships

All the relationships given as examples so far have been *binary*, in that they have connected together precisely two entity types. A relationship that connects three entity types is called a *ternary* relationship. For example, Figure 2.10 shows a ternary **shipped** relationship, between the **order** that was delivered, the **address** it was delivered to, and the **staff member** who delivered it. Cardinality and participation conditions can apply as with the binary case, but with extra complexity. In the Nutty Nuggets example, we might assume: each order must be shipped to one address by one delivery staff member; but each staff member can deliver zero or many orders to multiple different addresses; and each address might receive zero, one, or many orders, in the latter case possibly delivered by different staff.

ternary relationship

2.2.3 *Example E-R model*

The E-R model can now be applied to the example of the Nutty Nuggets customer database. Imagine you are an owner-manager at the analysis stage of system development. There is a need to hold information about customers and orders, stock and addresses. Choosing entities and attributes for such a model is a matter of judgment, often with more than one acceptable solution. Sometimes it is difficult to decide whether to characterize something as an attribute or an entity. In the modeling of the Nutty Nuggets database, there

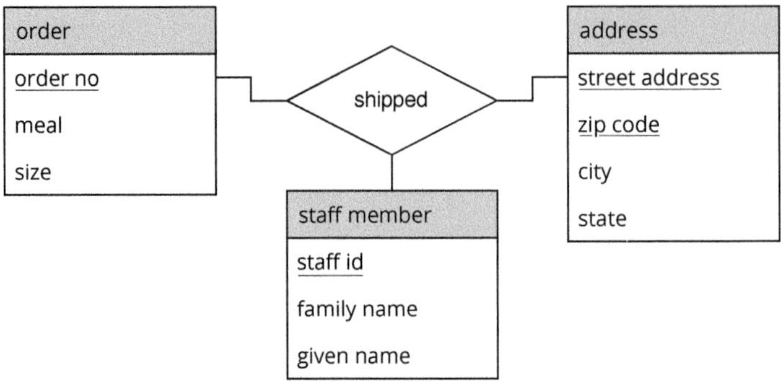

Figure 2.10: A ternary relationship, with cardinality conditions omitted

are several such choices. For example, should we make the city in which an address is located an entity or attribute? Some guidelines include:

- If the data type is relatively independent and identifiable, with its own attributes, then it is probably an entity; if it is just a property of something, then it is an attribute.
- If the data type enters into relationships with other entities (apart from being a property of something), then it is probably an entity.

In our system, the **city** is just a name and just one line of the address, so we choose to make it an attribute of **address**. If **city** had possessed its own attributes, such as **population**, then we would probably have made it an entity and constructed a relationship between entities **city** and **address**. Initial investigation reveals that the following entity types and their attributes are needed:

- **order (order no, meal, size)**
- **customer (given name, family name, contact number)**
- **address (street address, zip code, city, state)**
- **stock (ingredient name, quantity, units)**

Most of the attribute type names are self-explanatory, except that attribute types **meal** and **size** of entity type **order** capture the specific meal scheme ordered from the delivery menu (e.g., regular, low calorie, low carbohydrate, vegan) and the number of people the meal kit needs to feed (individual, couple, small or large family). The attributes **quantity** and **units** of **stock** denote the amount of that ingredient in stock (e.g., 10kg, 6l, or 250g). We have mentioned above that **contact number** is taken to be the identifier of the entity **customer**. By adopting this identifier, you will not be able to take business from customers who don't have a phone or who share their phone with other customers. Identifiers for the other entities require further analysis. For the purposes of this system, you decide that stock can be identified by its ingredient name; that each address can be uniquely identified by a combination of the street address (e.g., '1407 Graymalkin Lane') and zip code (e.g., '10573'); and that each order will require a unique identifier to be created for it, **order no**.

As we have already discussed, the entities **customer** and **order** are connected by the many-to-one **placed** relationship, with each customer optionally placing orders, but each order requiring (i.e., mandatory participation) exactly one customer to have placed that order (Figure 2.8). Similarly, each **order** is required to have at least one but possibly more ingredients from **stock**; while **stock** may optionally be included in zero, one, or many different orders (Figure 2.9). You decide to keep the **shipped** relationship to a simple binary relationship between an **order** and an **address**, rather than additionally capturing the delivery staff in a ternary relationship (cf. Figure 2.10). Each order is required to have exactly one delivery address. Each address in the database, however, may have zero, one, or many orders that need to be shipped there.

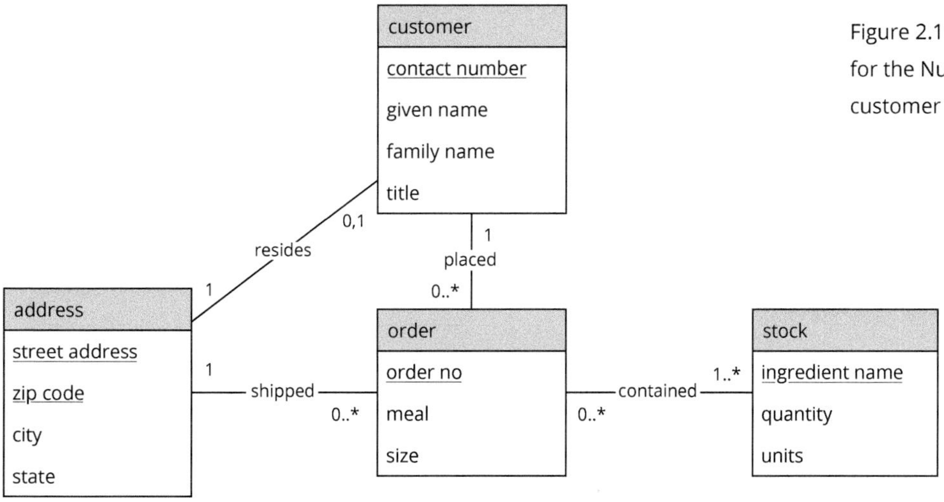

Figure 2.11: E-R diagram for the Nutty Nuggets customer orders database

Finally, you add a new one-to-one **resides** relationship to capture the home address of each customer. Each customer must nominate a single home address (mandatory participation); and if an address is a home address, you decide that there can be only one nominated customer at that address (i.e., optional participation). All this information can be modeled and represented by the E-R diagram shown in Figure 2.11.

Although originally developed in the 1970s, the E-R model remains an important and widely used modeling tool for database design. Measured against the three criteria of a modeling approach given at the beginning of Section 2.2.1, it scores highly. First, the method is based upon the intuitive notions of entity, attribute, and relationship, so it is easily grasped by non-specialists. E-R modeling therefore provides an excellent means of communication between systems analysts and users during the specification of the system. On the third criterion, we shall see later in the chapter how readily a conceptual data model summarized in our E-R diagram may be translated into a practical database implementation.

It is the second criterion, regarding the existence of sufficiently powerful modeling constructs, where the shortcomings of the basic E-R model are sometimes said to lie. This is particularly the case when the application domain to be modeled does not fit into the standard pattern and requires complex data types and relationships. Since it is precisely such systems that are the concern of this book, we will encounter later on more sophisticated modeling approaches that set out to remedy this deficiency. The price to be paid for any increase in modeling power that such extensions can provide is the slight loss of the natural feel and simple diagrams that the basic E-R model possesses.

2.3 *Relational databases*

The most widely used model for the overall structure of a database system is the *relational* model. The most important alternative model is the *graph* database model, covered in more detail in Section 2.4. The *object-oriented* model is less widely used in databases, but it involves some important concepts with direct relevance to modeling in GIS, returned to in Section 4.1.2. The focus of this section, however, is the relational model, introduced in a classic 1970 paper by Ted Codd.

2.3.1 *The relational model*

In the Nutty Nuggets example, we have seen that a database holds not only primary data about entities and their attributes, but also connections between entities as relationships. These connections are at the heart of the relational model. No single piece of data provides information on its own: it is the relationships between data items that provide much of the context for the data.

The structure of a relational database is very simple: this is what makes it so powerful. A relational database is a collection of tabular *relations*, often just called *tables*. It is unfortunate that the terms "relationship" (connection between entities in an E-R model) and "relation" (table in a relational database) are so similar. As we shall see, the concepts behind them are closely related, but it is still important not to confuse them.

Table 2.2 shows part of a relation called CUSTOMER (from the NUGGETS database given in full in Appendix A) containing data about some customers, their title, given and family names, and their contact phone number. A relation has associated with it a set of *attributes*: in Table 2.2 the attribute names are TITLE, GNAME, FNAME, and TEL, labeling the columns of CUSTOMER. The data in a relation is structured as a set of rows. A row, or *tuple*, consists of a list of values, one for each attribute. Each cell contains a single attribute occurrence, or *value*. The tuples of the relation are not assumed to have any particular order.

attribute

tuple

We make a distinction between *relation scheme*, which does not include the data but gives the structure of the relation, and *relation*, which includes the data. Data items in a relation are taken from *domains* (akin to data types

TITLE	GNAME	FNAME	TEL
Mr	Roberto	Da Costa	213-555-0506
Ms	Lorna	Dane	610-555-0195
Dr	Jane	Foster	939-555-0177
Mr	Bobby	Drake	757-555-0112
Mx	Loki	Laufeyson	785-555-0189
Ms	Joanna	Cargill	202-555-0125
...	...	...	...

Table 2.2: Part of the relation CUSTOMER

in programming). Each attribute of the relation scheme is associated with a particular domain. In basic database systems, the possible domains are often quite limited, comprising character strings, integers, floats, dates, etc. In our example, the attribute TEL might be associated with character strings of length 12. We may now give some definitions.

- A *relation scheme* is a set of attribute names and a mapping from each attribute name to a domain. (relation scheme)
- A *relation* is a finite set of tuples associated with a relation scheme in a relational database such that: (relation)
  - each tuple is a labeled list containing as many data items as there are attribute names in the relation scheme; and
  - each data item is drawn from the domain with which its attribute type is associated.
- A *database scheme* is a set of relation schemes, and a *relational database* is a set of relations. (database scheme; relational database)
- The database software that manages a relational database model is termed a *relational database management system* (RDBMS). (RDBMS)

Relations have the following properties:

- The ordering of tuples in the relation is not significant.
- Tuples in a relation are all distinct from one another.
- Columns are ordered so that data items correspond to the attribute in the relation scheme with which they are labeled.

Most relational systems also require that the data items are themselves *atomic*; i.e., they cannot be decomposed as lists of further data items. Thus a single cell cannot contain a set, list, or array. Such a relation, which contains only atomic attributes, is said to be in *first normal form* (1NF). In the example of the CUSTOMER relation, 1NF means that customers are not allowed to have multiple phone numbers or alternative names. The *degree* of the table is the number of its columns. The *cardinality* of the table is the number of its tuples. As tuples come, go, and are modified, the relation will change, but the relation scheme is relatively stable. The relation scheme is usually declared when the database is set up and then left unchanged through the lifetime of the system, although there are operations that will allow the addition, subtraction, and modification of attributes.

1NF

relation degree

cardinality

The theory of a relational database so far described has concerned the structuring of the data into relations. The other aspects of the relational model are the operations that may be performed on the relations (database manipulation) and the integrity constraints that the relations must satisfy. The manipulative aspects will be considered next, after we have described our working example.

2.3.2 *Relational database design*

We saw in Section 2.2 how E-R modeling can help in building a conceptual model of a database. An important criterion for the choice of E-R modeling as our conceptual modeling tool was its ability to translate from an E-R model into a practical database implementation. This section shows how an E-R model can aid in relational database design, and it considers some of the principles upon which a good design is based.

Database design, in the case of a relational database, concerns the construction of a relational database scheme. The central question is, "What characterizes a good set of relations for the target application?" Two advantageous features are:

redundancy

- lack of *redundant* data in relations (redundant data wastes space in the database and causes integrity problems); and
- fast access to data from relations.

These two features essentially trade off space (and integrity) against time. As we shall see later in this chapter, one of the most expensive of relational database operations involves combining information from multiple different relations, termed *relational join*. Fewer joins mean faster data access, which in turn implies fewer relations. Consequently, it is not usually efficient to have many small relations that will need to be joined in order to respond to common queries. On the other hand, while using fewer relations leads to fewer joins, it leads to other problems, shown by the following example from the NUGGETS database.

relational join

Suppose that we decide to have all the information about all the customers and their orders in a single relation in the database. The relation scheme could look something like:

CUSTOMER_ORDERS (TITLE, GNAME, FNAME, TEL, ORDER_NO, MEAL, SIZE)

This relation scheme of large degree might appear suitable if there were likely to be many retrievals requiring customer and order details together. The problem is that this scheme can result in redundant duplication of data, where a customer makes multiple orders. Table 2.3 shows an example of the problem. The relation includes redundant data. Each new order duplicates all of the customer data, including name and contact phone number. Such redundancy not only wastes space, but can also cause integrity problems (for example, if Lorna Dane's phone number changes, but only one of the three cells in the table in which it appears is updated).

TITLE	GNAME	FNAME	TEL	ORDER_NO	MEAL	SIZE
Mr	Roberto	Da Costa	213-555-0506	M315-22-06	Lo carb	4
Ms	Lorna	Dane	610-555-0195	M066-22-06	Regular	2
Ms	Lorna	Dane	610-555-0195	M066-22-07	Regular	1
Mr	Roberto	Da Costa	213-555-0506	M315-22-07	Lo carb	4
Dr	Jane	Blake	939-555-0177	M113-22-09	Vegan	1
Ms	Lorna	Dane	610-555-0195	M066-22-08	Regular	2
...	...	...	...	...	...	...

Table 2.3: Some rows and columns of the CUSTOMER_ORDER relation

The specific problems in the CUSTOMER_ORDERS relation above can be solved by splitting the scheme so that the customer data is held in one relation, and the order data is held in another. A first pass at this relation scheme is below:

CUSTOMER (TITLE, GNAME, FNAME, TEL)

ORDERS (ORDER_NO, MEAL, SIZE)

However, in designing this relation scheme we have lost the connection between the two tables, captured by the **placed** relation in our E-R model in Figure 2.11. Recovering this relationship requires that we first clarify the identifiers for each relation. In relational databases, a *candidate key* is an attribute or minimal set of attributes that will serve to uniquely identify each tuple of the relation. There may be several such candidate keys for a relation. The identifier in an E-R model will usually be a potential candidate key. One candidate key is chosen as the *primary key*.

candidate key

primary key

Following the E-R model in Figure 2.11, we can use TEL and ORDER_NO attributes as primary keys of the CUSTOMER and ORDERS relations, respectively. Armed with our primary keys, we could choose to add a new relation to our relation scheme called PLACED containing just two attributes TEL and ORDER_NO. This relation would then store the data about the relationship between customers and their orders. However, it is instead simpler to add the customer primary key TEL to the ORDERS relation. This construction is called *posting the foreign key*. In technical language, we have posted the identifier of CUSTOMER as a foreign key into ORDERS. Note that the converse—posting ORDER_NO as a foreign key to the CUSTOMER relation—will not work, as it will lead to the same duplication as in Table 2.3.

posting the foreign key

Posting a foreign key, rather than adding another relation, is usually an option if the relationship between entities is many-to-one. The resulting relation scheme is shown below. Just as identifiers for each entity type are underlined in our E-R model in Figure 2.11, so there is a convention that the set of attributes constituting the primary key of the relation is underlined in a relation scheme.

CUSTOMER (TITLE, GNAME, FNAME, TEL)

ORDERS (ORDER_NO, CUS_TEL, MEAL, SIZE)

The relations with the tuples from Table 2.3 appropriately distributed are shown in Table 2.2 above (CUSTOMER relation) and Table 2.4 below (ORDERS). There is now no redundant repetition of data.

Table 2.4: The ORDERS relation, together with the CUSTOMER relation in Table 2.2, capturing all the information in CUSTOMER_ORDER relation Table 2.3 without redundancy

ORDER NO	CUS_TEL	MEAL	SIZE
M315-22-06	213-555-0506	Lo carb	4
M066-22-06	610-555-0195	Regular	2
M066-22-07	610-555-0195	Regular	1
M315-22-07	213-555-0506	Lo carb	4
M113-22-09	939-555-0177	Vegan	1
M066-22-08	610-555-0195	Regular	2
...	...	...	...

Apart from avoiding redundant duplication of information, decomposition of relation schemes has the advantage that smaller relations are conceptually more manageable and allow separate components of information to be stored in separate relations. Of course, relations cannot be split arbitrarily. Relations form connections between data in the database and inappropriate decomposition can destroy these connections. One important guideline for appropriate decomposition has already been introduced in Section 2.3.1, with relations in 1NF having atomic attributes. In fact, there exists a hierarchy of normal forms for relational databases; higher normal forms require higher levels of decomposition of the constituent database relations (see Box 2.4 on page 58). The process of appropriately decomposing relations into normal form is termed *normalization*. Normal forms are useful guidelines for database design. However, in any logical data model, the level of normal form (i.e., degree of relation decomposition) must be balanced against the decrease in performance resulting from the need to reconstruct relationships by join operations.

normalization

In this way, an E-R model may be transformed into a set of relation schemes. As a first pass, the general principle is that to begin:

- each entity in the E-R model maps to a relation in the relation scheme;
- each attribute in the E-R model maps to an attribute in the corresponding relation;
- each identifier maps to a primary key;
- each relationship maps to a relation combining the identifiers/primary keys of the two entities/relations it connects.

In some cases it is also necessary to modify this first pass in order to maintain the information structure, while balancing the redundancy that accompanies too few relations and the inefficiency that arises from too many. The discussion above of the relations CUSTOMER and ORDERS provides one example of modifications needed to strike that balance. Posting the foreign key from CUSTOMER to ORDERS ensures the database scheme avoids the redundancy inherent in a single relation, Table 2.3, without the need for a third PLACED relation.

The database scheme below gives the final relational database scheme for our NUGGETS database.

CUSTOMER (<u>TEL</u>, HID, TITLE, GNAME, FNAME)
ORDERS (<u>ORDER NO</u>, CUS_TEL, AID, MEAL, SIZE)
STOCK (<u>INGREDIENT</u>, QUANT, UNITS)
CONTENT (<u>OID</u>, <u>SID</u>, WEIGHTG)
ADDRESS (<u>ADDRESS ID</u>, STREET_AD, ZIP, CITY, STATE)

With reference again to Figure 2.11, the following summarizes the decisions made in arriving at this database scheme:

- The CUSTOMER relation required the further addition of the foreign key HID as an attribute to identify the home address of each customer, necessitated by the **resides** relation.
- The ORDERS relation likewise includes the foreign key AID to capture the shipping address of each order (relation **shipped**).
- The STOCK relation with its attributes is mapped directly from the **stock** entity.
- The CONTENT relation captures the data associated with the **contained** relationship. Posting the foreign key (from ORDERS to STOCK or vice versa) is not an option for this many-to-many relationship without resulting in the kinds of redundancy already seen in Table 2.3. Hence, the primary key of the CONTENT relation becomes the combination of the foreign keys of ORDERS and STOCK: <u>OID</u> and <u>SID</u>, respectively. A primary key that is made up of two or more foreign keys is called a *compound key*. compound key
- It was further decided that it would be useful to know the quantity of each ingredient in the order (and not simply in the stockroom, already stored in the STOCK relation). Although this feature was not captured in the design of the E-R model in Figure 2.11, it is not unusual to need to tweak and refine designs during the implementation process. Consequently the weight in grams of each ingredient in each order is captured through the attribute WEIGHTG.

Finally, the ADDRESS relation has acquired a new primary key <u>ADDRESS ID</u>. An alternative candidate key is suggested by the identifiers of the **address** entity in Figure 2.11: the combination of attributes STREET_AD and ZIP. A candidate key that is made up of two or more attributes is called a *composite key*. (A compound key, above, is therefore a special case of a composite key.) composite key
Three reasons led to this decision. First, posting a composite foreign key from ADDRESS to CUSTOMER and ORDERS is possible, but it is arguably less clear for database users than posting a singleton foreign key. Second, the STREET_AD attribute is expected to be a relatively long text string in many cases (e.g., '1407 Graymalkin Lane'). Using such attributes as keys, and especially foreign keys, increases the chances of errors, for example, where a mistype in one table breaks the connection between the tuples. Hence, creating a new, simplified address code in <u>ADDRESS ID</u> as the primary key in ADDRESS was

Box 2.4: Second normal form (2NF)

As introduced above, normalization is the process of decomposing relations so that they conform to certain standard principles used to reduce redundancy and increase integrity of stored data, called *normal forms*. As we have seen, *first normal form* (1NF) requires all relations in a relational database to have atomic attributes. Each higher normal form builds on the lower normal form, such that the first condition of *second normal form* (2NF) is that a relation is in 1NF. 2NF further requires that all non-key attributes are "fully functionally dependent" on the primary key alone. "Full functional dependency" means that an attribute is dependent on the whole primary key, and not on a subset of that key. For example, the original ADDRESS table with the relation scheme ADDRESS (STREET AD, ZIP, CITY, STATE) is arguably not in 2NF. A violation of 2NF arises if the CITY or STATE attributes are functionally dependent on the ZIP attribute (i.e., if I know the zip code, then I can deduce the city or state). ZIP is only a part of the primary key for ADDRESS, so in that case CITY and STATE are not fully functionally dependent on the primary key, only part of the primary key, and so the relation is not in 2NF. Regaining 2NF would require that the relation be split into two, with each part of the compound key becoming the primary key of a new, decomposed relation. So why "arguably"? While this partial dependency holds in most cases, some US zip codes do straddle multiple cities and even states (such as 42223 that covers addresses on both sides of the border between Kentucky and Tennessee). Consequently, it could be argued that in order to allow for this eventuality, the relation is in 2NF.

considered to be worth the small additional redundancy it entailed (and see also the inset on second normal form, Box 2.4).

The full NUGGETS relational database can be found in Appendix A.

2.3.3 Operations on relations

A relation is nothing more than a structured table of data items. Each column of a relation is named by an attribute, and it has all its data items taken from the same domain. The basic operations supported by a relational database are therefore simple. There are five fundamental *relational operators*: *union*, *difference*, *product*, *project*, and *restrict*.

relational operator

The first three of these are traditional set-based operators, introduced in the next chapter. The project and restrict operators are described below. Three further relational operators *intersection*, *divide*, and *join*, termed *derived* relational operators, can be expressed using different combinations of the fundamental five operators.[1] Of these, intersection is another set-based operator introduced in the next chapter, join is described below, and divide is a less commonly used operator not discussed further here. The structure of these operations and the way that they can be combined is called *relational algebra*.

[1] Can you deduce how to define the three derived relational operators in terms of the five fundamental operators? Which two fundamental operators when combined can result in the join operator, described below, for example?

relational algebra

The relational model is *closed* with respect to all the above relational operations, because they each take one or more relations as input and return a relation as a result. The set operations union, intersection, product, and difference work on the relations as sets of tuples. Thus, if we have two relations, one holding all Californian and one all Texan addresses, then their union will hold the addresses in both states and their intersection will be empty. For all the set operations except product, the relations must be *compatible*, in that they must have the same attributes; otherwise, the new relation will not be well formed.

Project operator The *project* operation is unary, applying to a single relation. It returns a new relation that has a subset of attributes of the original. The relation is then modified so that any duplicate tuples formed are coalesced. The project operator π has the following syntax:

project operator

$$\pi_{<attribute\ list>}(relation)$$

For example, $\pi_{\text{CITY,STATE}}(\text{ADDRESS})$ returns the relation shown in Table 2.5a, and $\pi_{\text{STATE}}(\text{ADDRESS})$ returns the relation shown in Table 2.5b. Note that in the second case the four identical tuples containing the value "NY" have been coalesced into a single tuple.

Table 2.5: Results of relational projections and restrictions

(a)

CITY	STATE
New York	NY
North Salem	NY
Philadelphia	PA
Washington	DC
Fort Washington	NY
Broxton	OK
Brooklyn	NY
San Francisco	CA

(b)

STATE
NY
PA
DC
OK
CA

(c)

ORDER_ID	CUS_TEL	AID	MEAL	SIZE
M113-22-09	939-555-0177	74012-RA01	Vegan	1
M315-22-06	213-555-0506	94110-MS01	Lo carb	4
M315-22-07	213-555-0506	94110-MS01	Lo carb	4

(d)

ORDER_ID
M113-22-09
M315-22-06
M315-22-07

Restrict operator The *restrict* operation is also unary. The restrict operator works on the tuples of the table rather than the columns, and it returns a new relation that has a subset of tuples of the original. A condition specifies those tuples required. The restrict operator is often referred to as the *select* operator, and consequently it is denoted with the Greek symbol σ (sigma). The syntax used here is:

restrict operator

$$\sigma_{<condition>}(relation)$$

For example, the list of meal orders other than 'Regular' can be retrieved from the database using the expression $\sigma_{\text{NOT(MEAL='Regular')}}(\text{ORDERS})$. This will return the relation shown in Table 2.5c. Operations can be combined, for example:

$$\pi_{\text{ORDER\_ID}}(\sigma_{\text{NOT(MEAL='Regular')}}(\text{ORDERS}))$$

returns the order numbers of non-'Regular' meal orders, as shown in Table 2.5d.

join operator

natural join

Join operator With the *join* operation, the relational database begins to merit the term "relational." Join is a binary operator that takes two relations as input and returns a single relation. The join operation allows connections to be made between relations. There are several different kinds of relational join but we describe only the *natural join* of two relations, defined as the relation formed from all combinations of their tuples that agree on a specified common attribute or attributes. The join operator ⋈ has the following syntax:

$$\bowtie_{attribute_1 = attribute_2}(relation_1, relation_2)$$

to indicate that $relation_1$ and $relation_2$ are joined on attribute combinations $attribute_1$ of $relation_1$ and $attribute_2$ of $relation_2$. For example, to relate details of orders to the customers who placed each order on which they are showing, relations CUSTOMER and ORDERS are joined on the film title attribute in each relation. The expression is:

$$\bowtie_{\text{TEL=CUS\_TEL}}(\text{CUSTOMER, ORDERS})$$

The resulting relation shown in Table 2.6 combines tuples of CUSTOMER with tuples of ORDERS, provided that the tuples have the same customer phone number. Notice that the join has not repeated the duplicate attribute.

TEL	GNAME	FNAME	HID	TITLE	ORDER_ID	AID	MEAL	SIZE
213-555-0506	Roberto	Da Costa	94110-MS01	Mr	M315-22-06	94110-MS01	Lo carb	4
213-555-0506	Roberto	Da Costa	94110-MS01	Mr	M315-22-07	94110-MS01	Lo carb	4
610-555-0195	Lorna	Dane	10560-GL01	Ms	M066-22-06	19104-CS01	Regular	2
610-555-0195	Lorna	Dane	10560-GL01	Ms	M066-22-07	19104-CS01	Regular	1
610-555-0195	Lorna	Dane	10560-GL01	Ms	M066-22-08	19104-CS01	Regular	2
939-555-0177	Jane	Foster	74012-RA01	Dr	M113-22-09	74012-RA01	Vegan	1

Table 2.6: Result of relational join

If we only require the customer contact numbers, given names, and order numbers of non-'Regular' meal orders, we may again restrict and project the relation in Table 2.6 to only those desired attributes and rows. The combined relational algebra statement is below, with the resulting table shown in Table 2.7:

$$\pi_{\text{TEL,ORDER\_ID,GNAME}}(\sigma_{\text{NOT(MEAL='Regular')}}(\bowtie_{\text{TEL=CUS\_TEL}}(\text{CUSTOMER, ORDERS})))$$

Table 2.7: Result of combined relational join, restriction, and projection

TEL	GNAME	ORDER_ID
213-555-0506	Roberto	M315-22-06
213-555-0506	Roberto	M315-22-07
939-555-0177	Jane	M113-22-09

The last example may be used to demonstrate an important property of relation operations: the order in which operations are performed will affect performance. The join operation is the most time-consuming of all relational operations, because it needs to compare every tuple of one relation with every tuple of another. To extract data for Table 2.7 we performed operations join, project, and restrict. In fact, it would have been more efficient to have first done a restrict operation on the ORDERS table, then joined the resulting smaller table to CUSTOMERS, and then projected. The result would be the same, but the retrieval would perform better, because the join involves smaller tables.

In general, reordering the elements of a relational algebra expression may not lead to an equivalent expression. For example, the relational algebra expression $\pi_{\text{ORDER\_ID}}(\sigma_{\text{NOT(MEAL='Regular')}}(\text{ORDERS}))$ given above is not equivalent to $\sigma_{\text{NOT(MEAL='Regular')}}(\pi_{\text{ORDER\_ID}}(\text{ORDERS}))$, because MEAL is not a valid attribute of the relation $\pi_{\text{ORDER\_ID}}(\text{ORDERS})$. The topic of *query optimization* is a critical study for high-performance databases, concerned with processing queries as efficiently as possible. An important component of query optimization involves performing transformations (such as reordering) upon queries, to produce equivalent queries that can be processed more efficiently.

query optimization

2.3.4 *Structured query language*

The *structured query language* (SQL) provides users of relational databases with the facility to define the database scheme (data definition), and then insert, modify, and retrieve data from the database (data manipulation). The language may either be used on its own, as a means of direct interaction with the database, or embedded in a general-purpose programming language. The aim of this section is to provide an introduction to SQL, highlighting its close relationship with the relational algebra, without making any attempt to be a complete SQL reference.

SQL

Data definition using SQL The *data definition language* (DDL) component of SQL allows the creation, alteration, and deletion of relation schemes. Normally, a relation scheme is altered only rarely once the database is operational. A relation scheme provides a set of attributes, each with an associated data domain. SQL allows the definition of a domain by means of the expression below (square brackets indicate an optional part of the expression).

data definition language

```
CREATE DOMAIN name datatype
 [DEFAULT definition]
 [CHECK constraint]
```

The user specifies the name of the domain and associates that name with a predefined data type, such as a character string (VARCHAR), or an integer, float, date, time. The DEFAULT definition allows the user to specify a default value for a tuple; a common default value is NULL. The domain CHECK defines integrity constraints by restricting the domain to a set of specified values. An example of the definition of a domain for the attribute MEAL is as follows:

```
CREATE DOMAIN MEALTYPE VARCHAR(10)
  CHECK (VALUE IN ('Regular', 'Vegan', 'Lo carb', 'Lo fat', 'No gluten'));
```

Box 2.5: SQL and CRUD

SELECT statements covered in the main text are at the heart of SQL, and they provide access to the relational database retrieval operations. In addition, basic data creation, update, and deletion operations (found in CRUD) are provided by INSERT, UPDATE, and DELETE statements. For example:

```
INSERT INTO STOCK VALUES ('Flour', 37, 'kg');
```

will record a new entry in the STOCK table. In addition, the SELECT INTO statement makes a new table that is a copy of part or all of an existing table, such as:

```
SELECT * INTO STOCKBACKUP FROM STOCK;
```

The INSERT INTO can similarly copy data across from one table into another, suitably structured existing table. UPDATE and DELETE statements have a WHERE clause to specify which records to alter. For example:

```
DELETE FROM STOCK WHERE INGREDIENT='Oil';
```

deletes the oil from the STOCK table, and:

```
UPDATE STOCK SET QUANT=8 WHERE INGREDIENT='Corn';
```

updates the stored corn stock quantity.

A relation scheme is created as a set of attributes, each associated with a domain, with additional properties relating to keys and integrity constraints. For example, the relation scheme ORDERS can be created by the command:

```
CREATE TABLE ORDERS
  ORDER_NO CHAR(10),
  CUS_TEL CHAR(12),
  AID CHAR(10),
  MEAL MEALTYPE,
  SIZE INT(1),
  PRIMARY KEY (ORDER_NO),
  FOREIGN KEY (CUS_TEL) REFERENCES CUSTOMER(TEL),
  FOREIGN KEY (AID) REFERENCES ADDRESS(ADDRESS_ID),
  CHECK (SIZE < 7 AND SIZE > 0);
```

This statement begins by naming the relation scheme (called a table in SQL) as ORDERS. The attributes are then defined by giving their name and associated domain. It is optionally possible to identify to the database any keys, including the foreign keys posted from CUSTOMER and ADDRESS. Doing so enables the database to maintain referential integrity, for example, ensuring that if an address is deleted (or updated) from the ADDRESS relation, then any reference to it is also deleted (or updated) in ORDERS. Finally, one further integrity check is added to limit the meal size of an order to between 1 and 6 people, inclusive. Any attempt to insert a row with an order for 8 people, say, will be disallowed.

Data manipulation using SQL Having defined the relation schemes, the next step is to insert data into the relations. These SQL commands are quite straightforward, allowing insertion of single or multiple tuples, update of tuples in tables, and deletion of tuples (see Box 2.5). Data retrieval forms the most complex aspect of SQL: a large book could be written on this topic alone. Our treatment is highly selective, giving the reader a feel for SQL in this respect. The general form of the retrieval command is:

```
SELECT item-list
  FROM reference-list
```

```
[WHERE condition]
[GROUP BY attribute-list]
[HAVING condition]
```

A simple example of data retrieval, already considered in the relational algebra section, is to find the order number of all orders for meals other than 'Regular'. The corresponding SQL expression is:

```
SELECT ORDER_NO
  FROM ORDERS
  WHERE NOT(MEAL='Regular');
```

The SELECT clause serves to project (how confusing!) on the required ORDER_NO attribute. The FROM clause tells us from which table the data is coming, in this case ORDERS. The WHERE clause provides the restrict condition.

Relational joins are effected by allowing more than one relation (or even the same relation called twice with different names) in the FROM clause. For example, to find details of orders and where they are shipping to, we could give the following SQL command:

```
SELECT ORDER_NO, MEAL, SIZE, STREET_AD, ZIP, CITY, STATE
  FROM ORDERS, ADDRESS
  WHERE ORDERS.AID = ADDRESS.ADDRESS_ID;
```

In this case, the WHERE clause provides the *join condition* by specifying that tuples from the two tables are to be combined only when the values of the attributes AID in ORDERS (indicated by ORDERS.AID) and ADDRESS_ID in ADDRESS (ADDRESS.ADDRESS_ID) are equal. A more complex case, using all the clauses of the SELECT expression, is the following expression, which retrieves the total weight of ingredients in all orders with more than three items.

```
SELECT ORDER_NO, SUM(WEIGHTG)/1000 AS 'TOTAL (kg)'
  FROM ORDERS, CONTENT
  WHERE ORDERS.ORDER_NO = CONTENT.OID
  GROUP BY ORDER_NO
  HAVING COUNT(*) > 3;
```

With the exception of the term SUM(WEIGHTG)/1000 AS 'TOTAL (kg)', the first three lines of code act to retrieve the orders and associated ingredients, using the join of ORDERS and CONTENT. The GROUP BY clause serves to logically construct a table where the tuples are in groups, one for each order. The HAVING clause comes into play to operate as a filter condition on the groups in the grouped relation. It selects only groups that have a tuple count of at least three. This intermediate table (Table 2.8) is not a legal first normal form relation because values in most cells are not atomic (i.e., there are mini-tables within most cells). But we are concerned with the total weight for each order and will eventually project out unused attributes.

ORDER_ID	CUS_TEL	AID	MEAL	SIZE	SID	WEIGHTG
M066-22-06	610-555-0195	19104-CS01	Regular	2	Chickpeas	400
	610-555-0195	19104-CS01	Regular	2	Chili	10
	610-555-0195	19104-CS01	Regular	2	Rice	400
	610-555-0195	19104-CS01	Regular	2	Tomato	400
M066-22-08	610-555-0195	19104-CS01	Regular	2	Carrot	300
	610-555-0195	19104-CS01	Regular	2	Corn	200
	610-555-0195	19104-CS01	Regular	2	Potato	300
	610-555-0195	19104-CS01	Regular	2	Vegetable stock	500
M113-22-09	939-555-0177	74012-RA01	Vegan	1	Chickpea	400
	939-555-0177	74012-RA01	Vegan	1	Garlic	20
	939-555-0177	74012-RA01	Vegan	1	Lemon	200
	939-555-0177	74012-RA01	Vegan	1	Mushroom	600
M315-22-06	213-555-0506	94110-MS01	Lo carb	4	Chickpea	800
	213-555-0506	94110-MS01	Lo carb	4	Chili	20
	213-555-0506	94110-MS01	Lo carb	4	Corn	600
	213-555-0506	94110-MS01	Lo carb	4	Tomato	600

Table 2.8: Evaluation of an SQL query to the NUGGETS database: intermediate stage

The final stage of the retrieval, after projecting out the unwanted attributes, is shown in Table 2.9. The weights in the WEIGHTG column in Table 2.8 are aggregated by summation. SUM and COUNT are built-in SQL functions, along with a range of other aggregation functions including AVG, MAX, and MIN. Failing to specify any aggregate function for any attributes in the final projection (except the already grouped-by ORDER_ID) will result in an SQL error. In constructing the final result, dividing the SUM(WEIGHTG) attribute by 1000 in the projection converts the attribute values from grams to kilograms. The SQL AS keyword renames the column in the final output to make it more easily understandable.

ORDER_ID	TOTAL (kg)
M066-22-06	1.210
M066-22-08	1.300
M113-22-09	1.220
M315-22-06	2.020

Table 2.9: Evaluation of an SQL query to the NUGGETS database: final result

There is much more to SQL than could be covered in a text such as this, but a wealth of books and online resources are available to assist the interested reader in learning SQL. In addition to giving a flavor of the main capabilities of language, the brief introduction to SQL provided here is also designed to highlight the close relationship between the SELECT statement at the heart of SQL and the relational algebra operations that underpin it. Other database query languages, such as queries to the graph database model, cannot boast such rigorous formal foundations, as we shall discover in the following section.

2.4 *Graph databases*

The second main model of the overall structure of a database is the *graph* model. Where the relational model is founded on the table as its fundamental data structure, the graph model uses *abstract graphs*. Abstract graphs—referred to simply as "graphs" where there is no possibility of confusion—offer much greater flexibility in how data is structured than the more rigid tables we have seen in a relational databases. This additional flexibility comes at the cost of additional technical challenges in querying stored data efficiently. As a result, although graph DBMS have existed since the 1960s, it is only since around the mid-2000s that scalable graph DBMS have become widely available.

2.4.1 *Abstract graphs*

A *graph* is a highly abstracted model of connectivity between elements. Graphs are important models underpinning many spatial concepts, not only graph databases. So in this section we will take the time to introduce abstract graphs systematically.

graph

A graph is made up of a set of distinct *nodes* together with a set of *edges* that connect pairs of nodes. It is usual to summarize a graph by means of a diagram. The graph in Figure 2.12 consists of the five nodes *a*, *b*, *c*, *d*, and *e* and five edges. A node is said to be *adjacent* to another node if both nodes are connected by an edge. In Figure 2.12, nodes *a* and *b* are adjacent, but nodes *a* and *c* are not. An edge is said to *join* or to be *incident on* the nodes it connects. Likewise, a node is said to be *incident* with an edge that connects to it.

node

edge

adjacency

incident

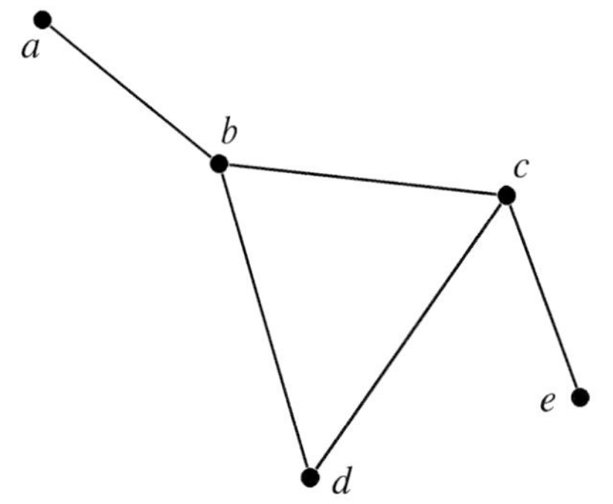

Figure 2.12: A connected, undirected graph with order 5

The *order* of a graph is the number of nodes it contains. The *degree* of a node is the number of edges with which it is incident. For example, the degree of nodes *b* and *c* in Figure 2.12 is three; the degree of node *d* is two; and the degree of nodes *a* and *e* is one. A *path* through a graph is a sequence of nodes in the graph where each consecutive pair of nodes is adjacent, i.e., connected by an edge in the graph. Examples of paths between nodes *a* and *e* in Figure 2.12 are *abce*, *abdce*. In contrast, the sequence *ace* is not a path in the graph because nodes *a* and *c* are not connected by an edge. A *connected graph* is one such that there exists a path between any two of its nodes. The

order

degree

path

connected graph

graph in Figure 2.12 is connected.

cycle

A path from a node to itself traversing at least one edge is called a *cycle*. There are two cycles in Figure 2.12: ***bcdb*** (or equivalently ***cdbc*** or ***dbcd***) and ***bdcb*** (or ***cbdc*** or ***dcbd***). A graph that has no cycles is called an *acyclic graph*.

acyclic graph

Two graphs may show exactly the same connectivity relationships. Such graphs are said to be *isomorphic*. Thus, graphs Figure 2.12 and Figure 2.13 are isomorphic, since both have precisely the same nodes and edges. In this case, we have made the case clearer by labeling the nodes to show the isomorphism.

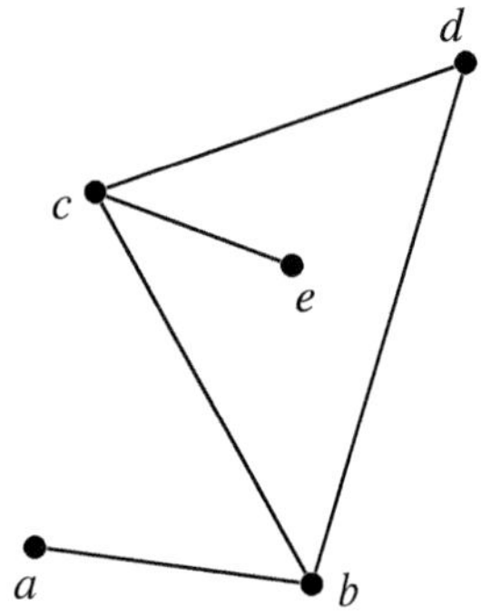

Figure 2.13: A graph isomorphic to that in Figure 2.12

undirected graph

The abstract graphs above are *undirected*, meaning that the direction of the edges is not significant. Thus, an undirected edge from a to b, written $\{a, b\}$, is the same as that from b to a (i.e., $\{a, b\}$ is equivalent to $\{b, a\}$). A *directed graph* or *digraph* is a graph in which each edge is assigned a direction. Thus in a directed graph an edge from a to b, written ***ab***, is distinct from an edge from b to a, written ***ba***. In diagrams, directed edges are usually indicated by arrowed lines.

directed graph

Figure 2.14 shows a directed graph of order 5. Because the direction of edges is significant, edge ***ba*** is an edge of the graph in Figure 2.14, but ***ab*** is not. Hence, in Figure 2.14 node b is adjacent to node a, but a is not adjacent to b (since there exists no directed edge *from* a *to* b). The asymmetry introduced by directed graphs means we have to refine our undirected definition of connected graphs above. A directed graph that satisfies the definition of a connected graph above—that there exists a path between every pair of nodes—is termed *strongly* connected. The graph in Figure 2.14 is not strongly connected. Replacing all the directed edges with undirected edges, as in Figure 2.12, does lead to a connected graph though. As a result, the graph in Figure 2.14 is said to be *weakly* connected. Closely related, a directed graph where there exists a path either from x to y *or* from y to x for any pair of nodes x and y is termed *semiconnected*.

[2] Can you construct your own example of a weakly connected, but not semiconnected graph? Hint: it is possible to achieve this by reversing the direction of just one of the directed edges in Figure 2.14.

strongly connected

weakly connected

semiconnected

The graph in Figure 2.14 is both weakly connected and semiconnected. Indeed, every semiconnected graph is also weakly connected; not every weakly connected graph is semiconnected, however.[2]

Four further types of graph are especially useful in representing data:

labeled graph

- A *labeled graph* is a graph in which each edge and/or node is assigned

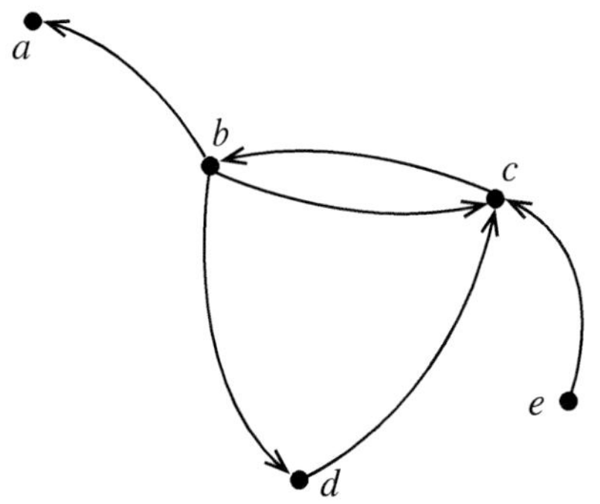

Figure 2.14: A directed, semiconnected graph of order 5, not isomorphic to that in Figure 2.12

a label (maybe a number or string). Labels are usually indicated on a diagram near to the corresponding edges or nodes.

- A *weighted graph* is a labeled graph in which each label is a (usually non-negative) number, or weight. Weighted graphs are of particular importance in GIS, as we shall see in the next chapter.
- A *multigraph* is a graph that supports multiple distinct edges between a pair of nodes, whereas ordinarily there can be at most one edge between any pair or nodes.
- A *hypergraph* is a graph where a single edge may connect more than two nodes, whereas ordinarily an edge connects exactly two nodes. Hypergraphs are mentioned here purely for completeness, but are not encountered again in this book.

weighted graph

multigraph

hypergraph

We will return to the topic of graphs in more detail in later chapters, as they are also important tools for representing spatial networks and spatial relationships. For example, one can imagine the usefulness of a graph in modeling the road network in a city center. In that case, edges can represent roads, with nodes representing road intersections. A directed graph could be used to additionally capture information about one-way streets. A labeled graph can be useful for storing road names, road lengths, or travel times.

However, for now we have enough knowledge of abstract graphs to introduce graph databases.

2.4.2 *The graph database model*

The most common graph databases are founded on three basic structures:

- Nodes are used to represent entities in a domain. Nodes are approximately equivalent to a record or a tuple in a relational database.
- Edges are used to represent relationships between entities. Edges are central to the difference between graph and relational databases. The relational model contains no comparable structure.
- *Properties* are data associated with nodes or edges. Properties are akin to attribute values in relational databases.

graph property

The most common graph databases structure their data as a directed, labeled multigraph of nodes, edges, and associated properties, known as the property graph model

property graph model. Another closely related graph database model which adopts a slightly different structure is called the *triple store.* Triple stores are particularly important in the context of linked open data and the Semantic Web, which we return to later in this book in Chapter 9.

The property graph model is an incredibly flexible way to structure and store an enormous range of different data. Indeed, this description of the property graph model may sound already familiar to you: the simpler E-R diagrams encountered earlier in this chapter can themselves be thought of as property graphs.[3]

[3] Indeed, edges in a graph database are often also called *relationships*, just like the corresponding structure in the E-R model. However, to avoid confusion with E-R model "relationships" (not to mention relational database "relations"!), we will continue to use the term *edge* in this book, highlighting the underlying connection to abstract graphs.

Figure 2.15 shows the Nutty Nuggets customer data within a graph database, displaying the same data encountered earlier in Table 2.2. Each tuple in the relational database table corresponds to a node in the graph database. Each data item in the relational database table can be found as a property of the corresponding node.

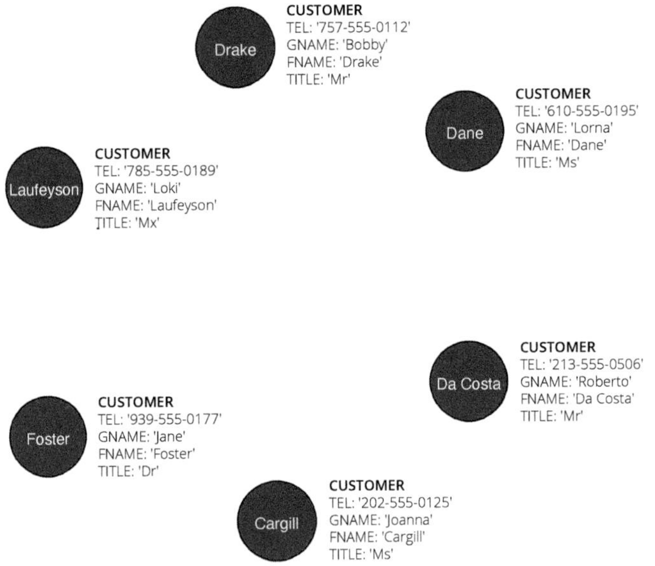

Figure 2.15: A view of the Nutty Nuggets graph database showing the **customer** data from Table 2.2

The graph in Figure 2.15 contains only nodes and properties, no edges. Adding in more data from the database into the view, we can see our first edges in Figure 2.16. As highlighted above, the relational model has no direct correspondent to edges; edges are a distinguishing feature of graph databases. However, an important consequence of this difference is that the Nutty Nuggets graph database needs no foreign keys. The customers in Figure 2.16 need no HID property, unlike the CUSTOMER relation scheme developed in Section 2.3.2. Instead, a RESIDES edge connects each customer with a home address. Additionally that edge is directed, indicating that it is the customer that resides at the address, not the address that resides in the customer. This support for relationship directionality is absent from the relational data model.

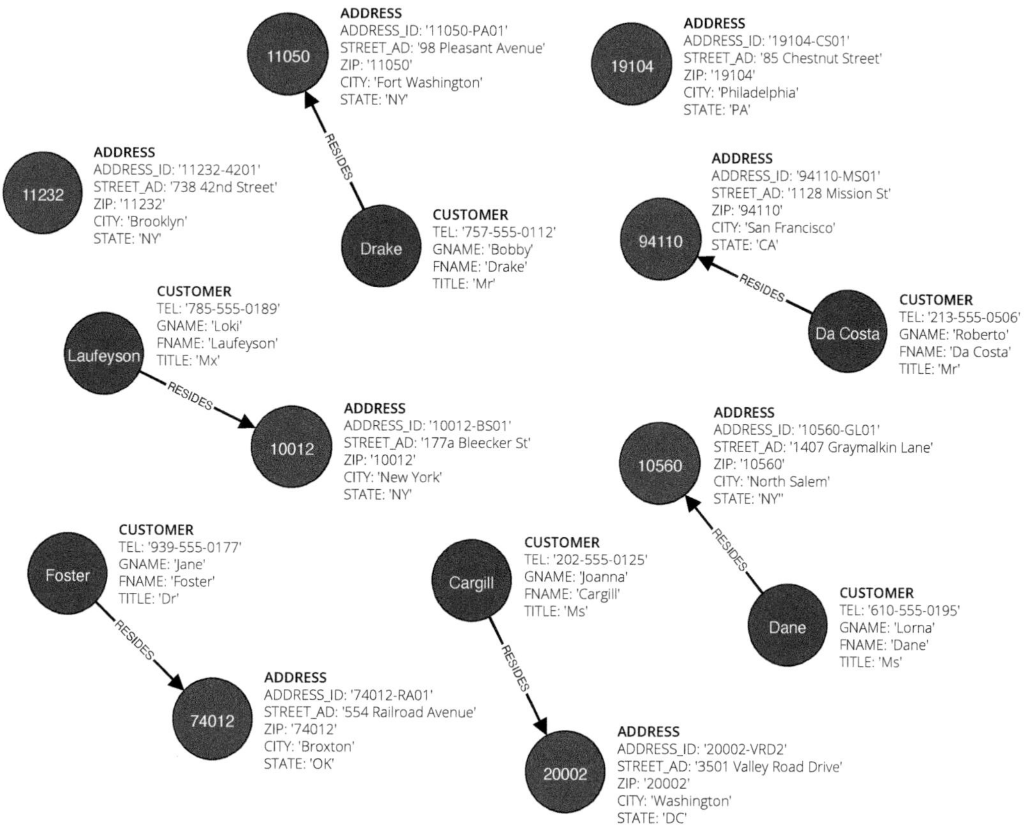

Figure 2.16: A view of the Nutty Nuggets graph database showing both the **customer** and **address** data

The resulting graph database view in Figure 2.16 also contains two isolated address nodes with no incident edges. These nodes relate to delivery addresses in the database that are not customer home addresses.

2.4.3 *Graph database design*

The process of database design begins in exactly the same way for a graph database as we have already seen for a relational database. Indeed, here we see emphasized again the most important property of any conceptual data modeling: it must be independent of any specific implementation. Hence, the E-R conceptual model developed in Section 2.2.1 and summarized in Figure 2.11 concerns the entities and relationships of interest in our application, but it is agnostic on the structure of the data itself.

Consequently, we have already completed the first and most important step in designing our graph database: designing the E-R model in Section 2.2.1. The next step is to again translate that model into a database scheme, but this time a graph database rather than relational database scheme. This step highlights an appealing feature of graph databases that has contributed greatly to their popularity. Whereas relational database design is a careful and methodical process with important implications for subsequent database performance, graph database design typically involves little more than identifying what will be represented as a node, as an edge, and as a property.

Some natural mappings immediately suggest themselves as a first pass, namely:

- each entity in the E-R model maps to a set of nodes in the graph database;
- each attribute in the E-R model maps a property found on each corresponding node;
- each relationship maps to a set of edges connecting corresponding nodes.

Unlike a relational database, a graph database does not usually require primary keys to be identified from the conceptual model. Most graph DBMS will instead automatically assign a unique system identifier to every node and edge, but they may also allow an identifying attribute from the data to be used in place of the system identifier. Without primary keys, a graph database also needs no foreign keys, as we have seen. Instead, we must be careful to capture explicitly any relationships in the E-R model as edges connecting pairs of nodes.

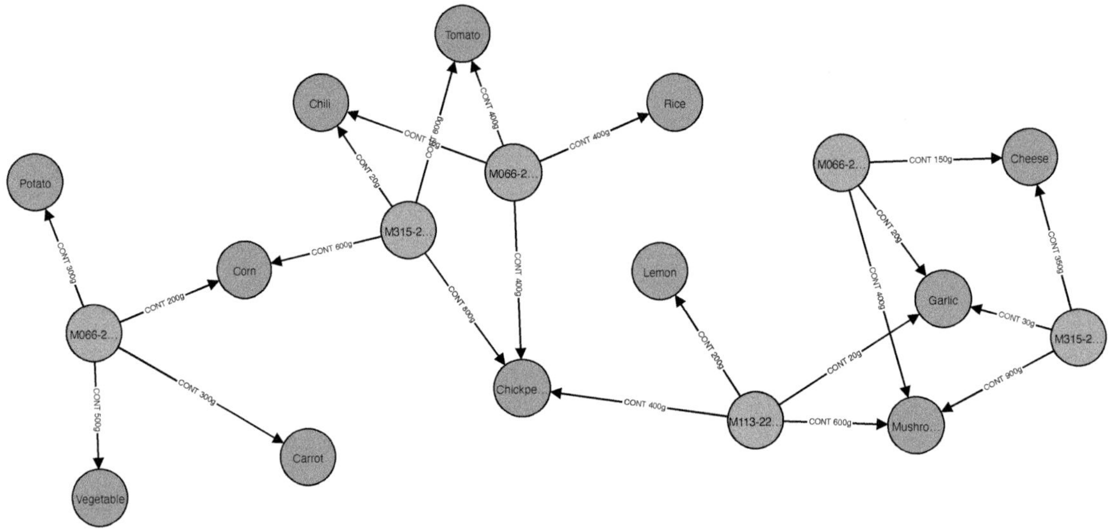

Figure 2.17: A view of the Nutty Nuggets graph database showing the WEIGHTG property of the CONT edge connecting STOCK and ORDERS nodes, node properties omitted

In a graph database, edges can have properties just as nodes can. These properties can be used to capture directly any attributes of relationships in the E-R model. For example, the view of the Nutty Nuggets graph database in Figure 2.17 shows the edges that capture the **contained** relationship from our E-R model in Figure 2.6. Where our relational database scheme required a CONTENT relation with a WEIGHTG attribute, as well foreign keys posted from STOCK and ORDERS, the relationship is captured in the graph database in Figure 2.17 by the CONT edges labeled with their WEIGHTG property.

The remarkable flexibility of the graph database model does come with some drawbacks, particularly for those database designers used to working within the structures provided the relational database model. At its most flexible, the graph database places no constraints upon the structure of each node and edge. There is no explicit requirement for any two nodes to share any particular properties in common, unlike that requirement enforced on tuples by the tabular structure of relations. The database developer is free to add a HAIR_COLOR: GREEN property only to the Lorna Dane customer node in Figure 2.15 or a CANS: 2 property only to the edge from order node 'M315-22-06' to stock node 'Chickpeas' Figure 2.17 without reference to the E-R model. Such *ad hoc* changes are not possible in the relational model, requiring first an update to the relation scheme and second the inclusion of corresponding hair color or cans data to other tuples in the relation (even if only NULL values).

Indeed, the different "types" of nodes in Figures 2.15–2.17, also shown in the complete database view in Figure 2.18, are colored for the purposes of graphical clarity only. The basic property graph model makes no such distinctions between different node or edge types. An extension to the property graph model, called the *labeled property graph* model, does however require nodes and edges to be grouped into labeled "types," but it still makes no restrictions on whether two nodes with the same label share any properties in common. The ability to have at least the minimal structure provided by node and edge labels is still practically useful. Accordingly, many graph DBMS today do adopt the labeled property graph model.

labeled property graph

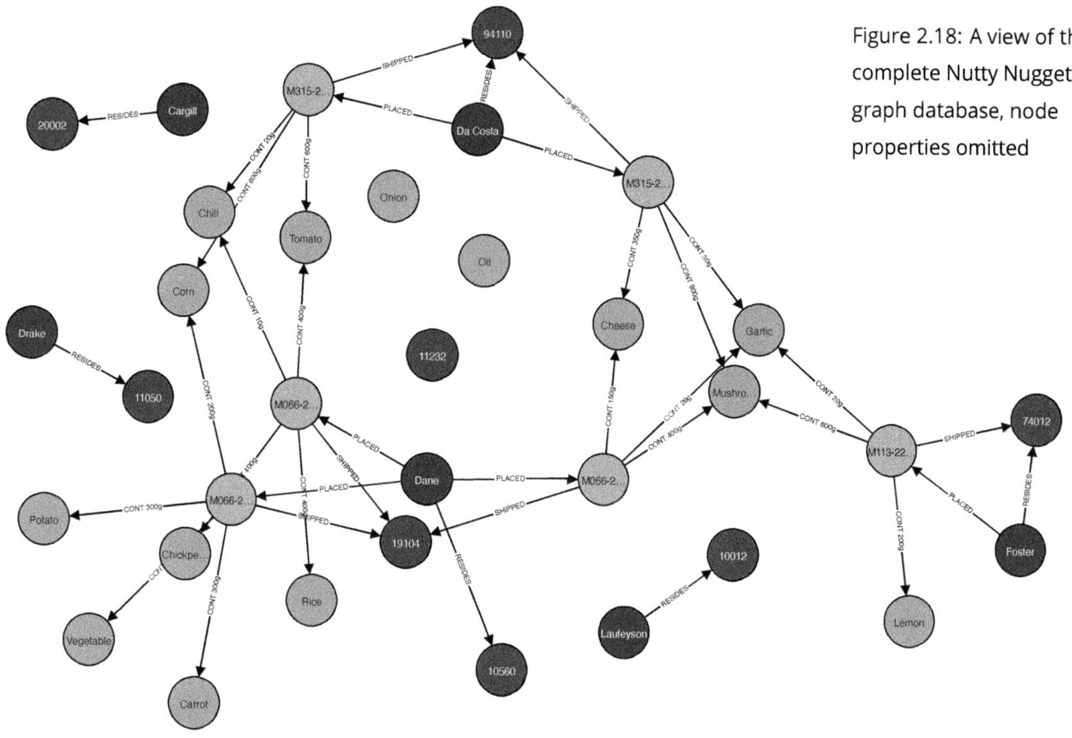

Figure 2.18: A view of the complete Nutty Nuggets graph database, node properties omitted

This extreme flexibility to not only add data but evolve structure has led to graph databases sometimes being referred to as "NoSQL" databases. The

NoSQL

term is somewhat misleading, suggesting that graph databases do need query languages or never accept SQL, neither of which is the case. Instead, a more accurate encoding of the idea would be a "not only SQL." As we shall see, a number of different languages are used to query graph databases, sometimes including extensions to SQL.

2.4.4 *Graph query languages*

Whereas Section 2.3.4 was entitled "Structured query *language*," this corresponding section for graph databases is intentionally titled "Graph query *languages*." Relational databases, founded on Codd's relational algebra introduced in Section 2.3.3, enjoy a standard query language, SQL, supported universally with only minor syntactic variations across different relational DBMS. Unfortunately, at the time of writing there exists no comparable consensus on a standard query language for graph databases.

The establishment in 2019 of an ISO (International Standards Organization) initiative to develop a standard GQL (Graph Query Language) makes it more likely one may exist when you read this. From the graph query languages available today[4], Cypher (and its open-source sibling openCypher) is used here, as it is amongst the most popular languages today and certain to be a major influence on any future standards.

[4] Common graph database query languages today include *Cypher* and *openCypher*; *GraphQL* and *GSQL*, two entirely different languages that should not be confused with the proposed standard *GQL*; *PGQL* (property graph query language); and *SPARQL*, which is a de facto standard for triple stores, the second major class of graph database discussed further in later chapters.

Cypher is an SQL-like language for querying labeled property graph databases. As a result, Cypher is built around nodes and edges and their associated properties and labels. As for SQL, the basic Cypher commands for creating, deleting, and updating data in the graph database are quite straightforward and not covered further here. Unlike SQL, querying involves searching for matching patterns rather than combinations of relational restrictions, projections, and joins. For example:

```
MATCH (n:CUSTOMER) RETURN n
```

retrieves all nodes with the label ("type") CUSTOMER (as shown in Figure 2.15), whereas:

```
MATCH e=()-[r:RESIDES]->() RETURN e
```

retrieves all edges with the label RESIDES together with their incident nodes (as shown in Figure 2.16). Many Cypher queries are similarly constructed and involve matching data in the database against various forms of the basic (node)-[directed edge]->(node) pattern.

While the two queries above reflect the underlying node and edge storage structures in a graph database, it is often still convenient to organize query results in the form of a table. For example, the query:

```
MATCH (c: CUSTOMER)-[p:PLACED]->(o: ORDERS)
  RETURN c.TEL, c.GNAME, c.FNAME, c.TITLE, o.ORDER_ID, o.MEAL, o.SIZE
```

will return the table of customer orders as per the relational join $\bowtie_{TEL=CUS\_TEL}$ (CUSTOMER, ORDERS) (already seen in Table 2.6) or equivalently from the SQL join query:

```
SELECT * FROM CUSTOMER, ORDERS WHERE CUSTOMER.TEL = ORDERS.CUS_TEL
```

Note that the explicit link between customers and their orders, in the form of an edge in the graph database, negates any need to specify the implicit connection between the primary keys of the corresponding tables, found in the relational database query.

The power and efficiency of graph pattern-matching starts to become most evident when queries responses require searching for paths through the graph. For example, the Cypher query:

```
MATCH (a)<-[:RESIDES]-(c)-[:PLACED]->(o)-[:SHIPPED]->(b)
  WHERE b<>a RETURN DISTINCT c.GNAME, c.FNAME,
    a.STREET_AD AS HOME, b.STREET_AD AS SHIP
```

identifies the customers who have a different home address to their shipping address. The query searches for non-cyclic paths from address nodes, through customers' home addresses and orders, and finally back to shipping addresses, using the pattern (a)<-[:RESIDES]-(c)-[:PLACED]->(o)-[:SHIPPED]->(b). The equivalent SQL query, below, requires a more complex four-way join across three tables to achieve the same result.

```
SELECT CUSTOMER.GNAME, CUSTOMER.HNAME,
    HADDRESS.STREET_AD AS HOME, SADDRESS.STREET_AD AS SHIP
  FROM ADDRESS AS HADDRESS, ADDRESS AS SADDRESS, CUSTOMERS, ORDERS
  WHERE HADDRESS.ADDRESS_ID = CUSTOMERS.HID AND
    CUSTOMER.TEL = ORDERS.CUS_TEL AND
    ORDERS.ORDER_ID = SADDRESS.ADDRESS_ID AND
    NOT(HADDRESS.ADDRESS_ID = SADDRESS.ADDRESS_ID)
```

Table 2.10 gives the final result for both Cypher and SQL versions of the query.

GNAME	FNAME	HOME	SHIP
Lorna	Dane	1407 Graymalkin Lane	85 Chestnut Street

Table 2.10: Evaluation of a graph-based path matching query or four-way relational join

Cypher also needs no GROUP BY statement, simplifying many aggregate queries. For example, the Cypher query below generates the total weight of orders containing more than three items, equivalent to the SQL query used to generate Table 2.9:

```
MATCH (o:ORDERS)-[c:CONTAINED]->()
  WITH o.ORDER_ID AS ORDN, COUNT(o.ORDER_ID) AS ORDC, SUM(c.WEIGHTG) AS WG
  WHERE ORDC > 3
  RETURN ORDN, ORDC, WG
```

Cypher also requires no comparable data definition commands to SQL, as there is no comparable need to define the schema structure in a graph database. Hence, SQL commands such as CREATE TABLE have no analog within Cypher. The one exception is that Cypher does allow the creation of integrity constraints to mandate the existence or uniqueness of properties on nodes or edges. For example, to require that the contact telephone number property is not null for customer nodes, we can issue the Cypher query:

```
CREATE CONSTRAINT ON (c:CUSTOMER) ASSERT c.TEL IS NOT NULL
```

This balance between structure and flexibility is at the root of the practical implications of choosing a relational versus a graph database. Graph databases are without question more flexible than relational databases, making it easy to change or adapt stored data structures in response to evolving needs. The rigid tabular structure of relational databases, however, provides in-built constraints that can improve consistency, for example, between the conceptual model and the implemented database.

2.4.5 Databases for spatial data handling

In an unmodified state, neither relational nor graph databases can claim to be well suited to spatial data management. In both cases, there can be difficulties when the technology is applied to spatial data. The main issues are:

Structure of spatial data Spatial data has a structure that does not naturally fit with tabular structures. Vector areal data is typically structured as boundaries composed of sequences of line segments, each of which is a sequence of points. Such sequences of arbitrary length violate first normal form. We shall see that some spatial data, in particular data about networks, does fit well with graph database structures. However, other spatial data—such as vectors and rasters—does not fit neatly into a graph structure either. So the need for special structures to store spatial data remains for graph databases too.

Performance The structure of relational databases can ensure simple and efficient queries about logically related data, in particular data contained within a single table. However, as queries become more complex, requiring joins across multiple tables, so the benefits of graph databases come to the fore, both in terms of performance and ease of formulation of queries. Reconstructing spatial objects and relationships is typically complex, usually requiring joining multiple tables stored in a relational database, with resulting performance overheads. Graph databases tend to be better adapted for more complex queries, and so they might be expected to be better suited to spatial queries. However, the underlying mismatch between graphs and some spatial data structures means the relationship is more complex than it might appear at first glance.

Indexes Indexing questions are the focus of Chapter 6. An index aims to increase the speed of queries by storing some additional data in order to aid access and retrieval. Both relational and graph databases need specialized spatial indexes if they are to perform well with spatial data structures. Happily, in many cases these index structures work well for spatial data stored in both relational and graph databases, as we shall see.

Reflections

Databases are a core technology for GIS; equally, databases are a core expertise for geographic information scientists. A solid grounding in databases often serves as a marker of advanced technical skills for geographic information scientists in the GIS industry. For those wishing to deepen their knowledge in databases, Elmasri & Navathe (2016) and Connolly & Begg (2014) are two venerable and trusted database texts with recently updated editions. Garcia-Molina, Ullman, & Widom (2013) is also recommended as a shorter yet respected, authoritative text.

All three classic database texts are firmly founded in relational databases, and less oriented towards graph databases specifically. In writing this third edition, it has been exciting to reflect on how much graph databases (mentioned only in passing in the second edition) have changed the landscape in databases and in GIS over recent years. Of the introductory database texts with an explicit graph database focus, Sullivan (2015) is recommended. Zhang, Song, & Liu (2014) and Baralis, Dalla Valle, Garza, Rossi, & Scullino (2017) provide some brief comparisons of the technical aspects of relational and NoSQL databases for spatial data, but no other introductory texts explicitly on the topic existed at the time of writing this book.

Practical mastery of databases, whether relational or graph, is best achieved by mastering a database query language.[5] Database query languages are among the most rewarding of computing languages to learn and use. There is great satisfaction to be gained from finding compact and elegant expression to capture a complex query using SQL, Cypher, or other graph database languages. While there are many excellent online resources for learning database query languages, Date (2015) is highly recommended as a resource for truly mastering SQL.

[5] The website to accompany this book contains downloadable relational and graph versions of the NUGGETS database in Appendix A for readers who would like to start learning to query there.

Finally, in introducing abstract graphs, set operations, and (relational) algebra, this chapter has taken a first step into the topic of discrete mathematics—an important topic in its own right that later chapters will build upon. As for databases, a foundation in discrete mathematics can be rewarding intellectually and provide the foundations of a deeper understanding of the GIS, and computing more generally. Lipschutz & Lipson (2021) and Epp (2018) are both excellent, accessible, and thorough introductions to the topic, replete with practical questions and exercises to support the learner. A brief discrete mathematics primer is also included in Appendix B of this book for those who only need a refresher.

Fundamental Spatial Concepts

3

Sections

Many different representations of space are needed for a GIS. This chapter explores the fundamental concepts and relationships underlying these different spaces. The topics covered in this chapter provide a guide for learning to:

- manipulate **Euclidean geometry**, in which **distances, angles,** and **coordinates** may be defined;
- understand **topology**, and its construction from the basic concept of a **neighborhood**;
- construct simple representations of space using **sets** and **graphs**; and
- appreciate how a **metric space** formalizes the concept of distance between points in space.

SPACE is a term that is difficult to precisely define (see Box 3.1 on the next page). We all have an intuitive idea about the concrete space in which our bodies move. In the context of GIS, we normally use the term "space" to refer to "geographic space": the structure and properties of the relationships between locations at the Earth's surface. In this chapter we examine "space" more carefully by considering the different ways of representing and reasoning about geographic space.

A fundamental concept underlying these different representations is that of *geometry*. A *geometry* provides a formal representation of the abstract properties and structures within a space. Modern treatments of geometry are founded on the notion of *invariance*: geometries can be classified according to the group of transformations of space under which their propositions remain true. This idea was first proposed in 1872, by the German mathematician Felix Klein in his inaugural address to the University of Erlangen (known as the *Erlangen Program*).

geometry

invariance

To illustrate, consider a space of three dimensions and our usual notion of distance between two points. Then a geometry is formed by the set of all transformations that preserve distances (that is, for which distance is an *invariant*). Into this set would fall translations and rotations, because the distance between two points is the same before and after a translation or rotation is effected. Scalings (enlargements) would not be members because they usually change distance. Looking at this the other way around provides us with a definition of a geometry as the study of the invariants of a set of

Box 3.1: What is space?

The notion of space is not easy to define. Gatrell (1991) defines space as "a relation defined on a set of objects," which includes just about any structured collection. This definition is too general to describe the spaces of interest in a GIS. A distinction is often made between the space we can apprehend using our visual perception, termed *perceptual*; and space that is too big for humans to observe all at once, termed *transperceptual* (see Kuipers, 1978; Montello, 1993). Zubin (1989) provides more detail and distinguishes four types of space: A-space contains manipulatable everyday objects, like phones and books; B-space contains objects larger than humans, but still observable from a single perspective, like buildings and buses; C-space contains geographic scenes that are too large to apprehend at one time, like landscapes; D-space contains objects that are too large for any human to truly experience, like the solar system or the galaxy. Freundschuh & Egenhofer (1997) give a full overview and synthesis of the different classifications of space that have been proposed. In GIS, we are primarily interested in Zubin C-spaces, although there is also growing interest in integrating information from B- and even A-spaces, as well as other types of virtual spaces, such as the Internet and cyberspace (an area of study termed "cybergeography," see Dodge & Kitchin, 2002; Fabrikant & Buttenfield, 2001; Kwan, 2001).

transformations. Thus the invariants of the set of translations, rotations, and scalings include angle and parallelism, but not distance.

Coordinatized Euclidean geometry provides a view of space that is intuitive, at least in Western culture, so this is our starting point (Section 3.1). In later sections we discuss the most primitive space of all—just collections of objects with minimal structure (Section 3.2)—and proceed to build up to richer geometries (Sections 3.3–3.4). Because of the nature of the topic, the treatment will of necessity sometimes be abstract and formal, but examples are provided along the way.

3.1 Euclidean space

Euclidean space

origin

axis

Cartesian plane

Spatial phenomena are most commonly modeled as embedded in a physical 2- or 3D space, called *Euclidean space*. We assume here for simplicity a 2D model, although all the concepts in this section can be generalized to higher dimensional spaces. For the 2D Euclidean plane, we can set up a coordinate frame consisting of a fixed, distinguished point (*origin*) and a pair of orthogonal lines (*axes*), intersecting in the origin. This coordinatized model of the Euclidean plane, known as the *Cartesian coordinate plane* or simply the *Cartesian plane*, transforms spatial properties into properties of tuples of real numbers and is the primary focus of this section.

3.1.1 Point objects

point

vector

A *point* in the Cartesian plane has associated with it a unique pair of real numbers (x, y) measuring its distance from the origin in the direction of each axis, respectively. The collection of all such points is often written as $\mathbb{R}^2$. It is often useful to view Cartesian points (x, y) as *vectors*, measured from the origin to the point (x, y), having direction and magnitude and denoted by a directed line segment (Figure 3.1). Vectors may be added, subtracted, and multiplied by scalars according to the rules:

$$(x_1, y_1) + (x_2, y_2) = (x_1 + x_2, y_1 + y_2)$$
$$(x_1, y_1) - (x_2, y_2) = (x_1 - x_2, y_1 - y_2)$$
$$k(x, y) = (kx, ky)$$

To distinguish clearly between points as locations and vectors as displacements, we use the over-arrow notation $\vec{a}$ to identify vectors (see Figure 3.1). Given a point vector, $\vec{a} = (x, y)$, we may form its *norm*, defined as follows:

norm

$$\|\vec{a}\| = \sqrt{(x^2 + y^2)}$$

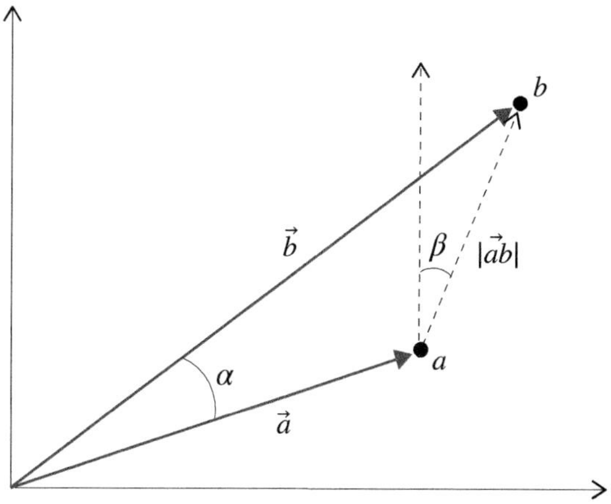

Figure 3.1: Distance $|\vec{ab}|$, angle α, and bearing β between points (vectors) $\vec{a}$ and $\vec{b}$ in the Euclidean plane

In a coordinatized system, measures of distance may be defined in a variety of ways (see Section 3.4.2 on metric spaces). A *Euclidean plane* is a Cartesian coordinate plane with the particular measures of distance and angle given below.[1] These measures form the foundation of most school geometry courses and refer to the "as the crow flies" concept of distance. Given points (vectors) $\vec{a}, \vec{b}$ in $\mathbb{R}^2$, the distance from $\vec{a}$ to $\vec{b}$, $|\vec{ab}|$, is given by:

Euclidean plane

[1] Conversely, a Cartesian plane is a Euclidean plane parameterized with coordinates.

$$|\vec{ab}| = \|\vec{b} - \vec{a}\|$$

Suppose that the points $\vec{a}, \vec{b}$ in $\mathbb{R}^2$ have coordinates (x_a, y_a) and (x_b, y_b), respectively. Then the distance $|\vec{ab}|$ is precisely the Pythagorean distance familiar from school days, given by:

$$|\vec{ab}| = \sqrt{(x_b - x_a)^2 + (y_b - y_a)^2}$$

The *angle* α (see Figure 3.1) between vectors $\vec{a}$ and $\vec{b}$ is given as the solution of the trigonometrical equation:

angle

$$\cos\alpha = \frac{x_a x_b + y_a y_b}{\|\vec{a}\| \times \|\vec{b}\|}$$

bearing

The *bearing* β (see Figure 3.1) of $\vec{b}$ from $\vec{a}$ is given by the unique solution in the interval $0 \leq \theta < 360$ of the simultaneous trigonometrical equations:

$$\sin\theta = \frac{x_b - x_a}{|\vec{ab}|}$$
$$\cos\theta = \frac{y_b - y_a}{|\vec{ab}|}$$

3.1.2 *Line objects*

Line objects are very common spatial components of a GIS, representing the spatial attributes of objects and their boundaries. The definitions of some commonly used types of line are introduced below in a parameterized form. The parameter λ is a real number constrained to vary over a given range (the range depending upon the object type). As λ varies, so the set of points constituting the linear object is defined.

line

- Given two distinct points (vectors) $\vec{a}$ and $\vec{b}$ in $\mathbb{R}^2$, the *line* incident with $\vec{a}$ and $\vec{b}$ is defined as the point set $\{\lambda\vec{a} + (1-\lambda)\vec{b}\}$.

line segment

- Given two distinct points (vectors) $\vec{a}$ and $\vec{b}$ in $\mathbb{R}^2$, the *line segment* between $\vec{a}$ and $\vec{b}$ is defined as the point set $\{\lambda\vec{a} + (1-\lambda)\vec{b} \,|\, 0 \leq \lambda \leq 1\}$.

half line

- Given two distinct points (vectors) $\vec{a}$ and $\vec{b}$ in $\mathbb{R}^2$, the *half line* radiating from $\vec{b}$ and passing through $\vec{a}$ is defined as the point set $\{\lambda\vec{a} + (1-\lambda)\vec{b} \,|\, \lambda \geq 0\}$.

Figure 3.2 shows some examples.

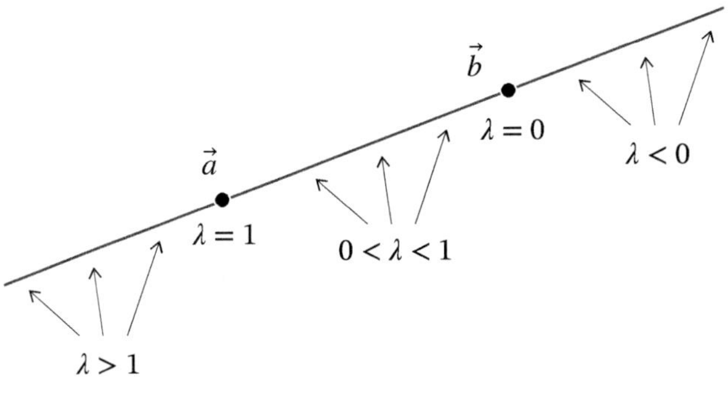

Figure 3.2: Parameterized representation of linear objects

Straight lines may also be specified by a single bivariate polynomial equation of degree one ($ax + by = k$). Not only straight lines are of interest for GIS, though. Higher-degree bivariate polynomials specify further classes of one-dimensional (1D) objects. Thus, polynomials of degree two (quadratics of the form $ax^2 + bxy + cy^2 = k$) specify conic sections, which could be circles, ellipses, hyperbolas, or parabolas.

3.1.3 *Polygonal objects*

polyline

A *polyline* in $\mathbb{R}^2$ is defined to be a finite set of line segments (called *edges*) such that each edge end-point is shared by exactly two edges, except possibly for two points, called the *extremes* of the polyline. If, further, no two edges intersect at any place other than possibly at their end-points, the polyline is called a *simple* polyline. A polyline is said to be *closed* if it has no extreme points. A (simple) *polygon* in $\mathbb{R}^2$ is defined to be the area enclosed by a simple closed polyline. The polyline forms the *boundary* of the polygon. Each end-point of an edge of the polyline is called a *vertex* of the polygon. Some possibilities are shown in Figure 3.3. An extension to the definition would allow a general polygon to contain holes, islands within holes, etc.

simple

polygon

boundary

vertex

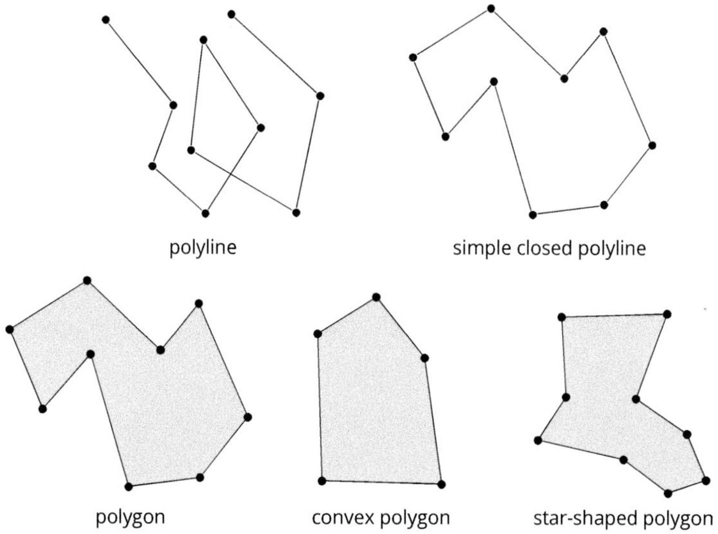

Figure 3.3: Polylines and types of polygons

Many different types of polygon have been defined in the computational geometry literature. A useful category is the set of *convex polygons*, each of which has an interior that is a convex set and thus has all its internal angles not greater than 180°. For a convex polygon, every interior point is visible from every other interior point in the sense that the line of sight lies entirely within the polygon. A *star-shaped polygon* has the weaker property that there exists at least one point which is visible from every point of the polygon (see Figure 3.3). Convexity is discussed further in Section 3.2.4.

convex polygon

star-shaped polygon

The definition of *monotone polygons* depends upon the concept of *monotone chains*. Let chain $C = [p_1, p_2, ..., p_n]$ be an ordered list of n points in the Euclidean plane. Then C is *monotone* if and only if there is some line in the Euclidean plane such that the projection of the vertices onto the line preserves the ordering of the list. Figure 3.4 shows monotone and non-monotone chains.

monotone

A polygon is a monotone polygon if its boundary can be split into two polylines, such that the chain of vertices of each polyline is a monotone chain. Clearly, every convex polygon is monotone. However, the converse of this statement is not true. It is not even true that every monotone polygon is star shaped, as may be seen from the example in Figure 3.5.

monotone polygon

A *triangulation* of a polygon is a partition of the polygon into triangles

triangulation

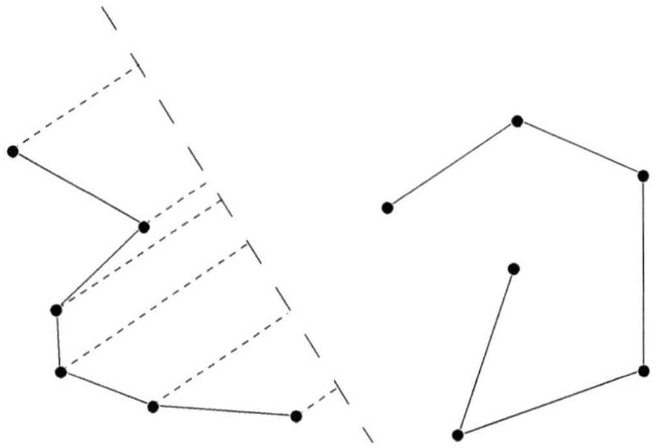

Figure 3.4: Monotone and non-monotone chains

that intersect only at their mutual boundaries. It is not too hard to show that a triangulation of a simple polygon with n vertices that introduces an extra m internal vertices (sometimes called *Steiner points*) will result in exactly $n + 2m - 2$ triangles.

Steiner point

3.1.4 *Transformations of the Euclidean plane*

transformation

This section describes some common transformations of the Euclidean plane. A *transformation* of $\mathbb{R}^2$ is a function from $\mathbb{R}^2$ to itself. Thus every point of the plane is transformed into another (maybe the same) point. Some transformations preserve particular properties of embedded objects:

congruences

Euclidean transformations (also termed *congruences*) preserve the shape and size of embedded objects. An example of a Euclidean transformation is a translation. For example, a translation of the point (x, y) through real constants a and b has the formula $(x + a, y + b)$.

similarity transformation

Similarity transformations preserve the shape but not necessarily the size of embedded objects. All Euclidean transformations are also similarities. An example of a similarity transformation is a scaling. Scaling a point (x, y) by real constants a and b has the formula (ax, by).

affine transformation

Affine transformations preserve the affine properties of embedded objects, such

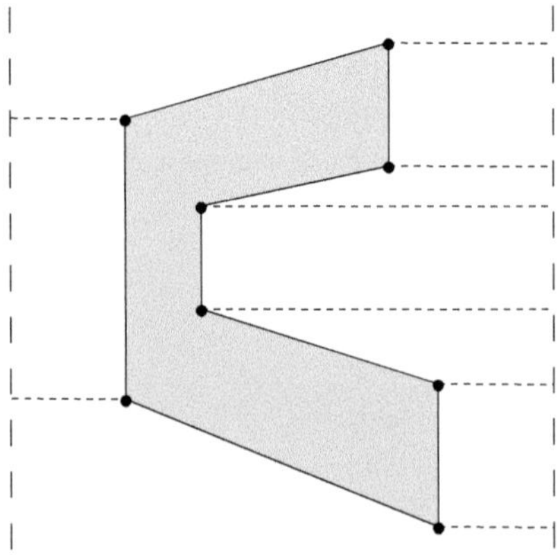

Figure 3.5: A monotone but not star-shaped polygon

as parallelism. All similarity transformations are also affine. Examples of affine transformations of the point (x, y) include *rotations*, e.g., through angle θ about the origin with the formula $(x \cos\theta - y \sin\theta, x \sin\theta + y \cos\theta)$; *reflections*, e.g., reflected in the line through the origin at angle θ to the x-axis with the formula $(x \cos 2\theta + y \sin 2\theta, x \sin 2\theta - y \cos 2\theta)$; and *shears*, e.g., parallel to the x-axis with real constant a and formula $(x + ay, y)$.

rotation

reflection

shear

Projective transformations preserve the projective properties of embedded objects. An intuitive idea of a central projection is the action of a point light source, sending a figure to its projection upon a screen. For example, the projection of a circle may result in an ellipse. All affine transformations are also projective.

projective transformation

Topological transformations (homeomorphisms or bicontinuous maps) preserve topological properties of embedded objects. We shall study this class in detail later.

3.2 Set-based spaces

The Euclidean plane is a highly organized kind of space, with many well-defined operations and relationships that can act upon objects within it. This section retreats to the much more rarefied realm of set-based space, and then gradually builds up more structure with topological and metric spaces.

3.2.1 Sets

The set-based model of space does not have the rich set of constructs of the Euclidean plane. The set-based model simply involves:

- The constituent objects to be modeled, called *elements* or *members*.
- Collections of elements, called *sets*. For computer-based models, such collections are usually finite, or at least countable.
- The relationship between the elements and the sets to which they belong, termed *membership*. We write $s \in S$ to indicate that an element s is a member of the set S.

element

set

membership

The set-based model is abstract and provides little in the way of constructions for modeling spatial properties and relationships. Nevertheless, sets are a rich source of modeling constructs (see "Russell's paradox," Box 3.2 on the following page) and are fundamental to the modeling of any spatial information. For example, relationships between different base units of spatial reference may be modeled using set theory. States or counties may be contained within (i.e., members of) a country; countries may themselves be members of continents; continents may be elements of the set of landmasses. Such hierarchical relationships are adequately modeled using set theory. Sometimes, areal units are not so easily handled. For example, in the US there is no simple set-based relationship between zip codes, used for the distribution of mail, and

Box 3.2: Russell's paradox

Even though the set concept expresses the most rudimentary relationships and structuring of space, it is surprisingly difficult to capture the essence of a set in a few words. A first attempt might be "a set is any collection of objects," but this lays itself open to Russell's paradox (after the British mathematician Bertrand Russell). Russell's paradox is sometimes explained in terms of a library index. Imagine that an assiduous librarian wishes to index every book in the library that does not contain a reference to itself (unlike this book, Duckham, Sun, & Worboys, 2023). The librarian can write down the names of books that do not refer to themselves in a new index book. To be complete, this index should list itself, as it does not contain a self-reference. But as soon as the librarian adds a self-reference to the index, that reference should be removed, because now the index book does cite itself! We can express this tortuous problem mathematically by considering the set *S* of all sets that are not members of themselves. If *S* is a member of itself, then by definition it is not a member of itself; on the other hand, if *S* is not a member of itself, then it must be a member of itself. Either way, we arrive at a contradiction. This paradox spurred considerable efforts in the 20th century to find a more adequate definition of a set, but no definition has proved completely satisfactory.

the administrative units such as city, county, and state (as already noted in Box 2.4 on page 58). This mismatch causes considerable problems when data referenced to one set of spatial units is compared or combined with spatial data referenced to a different set (discussed further in Box 8.5 on page 321).

In classical set theory an object is either an element of a particular set or it is not. There is no halfway house or degree of membership. If the binary on-off nature of the membership condition is relaxed, then it is possible to arrive at some more expressive models. Some of these models are important for modeling uncertainty in spatial information, a topic tackled in Chapter 10.

From the basic constructs of element, set, and membership, a large number of modeling tools may be constructed. We shall consider just a few here.

- equality — *Equality* is defined as a relationship between two sets that holds when the sets contain precisely the same members.
- subset — The relationship between two sets where every member of one set is a member of the second is termed *subset*. The relationship that set S is a subset of set T is denoted $S \subseteq T$.
- power set — A *power set* is the set of all subsets of a set. The power set of set S is denoted $\mathcal{P}(S)$.
- partition — A *partition* of a set is a set of subsets such that each and every member of the original set appears in exactly one subset in the partition.
- empty set — The *empty set* is the set containing no members, denoted $\emptyset$.
- cardinality — The number of members in a set is termed *cardinality*. The cardinality of set S is denoted $\#S$ or $|S|$.
- intersection — The *intersection* operation is a binary operation that takes two sets and returns the set of elements that are members of both the original sets. The intersection of sets S and T is denoted $S \cap T$.
- union — The *union* operation is a binary operation that takes two sets and returns the set of elements that are members of at least one of the original sets. The union of sets S and T is denoted $S \cup T$.
- difference — The *difference* operation is a binary operation that takes two sets and re-

turns the set of elements that are members of the first set but not the second set. The difference of sets S and T is denoted $S \backslash T$.

- The *complement* operation is a unary operation that when applied to a set returns the set of elements that are not in the set. The complement is always taken with reference to an (implicit) universal set. The complement of set S is denoted S'.

complement

Figure 3.6 shows by shading on set diagrams some of the Boolean set operations described above.

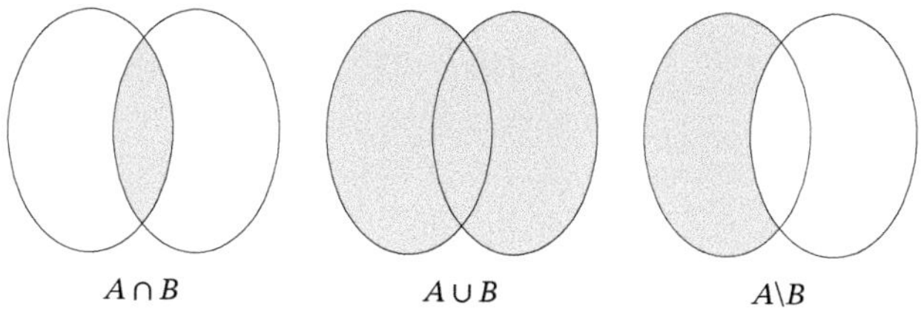

Figure 3.6: Set intersection, union, and difference

Some sets, particularly sets of numbers, are used so often that they have a special name and symbol. Some of these are listed in Table 3.1. The Boolean set $\mathbb{B}$ is used whenever there is a two-way choice to be made. This choice may be viewed as between on and off, true and false, or one and zero, depending upon the context. The set of integers $\mathbb{Z}$ is used in discrete models. Sometimes, only the positive integers are needed, written $\mathbb{Z}^+$.

For continuous models, the real numbers $\mathbb{R}$ are required. In fact, it is provably impossible to capture the reals completely using a computer; therefore, rational numbers are used in practice. Rational numbers are real numbers that can be expressed as a ratio of integers, e.g., 123/46. Like real numbers, rational numbers have the property that they are *dense*: that is, between any two rationals a and b, where a is greater than b, no matter how close a and b are, it is always possible to find a third rational c such that $a < c < b$ (see Figure 3.7). In this way, the rationals are a useful approximation for modeling continuous processes. Any particular computer implementation will place a restriction upon the precision of the rationals.

dense set

Name	Symbol	Description
Booleans	$\mathbb{B}$	Two-valued set of true/false, 1/0, or on/off
Integers	$\mathbb{Z}$	Positive and negative numbers, including zero
Rationals	$\mathbb{Q}$	Ratios of two integers
Reals	$\mathbb{R}$	Measurements on the number line
Real plane	$\mathbb{R}^2$	Ordered pairs of reals, the Cartesian plane
Closed interval	$[a, b]$	All reals between a and b (including a and b)
Open interval	$]a, b[$	All reals between a and b (excluding a and b)
Semi-open interval	$[a, b[$	All reals between a and b (including a and excluding b)

Table 3.1: Some distinguished sets

Some subsets of the reals are particularly useful. *Intervals* are connected sets of real numbers. They may or may not contain their end-points and are then called *closed* or *open*, respectively. It is also possible for intervals to be

interval

Figure 3.7: The rationals are dense

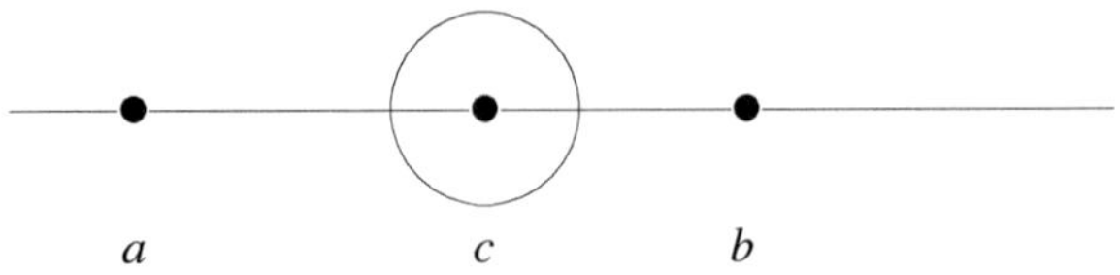

closed at one end and open at the other. Such intervals are called *semi-open* (or *semi-closed*). Thus the closed interval $[2, 5]$ denotes the set of all real numbers not less than two and not greater than five. The semi-open interval $]2, 5]$ denotes the set of all real numbers greater than two but not greater than five.

3.2.2 Relations

Sets on their own are limited in their application to modeling. Life becomes more interesting when relationships between two or more sets are modeled. In order to provide these tools, a further set-based operation is defined.

product

- The binary operation *product* returns the set of ordered pairs, whose first element is a member of the first set and second element is a member of the second set. The product of sets S and T is denoted $S \times T$.

An example of a product set is the set of points in the Cartesian coordinate plane, introduced in Section 3.1.1. Each point in the Cartesian plane is represented as an ordered pair of real numbers, measuring the point's distance from a given origin in the direction of the two axes. In set-theoretic terms, the collection of all such points is a product set, being the product of the set of real numbers with itself. This set is denoted by $\mathbb{R} \times \mathbb{R}$ or $\mathbb{R}^2$. This notion may be generalized to Cartesian coordinate 3-space $\mathbb{R}^3$ or indeed Cartesian n-space $\mathbb{R}^n$. A second example would be the set of points in the *unit square*, the square with vertices $(0, 0)$, $(0, 1)$, $(1, 0)$, and $(1, 1)$ in the Cartesian plane (Figure 3.8). This set is the product of two intervals, $[0, 1] \times [0, 1]$. It is the case that $[0, 1] \times [0, 1] \subseteq \mathbb{R}^2$.

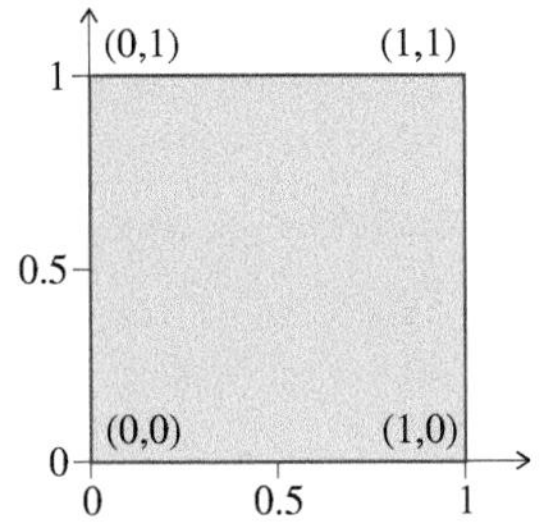

Figure 3.8: The unit square

Product spaces provide a means of defining relationships between objects. Hence the following construction:

binary relation

- A *binary relation* is a subset of the product of two sets, whose ordered pairs show the relationships between members of the first set and members of the second set.

Suppose that S = {Fred, Mary} and T = {apples, oranges, bananas}. Then the formal way of expressing the relationship "likes" between people and fruit is by means of a relation, constructed as a set of ordered pairs {(Fred, apples), (Fred, bananas), (Mary, apples)}. This set of ordered pairs is a subset of the entire product space $S \times T$, containing six pairs.

Relations may in general apply to the product of more than two sets (as in the relations in the relational database model). However, even binary relations have many modeling applications. For example, suppose that we are given two sets: the places of interest in The Potteries P and town centers in The Potteries T. A relation $R_1 \subseteq P \times T$ might provide for each a place of interest $p \in P$ the nearest town center $t \in T$ to p. Figure 3.9 shows the relation R_1 diagrammatically. The fact that Hanley is the nearest town center to the City Museum is indicated in Figure 3.9 with an arrow from the City Museum to Hanley, and would be written (City Museum, Hanley) $\in R_1$.

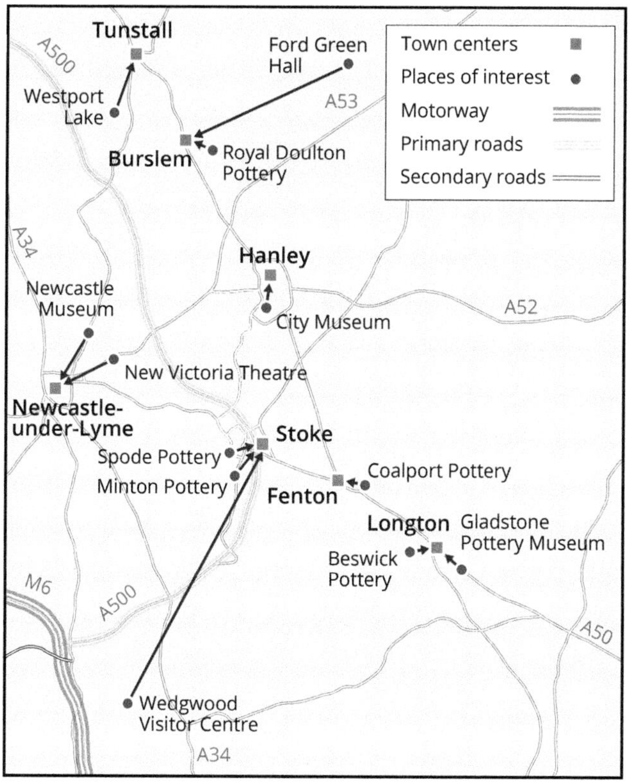

Figure 3.9: Binary relation on the sets of Potteries places of interest and town centers

As we have seen with the relation $\mathbb{R}^2$, a relation may act upon a single set. Suppose for the set of places P, a place of interest $p \in P$ is related to $p' \in P$ if p' is the nearest place of interest to p. This relation $R_2 \subseteq P \times P$ is shown in Figure 3.10. The fact that Spode Pottery is the nearest place of interest to the City Museum is indicated with an arrow from the City Museum to Spode Pottery, and written (City Museum, Spode Pottery) $\in R_2$.

Some binary relations between objects of the same set have special properties.

- A binary relation R on X where every element of the set $x \in X$ is related to itself, $(x, x) \in R$, is termed a *reflexive* relation. A relation is *irreflexive* if every element of the set is not related to itself, $(x, x) \notin R$ for all $x \in X$.
- A binary relation R where if x is related to y, then y is related to x is termed a *symmetric* relation. A relation is *antisymmetric* if for every pair of elements, x and y in X, if x related to y then y is not related to x.
- A binary relation R where if x is related to y and y is related to z, then x is related to z is termed a *transitive* relation.

reflexive
irreflexive
symmetric
antisymmetric
transitive

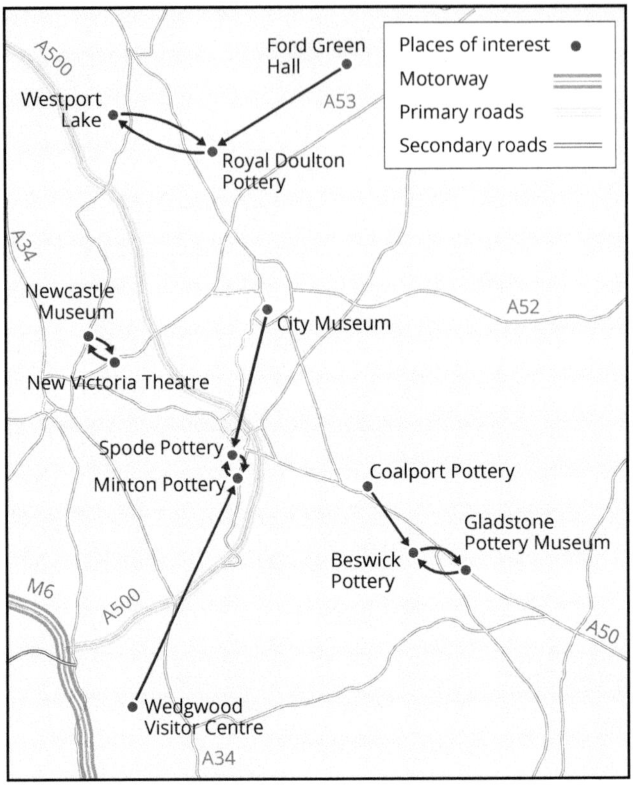

Figure 3.10: Binary relation on the sets of Potteries places of interest

In the example of Figure 3.10, there is an implicit assumption that a place cannot be related to itself; therefore, the relation is not reflexive and is also irreflexive. It is less obvious whether or not the relation "has nearest place of interest" is symmetric. For example, Spode Pottery is nearest to Minton Pottery and vice versa. However, the nearest place to the City Museum is also Spode Pottery while the converse is not true, so the relation "has nearest place of interest" cannot be symmetric. Also, the relation is not transitive because if p has nearest place of interest q and q has nearest place of interest r, then p cannot have nearest place of interest r unless $q = r$ or $q = p$ (and we have already disallowed a place being nearest to itself).

equivalence relation

A binary relation that is reflexive, symmetric, and transitive is termed an *equivalence relation*. Another useful class of relations is order relations, which satisfy the transitive property and are also irreflexive and antisymmetric. These further properties and partial orders are discussed in Section 3.4.1.

3.2.3 *Functions*

function

A *function* is a special type of relation which has the property that each member of the first set relates to exactly one member of the second set. Thus, a function provides a rule that transforms each member of the first set, called the *domain*, into a member of the second set, called the *codomain*. We use the notation:

domain

codomain

$$f : S \to T$$

to mean that f is a function, S is the domain, and T is the codomain. If the result of applying function f to element x of S is y, we write $y = f(x)$ (read "y equals f of x") or $f : x \mapsto y$ (read "f maps x to y"). Figure 3.11 shows schematically the relationship between function, domain, codomain, and image (defined shortly).

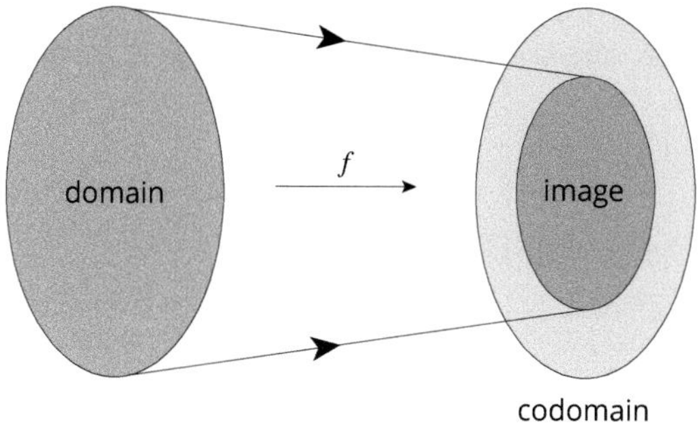

Figure 3.11: An abstract function f

For example, suppose that S is the set of points on a spheroid and T the set of points in the plane. A map projection is essentially a function with domain S and codomain T. Usually, map projection functions are constructed to leave some properties of the sets invariant, for example, lengths, angles, or areas. The familiar Universal Transverse Mercator (UTM) projection is a function that preserves angles. The UTM projection has a further property that any two different points in the domain are transformed to two distinct points in the codomain. Such a function is said to be an *injection*. Map projections are discussed further in the context of cartography, in Section 8.2.3.

injection

Not all functions are injections. Consider, for example, the function that computes the square of any integer. Then, because both 2 and −2 square to the same number, 4, the square function is not an injection. The two relations described in the previous section (Figures 3.9 and 3.10) are also functions (i.e., each member in the first set relates to exactly one member in the second set) but not injections.

The set of outputs from the application of a function to elements in its domain will form a subset of the codomain. Thus the UTM map projection will project a spheroid onto a finite subset of the plane. The set of all possible outputs is called the *image* or *range* of the function. If the image actually equals the codomain, then the function is termed a *surjection*. The UTM projection is not a surjection because the spheroid projects onto only a finite portion of the plane and not the whole plane. A function that is both a surjection and an injection is termed a *bijection*.

image

surjection

bijection

Injective functions have the special property that they have inverse functions. Consider again the UTM projection from the spheroid to the plane. Given a point in the plane that is part of the image of the transformation, it is possible to reconstruct the point on the spheroid from which it came. This reversal of the process allows us to form a new function whose domain is the image of the UTM, and which maps the image back to the spheroid. This is called the *inverse* function.

inverse function

3.2.4 Convexity

Convexity has already been discussed for polygons and is now generalized to the same property for arbitrary point-sets in the Euclidean plane. The same notion is also meaningful in Euclidean 3-space, and the definition is easily extensible to this case. The essential idea is that a set is convex if every point is visible from every other point within the set. To make this idea precise, we define visibility and then convexity. Let S be a set of points in the Euclidean plane. Then:

- the point x in S is *visible* from point y in S if either $x = y$ or it is possible to draw a straight-line segment between x and y that consists entirely of points of S;
- the point x in S is an *observation point* for S if every point of S is visible from x;
- semi-convex — the set S is *semi-convex* (*star-shaped* if S is a polygonal region) if there is some observation point for S; and
- convex — the set S is *convex* if every point of S is an observation point for S.

Figure 3.12 shows the visibility relation within a set between three points x, y, and z. Points x and y are visible from each other, as are points y and z. But, points x and z are not visible from each other. The visibility relation is reflexive and symmetric, but not transitive. Also, observe that any convex set must be semi-convex (although the converse is not necessarily true). Figure 3.13 gives some examples of sets that are not semi-convex, semi-convex but not convex, and convex, respectively.

Figure 3.12: Visibility between points x, y, and z

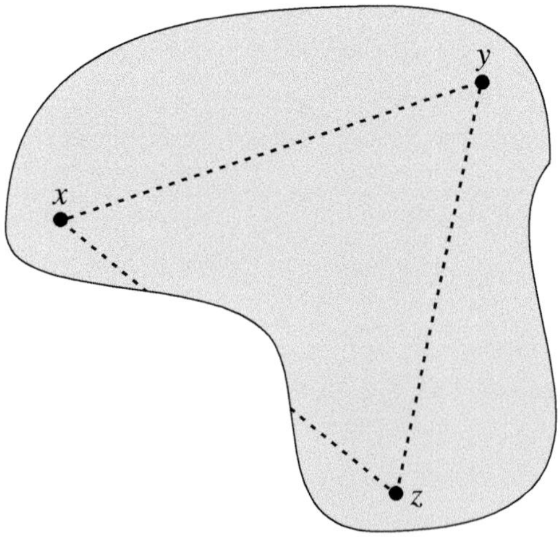

convex hull

The intersection of a collection of convex sets is also convex, and therefore any collection of convex sets closed under intersection has a minimum member. This leads to the definition of a *convex hull* of a set of points S in $\mathbb{R}^2$ as the intersection of all convex sets containing S. From above, the convex hull must be the unique smallest convex set that contains S (Figure 3.14). A convex hull of a finite set of points is always a polygonal region.

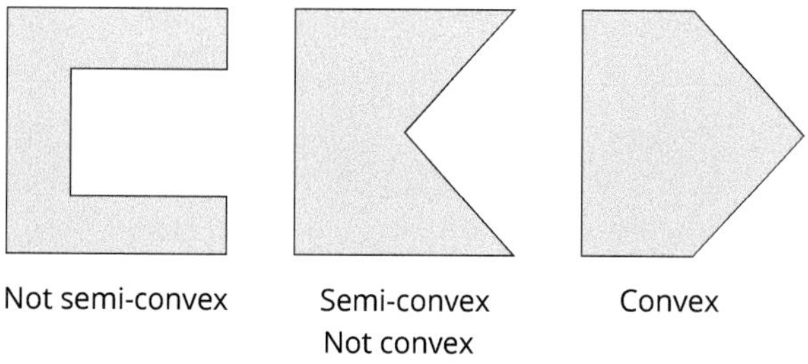

Figure 3.13: Degrees of convexity in point sets

3.3 *Topology of space*

The word "topology" derives from Greek and literally translates to "study of form." This may be contrasted with the word "geometry," which translates to "measurement of the Earth." *Topology* is a branch of geometry concerned with a particular set of geometrical properties—those that remain invariant under topological transformations. So, what is a topological transformation? In fact, topology is a general notion that can be applied to many different kinds of space.

topology

3.3.1 *Topological spaces*

To gain some topological intuition, imagine the Euclidean plane to be an unbounded sheet of fine-quality rubber that has the ability to stretch and contract to any desired degree. Imagine a figure drawn upon this rubber sheet. Allow this sheet to be stretched but not torn or folded. Certain properties of the original figure will remain, while others will be lost. For example, if a polygon were drawn upon the sheet, and a point was drawn inside the polygon, then after any amount of stretching the point would still be inside the polygon; on the other hand, the area of the polygon may well have changed. The property of "insideness" is a topological property (because it is invariant under rubber sheet transformation), while "area" is not a topological property. The transformation induced by stretching a rubber sheet is called a *topological transformation* or *homeomorphism*. Thus we have the following definitions:

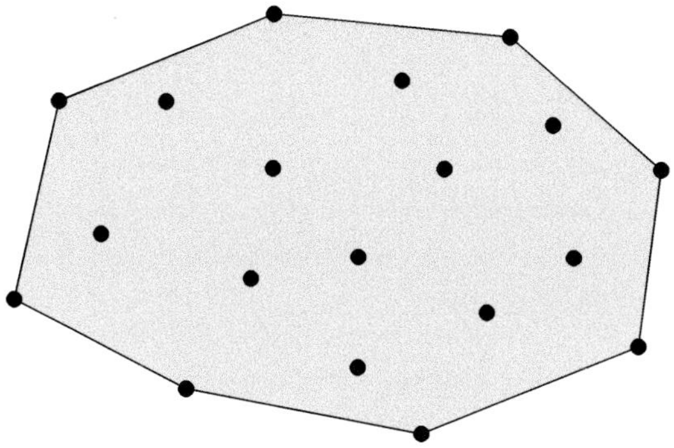

Figure 3.14: Convex hull of a point set

Topological properties properties that are preserved by topological transformations of the space.
Topology the study of topological transformations and the properties that are left invariant by them.

Table 3.2 lists some topological and non-topological properties of objects embedded in the Euclidean plane. For object types we take points, arcs, and area. Later, these types will be defined more carefully; for now, assume the obvious meanings to arc (possibly curved linear object) and area (2D piece of plane, possibly with holes and islands).

Table 3.2: Topological and non-topological properties of objects in the Euclidean plane with the usual topology

Topological	A point is at an end-point of an arc
	A point is on the boundary of an area
	A point is in the interior/exterior of an area
	An arc is simple
	An area is open/closed/simple
	An area is connected
Non-topological	Distance between two points
	Bearing of one point from another point
	Length of an arc
	Perimeter of an area

point-set topology

combinatorial topology

The discussion will cover two branches of topology: point-set (or analytic) topology and combinatorial (or algebraic) topology. In *point-set topology*, the focus, as one would expect, is on sets of points and in particular on the concepts of neighborhood, nearness, and open sets. We shall see that several important spatial relationships, such as connectedness and boundary may be expressed in point-set topological terms. The other important branch of topology, which has been applied to spatial data modeling, is *combinatorial topology*, in particular the theory of simplicial complexes. Even though these ideas may at times seem rarefied and far removed from spatial databases, in fact they do form the basis of several prominent conceptual models used in GIS. It is certainly true that the construction of sound and lasting generic spatial models relies on knowledge of the material that is introduced here.

3.3.2 General point-set topology

It is possible to define a topological space in several different ways. The definition below is based upon a single primary notion, that of *neighborhood.* A set upon which a well-defined notion of neighborhood is provided is then a topological space. It turns out that all the familiar topological properties are definable in terms of the single concept of neighborhood. Given any set, the approach is to define a collection of its subsets, constituting the neighborhoods, and thus provide a neighborhood topology on the set. The formal definition is now given.

topological space

neighborhood

- Let S be a given set of points. A *topological space* is a collection of subsets of S, called *neighborhood*, that satisfy the following two conditions.

T1 Every point in S is in some neighborhood.

T2 The intersection of any two neighborhoods of any point x in S contains a neighborhood of x.

Figure 3.15 shows the two conditions of a topological space in action. Neighborhoods are shown surrounding each point in the set, and two neighborhoods are shown overlapping and containing in their intersection another neighborhood.

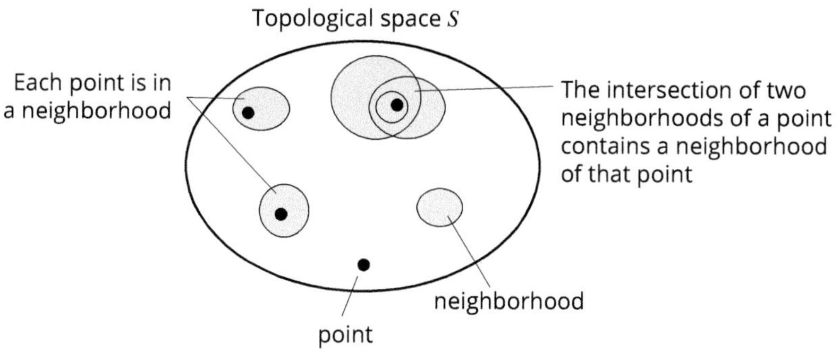

Figure 3.15: Points and neighborhoods in a topological space

By far, the most important example of a topological space, for our purposes, is the *usual topology* for the Euclidean plane. The usual topology is so called because it is the topology that naturally comes to mind with the Euclidean plane, and it corresponds to the rubber-sheet topology introduced earlier. It is possible to define other (unusual?) topologies on the Euclidean plane (see Box 3.3 on the next page), and one important example of these, travel time topology, follows the usual topology.

The usual topology of the Euclidean plane Define an *open disk* to be a set of points bounded by a circle in the Euclidean plane, but not including the boundary. An example is given in Figure 3.16. The convention is that a hatched line at the boundary indicates that the boundary points are excluded, whereas a continuous line indicates that boundary points are included.

open disk

Figure 3.16: An open disk in the Euclidean plane with the usual topology

Define a neighborhood of a point x in $\mathbb{R}^2$ to be any open disk that has x within it (see Figure 3.17). We now show that, under this definition of neighborhood, $\mathbb{R}^2$ is a topological space. To check that condition *T1* for a topological space holds, it is sufficient to observe that every point in $\mathbb{R}^2$ can certainly be surrounded by an open disk. For *T2*, take any point x in $\mathbb{R}^2$ and surround it by two of its neighborhoods (open disks with x inside), N_1 and

Box 3.3: Topological spaces

There are many common topological spaces other than the usual topology. For example, let S be any set, and define the neighborhoods to be *all* the subsets of S. It is easy to confirm that this neighborhood structure defines a topological space (by checking conditions $T1$ and $T2$). The space is called the *discrete topology*, because the smallest neighborhood of each point x in S is $\{x\}$, so each point in S is separated by a neighborhood from every other point. Another example of an extreme topology occurs if we let S be any set, and define the only neighborhood to be the set S itself. Again, this may easily be verified to be a topological space, called the *indiscrete topology*. The usual topology of the Euclidean plane may be scaled up or down to Euclidean space of any dimension. This is illustrated with the usual topology on the Euclidean line. For any real number $x \in \mathbb{R}$, define a neighborhood of x to be any open interval containing x. This is the 1D equivalent of the usual topology on the Euclidean plane, and it can be shown to satisfy properties $T1$ and $T2$ in a similar way.

N_2. Now, x will lie in the intersection of these two neighborhoods, and it is always possible to surround x with an open disk entirely within this intersection. To see this, let d_1 be the minimum distance of x from the boundary of N_1 and d_2 be the minimum distance of x from the boundary of N_2. Then the open disk with center x and radius the minimum of d_1 and d_2 will contain x and lie entirely in the intersection of N_1 and N_2 (see Figure 3.18). Thus $T1$ and $T2$ are satisfied, and under this definition of neighborhood, $\mathbb{R}^2$ is a topological space, called the *usual topology* for $\mathbb{R}^2$.

usual topology

Figure 3.17: A neighborhood of x in $\mathbb{R}^2$ with the usual topology

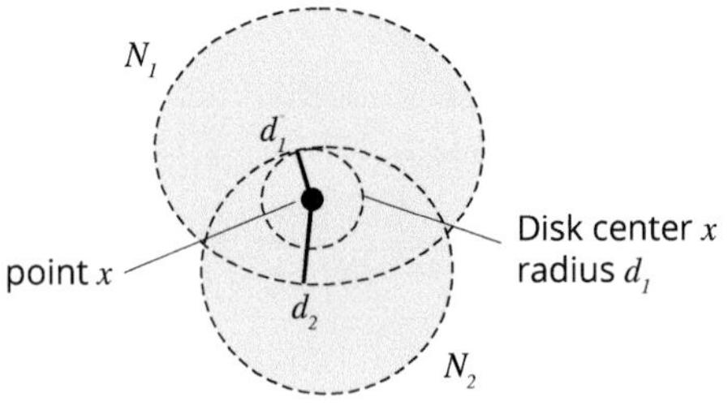

Figure 3.18: Condition $T2$ is satisfied for $\mathbb{R}^2$ with the usual topology

Travel time topology Another important topology for spatial information is *travel time topology*. Let S be the set of points in a region of the plane. Suppose that the region contains a transportation network and that we know the average travel time between any two points in the region using the network, following the optimal route. For the purposes of this example, we need to assume that the travel time relation is symmetric: that is, it must always be the case that the travel time from x to y is equal to the travel time from y to x. For each time t greater than zero, define a t-zone around point x to be the set of all points reachable from x in less than time t. As an illustration, Figure 3.19 shows a 5-zone, 10-zone, and 15-zone around the Spode Pottery.

travel time topology

Let the neighborhoods be all t-zones (for all times t) around all points. Then, clearly $T1$ is satisfied, since each point will have some t-zones surrounding it. The argument that $T2$ is satisfied is similar to that used for the usual topology of the Euclidean plane and is omitted. However, the symmetry of the travel time relation is required at a critical stage in this argument.

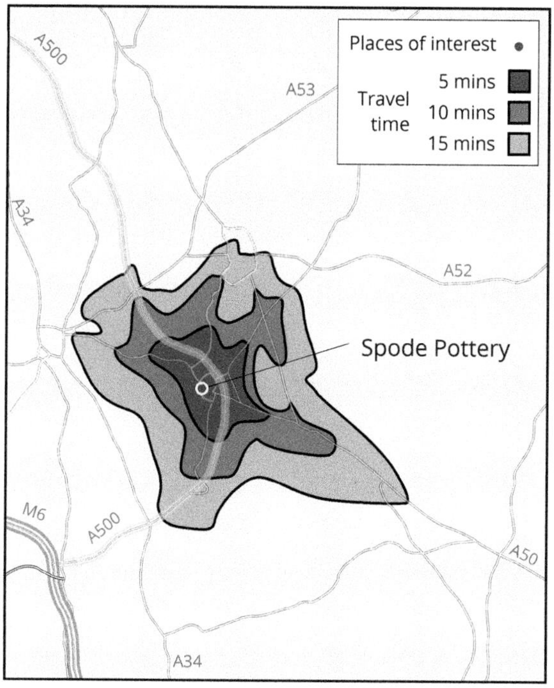

Figure 3.19: Travel-time topology example, showing some neighborhoods (t-zones)

The symmetric travel time measure between two points is an example of a *metric* (discussed later in Section 3.4.2). The travel time topology is a special case of the topology that can be induced by any metric on a space.

3.3.3 *Properties of a topological space*

It is surprising that out of the single primitive notion of neighborhood it is possible to construct all the features and properties of a topological space. This section describes some of these constructions, in a similar way to Henle (1979), beginning with the definition of "nearness." Many topologists use the phrase "limit point" to replace our use of "near point."

- Let S be a topological space. Then S has a set of neighborhoods associated with it. Let X be a subset of points in S and x an individual point in S. Define x to be *near* X if every neighborhood of x contains some point of X.

For example, in the Euclidean plane with the usual topology, let C be the open *unit disk*, centered on the origin, $C = \{(x, y)|x^2 + y^2 < 1\}$. Then the point $(1, 0)$, although not a member of set C, is near to C, because any open disk (no matter how small) that surrounds $(1, 0)$ must impinge into C. In fact, any point on the circumference of C is near to C, as indeed is any point inside C.

However, any point exterior to C and not on the circumference is not near C, since it will always be possible to surround it with a neighborhood that separates it from C (see Figure 3.20).

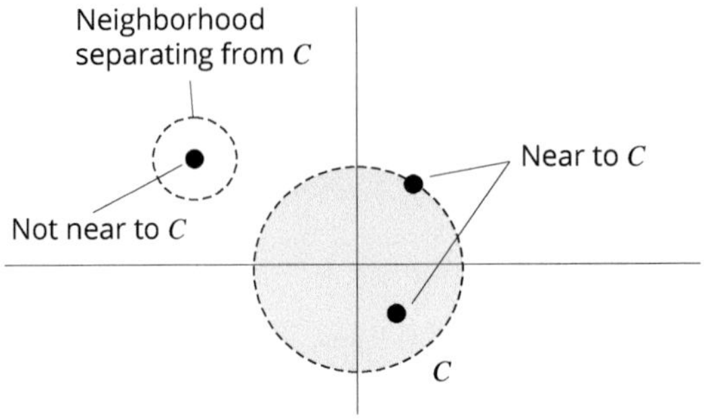

Figure 3.20: Points near and not near to the open unit disk C in the Euclidean plane with the usual topology

A fundamental topological invariant of any set is its boundary. This notion can be constructed out of our primitives as follows. First, open and closed sets are introduced using the neighborhood idea. It will emerge that an *open set* is a set that does not contain its boundary, whereas a *closed set* is a set that contains all its boundary.

open set

closed set

- Let S be a topological space and X be a subset of points of S. Then X is *open* if every point of X can be surrounded by a neighborhood that is entirely within X.
- Let S be a topological space and X be a subset of points of S. Then X is *closed* if it contains all its near points.

The open unit disk C above is (obviously) open, because any point, no matter how close to the circumference, may be surrounded by a neighborhood made sufficiently small that it is entirely within C. C is not closed, because points on the circumference are near points to C but not contained in C. For C to be closed, it would have to include its circumference. Note that it is possible for a topological space to contain sets that are both open and closed or neither open nor closed (see Box 3.4 on the facing page).

This leads us to the related definition of *closure*.

closure

- Let S be a topological space and X be a subset of points of S. Then the *closure* of X is the union of S with the set of all its near points. The closure of X is denoted X^-.

Clearly the closure of a set is itself closed; in fact the closure of set X is the smallest closed set containing X. In our example, the closure of the open unit disk C is formed by annexing its circumference. Thus $C = \{(x,y)|x^2 + y^2 \leq 1\}$. This set is called the *closed unit disk*. We may also force a set to be open by stripping away unwanted near points, constructing the *interior* of a set as follows:

interior

- Let S be a topological space and X be a subset of points of S. Then the *interior* of X consists of all points which belong to X and are not near points of X', the complement of X. The interior of set X is denoted X°.

Box 3.4: Topological spaces (open and closed sets)

Recall from the previous "Topological spaces" Box 3.3 on page 94, that the discrete topology on a space S defined the neighborhoods to be all the subsets of S. Let X be any subset of S, then the only near points of X are the points of X itself (since any point not in X may be surrounded by the neighborhood containing just that point). Thus X is closed. X is also open, because we can surround any point of X with a neighborhood entirely in X containing just that point. The discrete topology is therefore odd in that every set within it is both open and closed. This is not the case for the indiscrete topology.

Let S be a set upon which the indiscrete topology is defined and let X be any subset of S. Then, the only neighborhood in S is S itself. Therefore, every point in S is a near point of X. Thus, unless X is either empty or equal to S, it is neither open nor closed. For the travel time topology, an example open set is the set of all points less than 1 hour's traveling time from a specified point, say, Spode Pottery. This set has as its boundary the set of all points that are exactly 1 hour from Spode Pottery, and as its closure the set of all points having travel time from Spode Pottery not greater than 1 hour.

Notice that for a point x to be near to the complement of set X, it must be the case that each neighborhood of x impinges upon X'. Therefore, a point in X which is not a near point of X' has at least one neighborhood of it that is entirely within X. Thus, the interior of a set is open. In fact, the interior of a set X is the largest open set contained in X. As an example, if D is the closed unit disk, then D° is the open unit disk. Figure 3.21 shows a further example of an open and closed set in the Euclidean plane with the usual topology.

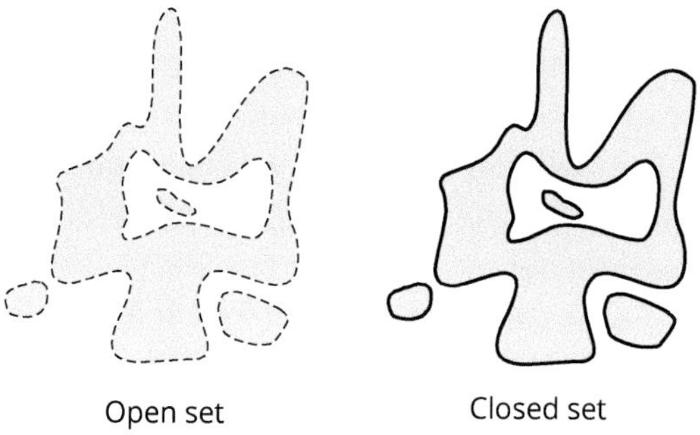

Figure 3.21: Open and closed sets in the Euclidean plane with the usual topology

We are now sufficiently prepared to define the *boundary* of a set in purely topological terms.

boundary

- Let S be a topological space and X be a subset of points of S. Then the boundary of X consists of all points which are near to both X and X'. The boundary of set X is denoted ∂X.

Let point x be a member of ∂X. Since x is near to X, then x must be in X^-. Since x is near to X', then x cannot be in X°. Thus ∂X is the set difference of X^- and X°. In the case of the unit disk C in the Euclidean plane with the usual topology, ∂X is the circumference of C, as we would expect. As a further example using the Euclidean plane with the usual topology, suppose that S is the connected region of the plane containing a single hole shown in Figure 3.22. We see that the outer boundary of S is excluded from S, but the inner boundary of S (i.e., the boundary of the hole) is contained in S. The interior, closure, and boundary of S are as shown in Figure 3.22. S° contains all the points of S excluding its inner boundary. S^- includes both inner and outer boundaries. ∂S is the union of the inner and outer boundaries of S.

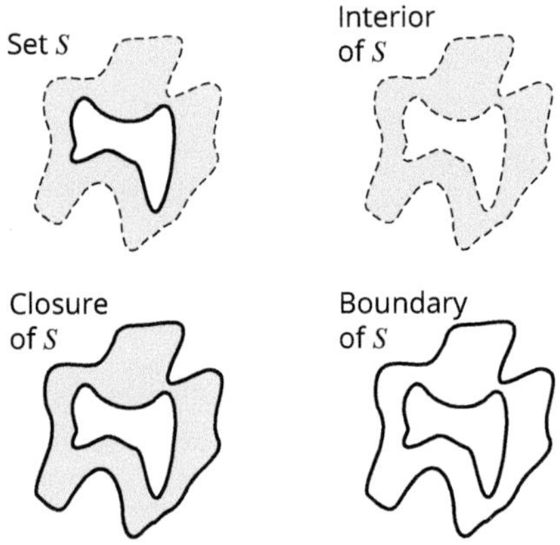

Figure 3.22: Interior, closure, and boundary of a region in the Euclidean plane with the usual topology

It is important to realize that it is not possible to consider the topological properties of sets in exclusion from the larger spaces in which they are embedded. To illustrate this point, consider a finite length of straight line. If we take the line to be embedded in a 2D (or higher-dimensional) Euclidean space, as shown in Figure 3.23a, then its interior is the empty set, and its boundary and closure are both the line itself. On the other hand, if the same line is embedded in a 1D Euclidean space, that is, in the real number line, as shown in Figure 3.23b, then its interior is the line excluding its end-points, its closure is the line itself, and its boundary consists of the end-points of the line.

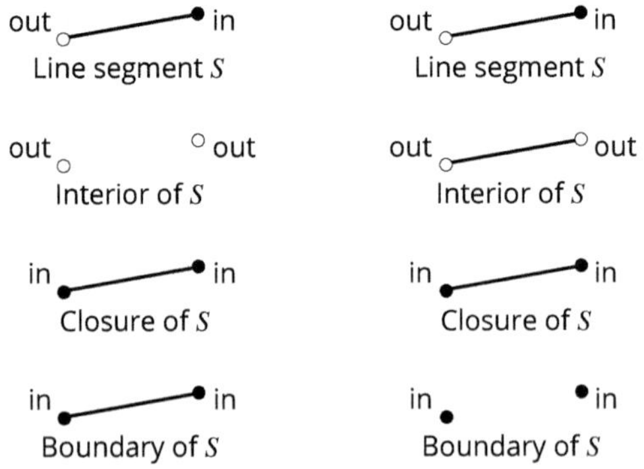

a. Topology of a line in Euclidean 2-space

b. Topology of a line in Euclidean 1-space

Figure 3.23: Interior, closure, and boundary of a line segment in Euclidean 2-space and 1-space with the usual topology

We conclude this brief excursion into point-set topology by considering the notion of *connectedness*. In fact, point-set topology recognizes several different kinds of connectedness. This section defines a simple form based directly upon the neighborhood properties of a topological space. The next section includes a description of further forms, including weak, strong, and path-connectedness.

- Let S be a topological space and X be a subset of points of S. Then X is *connected* if whenever it is partitioned into two non-empty disjoint subsets, A and B, then either A contains a point near B, or B contains a point near A, or both.

connected

Consider the three sets shown in Figure 3.24. In the case of Figures 3.24a and b, no matter how we choose to divide them into two, the partition will always satisfy the condition of the definition of connectedness above. Even if the set in Figure 3.24a is partitioned into its upper and lower disks, with the point of intersection included in the upper disk, then this point is certainly a near point of the lower disk. Therefore, the sets in Figures 3.24a and b are connected. Figure 3.24c shows a set that is not connected. To see this, partition the set in Figure 3.24c into its upper and lower component disks. Then no point of the upper disk is a near point of the lower disk, and no point of the lower disk is a near point of the upper disk. This example shows that the topological definition of connectedness accords with our intuition.

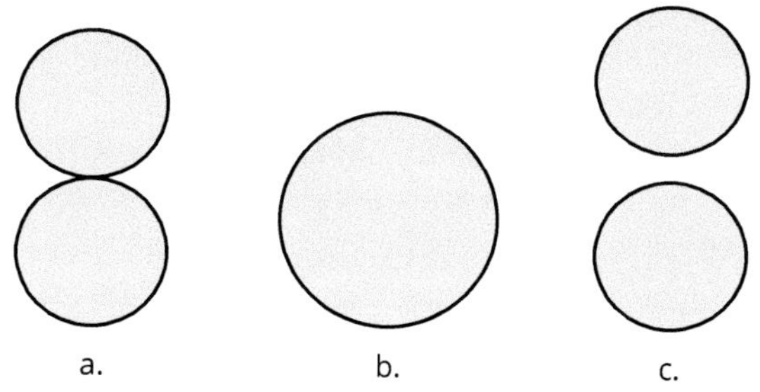

Figure 3.24: Connected (a, b) and disconnected (c) sets in the Euclidean plane with the usual topology

3.3.4 *Point-set topology of the Euclidean plane*

The Euclidean plane with the usual topology provides by far the most important example of a topological space, for the purposes of GIS. A *homeomorphism* (or *topological transformation*) of $\mathbb{R}^2$ is a bijection of the plane that transforms each neighborhood in the domain to a neighborhood in the image. Furthermore, any neighborhood in the image must be the result of the application of the transformation to a neighborhood in the domain. Put more simply and intuitively, a homeomorphism corresponds to the notion of a rubber sheet transformation, which stretches and distorts the plane without folding or tearing.

homeomorphism

If the result of applying a homeomorphism to a point-set X is point-set Y, we say that X and Y are *topologically equivalent*. Thus, in Figure 3.25, the disk S and the area T are topologically equivalent, but neither is topologically equivalent to the area U. From an intuitive point of view, it is clearly possible (at least, in the mind's eye, with a lot of stretching) to transform S to T (and back again) by stretching and contracting a rubber sheet. However, the only option to arrive at the area U from S is to tear the sheet, so as to form the hole in U. Tearing is not allowed and therefore S and U are not topologically

topological equivalence

equivalent. Set V, which is formed by gluing the disk to itself at a single point, is homeomorphic to none of the sets S, T, or U.

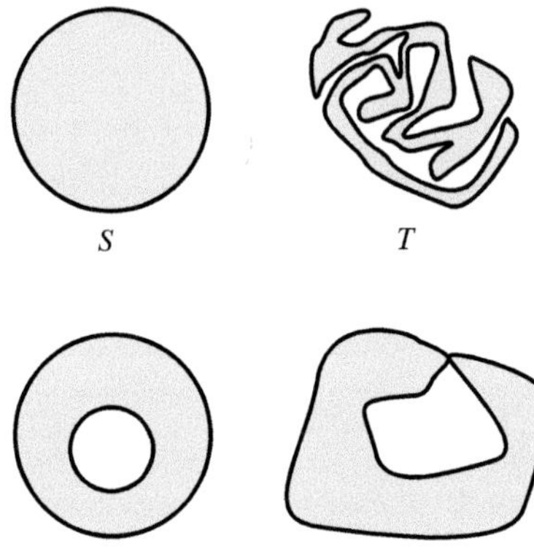

Figure 3.25: Topologically equivalent and inequivalent planar objects

Many mapping ideas are based upon the idea of homeomorphism. For example, Figures 3.26 and 3.27 show two maps of The Potteries' bus routes, first (Figure 3.26) as a similarity transformation of the actual routings (neglecting the Earth's curvature), and second (Figure 3.27) as a topological transformation of the actual routings. The two bus route maps are topologically equivalent.

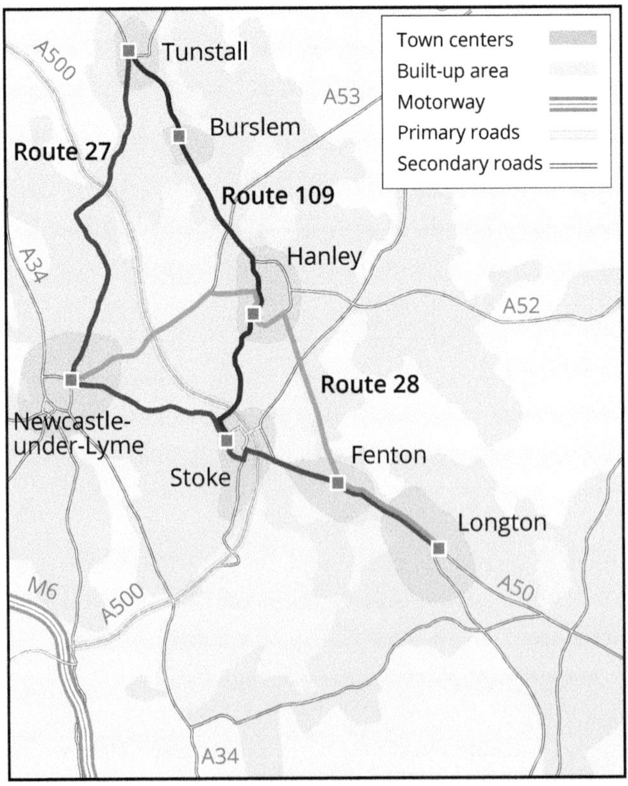

Figure 3.26: Map (similarity transformation) of Potteries' bus routes

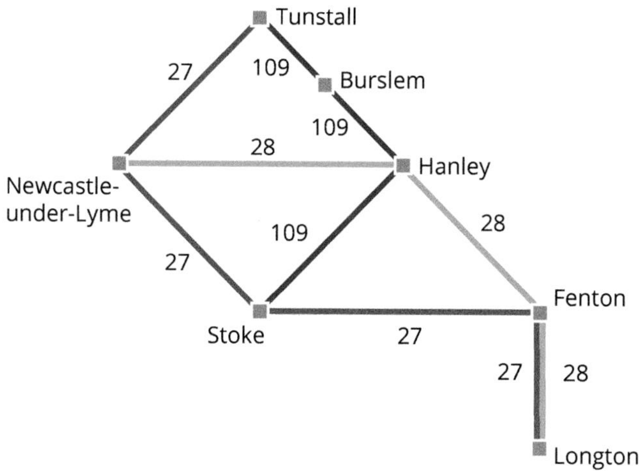

Figure 3.27: Map of Potteries' bus routes, topologically equivalent to Figure 3.26

Properties that are preserved by homeomorphisms are called *topological invariants.* From the definition of a homeomorphism, it is clear that the configuration of neighborhoods of a space is a topological invariant. Other constructs, such as open set, closed set, and boundary, are also topological invariants because they are defined purely in terms of neighborhoods.

topological invariance

The paradigm for all open sets in the Euclidean plane is the unit open disk, with its center at the origin. Therefore, all topologically equivalent areas to this are open as well. Any set homeomorphic to the unit open disk we term an *open cell.* Any set homeomorphic to the closed unit disk is a *closed cell* (see also Box 3.5 on the following page). An open cell is clearly an open set, and similarly a closed cell is clearly a closed set. In general, if a set of points owns its entire boundary then it is closed, and if it owns none of its boundary then it is open.

open cell

closed cell

Connectedness has already been defined in the context of a general topological space. Connectedness is defined purely in terms of the topology, so it is a topological invariant. Essentially, a connected set is "all of a piece," such as a line, a cell, or an annulus (a ring). Connected sets are not necessarily homeomorphic to each other; a cell is not homeomorphic to an annulus because the rubber sheet would need to be torn to make the hole. Connected sets that do not have holes are called *simply connected.* Being simply connected is a topological invariant. Cells are simply connected; annuli are not simply connected.

simply connected

Two basic 1D object types are the straight-line segment and the circle (a circle is the boundary of a disk). Using the notion of homeomorphism, we may generalize these two basic object types as follows. A simple arc is topologically equivalent to a straight-line segment: it is clearly connected. The homeomorphism from a straight-line segment to a simple arc is a bijection, so it cannot be possible for a simple arc to cross over itself or for its end-points to be coincident. If the condition on no self-crossings is relaxed, then the resulting 1D object is termed an *arc.* If the end-points are coincident and self-crossings are not allowed, then the object is termed a *simple loop.* If the

simple arc

arc

simple loop

Box 3.5: Brouwer's fixed point theorem

Cells, though basic to topological work, are the subject of one of the most intriguing of results in introductory topology. Imagine a cell laid out as a rubber shape on the plane. Now deform the cell by twisting, stretching, even folding, but not tearing, and replace the cell on the plane in such a way that it is entirely within its original outline. Then, there will always be at least one point that has not moved from its original position! Such a point is called a *fixed point* and the result is known as *Brouwer's fixed point theorem*. Brouwer's fixed point theorem holds for Euclidean n-space and has many intriguing and counter-intuitive implications. As a 3D example, assume the water within a lake is homeomorphic to the closed unit sphere. Assume further that the currents within the lake operate as a continuous rubber sheet transformation of the water. Then according to Brouwer's fixed point theorem, for any two points in time there must always be some water that is in exactly the same location within the lake at both points in time (although it may have moved in between). Fixed point theorems have played an important role in computer science, in particular underpinning the theory of recursion.

loop

end-points are coincident and self-crossings are allowed, then the object is termed a *loop*. A simple loop is topologically equivalent to a circle. Examples of some 1D planar objects are shown in Figure 3.28.

Figure 3.28: Some 1D planar objects

Moving up a dimension, we have already defined the cell as the primary 2D topological object, topologically equivalent to the disk and simply connected. Another notable class of areal objects is the annuli (cells with a single hole). We can also allow the holes to be occupied by further objects (islands) and so on (see Figure 3.29).

Figure 3.29: Holes within holes within holes

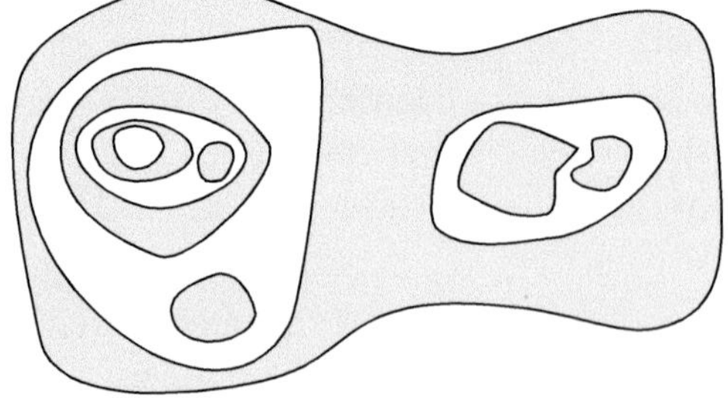

The next paragraphs explore further the concept of connectedness, particularly as it relates to areal objects. The topological property of connectedness was defined earlier, in Section 3.3.2. We begin by providing a different definition.

path-connected

- A set in a topological space is *path-connected* if any two points in the set can be joined by a path that lies wholly in the set.

A natural first question is: Given that there are two notions of connectedness, namely, connectedness (as defined earlier) and path-connectedness, do these notions define the same property? The answer to this question is, in general, no. Although it is possible to show that every path-connected set is connected, there are sets that are connected but not path-connected. However, in the special case of the Euclidean plane with the usual topology, each example of a connected but not path-connected set is pathological, involving an infinite number of twists and turns. Therefore, for practical purposes, we may identify notions of connectedness and path-connectedness, certainly for the areal objects that we will define shortly. Path-connectedness is a more intuitive notion than pure connectedness, and can therefore be used as a test for connectedness in practical cases. To summarize, test for connectedness by asking the question: Given any two points in the set, is it possible to move from one point to the other along a path entirely within the set?

Many applications of spatial analysis require classes of planar objects that are purely areal, that is, not mixtures of points, lines, and areas. Also they do not have isolated missing points (*punctures*) or arcs (*cuts*). Interestingly, it is possible to define the notion of a purely areal object using only topological notions.

regularization

- Let X be a set of points in the Euclidean plane under the usual topology. Then define the *regularization* of X to be the closure of the interior of X, that is, $reg(X) = X^{\circ-}$.

The regularization process has the effect of eliminating from a set any pathological and non-areal features. Consider the example shown in Figure 3.30a, which is an amalgamation of a punctured and cut cell with some arcs and isolated points. The regularization of the set in Figure 3.30b removes all cuts, punctures, extraneous arcs, and isolated points. Regularization first finds the interior of the object, which will remove exterior arcs and points. Taking the closure will then remove cuts and punctures. What remains is always a closed, purely areal object. In our example, shown in Figure 3.30b, the result is a cell.

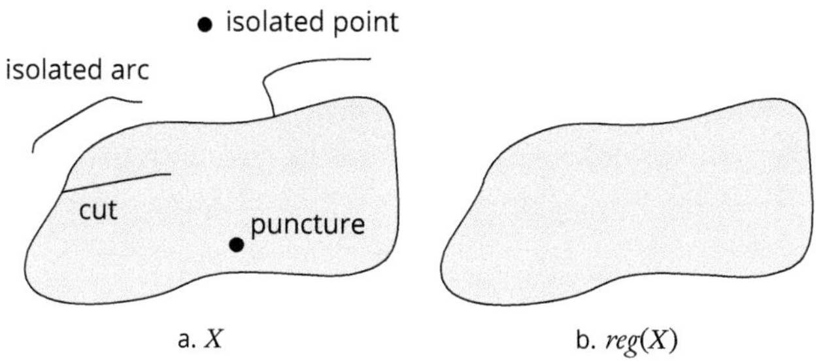

Figure 3.30: Spatial object X comprising an area with cut, puncture, arcs, and points, and $reg(X)$, its regularization

The regularization concept can now be used to characterize pure area. If an object is already purely areal, then regularizing it will have no effect. An object for which regularization has no effect is termed *regular closed*. The regular closed sets are exactly the purely areal objects that we require. The formal definition follows.

regular closed

- Let X be a set of points in the Euclidean plane under the usual topology. Then X is *regular closed* if and only if $X^{\circ -} = X$.

Having finally arrived at a topological characterization of objects that are purely areal, we now reconsider in more detail the notion of connectedness. Figure 3.31 shows three connected sets.

Figure 3.31: Three connected sets

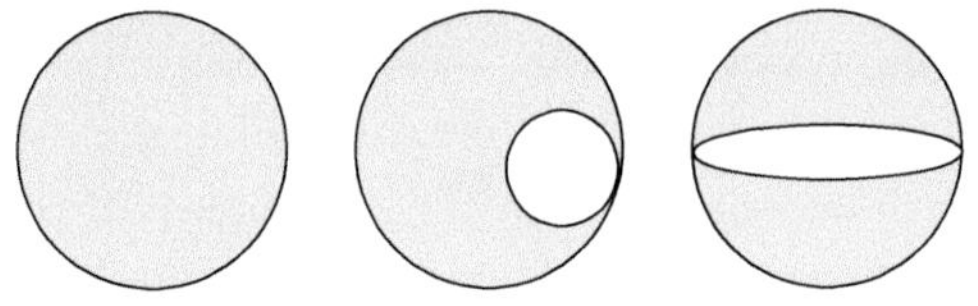

There are some clear differences in the kinds of connectedness here. In the first two cases X and Y, given any two points in each set, there are few constraints upon the path of connection from the first point to the second. All that is required is that the path starts at the first point, stays within the set, and ends at the second point. However, in the case of set Z, if the two points are in the upper portion and lower portion, respectively, then the path is constrained to pass through one of the two points on the horizontal diameter. This difference is expressed by saying that X and Y are *strongly connected*, but that Z is *weakly connected*. To arrive at a formal definition, note that Z may be made disconnected by removing a finite number of points (in fact, the two points on its horizontal diameter). However, no matter how large a finite number of points we remove from X and Y, they will remain connected.

weakly connected

strongly connected

- A connected set X in the Euclidean plane with the usual topology is *weakly connected* if it is possible to transform X into an unconnected set by the removal of a finite number of points.
- A connected set X in the Euclidean plane with the usual topology is *strongly connected* if it is not weakly connected.

Figure 3.32 shows some more strongly and weakly connected sets. While the sense of strongly and weakly connected sets will resonate with the definitions of strongly and weakly connected directed graphs introduced in the previous chapter (Section 2.4.1), the two ideas are quite distinct as can be seen from their definitions. The notions of strong and weak connectedness in sets play an important role later in the categorization of planar objects in the object-based approach to spatial modeling.

The 4-intersection model (4IM) We have seen above how point-set topology leads directly to formal of *boundary*, *interior*, and *closure*. An important characterization of the topological relationships between spatial objects—more widely cited in the literature than any other—is based on the question: What can be deduced about the topological relationship between two spa-

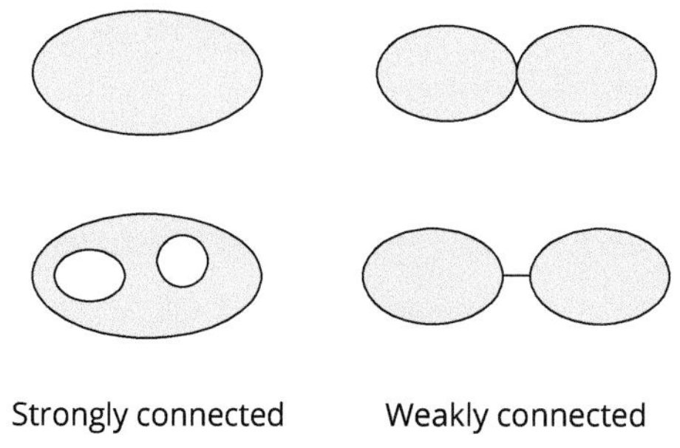

Figure 3.32: More strongly and weakly connected sets

tial regions from the set-theoretic relationship of intersection between their interiors and boundaries?

To be more precise, let X and Y be the spatial cells, and assume that the boundaries, ∂X and ∂Y, and interiors, X° and Y°, are known. In order to find the topological relationship between X and Y, compute the set-oriented relationships between ∂X, X°, ∂Y, and Y°. In fact, for a first-pass determination of the topological relationships, we consider the following four sets:

$$\partial X \cap \partial Y \qquad \partial X \cap Y^\circ$$
$$X^\circ \cap Y^\circ \qquad X^\circ \cap \partial Y$$

Each of these point sets may either be empty (written $\emptyset$, the empty set) or non-empty ($\neg\emptyset$). There are $4 \times 4 = 16$ different mutually exclusive combinations of possibilities. Each possibility is a condition on the boundaries and interiors of sets, so each possibility will lead to a relationship between the sets that is preserved under topological transformations (homeomorphisms). For general sets in a point-set topology, each of the 16 combinations can exist and lead to a distinct topological relationship between the two sets X and Y. We are, however, concerned only with spatial cells embedded in the Euclidean plane. In this case, only *eight* of the 16 combinations can occur. Table 3.3 shows the eight possibilities, together with their common names.

Table 3.3: Eight relations between cells in the Euclidean plane

$\partial X \cap \partial Y$	$X^\circ \cap Y^\circ$	$\partial X \cap Y^\circ$	$X^\circ \cap \partial Y$	Name	Alternative
$\emptyset$	$\emptyset$	$\emptyset$	$\emptyset$	X disjoint Y	
$\neg\emptyset$	$\emptyset$	$\emptyset$	$\emptyset$	X meets Y	X touches Y
$\neg\emptyset$	$\neg\emptyset$	$\emptyset$	$\emptyset$	X equals Y	
$\emptyset$	$\neg\emptyset$	$\neg\emptyset$	$\emptyset$	X inside Y	Y contains X
$\neg\emptyset$	$\neg\emptyset$	$\neg\emptyset$	$\emptyset$	Y covers X	X covered by Y
$\emptyset$	$\neg\emptyset$	$\emptyset$	$\neg\emptyset$	Y inside X	X contains Y
$\neg\emptyset$	$\neg\emptyset$	$\emptyset$	$\neg\emptyset$	X covers Y	Y covered by X
$\neg\emptyset$	$\neg\emptyset$	$\neg\emptyset$	$\neg\emptyset$	X overlaps Y	

Any pair of spatial cells in the Euclidean plane must satisfy one and only one of these possibilities. In technical language, the eight topological relationships are JEPD: jointly exhaustive (i.e., no other possibilities) and pairwise disjoint (i.e., a pair of regions cannot satisfy two possibilities simultaneously). These eight possibilities are summarized graphically in Figure 3.33. Because four intersections are tested for empty in this method, it is usually known as the *4-intersection* model (4IM).

4-intersection model

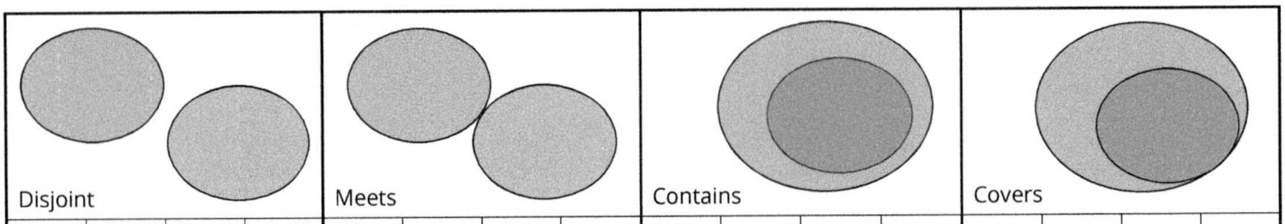

$\emptyset$	$\emptyset$	$\emptyset$	$\emptyset$	$\neg\emptyset$	$\emptyset$	$\emptyset$	$\emptyset$	$\emptyset$	$\neg\emptyset$	$\emptyset$	$\neg\emptyset$	$\neg\emptyset$	$\neg\emptyset$	$\emptyset$	$\neg\emptyset$
$\partial X\cap\partial Y$	$X^\circ\cap Y^\circ$	$\partial X\cap Y^\circ$	$X^\circ\cap\partial Y$	$\partial X\cap\partial Y$	$X^\circ\cap Y^\circ$	$\partial X\cap Y^\circ$	$X^\circ\cap\partial Y$	$\partial X\cap\partial Y$	$X^\circ\cap Y^\circ$	$\partial X\cap Y^\circ$	$X^\circ\cap\partial Y$	$\partial X\cap\partial Y$	$X^\circ\cap Y^\circ$	$\partial X\cap Y^\circ$	$X^\circ\cap\partial Y$
$\neg\emptyset$	$\neg\emptyset$	$\emptyset$	$\emptyset$	$\neg\emptyset$	$\neg\emptyset$	$\neg\emptyset$	$\neg\emptyset$	$\emptyset$	$\neg\emptyset$	$\neg\emptyset$	$\emptyset$	$\neg\emptyset$	$\neg\emptyset$	$\neg\emptyset$	$\emptyset$

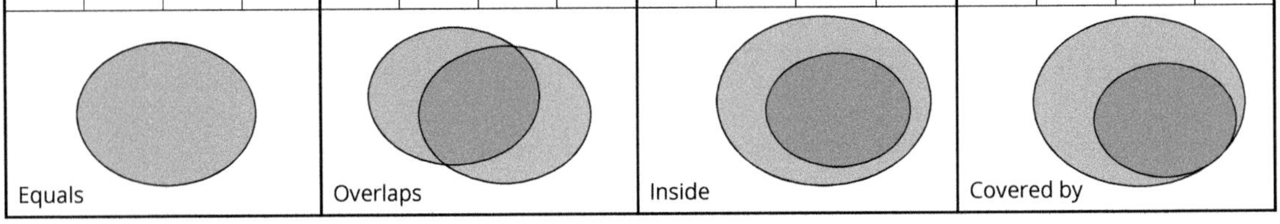

Figure 3.33: The eight JEPD topological relationships of the 4-intersection model (4IM)

It can now be seen that 4IM relations meets and overlaps are topological refinements of the set-oriented operation *intersection*, while inside and covers are refinements of the *subset of* relationship.

co-dimension

This work has been extended for higher-dimensional spaces, and where the *co-dimension* is non-zero (i.e., where the dimension of the spatial objects is less than the dimension of the space in which they are embedded). In this case, three topological operations are used: boundary, interior, and set complement. The binary spatial relation between two spatial objects X and Y is classified by checking for emptiness/non-emptiness of the nine combinations of the operations applied to X and Y, known as the *9-intersection* model (9IM). The 9 combinations are[2]:

9-intersection model

[2] Recall that X' is the set complement or *exterior* of X.

$$\begin{array}{ccc} \partial X\cap\partial Y & \partial X\cap Y^\circ & \partial X\cap Y' \\ X^\circ\cap\partial Y & X^\circ\cap Y^\circ & X^\circ\cap Y' \\ X'\cap\partial Y & X'\cap Y^\circ & X'\cap Y' \end{array}$$

3.3.5 *Combinatorial topology of the Euclidean plane*

The other topological area considered in this chapter is combinatorial topology. In some ways, combinatorial topology is more pertinent to computer-based models than point-set topology, because the often finite and discrete structures that arise in combinatorial topology are highly suitable for representation in computer-based data structures. A typical result of combinatorial topology is Euler's famous formula:

Euler's formula

- Given a polyhedron with f faces, e edges, and v vertices, then $f - e + v = 2$.

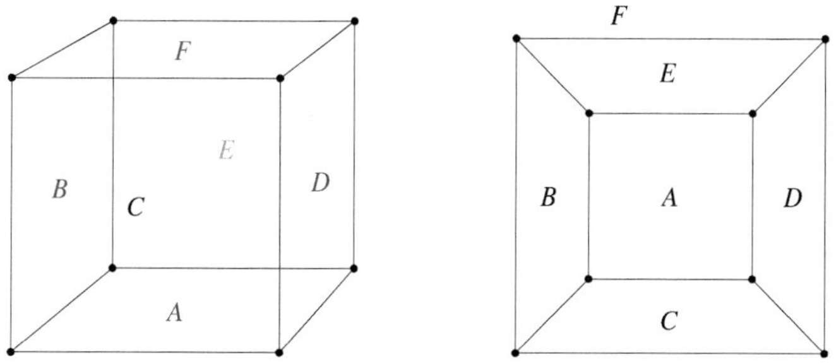

a. Cube

b. Planar configuration homeomorphic to a cube

Figure 3.34: Example polyhedron and 3-space homeomorphism

For example, Figure 3.34a shows a cube (six faces, 12 edges, and eight vertices). A very similar formula applies to an arrangement of cells in the plane. Remove a single face from a polyhedron (for example, in Figure 3.34a remove face F) and apply a 3-space homeomorphism to flatten the shape onto the plane. What results is a configuration of cells, with arcs forming common boundaries and nodes forming the intersection points of the arcs. Flattening the cube in Figure 3.34a results in the planar configuration in Figure 3.34b. Since we have removed a face from the polyhedron (it has actually become the exterior to the cellular configuration, as in Figure 3.34b), we may simply modify Euler's formula for the sphere to derive Euler's formula for the plane.

- Given a cellular arrangement in the plane, with f cells, e edges, and v vertices, $f - e + v = 1$.

Figure 3.35 shows an example of a planar configuration of cells, with 25 faces, 72 edges, and 48 nodes (inspired by the configuration of the 25 wards of the City of London).

The topological content of these results becomes clear when we observe that no matter how the surface of the sphere is divided into polyhedral arrangements, the result of $f - e + v$ is always two, and for planes the result is always one. Thus the number two characterizes a sphere and distinguishes it from a plane. If we were to perform the same exercise on the surface of a torus (doughnut shape), the result is always zero. The result of $f - e + v$ is called the *Euler characteristic* of a surface.

Euler characteristic

Simplexes and complexes As we will see in later chapters, many of the models of space found in a GIS use variations of the planar cellular arrangements described above. The most fundamental formal model of such cellular arrangements uses the notion of a *simplicial complex*. In the two-dimensional case, simplicial complexes are simple triangular network structures in the Euclidean plane.

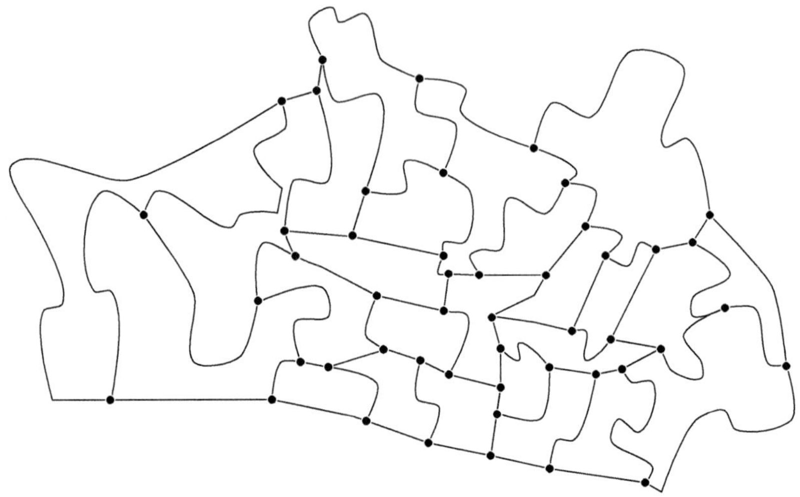

Figure 3.35: Planar cellular arrangement where $f = 25, e = 72$, and $n = 48$

- A *0-simplex* is a set consisting of a single point in the Euclidean plane.
- 1-simplex — A *1-simplex* is a closed finite straight-line segment.
- 2-simplex — A *2-simplex* is a set consisting of all the points on the boundary and in the interior of a triangle whose vertices are not collinear.

The *vertices* of a simplex are defined as: for a 0-simplex the point itself; for a 1-simplex its end-points; and for a 2-simplex the vertices of the triangle. A *face* of a simplex S is a simplex whose vertices form a proper subset of the vertices of S. Figure 3.36 shows examples of 0-, 1-, and 2-simplexes. The *boundary* of a simplex S, written ∂S, is the union of all its faces. For example, suppose that a 2-simplex S has vertices x, y, and z. Then the faces of S are the three 1-simplexes xy, xz, yz and the three 0-simplexes x, y, z. The boundary of S is the union of these faces, thus corresponding to the usual point-set topological definition of boundary.

Figure 3.36: Examples of 0-, 1- and 2-simplexes

Simplexes are the building blocks of larger structures, called simplicial complexes. Complexes are built out of simplexes in a way that is now made precise. simplicial complex — A *simplicial complex* C is a finite set of simplexes satisfying the properties:

1. A face of a simplex in C is also in C.
2. The intersection of two simplexes in C is either empty or is also in C.

Figure 3.37 shows examples of configurations, two of which are simplicial complexes and two are not. In fact, the complexes on the right are formed by adding sufficient nodes and edges to the configurations on the left to make them satisfy the simplicial complex formation rules, a form of "completing the topology." In the case of the 2-simplexes ***abc*** and ***def***, their intersection is not a face of either simplex. This is rectified by adding nodes ***k*** and ***l*** and decomposing the original simplexes ***abd*** and ***def*** into simplexes ***akf***, ***afb***, ***bfc***, ***cfl***, ***fkl***, ***dkl***, and ***del***. These simplexes, along with their faces ***ab***, ***af***, ***ak***, ***bc***, ***bf***, ***cf***, ***cl***, ***dk***, ***dl***, ***de***, ***el***, ***fl***, ***fk***, ***kl*** and vertices ***a***, ***b***, ***c***, ***d***, ***e***, ***f***, ***k***, ***l***, form a simplicial complex. The 1-simplexes in the lower part of the figure are enhanced in a similar way by adding vertex ***m***.

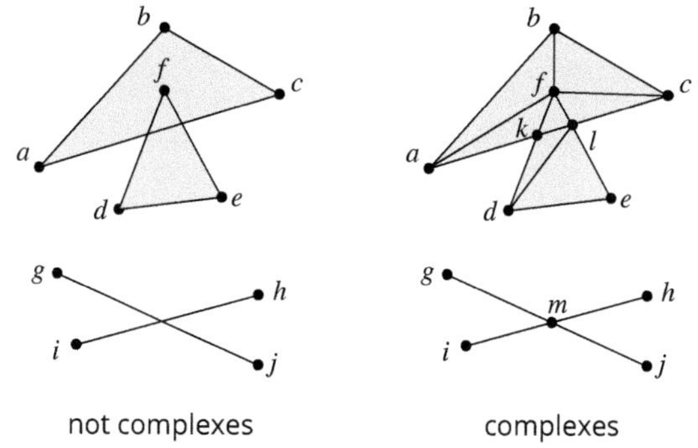

Figure 3.37: Aggregations of simplexes

The set of points contained in the constituent simplexes of a simplicial complex is a set in the Euclidean plane, called the *planar embedding* of the complex. Figure 3.38 shows the planar embedding of the upper complex in Figure 3.37.

planar embedding

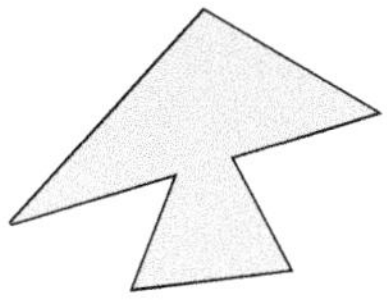

Figure 3.38: Planar embedding of a simplicial complex from Figure 3.37

Even though we have given an abstract presentation of simplicial complexes, such structures are common in GI science. An n-simplex is said to have *dimension* n. The dimension of a simplicial complex is the maximum dimension of its constituent simplexes. The constructions performed in this section are all planar, but the ideas can be generalized to higher dimensional structures. Thus a 3-simplex would be a tetrahedron.

For an n-complex, C, the boundary of C, ∂C, is a simplicial complex of dimension $n - 1$. 1D complexes are *graphs*, covered previously in Sections 2.4.1 and 3.4.1. 2D complexes may be used to model the triangulated irregular networks (TINs) used in terrain modeling, or indeed any areal objects.

Combinatorial map A different approach to combinatorial topology is known as the *combinatorial map*. Assume that the boundary of a cellular arrangement is decomposed into simple arcs and nodes, as in Figure 3.35. Next, distinguish the two directions of each arc, akin to a bidirected graph structure. A combinatorial map is then:

combinatorial map

- The set of directed arcs D, called *darts*. In Figure 3.39, the set of darts $D = \{ab, ag, ah, ba, bc, bi, cb, ci, cd, dc, de, di, ...\}$ and so on.
- A function $\sigma : D \rightarrow D$ that for each dart specifies the next dart in a counterclockwise direction at each node. For example, in Figure 3.39 $\sigma(ab) \mapsto ag$, $\sigma(ag) \mapsto ah$, $\sigma(ah) \mapsto ab$ and so on. Written as a relation, $\sigma = \{(ab, ag), (ag, ah), (ah, ab), (ba, bi), (bi, bc), (bc, ba), ...\}$. The function σ is a bijection (an injection and a surjection).
- A function $\alpha : D \rightarrow D$ that for each dart specifies the other dart that makes up that arc. For example, in Figure 3.39 $\alpha(ab) \mapsto ba$, $\alpha(ba) \mapsto ab$, $\alpha(ag) \mapsto ga$ and so on. The function α is also a bijection, as well as a symmetric, irreflexive function. Written as a relation, $\alpha = \{(ab, ba), (ag, ga), (ah, ha), (ba, ab), (bc, cb), (bi, ib), (cb, bc), ...\}$.

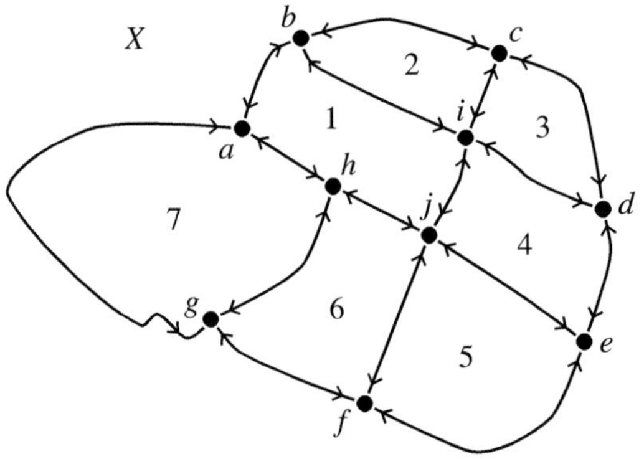

Figure 3.39: A planar cellular arrangement, augmented with a set of darts D as the basis for a combinatorial map

Intriguingly, this abstract structure of darts and functions provides a unique and unambiguous representation of a cellular arrangement, such as that in Figure 3.39. Adopting the rule that a dart bounds the face on its right as we traverse it, we can use the functions σ and α to reconstruct each face. For example, let us reconstruct the boundary of face 7 in Figure 3.39. We start with any one of its bounding darts, say *ah*. Next, alternating the functions α and σ will allow us to trace around face 7: $\alpha(ah) \mapsto ha$, $\sigma(ha) \mapsto hg$, $\alpha(hg) \mapsto gh$, $\sigma(gh) \mapsto ga$, $\alpha(ga) \mapsto ag$, and finally back to the start of the cycle with $\sigma(ag) \mapsto ah$. The choice of starting dart was arbitrary; we might have picked any of the darts that bound 7 on their right: *ah*, *hg*, or *ga*. Each of the other faces may be similarly traversed. Dart *ha*, for example, bounds face 1: $\alpha(ha) \mapsto ah$, $\sigma(ah) \mapsto ab$, $\alpha(ab) \mapsto ba$, $\sigma(ba) \mapsto bi$, ... and so on. For completeness, we must also include the external face, X, bounded by *ba*, $\alpha(ba) \mapsto ab$, $\sigma(ab) \mapsto ag$, $\alpha(ag) \mapsto ga$, $\sigma(ga) \mapsto gf$, $\alpha(gf) \mapsto fg$, $\sigma(fg) \mapsto fe$, ... and so on.

The combinatorial map is the foundation of the NAA (node-arc-area) and DCEL representations of planar configurations, discussed in more detail in Chapter 5.

3.4 *Further spaces*

The picture emerging over this chapter is that space is not just one thing, one reality; rather it is a family of related representations each of which foregrounds particular properties of space and offers particular characteristics and operations. This final section visits three further types of spaces: network spaces build on the foundations of abstract graphs introduced in the previous chapter; metric spaces formalize the concept of distance between objects in the space; and fractal geometry tackles spatial objects that exhibit self-similarity and scale dependence.

3.4.1 *Network spaces*

A great many spatial problems can be represented using a network. For example, a system of roads or rail links is often thought of as a network; the task of deciding what route to take when traveling by car is essentially the task of finding a suitable route through a network. Several famous spatial problems have been solved by representing the problem as a network, such as the Königsberg bridge problem (see Box 3.6 on the next page).

We have already encountered *abstract graphs* as a formal model of the network, in connection with graph databases in the previous chapter (Section 2.4.1). Building on these foundations, this section introduces some further graph concepts important for modeling geographic spaces, including *trees* and *planar graphs*.

Abstract graphs revisited Recall from Section 2.4.1 that a graph is a set of nodes together with a set of edges that connect pairs of nodes. Armed with our knowledge of sets, relations, and functions gained in the previous section, we can now more precisely define the following graph structures already encountered:

- An undirected graph $G = (N, E)$ is a set of nodes N together with a set E of unordered pairs of nodes, called edges.
- A directed graph $G = (N, E)$ is a set of nodes N together with a set of edges drawn from a relation on the set of nodes, $E \subseteq N \times N$.
- An (edge) labeled graph $G = (N, E, w)$ is a set of nodes N, edges E, together with an edge-labeling function $w : E \rightarrow L$ mapping each edge to a label in the set L.
- An (edge) weighted graph is a labeled graph $G = (N, E, w)$ where the labels are drawn from positive numbers, e.g., $w : E \rightarrow \mathbb{Q}^+$.
- A (directed) multigraph $G = (N, E, s, t)$ is a set of nodes N, a set of edges E, and two functions $s : E \rightarrow N$ and $t : E \rightarrow N$ mapping each edge to its start and end node, respectively.

Recall also from Section 2.4.1 that a path is an ordered sequence of nodes in a graph, where each consecutive pair of nodes is connected by an edge; a

Box 3.6: Königsberg bridge problem

The Königsberg bridge problem addresses the question of whether it is possible to walk a circuit that crosses each of seven bridges (shown below left) in the city of Königsberg once and only once. You might try to find such a route yourself. However, don't try for too long as in 1786 the mathematician Leonard Euler succeeded in proving that the task is impossible. Euler's proof is based on a model of the topological relationships between the Königsberg bridges (shown below right). The nodes, labeled w, x, y, and z, are abstractions of the regions of dry land. The edges between nodes are abstractions of the bridges connecting regions of dry land. Euler noted that apart from the start and end nodes, any path through a node must come in along one edge and out along another edge. So, if the problem is to be solvable, then the number of edges incident with each intermediate node must be even. However, in Königsberg *none* of the nodes is incident with an even number of edges. Thus, Euler proved that it was impossible to cross each bridge just once. A path through a graph that visits each edge exactly once is still termed an *Eulerian trail*.

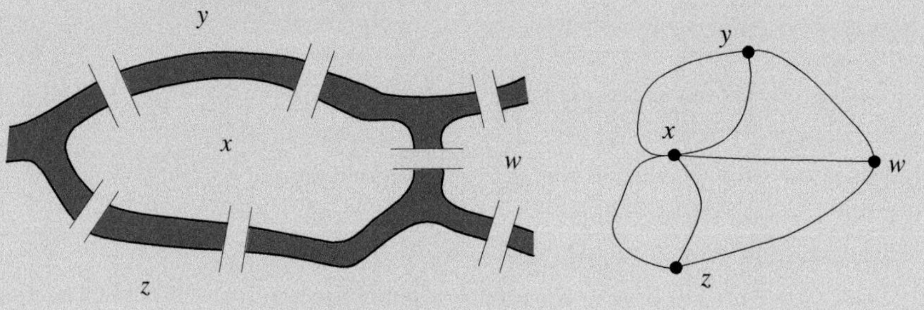

cycle is a path from a node to itself traversing at least one edge; a graph is connected if there exists a path between every pair of nodes.

tree

Trees and DAGs An especially useful class of graphs is the *tree*. A tree is a connected acyclic graph. Figure 3.40 shows the three non-isomorphic trees, each with five nodes.

rooted tree

A *rooted tree* is a tree that has one of its nodes, the *root*, distinguished from the others. Rooted trees (often we omit the word "rooted") are conventionally drawn with the root at the top and nodes occupying successive levels down, depending upon their distance (in terms of path) from the root. Nodes immediately below the root are termed *immediate descendants* of the root; they themselves may have descendants, and so on. A node with no descendants is termed a *leaf*. Figure 3.41 shows an example of a layered rooted tree. Trees

leaf

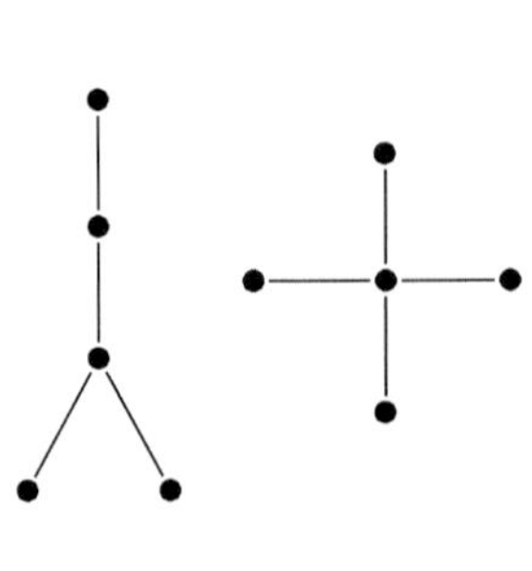

Figure 3.40: The three non-isomorphic trees with five nodes

provide some useful data structures for computational purposes, to which we return in later chapters.

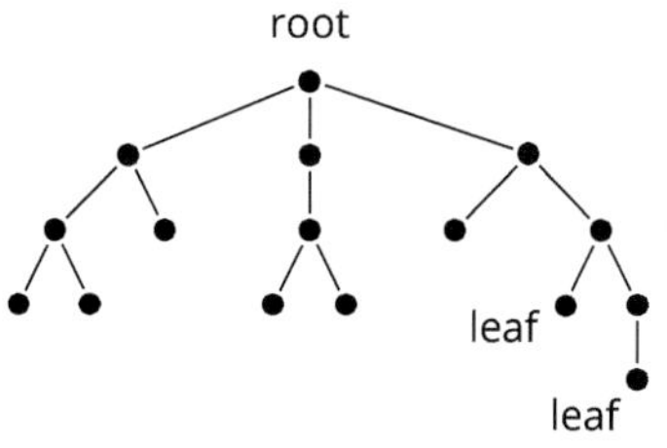

Figure 3.41: Rooted tree with five levels and eight leaves

The class of directed graphs that have no cycles is particularly useful for a wide range of applications. These graphs are termed *directed acyclic graphs* or DAGs. A DAG defines a *partial order* on its nodes. As introduced above, a (strict) partial order is a special form of relation on a set, which is irreflexive, antisymmetric, and transitive.

DAG

partial order

An example of a partial ordering is the relation "greater than," between two real numbers. No number can be greater than itself (irreflexive); if x is greater than y, then it cannot be that y is greater than x (antisymmetric); and x greater than y greater than z implies x greater than z (transitive). For a DAG, if we define a relation R where $(a, b) \in R$ means "there exists a path from a to b," then it can be shown that this relation is a partial order.

Planar graphs A further level of information may be added to the graph-theoretic model by considering the embedding of the graph in the Euclidean plane. A planar graph is a graph that can be embedded in the plane in a way that preserves its structure. In particular, a planar graph's edges are embedded as arcs that may only intersect at nodes of the graph. Figure 3.42a shows a planar graph while Figure 3.42b shows a non-planar graph. For the non-planar graph in Figure 3.42b, no rearrangement of arcs in the plane will preserve the original connectivity without leading to edges crossing somewhere other than at a node.

planar graph

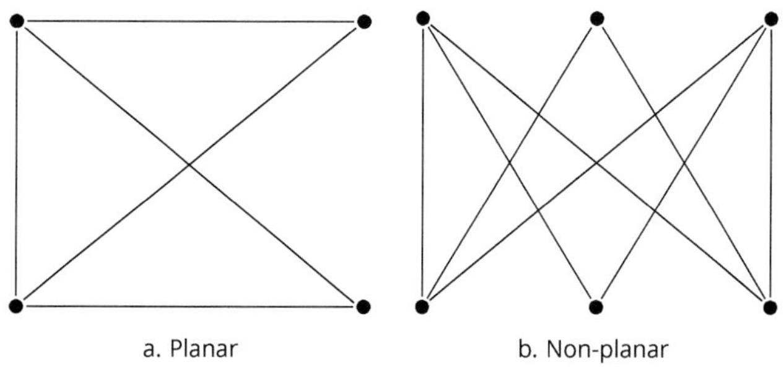

Figure 3.42: Planar and non-planar graphs

In general, there are many topologically inequivalent planar embeddings of a planar graph in the plane. A planar graph together with an embedding into the plane such that arcs intersect only at nodes of the graph is called a *plane graph*.

plane graph

Figure 3.43 shows three embeddings of the planar graph of Figure 3.42. The upper two embeddings are homeomorphic, but not homeomorphic to the lowest, with respect to the usual topology of the Euclidean plane. There is no way, using topological transformations (rubber-sheet geometry), to move the node of degree two inside the triangle to a position outside the triangle.

This last discussion raises an interesting consideration. Configurations that are taken to be equivalent in one model may very well be inequivalent in another. The second and third configurations are equivalent in a graph-theoretic sense, both being identically connected. However, when viewed as embeddings in the plane, they are inequivalent in the topological sense (and certainly in the metric sense). It all depends on the level of abstraction.

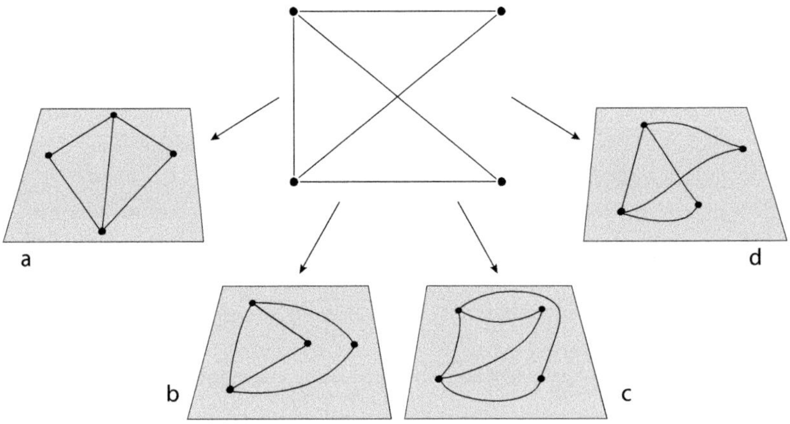

Figure 3.43: Three planar embeddings of a planar graph (plane graphs a, b, and c, where b and c are also homeomorphic) and one non-planar embedding of a planar graph (d)

A planar embedding of a planar graph determines a subdivision of the plane into faces or regions. A simple integrity constraint upon a planar graph discussed in an earlier section is given by the Euler formula for the plane, which provides a relationship between the number of nodes n, arcs a, and faces f, as $f - a + n = 1$.

dual graph

A useful concept associated with planar embedded (i.e., plane) graphs is that of *duality*. The *dual* G^* of a plane graph G is obtained by associating a node in G^* with each face in G. Two nodes in G^* are connected by an edge if and only if their corresponding faces in G are adjacent. Given an edge e in G, the dual edge e^* joins the nodes in G^* corresponding to the two faces in G incident with e. Figure 3.44 shows a plane graph (nodes filled in black, edges marked with continuous lines) and its dual (nodes shown filled in color, edges marked with dotted lines).

When the planar graph G is a diagonal triangulation of a polygon (with no Steiner points), then the dual graph G^* has the properties that the degree of each node is no more than three (because triangles have three sides), and that G^* is acyclic and connected: it is a tree.

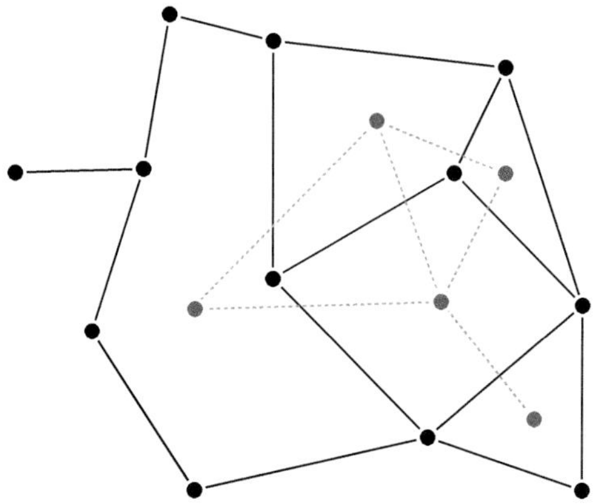

Figure 3.44: A planar graph and its dual

3.4.2 *Metric spaces*

This final section explores the kind of properties a model of space should have if it is to include the concept of distance between objects in the space. Such a space is called a *metric space*. The formal definition follows, but it should be noted that it does not always accord with our commonsense notion of distance.

metric space

A point-set S is said to be a *metric space* if there exists a distance function d that takes ordered pairs (s, t) of elements of S and returns a distance which satisfies the following three conditions:

1. for each pair s, t in S, $d(s, t) > 0$ if s and t are distinct points and $d(s, t) = 0$ if s and t are identical;
2. for each pair s, t in S, the distance from s to t is equal to the distance from t to s, $d(s, t) = d(t, s)$;
3. for each triple s, t, u in S, the sum of the distances from s to t and from t to u is always at least as large as the distance from s to u, that is: $d(s, t) + d(t, u) \geq d(s, u)$.

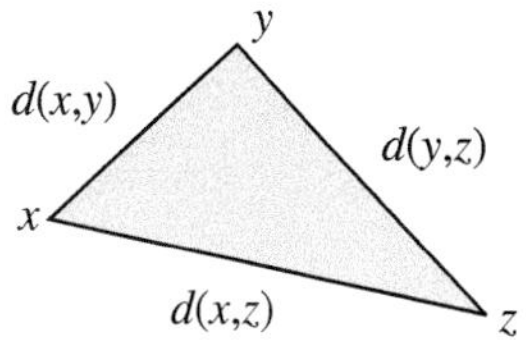

Figure 3.45: Triangle inequality

Put into more informal language, the first condition stipulates that the distance between points must be a positive number unless the points are the same, in which case the distance will be zero and the points will be *indiscernible* from each other. The second condition ensures that the distance between two points is independent of which way around it is measured, i.e., its distance measure is *symmetric*. The third condition states that it must always be at least as far to travel between two points via a third point rather than to travel directly. This third property is termed the *triangle inequality* (i.e., any two sides of a triangle must together be longer than the third side) due to the configuration shown in Figure 3.45.

triangle inequality

These three properties of a distance measure—indiscernibility, symmetry, and the triangle inequality—together define a *metric*. In order to motivate this definition, we give below some possible distance functions and consider them with respect to properties 1–3 above. Let S be a set of cities on the globe and distance between two cities in S defined as follows (see Figure 3.46):

metric

geodesic distance

Geodesic distance The distance "as the crow flies" is termed the *geodesic distance*. In our example, it is the distance along the great circle of the Earth passing through the two city centers.

Manhattan distance

Spherical Manhattan distance The name "Manhattan distance" arises because in Manhattan many of the streets are arranged in a grid-like configuration. In the plane, the *Manhattan distance* is simply the sum of the difference in x and y coordinates. Similarly, the spherical Manhattan distance is the difference in latitudes plus the difference in longitudes (see Box 3.7 on the facing page).

travel time distance

Travel time distance The minimum time required to travel from one city to the other (for example, using a sequence of scheduled airline flights) is the travel time distance.

lexicographic distance

Lexicographic distance The absolute value of the difference between the positions of cities in a fixed list of place names (*gazetteer*) is termed the *lexicographic distance*.

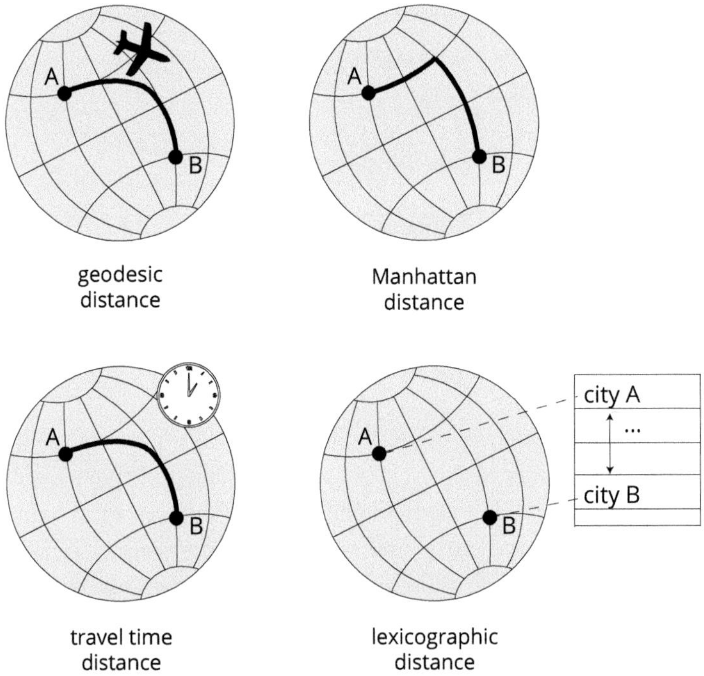

Figure 3.46: Distances defined on the globe

The first property of a metric space is quite uncontroversial, and satisfied by any self-respecting distance function. Sensible distances cannot be allowed to be negative. Also, the distance between an element and itself is always zero, whereas the distance between two distinct elements is always greater than zero. All the distance functions in our city example possess these properties. The third property, the triangle inequality, would also be a surprise if it did not hold for our distance measure. In plain English, the triangle inequal-

Box 3.7: Latitude and longitude

Many of the models in this chapter have been founded upon the plane. The plane is a useful approximation to the Earth's surface over small distances, but the Earth is not flat, so over larger distances other approximations are needed. The most common such approximation is the sphere. The surface of the sphere, although embedded in Euclidean 3-space, is 2D. Thus, any point on it may be uniquely specified by two numbers. A familiar system of coordinates for points on the surface of the sphere is latitude and longitude. In the diagram below, o is origin of the coordinate system, p is an arbitrary point on the sphere, and q is the projection of p onto the xy-plane. The angle between op and oq is the latitude of the point p. The angle between oq and the x-axis is the longitude of p. The great circle of the sphere in the xy-plane is the *equator* and the great circle in the xz-plane is the *meridian*. The usual topology of the surface of the sphere is similar to the usual topology of the Euclidean plane. Neighborhoods are sets of points with a constant distance from a fixed point, measured along a geodesic (great circle). Computing distances, angles, and areas on the sphere requires spherical trigonometry, with substantial differences to the planar geometry introduced in this chapter. On the sphere, straight lines meet at two points and the internal angles of a triangle sum to more than 180°, for example. A more detailed introduction to latitude and longitude may be found in Longley et al. (2015).

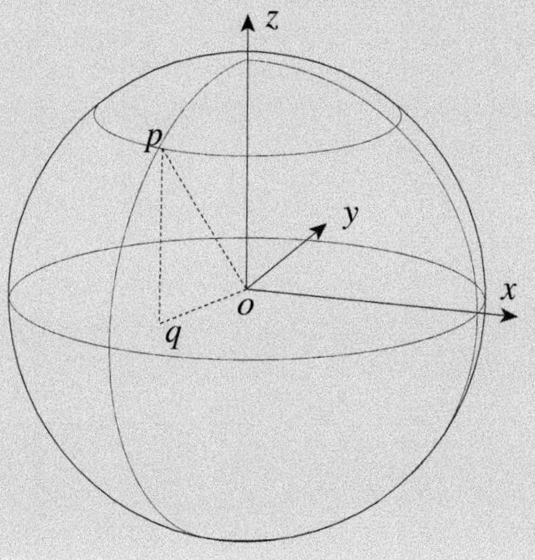

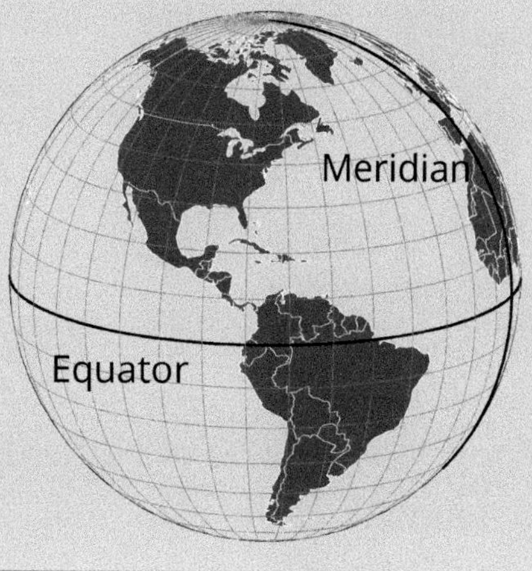

ity implies that it is never any farther to go by a direct route than an indirect route. All the above examples obey the triangle inequality.

Only the second property—of symmetry, that the distance from a to b is always the same as the distance from b to a—is a little more controversial. The geodesic, Manhattan, and lexicographic distances do all satisfy this condition. However, the travel time distance is not necessarily symmetric. For example, it is perfectly possible (and indeed usual due to prevailing winds) for the flight times between two cities to be different in each direction. A distance function that obeys properties 1 and 3 but not 2 is called a *quasimetric*.

quasimetric

So, the collection of cities together with the geodesic distance function, the Manhattan distance function, and the lexicographic distance function are all metric spaces. The collection of cities with the travel time function is a quasimetric but not a metric space. The archetypal example of a metric space is Euclidean space, where the distance between two points is defined by the Pythagorean formula given earlier (Section 3.1.1). This distance function can easily be extended to higher dimensions.

Topology of metric spaces It turns out that a metric space has a natural topology. Let S be a metric space with distance function d. For each point, x in S and each real number r, define the *open ball* $B(x, r)$ to be the set of points whose distance from the point x is less than the number r. Expressed formally:

$$B(x, r) = \{y | d(y, x) < r\}$$

Define the set of neighborhoods to be the set of open balls. It is not hard to verify that this defines a topology for S. In the case of the Euclidean metric, this example reduces to the usual topology of the Euclidean plane. In the case of the Manhattan metric, then the open ball $B(x, r)$ will contain all points for which the sum of the horizontal and vertical distances from x is less than r. In the case of the Manhattan metric applied to a flat planar space, the open balls $B(x, r)$ will be squares of diagonal $2r$, as shown in Figure 3.47.

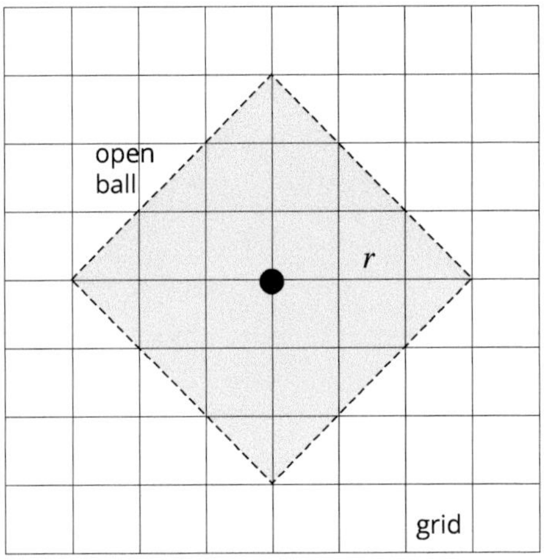

Figure 3.47: An open ball induced by the Manhattan metric

Travel time measures have some interesting properties. We have seen that, in general, they do not lead to metric spaces because they are not necessarily symmetric (for example, in a one-way traffic system). Let us for the moment make the simplifying assumption that distances are symmetric, so a topology may be defined as above. This is the travel-time topology introduced in Section 3.3.2. Computing topological neighborhoods can be illuminating in this case. Figure 3.48 shows a travel-time neighborhood (t-zone) of Liège, computed by Dussart and redrawn and considered by Tobler (1993).

The shaded region shows the area within 1 hour's travel time of the center of Liège in 1958 by the common means of travel available at that time. This "neighborhood" is not even connected, being the disjoint union of a finite number of cells. This type of pattern arises because of the discontinuous nature of travel on public transport, boarding and alighting at fixed and discrete points. As Tobler points out, such travel-time configurations are not amenable to modeling within the framework of Euclidean space. Such problems force us to use network models. Later chapters further develop this theme.

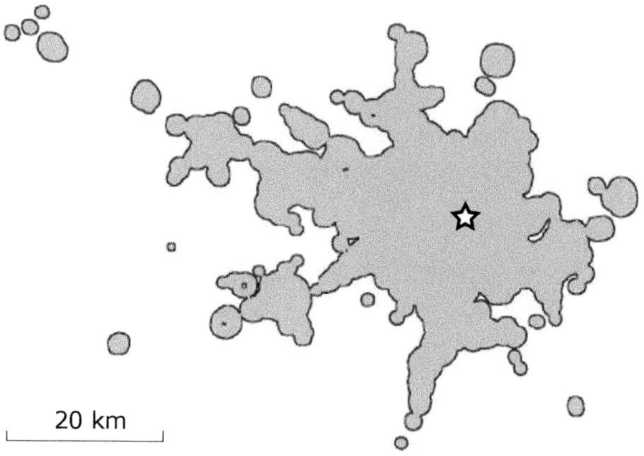

Figure 3.48: A travel-time neighborhood of Liège, computed by Dussart (Tobler, 1993). Shading indicates areas attainable within 1 hour from Liège center in 1958 by a combination of tramway, autobus, chemin de fer, and walking (5 km/h)

3.4.3 *Fractal geometry*

The appearance and characteristics of many geographic and natural phenomena, such as landscapes and coastlines, depend on the scale at which these phenomena are observed, termed *scale dependence*. For example, Figure 3.49 shows a satellite image of the Ganges River delta, Bangladesh, at three different scales. More detail is revealed at each finer scale, with detail revealed at finer scales tending to resemble details at coarser scales, termed *self-similarity*. The shape of the coastline and sinuosity of the rivers are similar at each scale, making it difficult to gauge the size of land forms in Figure 3.49. The scale has been deliberately omitted from Figure 3.49 to emphasize the effect; in fact, the first image is of an area about 60 km across, the final image approximately 15 km across. We can imagine repeating the process of zooming in over and over, perhaps until we reach the microscopic level. At every scale, new and self-similar detail would be revealed.

scale dependence

self-similar

The straight lines and smooth curves of Euclidean geometry are not well suited to modeling self-similarity and scale dependence. In a classic book, the Polish mathematician Benoit Mandelbrot (1982) argued that a fundamentally different type of geometry, which he dubbed *fractal geometry*, provides a more faithful representation of such natural and geographic phenomena. Fractal geometry concerns the study of shapes, called *fractals*, that are self-similar across all scales. Although the term "fractal" is relatively recent, fractal shapes and the problems they cause mathematicians have been known for hundreds of years (see Box 3.8 on page 121).

fractal geometry

fractal

True fractals are self-similar because they are defined recursively, rather than by describing their shape directly. For example, Figure 3.50 shows the first four stages in the construction of a famous fractal, the *Koch snowflake*, named after a Swedish mathematician, Helge von Koch. Building a Koch snowflake starts with an equilateral triangle, termed the *initiator* step. Then each straight line is divided into three equal parts. The middle part of each line is replaced with a new equilateral triangle, correctly scaled and with no base, termed the *generator* step. The Koch snowflake is the result of iterating

Koch snowflake

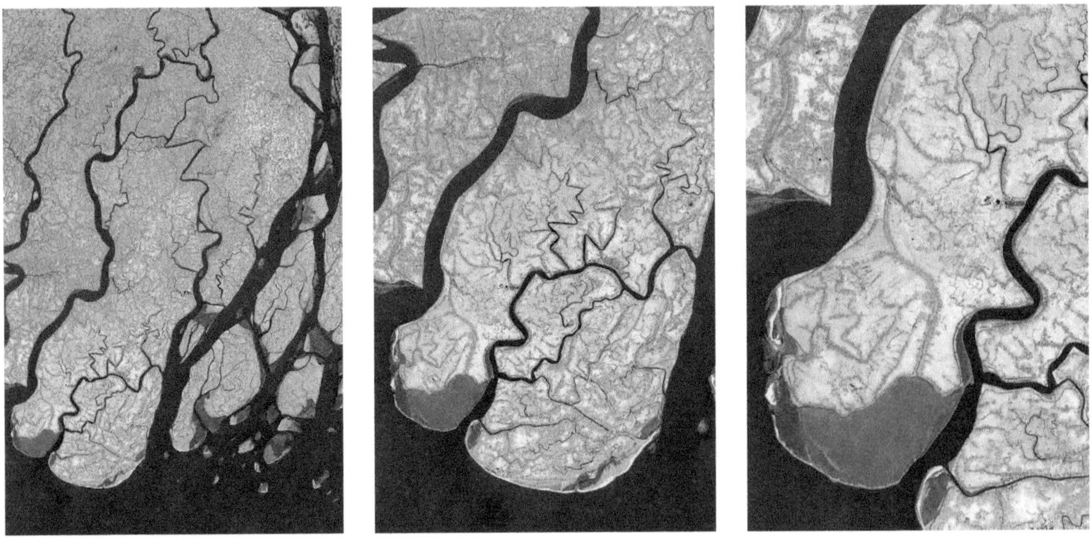

Figure 3.49: False-color composite using green, infrared, and blue wavelengths of Ganges River delta at three levels of detail (Source: NASA Landsat-7 image, February 2000)

the generator step an infinite number of times. Irrespective of the scale at which the Koch snowflake is viewed, it always exhibits the same level of detail.

Simple fractals like the Koch snowflake form a useful analogy to natural self-similar phenomena. More complex fractals can produce more "realistic" looking shapes. Figure 3.51 shows a synthetic "fractal landscape." Like the Koch snowflake, this landscape (including the clouds and water) is the result of recursively applying a feedback generator. The generator step of the Koch snowflake is essentially a similarity transformation, introducing at each iteration scaled copies of the original shape at one-third of the size of the previous iteration. In addition to being self-similar, fractals like those in Figure 3.51 are *self-affine*. Self-affine fractals can be constructed using affine transformations within the generator, so rotations, reflections, and shears can be used in addition to scaling (recall from Section 3.1.4 that all affine transformations are also similarity transformations).

self-affine

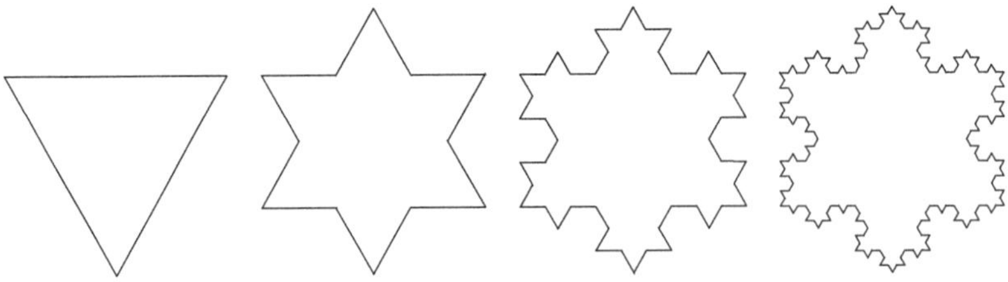

Figure 3.50: First four stages in the construction of the Koch snowflake

Aside from producing attractive pictures, fractal geometry has a number of more serious uses in GIS, for example, for map generalization (discussed further in Chapter 8). In order to produce maps at a range of different scales, it is often important to be able to decrease, and occasionally increase, the level of detail in the representation of spatial data while still retaining the essential characteristics of that line (for example, its "wiggliness"). Simplifica-

Box 3.8: Jordan curve theorem

In 1887, the French mathematician Camille Jordan formulated a famous theorem about simple loops, now known as the *Jordan curve theorem*. The theorem states that, given any simple loop, then the complement of the simple loop is not connected, but is partitioned into two connected components, one of which is bounded (called the *inside* of the loop) and one not bounded (called the *outside* of the loop). This proposition may seem so blindingly obvious as to be the sort of thing that gives mathematicians a bad name. In fact, it is quite difficult to prove convincingly. The problem lies in the wide variety of shapes, including some fractals, that qualify as loops. For example, the Koch snowflake in Figure 3.50 is a loop, but does not have a defined slope at any point on it. Technically, it is nowhere differentiable (see Chapter 4). The effort involved in proving the Jordan curve theorem resulted in the development of techniques that were instrumental in the birth of topology as a major branch of mathematics. The theorem has a practical application in the field of GIS, being the foundation of the point-in-polygon operation, discussed in more detail in Chapter 5.

tion requires less detail; enhancement requires more detail. By approximating the shape of a river, for example, as a fractal we can easily generate representations of that river at arbitrary levels of detail.

Figure 3.51: Entirely synthetic fractal landscapes

Other uses of fractals are often based on the concept of *fractal dimension*. Fractal dimension is an important property of fractals, which provides a measure of the degree to which new detail is revealed at different scales. The fractal dimension of a shape lies somewhere between the Euclidean dimensions of the shape and its embedding space. For example, the fractal dimension of the Koch snowflake in Figure 3.50 lies between 1, the dimension of a line, and 2, the dimension of the Euclidean plane.[3] Fractal dimension is an indicator of shape complexity: a shape with a high fractal dimension is complex enough to nearly fill its embedding space (e.g., a curve with a fractal dimension of 1.8 almost fills the plane). As a result, fractals are often referred to as *space-filling*. The space-filling characteristics of fractals are useful for indexing spatial information, as we shall see in Chapter 6.

fractal dimension

[3] The Koch snowflake, for example, has a fractal dimension of 1.26.

space-filling

Fractal dimension can also be a useful descriptor of a geographic shape. The fractal dimension of a river may be used as an indicator of the underlying geomorphological and hydrological processes involved in river formation. Similarly, fractal dimension analysis is used in landscape ecology to assess the complexity of geographic patches, such as plant or animal habitats. The fractal dimension of a "true" fractal shape, like the Koch snowflake, can be determined using theoretical analysis of the fractal generator. However, the fractal dimension of natural geographic phenomena, such as rivers, terrain surfaces, and animal habitats, must be determined empirically. Such empirical measurements of fractal dimension are notoriously unreliable and require careful analysis.

3.4.4 *Summary of operations*

This chapter has set out the fundamental spatial concepts that underpin most GIS. While other important spatial concepts exist, the Euclidean, geometric, topological, metric, and network spaces and concepts in this chapter are those most commonly encountered. We close this chapter by taking a slightly different perspective, summarizing some of the most important *operations* covered in this chapter in Table 3.4.

Table 3.4: Catalog of spatial operations

Group	Operation	Symbol	Operand(s)	Output
Set-oriented	equals	$=$	region, region	Boolean
	member of	$\in$	point, region	Boolean
	subset of	$\subseteq$	region, region	Boolean
	disjoint from		region, region	Boolean
	intersection	$\cap$	region, region	region
	union	$\cup$	region, region	region
	difference	$\setminus$	region, region	region
Network-	adjacent		node, node	Boolean
oriented	incident		node, edge	Boolean
	connected		node, node	Boolean
	degree		node	integer
	connected		graph	Boolean
	semi-connected		directed graph	Boolean
Topological	boundary	∂	cell	loop
	interior	$^\circ$	cell	open cell
	closure	$^-$	cell	closed cell
	meets		cell, cell	Boolean
	overlaps		cell, cell	Boolean
	inside		cell, cell	Boolean
	covers		cell, cell	Boolean
	connected		region	Boolean
	extremes		arc	set(points)
	within		point, loop	Boolean
	distance	$\|\|$	point, point	real
Euclidean,	bearing/angle	$\angle$	point, point	$[0, 2\pi[$
metric	length	$\|\|$	arc	real
	area		cell	real
	perimeter		cell	real

The operations in Table 3.4 are grouped into general, set-oriented, network-oriented, topological, and Euclidean and metric. The types for inputs to each operation (called *operands*) are shown along with the resulting output type. The spatial types used are *points*, of course, the basis of so many spatial structures; *regions* (arbitrary one- or two-dimensional point sets); *cells* (point sets homeomorphic to a disk); *arcs* (one-dimensional curves); *loops* (closed arcs with no self-intersections); as well as network types *nodes*, *edges*, and *graphs*, introduced in the previous chapter.

Some operations are unary (applying to a single operand); others are binary (applying to two operands). In addition to operations we have already seen in this chapter, the operation extremes acts on an object of type arc and returns the pair of points of the arc that constitute its end-points. Operation within provides a relationship between a point and a loop, returning true if the point is enclosed by the loop. This relationship is the often-used point-in-polygon operation, discussed in detail in Chapter 5.

The operations in Table 3.4 are a summary and guide, but by no means provide a complete typology.[4] For example, the operations distance, bearing, and angle are defined between the point elements of the space. In practice, it is often important to measure distances and angles between objects of different dimensions. For example, we might wish to know the distance of a town from a motorway. There are several ambiguities here. Do we mean the town center or the town as an area? Are we measuring distance along roads as the crow flies, or by some other means? If, for example, we are measuring as the crow flies, do we want the distance to the nearest point on the motorway or to the nearest intersection? These ambiguities must be resolved before the question can be answered properly. We return to some of the options in Chapter 5.

[4] It should be noted that not all operations in Table 3.4 are independent. For set theoretic operations, the usual set-theoretic constraints hold (for example, De Morgan's laws, $X\backslash(Y \cap Z) = (X\backslash Y) \cup (X\backslash Z)$ and $X\backslash(Y \cup Z) = (X\backslash Y) \cap (X\backslash Z)$). Other examples of interdependencies exist between the topological boundary, interior, closure, and set complement operations: for example, ∂X is the set difference of X^- and X°; and X^- is the set complement of $(X')^\circ$, the interior of the complement of X.

Reflections

Space is so much more than coordinates! While Euclidean geometry and the Cartesian plane are the most familiar and arguably most intuitive representations of space, a GIS employs multiple different representations in tandem. Geographic information scientists are experts in selecting the representation of space appropriate to the problem at hand, and they are also adept at switching between representations as required.

A wealth of mathematical texts exists to support those who wish to deepen their knowledge of the different types of spaces. Readers already more mathematically inclined may enjoy Coxeter (1961), which gives a fascinating overview of the variety that geometry offers, from basic work with triangles and polygons, through tessellations and 2D crystallography, to the platonic solids and golden section. A similarly stimulating tour of some of the most engaging ideas in graph theory can be found in Hartsfield & Ringel (2003), with a more systematic introduction to the topic in Saoub (2021).

Algebraic or combinatorial topology can be quite inaccessible to all but the mathematically gifted. A book that has withstood the test of time as a readable elementary introduction is Giblin (1977). Combinatorial maps were introduced in the Master's dissertation of distinguished computer scientist

Jack Edmonds (1960). Ohori, Ledoux, & Stoter (2015) explore the application of combinatorial structures, including simplicial complexes, to 3D and higher dimensional GIS. Zlatanova, Rahman, & Shi (2004) give a readable overview of some of the key concepts and challenges in capturing the topology in 3D GIS.

In connection with fractal geometry, Mandelbrot (1982) has become a modern classic. Burrough (1981) and Goodchild (1988) are two early examples of fractal geometry applied to spatial information, with further examples of fractal geometry applied to spatial information and GIS found in Lam & De Cola (1993) and Duckham, Drummond, & Forrest (2000).

MODELS OF GEOSPATIAL INFORMATION

4

SECTIONS

GIS do more than simply store and process spatial data: they embody high-level **models** of the geographic world around us. This chapter examines the models and modeling concepts embedded in a GIS. Specifically, the chapter is designed to aid in learning to:

- discuss the features of **object-based models** and explain how they can be used to model spatial objects;
- contrast with the **field-based model** and select an appropriate spatial model for specific applications; and
- integrate **time** and dynamic phenomena, such as **movement**, in a more complete description of space.

MODELS help to define the relationship between things in the world and the representation of those things within the computer. In a GIS, we are interested in modeling parts of the world at *geographic* scales: those larger than can be apprehended from any single vantage point—sometimes called *transperceptual spaces*—but smaller than cosmic scales (see also the discussion of Zubin C-spaces in Box 3.1 on page 78). This chapter's title, reintroducing the term "geospatial" first encountered in Chapter 1 but not in the intervening chapters, is a conscious reference to this renewed focus on the connection to our geographic world.

transperceptual space

Being clear about what things in the world we are modeling, and how they relate to the stored data and data structures in our GIS, is essential if we are to avoid serious and costly errors that can otherwise make nonsense of the outputs from a GIS. Modeling, then, begins by forming a clear picture of the underlying categories and kinds of things being modeled. The study of general classifications of, and relationships between, things is the business of *ontology*, a field of study within philosophy. So, for example, trees, a forest, the location of the forest, and the land-cover type "forested" each fall into distinct ontological categories (Figure 4.1). While we are not suggesting that all GIS modelers need to be philosophers, an understanding of basic ontological distinctions is fundamental to being a geographic information scientist.

ontology

Next, modeling selects the appropriate data structures and associated operations needed to store and process data about the things being modeled. The key distinction between ontology and data modeling is that the former aims to develop general taxonomies of what exists, while the latter aims to

Figure 4.1: A forest, the location of a forest, the trees that make up a forest, and the land-cover type "forested" are all instances of different ontological categories

develop classifications within a particular application domain. Thus, the distinction between a *fiat boundary* (such as a state boundary, national border, or cadastral parcel, which exists by virtue of human "fiat" or decree) and *bona fide boundary* (which reflects a physical discontinuity, such as a cliff or coastline, see Box 4.1 on the next page) is not a data modeling issue, although it may be useful in data modeling. Conversely, the decision as to whether to represent a road in a spatial database as a 1D polyline or as a 2D polygonal area is a data modeling question, not an ontological one.

fiat boundary

bona fide boundary

We have already started to become modelers, with our introduction to the modeling process in Section 1.3.1 and foray into conceptual data modeling for databases in Section 2.2.1. This chapter aims to deepen your modeling expertise by delving in more detail into the important models for a GIS. Models in a GIS can be classified into three major categories: *objects* in Section 4.1, *fields* in Section 4.2, and *events* (and processes) in Section 4.3. The effects of adopting one model over another have far-reaching implications for spatial data structures, for analysis operations, for the choice of implementation, and for how the outputs from GIS more broadly should be understood.

4.1 Object-based models

A basic ontological distinction is between entities in the world with identities that endure *through* time—such as my computer, the city of Kinshasa, and your shoes; and more ephemeral entities in the world that occur *in* time—such as my birthday party, a heatwave, and your life. Objects fall firmly into this first category, termed *continuants* or *endurants*. The second category, which includes *events* and *processes*, is termed *occurrents* or *perdurants* and is the focus of Section 4.3.

continuant

occurrent

object-based model

Object-based models decompose space and populate GIS with spatially referenced *objects* (e.g., cities and countries, roads and stations, rivers and forests) each with attributes (e.g., population, speed limit, names, and identifiers). Consider, for example, the Californian wildfire pictured in Figure 4.2.

Box 4.1: Fiat and bona fide boundaries

Philosopher Barry Smith introduced the ontological distinction between *fiat* and *bona fide* boundaries (see Smith, 1995; Smith & Varzi, 2000). Fiat boundaries are boundaries that exist as the result of human cognition, while bona fide boundaries exist as the result of physical discontinuities in the world, independent of human cognition. Geopolitical boundaries, such as the border between Scotland and England, are good examples of fiat boundaries. Coastlines, rivers, and firefronts are all examples of bona fide geographic boundaries. Sometimes fiat and bona fide boundaries may coincide. More than two millennia ago, the Great Wall of China (bona fide boundary) marked the northern boundary of the Imperial China (fiat boundary). Numerous changes in China's northern fiat boundary over the millennia mean that most of the complex of fortifications that make up the wall now lie well within the (fiat) boundary of modern China (including all of the Ming Dynasty fortifications, the most recognizable and photographed parts of the Great Wall, below). Ontological distinctions—such as that between fiat and bona fide boundaries—attempt to provide a categorization of things in the world without reference to any particular application domain. Indeed, the distinction between fiat and bona fide boundaries is very general and not restricted to geographic entities. Your skin forms a bona fide boundary to your body, while the boundary between your head and your neck is a fiat boundary. Examples of other ontological investigations of geographic boundaries, regions, and associated information include Mark, Smith, & Tversky (1999) and Couclelis (2010).

What geospatial objects might we wish to model in a GIS, for example, to assist firefighters with the containment and suppression of the fire?

Typical objects that we might choose include buildings and infrastructure, roads and access paths, and of course the firefront locations themselves. Figure 4.3 shows an object-based map of the image in Figure 4.2 (with the orientation rotated through 180 degrees). Figure 4.3 is populated with objects such as roads and rail, water features, and the highlighted firefronts. In general, objects in a GIS must be:

object

- identifiable,
- relevant (be of interest), and
- describable (have characteristics).

For example, the roads in Figure 4.3 are identifiable through their street names and road numbers; the roads are relevant to the wildfire application because they enable access by emergency responders and egress by local residents; and the roads are describable through their attributes, such as speed limit, width, surface material, road class as well as location. While entities such as roads are unambiguously treated as objects, the status of other entities

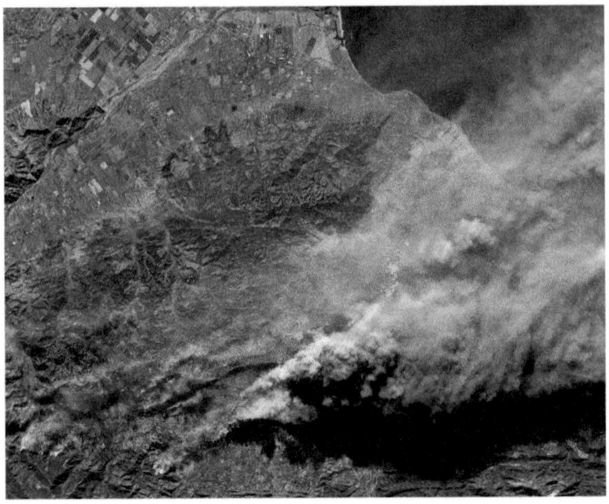

Figure 4.2: An active wildfire near Ventura, California

such as the fire itself may be less clear. The fire is clearly relevant and is describable, for example, through its location, size, and intensity. But is the fire identifiable? The answer to that question is a qualified "yes." For the purposes of firefighting at least, wildfires are usually treated as having an identity. The characteristics, parts, location, and even the materials that constitute the wildfire may change minute-to-minute. Nevertheless, a wildfire is commonly identified by emergency services and the impacted community as a single entity throughout its lifetime. In some cases, we even give names to wildfires (and some ontologically related entities such as storms and hurricanes, see Box 4.2 on the facing page).

Even though we may choose to treat our wildfire as an object for one application, it does not preclude other applications requiring a different ontological perspective. Fire spread and fire prediction models, for example, typically require a different ontological perspective of the wildfire as a *field*, covered in the next Section 4.2. In other applications, the highly dynamic nature of fire makes it preferable to treat wildfires as *events* or *processes*, explored in Section 4.3. However, in all cases, we must be careful not to conflate these

Figure 4.3: An object-based view of the wildfire in Figure 4.2, with transport infrastructure, waterways, and firefront locations marked (Map source: Open Street Map)

Box 4.2: The Ship of Theseus

The question of whether entities such as wildfires and hurricanes do or do not have an identity underlies a knotty ontological problem that philosophers have grappled with for millennia. Indeed, the problem arises not only for dynamic entities, such as wildfires and hurricanes, but for even the most object-like entities, such as axes and ships. The story of the Ship of Theseus, known to date back more than 2400 years to the time of the ancient Greek philosopher Heraclitus, is often used to illustrate the problem of identity over time. In legend, the ship that returned the king and hero Theseus to Athens, after his defeat of the Minotaur on Crete, was preserved in honor of the god Apollo. However, as the ship slowly decayed in harbor, rotted timbers were replaced with new planks to maintain the ship in seaworthy condition. Over time, none of the original timbers remained—decades later the ship was composed of entirely new and different materials from that which Theseus disembarked. The question is: Is it still the same ship, once every part has been replaced? And if it is, a further twist posed by 17th century philosopher Thomas Hobbes asked: What if the parts of the original ship were gathered up, preserved, and reconstructed in a new location? Which of the *two* ships then is the Ship of Theseus?

different perspectives, and we must retain clarity about what kinds of entities we are discussing.

In addition to characteristics (i.e., attributes or properties) and locations, objects may be aggregates of other objects or have parts that are also objects. The wildfire in Figure 4.3 may at any time be made up of multiple firefronts, each of which may itself be treated as an object. Similarly, the city of Ventura may be an aggregate of the buildings, roads, and infrastructure that make it up, as well as perhaps the citizens that people it.

In discussing objects and their characteristics, we can make an important distinction between objects and *literals*. Literals differ from objects in that they have an immutable state that cannot be created, changed, or destroyed. For example, the Californian city of Ventura may have many describable characteristics, such as its population, its location, and of course its name. However, the actual *values* of these characteristics are literals. For example, the name "Ventura" itself cannot be created, destroyed, or changed, even though we may change the name of the city.[1] Similarly, a future change to the official city boundary might result in a new polygon being used to represent this boundary. However, the underlying points and arcs that make up the boundary are literals and cannot be created, changed, or destroyed.

literal

[1] In fact, the Indigenous Chumash people have lived in the region of modern day Ventura for more than 10,000 years. The village upon which today's Ventura was built was known as "Shisholop" by the Chumash Traditional Owners.

4.1.1 Spatial objects

Spatial objects are called "spatial" because they have spatial extents that populate an *embedding space*. Several different types of space were introduced in Chapter 3, including Euclidean space, metric space, network space, and set-oriented space. The specification of a spatial object depends upon the structure of its embedding space. For example, it would not be meaningful to define a circle that is embedded in a network space.

embedding space

The most common situation is that the underlying space is Euclidean and each spatial object is specified by a set of coordinate tuples. Another approach is to specify a set of primitive objects, out of which all others in the application domain can be constructed, using an agreed upon set of operations. Primitive spatial object classes that have been suggested include closed half-planes (although these suffer from the counter intuitive property of being infinite in extent); simplicial complexes, introduced in Section 3.3.5 (the disadvantage here being computational expense in building rich geographies with such primitive objects); and point-line-polygon primitives (most common in existing systems, introduced in Chapter 3 and explored further in this chapter, below). Consider the following very simple analysis that may be required of a GIS.

- For Italy's capital city, Rome, calculate the total length of the River Tiber which lies within 2.5 km of the Colosseum.

Figure 4.4: Satellite image of Rome centered on Colosseum (Source: NASA ASTER image, May 2003)

Figure 4.4 shows a satellite image of Rome, centered on the Colosseum (highlighted with a star). Such Earth observation data is not well suited to answering our question: it does not explicitly represent objects like "the Colosseum" or "the River Tiber," and contains much more detail than we need. Instead, we need to model the relevant parts of Rome as objects. We might start by identifying a river object (with "Tiber" as its name), and a building object (named "Colosseum"). The spatial references for these objects might be an arc and point, respectively. The spatial object class "circle" is also important, as it describes the region within a 2.5-km radius of the Colosseum. A "length" operation will be needed to act on the river arc, returning a real number; and an "intersect" operation will be needed to construct the pieces of the river arc in common with the circle. The configuration of spatial objects is shown in Figure 4.5.

The analysis might proceed as in five stages below. Each stage of the analysis uses objects defined in the model and permitted operations applied to those objects. Specifically, the steps in the analysis are:

1. Find the arc representing the River Tiber and the point representing the Colosseum.

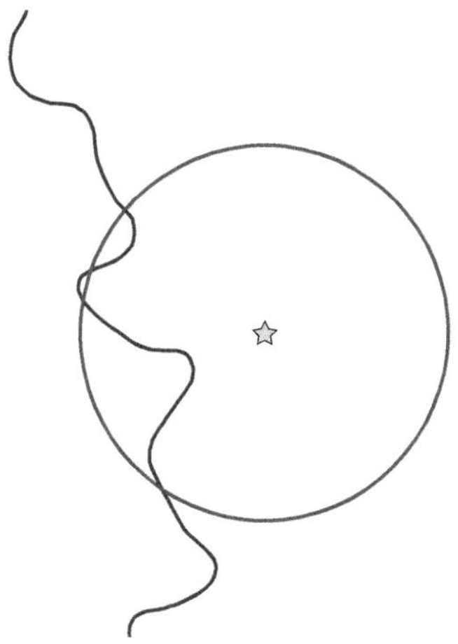

Figure 4.5: Continuous spatial objects in Rome: the River Tiber (blue) and a circle (red) of radius 2.5 km centered on the Colosseum (yellow)

2. Construct a circle with radius 2.5 km centered at the Colosseum.
3. Find the intersection of the circle constructed in stage 2 with the arc obtained in stage 1.
4. Find the length of each of the clipped arcs from stage 3.
5. Sum the lengths obtained in stage 4.

Unfortunately, the objects and operations described above are not directly amenable to computation, because they are continuous and infinite. A process of discretization must convert the objects to types that are computationally tractable. For example, a circle may be represented as a discrete polygonal area; arcs by chains of line segments; and points may be embedded in some discrete space. Operations such as "intersection" and "length" may then be computed using standard algorithms from computational geometry (explored further in the next chapter). Figure 4.6 shows a discretized version of our Roman example.

Figure 4.6: Discrete spatial objects in Rome: a discretized River Tiber (blue) intersected with a discretized circle (red) centered on the Colosseum (yellow)

4.1.2 Object orientation

Our next step into modeling spatial objects will require the introduction of a new modeling technique called *object orientation*. Object orientation extends the capabilities of the entity-relationship model, introduced in Section 2.2.1, with further constructs that will help us make the jump from the databases to geospatial objects.

object orientation

As might be expected from the name, the concept of an *object* is central to object orientation. In common with entity-relationship entities, objects in object orientation can model the static, attribute-oriented aspects of an entity. For example, a object-oriented (OO) city object might have name and population among its attributes. The totality of a given OO object's attribute values at any moment constitutes that object's *state*.

state

Unlike entities in the E-R model, however, object orientation can also model the dynamic *behavior* of an object. The behavior of an OO object is expressed as a set of operations that the object can perform. For example, calculating the area of the discretized polyline boundary of a city object or constructing a 2.5 km buffer around the coordinate point location of a building object might be examples of operations available to OO geometry objects. For the object-oriented approach, the key notion is that:

behavior

$$\text{object} = \text{state} + \text{behavior}.$$

Four further constructs are important in object orientation: *identity*, *classification*, *inheritance*, *encapsulation*, and *association*.

Identity Just as objects in the ontological sense have identity, discussed above, so OO objects have unique and immutable *identities*, independent of their attribute values. Once created, an object's identity is maintained even if all that object's attributes change their values. Immutable identities ensure that OO objects are never subject to the types of ambiguous identity problems that can plague real-world objects (see Box 4.2 on page 129).

object identity

Classification Just as it is important to distinguish between entity instances and entity types in E-R modeling, so must we distinguish between object instances and object classes in object orientation. Classification in object orientation is the process of collecting and organizing object instances into structured templates or blueprints called *classes*. Objects with the same structure are collected together into a class in the same way that similar entity *instances* in the E-R model are related to a specified entity *type*.

class

Consider the diagram in Figure 4.7, which captures our familiar Nutty Nuggets database model, introduced in Chapter 2, in a new object-oriented way. As for E-R models introduced in Chapter 2, the graphical notation for Figure 4.7 is based on the unified modeling language, UML. The diagram in Figure 4.7 is an example of a UML *class diagram*, which exhibits a number of new features when compared with E-R diagrams such as that in Figure 2.11.

class diagram

Four named classes appear in Figure 4.7: Customer, Address, HomeAddress, and DeliveryAddress. Beneath each class name, the associated attributes for that class are listed, much as for an entity type in an E-R diagram.

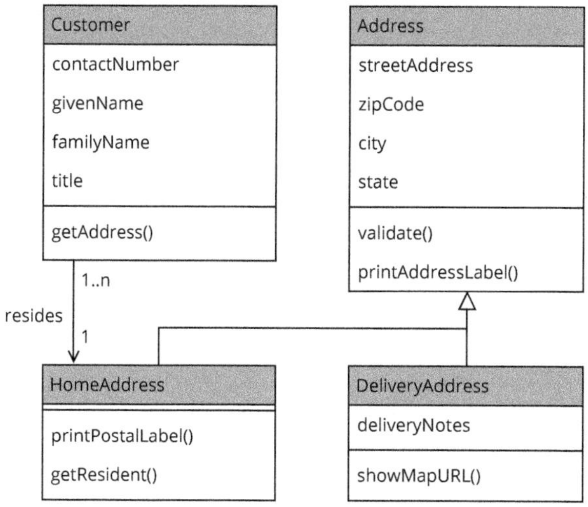

Figure 4.7: Simple UML class diagram derived from the Nutty Nuggets database example

However, unlike E-R entity types, classes also include *behaviors*—since as we have seen: object = state + behavior. Behaviors, such as validating an address or formatting a printable label, are listed in the second, lower compartment of each box. Behavior names are followed by parentheses to show they are operations or functions and to distinguish them from attributes. For example, the Customer class has a getAddress() behavior listed, which presumably retrieves the home address of a customer.

Encapsulation Encapsulation ensures that the internal mechanisms of an object's behavior are rigidly separated from external access to, and the effects of that behavior. Encapsulation helps the object-oriented model further reduce complexity by treating *what* behaviors an object exhibits separately from *how* an object achieves those behaviors.

encapsulation

In Figure 4.7, the details of *how* a behavior operates remain hidden. How a DeliveryAddress is validated with the validate() behavior, for example, is not important at this level of abstraction—only the *what*, not the *how*, is relevant to a class diagram.

Inheritance In an object-oriented model, *inheritance* allows classes (and so objects) to share common properties. Inheritance acts in two ways: *generalization* and *specialization*, one the inverse of the other. A generalized class will exhibit properties common to all its specialized *subclasses*. Conversely, classes specialize, adapt, and add to the properties of generalized *superclasses*.

inheritance

generalization

subclass

specialization

superclass

Inheritance allows specialization or generalization of behaviors as well as attributes. Superclasses (such as Address in Figure 4.7) can share their attributes and behaviors with subclasses (such as HomeAddress and DeliveryAddress). Subclasses, indicated with the arrow relationship pointing to the generalized superclass, inherit all the attributes and behaviors of the superclass, to which they may add their own specialized attributes and behaviors. A DeliveryAddress class, for example, might possess all the attributes and behaviors of a generalized Address, but additionally have specialized delivery instructions or access maps. Inheritance relationships are often called *"is a" relationships* since, for example, a DeliveryAddress "is an" Address.

"is a" relationship

association

Association Finally, *association* groups objects together in order to model phenomena with complex internal structure. Associations between classes operate in much the same way as relationships in the E-R model. For example, the resides association in Figure 4.7 has many-to-one cardinality conditions indication that a Customer resides at exactly one HomeAddress, but each HomeAddress may have one Customer or more residing there.

Aggregation relationships are a special type of association, sometimes called a "part of" relationship. Aggregation is *antisymmetric* (if A is part of B, it means B cannot be part of A). A hierarchy of aggregation relationships is termed a *partonomy* or sometimes a *mereological hierarchy* (in contrast to an inheritance hierarchy, also termed a *taxonomy*).

"part of" relationship

partonomy

taxonomy

From a modeling perspective, object orientation offers three main advantages for modeling spatial objects:

Reduces complexity Decomposing complex phenomena into objects with state and behavior helps make conceptual models simpler and easier to manage. Organizing groups of objects with similar behaviors into classes further helps reduce model complexity. Spatial objects often have complex structures, associations, and behaviors, so object orientation is especially useful in managing that additional complexity.

Increases metaphorical power Treating complex spatial objects as OO objects is a useful *metaphor* that leverages our familiarity with handling real-world objects, such as cars and phones, in everyday life. The term *impedance mismatch* is often used to describe the problems that arise through the gaps between how we conceptualize the world, how we represent those concepts in a model, and how we implement that model in a computer. Impedence mismatch is a particular problem when modeling spatial objects because of the complex data structures and algorithms needed to support sophisticated representations of space, as we shall see. Object orientation helps to reduce impedence mismatch by bringing our models closer to familiar concepts in the world. Further, object-oriented technology, in particular *object-oriented programming languages* (OOPLs), is the dominant approach to software development today, which further aids in closing the gap between model and implementation.

impedance mismatch

object-oriented programming languages

Promotes reuse Systems can be developed more efficiently if they can be constructed from collections of well-understood and well-specified subcomponents. Once a system has been successfully captured in an object-oriented model, encapsulation ensures it can easily be reused as a sub-component within more complex models. Reuse is especially important to sophisticated GIS and spatial data, which enjoys myriad different uses, both by experts and non-experts (in comparison to, say, genetic sequencing data which is typically subject to relatively specialized uses by expert users).

4.1.3 Geospatial objects with planar geometry

Figure 4.14 on page 139 shows an object-oriented class diagram for geospatial objects with geometries in the Euclidean plane with the usual topology. The class diagram is adapted from a widely used model of spatial objects: the *simple feature model*. The simple feature model is an important international standard, implemented in many GIS and underlying many object-based spatial data sets. Using the principles of object-oriented modeling introduced in the previous section, this section steps through the construction and key features of the simple feature model.

simple feature model

The Geometry class The most general class of geometric object, Geometry, is at the top of the simple feature model hierarchy. The Geometry class possesses attributes and behaviors common to all spatial objects, illustrated in the UML class in Figure 4.8. As indicated in the introduction to UML class diagrams above, the trailing parentheses distinguish behaviors from attributes. The class diagrams in this section additionally specify data types for attributes and behaviors. Specifically:

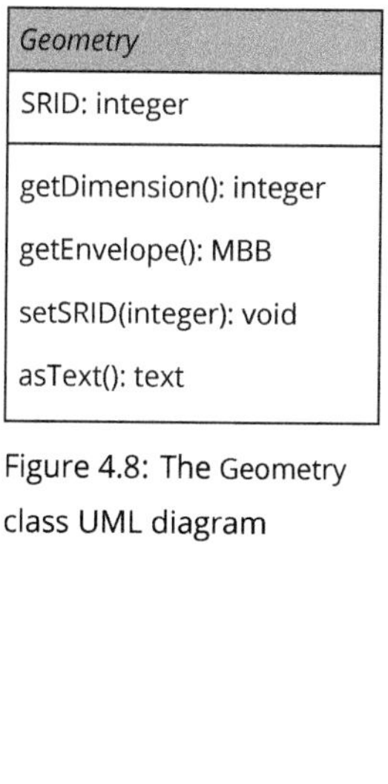

Figure 4.8: The Geometry class UML diagram

- The behavior getDimension() retrieves the dimensionality of a Geometry instance, for example, whether a zero-dimensional point or a 2D region. The getDimension() behavior outputs ("returns") an integer number (i.e., 0, 1, 2, ...).

- The behavior getEnvelope() returns the minimum bounding box (MBB) for a geometry object. The MBB is the smallest enclosing axes-parallel rectangle enclosing a geometry. The MBB is an important descriptor of spatial objects used in spatial algorithms and database indexes as we shall explore in more detail later in Chapters 5 and 6.

- The asText() behavior outputs a text-based representation of the spatial object in *well-known text* (WKT) format, a standard language for representing spatial object geometries.

well-known text

- Finally, the Geometry class also maintains information about the spatial reference system used to encode the geometry via the SRID attribute: the *spatial reference system identifier*. Such *accessor* behaviors are commonly used in object-oriented models to encapsulate attribute access behind sophisticated validation methods and consistency checks.

SRID

accessor

Box 4.3: Spatial reference systems

If everyone used the same system for specifying geographic coordinates, there would perhaps be no need for spatial reference systems. Any such universal coordinate system would have to be a spherical coordinate system, because the Earth's surface is *topologically* a sphere. Unfortunately, the Earth is not a sphere *geometrically*. The geometric shape of the Earth is the subject of the study of *geodesy*. Different geodetic approximations to the Earth's shape are built on different global bases (called *geodetic datums*). As a result, the same spherical latitude and longitude coordinates (see Box 3.7 on page 117) can refer to different locations on the Earth's surface, dependent on the global geodetic reference system used. Further complicating the situation is the fact that spherical coordinate systems are much less desirable than planar coordinates in many practical and computational situations. Basic spatial computations, such as distance and angle in Section 3.1.1, become much more complicated on the sphere than in the plane. Although planar coordinates can never accurately represent locations on a curved surface, the distortions are acceptable over small distances. Consequently, most GIS rely on planar coordinate systems, with many spatial data sets using locally optimized spatial reference systems. Map projections, such as Universal Transverse Mercator (UTM), have already been encountered in Chapter 3 and are revisited in more detail in Section 8.2.3.

The SRID spatial reference system enables stored geometries in a range of different projections to be related to locations on the Earth's surface (see Box 4.3). The setSRID() behavior provides controlled access to the integer SRID attribute, ensuring it can only be updated using a valid SRID. SRID codes can be specified according to any one of a number of international standards, most notably EPSG codes or ISO standard 19111:2007.

All other classes in the simple feature model are subclasses of Geometry, and consequently inherit all the attributes and behaviors of Geometry.

The Point class The Geometry class is useful because it captures some important common properties and behaviors of all spatial objects, even though the Geometry class itself is not expected ever to be instantiated as an object itself. By analogy, the biological class of "mammals" is useful in describing the common features of many different individual animals on Earth. However, any mammals we may meet—such as humpback whales, Tasmanian devils, or horseshoe bats—are instances of those specific species (i.e., subclasses of mammals). Such instances are *also* mammals (by the "is a" relationship), but not *only* mammals. In object-oriented terminology, such non-instantiated classes are usually called an *abstract class*. Abstract classes are indicated in UML using an italic name in the class diagram header (see Figure 4.8). Instantiated classes are also called *concrete* classes.

abstract class

The most basic concrete, instantiated Geometry subclass is the Point class (Figure 4.9). The Point class has two attributes, an x and a y coordinate. As we have seen, the continuous Euclidean plane is not computable and must be discretized in order that computation can take place. In a computer, the real numbers (Euclidean 1-space) are discretized by storing digits to only a finite level of precision. For example, using 32-bit precision floating-point numbers, up to six significant decimal digits can be stored (e.g., the number 0.123456789 would be rounded up to the 32-bit floating point number 0.123457). Thus, points can be discretized by storing their coordinates as fi-

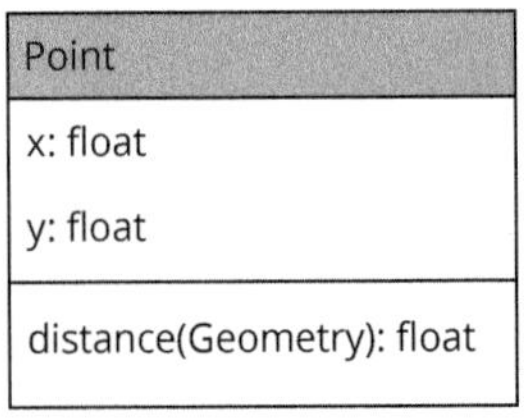

Figure 4.9: The Point class UML diagram

nite precision numbers, such as the float data type used for the Point class in Figure 4.9.

Points may have a range of behaviors, with one example illustrated in Figure 4.9. The distance() behavior computes the distance from the Point to another specified Geometry object. In the case of computing the distance to another point geometry, this behavior can encapsulate the distance measure already encountered in Section 3.1.1. However other geometry types may require more sophisticated approaches, covered in more detail in the next chapter on spatial algorithms, Chapter 5.

The Arc, Region, and Collection classes The Arc, Region, and Collection classes are abstract superclasses of polylines, polygons, and collections of points, polylines, and polygons. Each class possesses its own behaviors, inherited by those subclasses. The Arc class has behaviors to compute the length() of the arc, retrieve the startPoint() and endPoint() of an arc, and check if the arc isSimple() and isClosed() (see Figure 4.10).

Arc
length(): float
startPoint(): Point
endPoint(): Point
isClosed(): boolean
isSimple(): boolean

Figure 4.10: The Arc class UML diagram

The Region class has behaviors to compute the area() and centroid() (central point) of a polygon, discussed in more detail in the next chapter. The Collection class has accessor behaviors that retrieve the number of geometries in the collection (numGeometries()) and individual geometries within that collection (getGeometry()).

The Polyline and Ring classes A Polyline object is an aggregation of at least two Point objects—a start and end point—plus a number of internal points, which define the line segments that make up the polyline. In this way, the Polyline class provides a discrete approximation of a continuous curve as a finite sequence of two or more discretized points (see Figure 4.11).

The numPoints() and getPoint() behaviors retrieve the total number of points in the polyline or an individual constituent point, respectively. The Ring subclass is a special case of a simple Polyline where the first and last points are interpreted as being joined by a line segment.

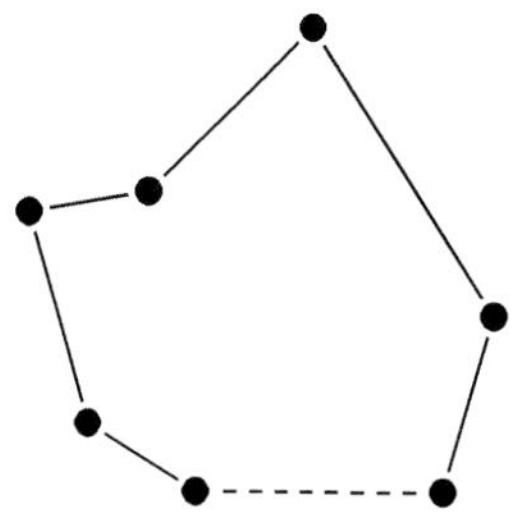

Figure 4.11: The Ring class is a Polyline with end points joined by a line segment

The Polygon class A Polygon object is an aggregation of at least one Ring object. At least one simple, closed polyline (a Ring) is required to define the exterior boundary to the polygon. In addition, a Polygon may have zero or more internal Ring objects to enable the representation of polygons with holes. By convention, polygon exteriors use a counterclockwise orientation, and holes a clockwise orientation, as illustrated in Figure 4.12. Interior holes may not intersect each other or the exterior, except at a tangent (as shown with the two right-hand holes in Figure 4.12).

The Polygon class also includes accessor methods to retrieve the exterior polyline (exteriorRing()) and each of the interior polylines (via numInteriorRings() and getRing()). Once again, the Polygon is discretized, comprising a finite number of discrete Point objects.

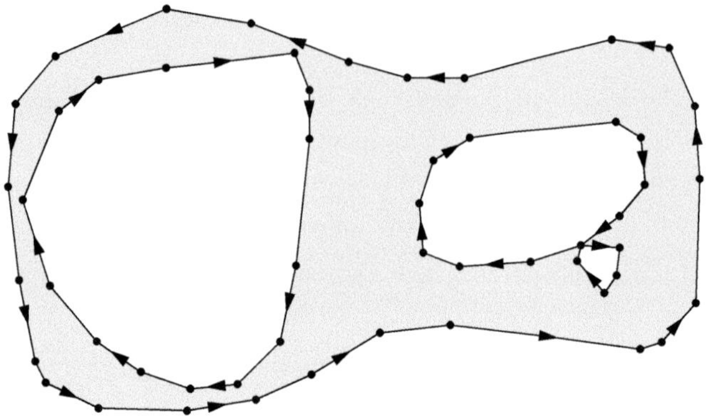

Figure 4.12: A simple feature Polygon with three holes, illustrating the clockwise orientation of points around each hole

The MultiPoint, MultiPolyline, and MultiPolygon classes Finally, the three concrete collections, MultiPoint, MultiPolyline, and MultiPolygon are formed from an aggregation of one or more Point, Polyline, or Polygon objects, respectively. The associations in Figure 4.14 show these aggregations. The MultiPolygon object again uses the orientation of points around a Ring boundary to distinguish boundaries of holes from the boundaries of region components. This enables the representation of more complex polygonal structures with nested holes and islands, as in Figure 3.29. Figure 4.13 shows a complex MultiPolygon with multiple levels of holes within islands. At each level the Ring orientation helps to distinguish apart boundaries to regions and to holes.

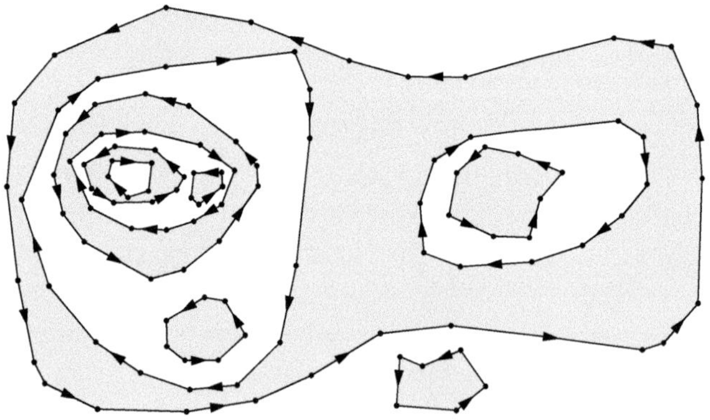

Figure 4.13: Ring orientation in the simple feature model for *MultiPolygon* objects. Holes use a clockwise orientation of points in the polyline; islands and polygon exteriors a counterclockwise orientation

The complete simple feature model class diagram is found in Figure 4.14. It is worth taking a little time to study, understand, and internalize the classes, attributes, behaviors, and relationships captured by the simple feature model. It is today one of the most widely used data models in GIS, and underpins a wide range of geospatial data, software, and analyses.

4.2 Field-based models

relative space

absolute space

Space may be conceptualized in two distinct ways: dependent on a set of objects with spatial properties, referred to as *relative space*; or existent in itself as a set of locations with properties, termed *absolute space*. As we have seen

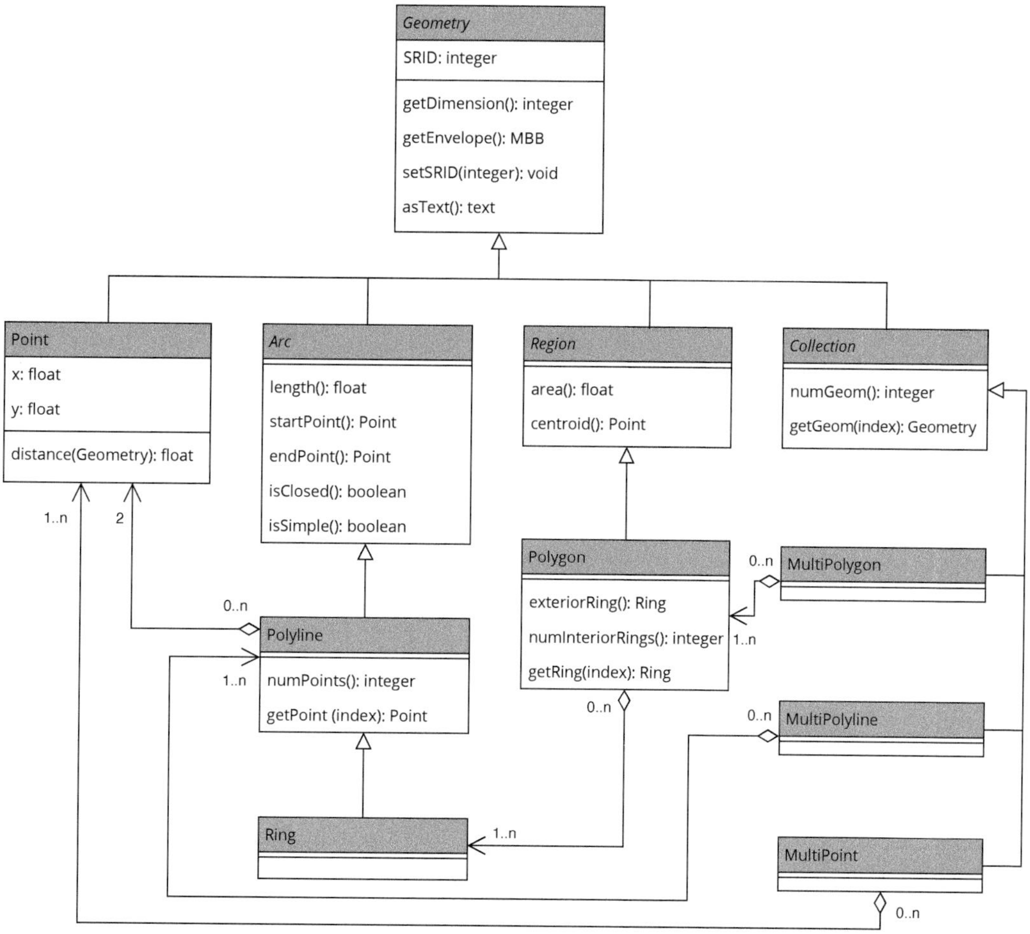

Figure 4.14: UML class diagram for spatial objects, based on the simple feature model

in the previous section, object-based models adopt the former perspective and treat space as populated by discrete, identifiable entities, each with a geospatial reference.

field-based model

By contrast, the *field-based model* is founded on the latter perspective, and treats geographic information as collections of spatial distributions, called *fields*. Each field defines the value of an attribute at locations across space. Figure 4.15 shows a field-based map aligned with the image in Figure 4.2 (again with the orientation rotated through 180 degrees). Figure 4.15 is populated primarily with land cover information in addition to the active wildfire location. Compared with the object-based map in Figure 4.3, Figure 4.15 is not concerned with objects *at* a location, such as towns, roads, or lakes. Instead, Figure 4.15 is concerned with the attributes and qualities *for* a location, such as whether it is urban, cropped, grassland, or forested.

field

spatial framework

In a field-based model, each field may be formalized as a mathematical function from a *spatial framework* to an attribute domain. A spatial framework is a *partition* of a region of space. Recall that a partition is a set of subsets such that every element in the set appears in exactly one subset of the partition. For example, a regular grid placed over a space can be used to form a parti-

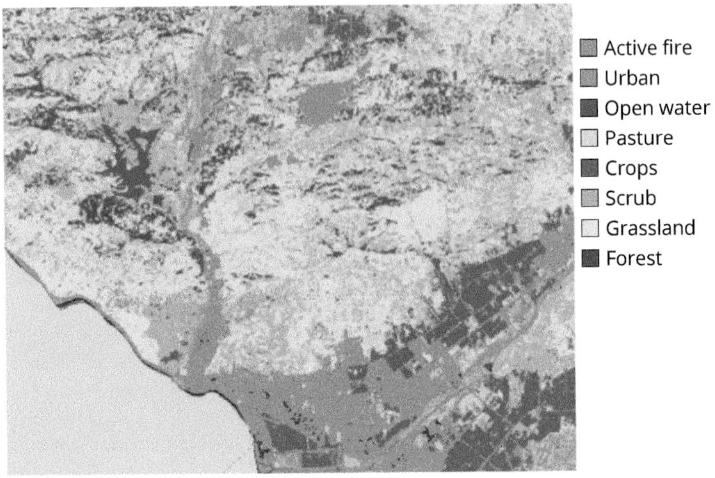

Figure 4.15: A field-based view of the wildfire in Figure 4.2, with urban and vegetation land cover shown together with fire locations (Map source: US NLCD 2019 Land Cover)

tion, such that each location in the space falls into one and only one grid cell. Indeed, this is the most common spatial framework for a field, called a *raster* (and introduced in Chapter 1).

Thus, in the field-based model, each field uses a fixed spatial framework, such as a regular grid, to capture the variation in attribute values with respect to this reference framework. Many different types of spatial data fit neatly into this view, such as patterns of topographic height, rainfall, temperature, as well as land cover in Figure 4.15. The spatial framework for a given field-based data set is sometimes referred to as the *support* of that data set. Note too that it is the *function* that is the field, not the set of values.

support

The field-based and object-based models are in a sense inverse to each other. With the field-based approach, the first-class entities in the model are functions (fields) from the spatial framework to the other attributes. On the other hand, the object-based approach constructs a population of entities with a spatial embedding, i.e., a function from entities to a spatial framework (see Figure 4.16, and Box 4.4 on the next page).

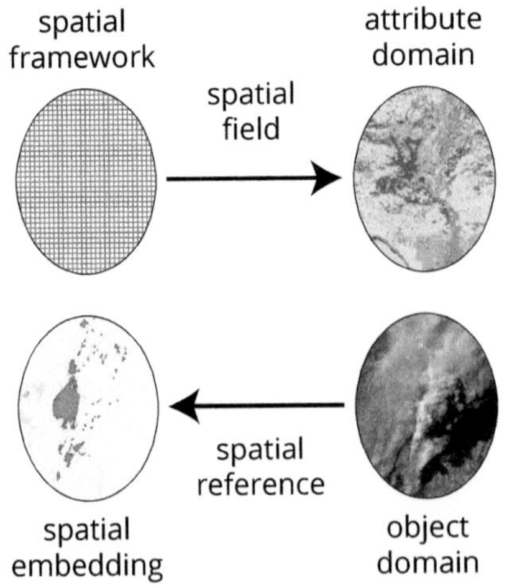

Figure 4.16: Spatial field and object are inverse constructs

Box 4.4: Object versus field

The dichotomy between objects and fields turns out to have far-reaching implications for spatial modeling, where absolute space is modeled as a set of fields and relative space as collections of spatially referenced objects (see Chrisman, 1978; Peuquet, 1984). Couclelis (1992) draws a parallel between the object versus field distinction and a more general distinction in the philosophy of science: that of *plenum* versus *atom* and *wave* versus *particle*. It is natural to think of continuously varying geographic phenomena, such as temperature or rainfall, as fields. Similarly, it is natural to think of discrete geographic phenomena, such as buildings or roads, as objects. However, we should never forget that field and objects are simply perspectives on the world, and not inherent in the world itself. We will always have a choice of how to model geographic phenomena. For example, we may choose to view cities as collections of objects, such as buildings, roads, parks, and paths. But the same cities may equally be viewed as fields connecting urban locations with land use and land cover types, such as residential, recreational, impermeable surfaces, and water bodies. Conversely, rainfall is commonly viewed as a field, with maximum, minimum, or average rainfall mapped at each location, for example. However, rainfall can equally be approached from an object-based perspective, for example, as zones of high, moderate, and low rainfall. It is always possible to view the properties-of-objects-in-space as a field, or view spatial aggregates of like-properties-of-a-field as objects.

The spatial framework for any field is a finite structure—it must be so in order to be tractable for computation. Often the application domain will not be finite, or it may be larger than can be practically accommodated in totality. This implies that the phenomena to be modeled will be *sampled* and imprecision will necessarily be introduced by the sampling process. Imprecision is one of the topics we return to in Chapter 10.

The combination of the spatial framework and the field that assigns values to each location in the framework, is termed a *layer*. In general, there may be many layers in a GIS, each with respect to the same or different underlying spatial frameworks. For example, Figure 4.17 shows two layers for the same spatial framework. The upper layer shows elevation, with lighter yellows indicating higher elevation and darker greens lower elevation. A second layer, partially occluded by the elevation layer, shows soil moisture content, with lighter blues indicating drier soils, darker blues wetter soils. The boundary of each area in Figure 4.17 connects locations with the same attribute value (elevation or soil moisture content). The locus of all points in a field with the same attribute value is termed an *isoline*.

layer

isoline

The field-based spatial model is now defined more precisely. For simplicity, assume that each field in the model has the same spatial framework F. In general, we would allow the spatial framework to be based upon any mathematical model of space. In practice, most spatial frameworks are based upon Euclidean space in two dimensions (the Euclidean plane), with distance and angle defined as in Chapter 3.

- A field-based model based upon a spatial framework F consists of a finite collection of n spatial fields $\{f_i : 1 \leq i \leq n\}$.
- For $1 \leq i \leq n$, each *spatial field* f_i is a computable function from set F to a finite attribute domain A_i, $f_i : F \to A_i$.

spatial field

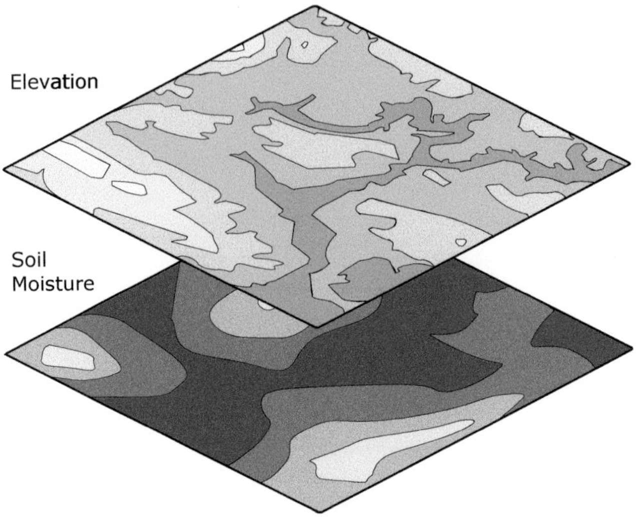

Figure 4.17: Two layers in a field-based model

In order that this model will work computationally, the number of fields, the *cardinality* of the domains A_i, and the spatial framework F are all stipulated to be finite. Also, the field functions f_i are defined to be computable (i.e., are capable of evaluation using a computer). In the special case where the spatial framework is a Euclidean plane and the attribute domain is a subset of the set of real numbers, then a field may be represented as a surface in a natural way. Then, the Euclidean plane plays the role of the horizontal xy-plane and the spatial field values give the z-coordinates, or "heights" above the plane.

To sum up the modeling method using a field-based approach:

1. Construct or use a given suitable model of the underlying space to act as the spatial framework, F.
2. Find suitable domains for the attribute(s) to act as the A_i $(1 \leq i \leq n)$.
3. Sample the phenomena under consideration at the locations in the spatial framework, so as to construct the spatial field functions f_i $(1 \leq i \leq n)$.
4. Perform analyses by computing with the spatial field functions.

A simple example of a field model would be regional climate variations. Imagine placing a square grid over a region and measuring aspects of the climate at each node of the grid. A more realistic model might require using an irregular spatial framework to reflect where it is practical to set up weather stations to record meteorological measurements. Different fields would then associate locations with values from each of the measured attribute domains.

Another example might be regional health variations. Here the spatial framework might be an irregular tessellation of the counties or districts in which people reside. For the population within each region, we might measure the incidence of an infectious disease. The model might then be used to analyze spatial patterns of specific diseases or relate such patterns to other spatial fields for the same spatial framework. A third example would be a *digital elevation model* (DEM), where terrain elevation is measured over a

DEM

Box 4.5: Digital elevation models

Digital elevation models (DEMs) have been an important application of the field model since the term was first coined in the 1950s (Miller & Laflamme, 1958). In a DEM, the spatial framework represents a set of locations within a portion of the Earth's surface, and the field values are the heights of the terrain surface at those locations. Since the field function is single-valued, overhanging terrain features, such as cliffs and caves, cannot be represented in a DEM. As a consequence, DEMs are sometimes referred to as "2.5-dimensional" objects. The spatial framework for a DEM is often in the form of a regular grid, which has the advantage of simplicity. However, irregular tessellations, such as a TIN, can allow the structure of the field to more closely parallel the structure of the terrain, e.g., following ridges and valleys. Many common GIS applications require a DEM as their basic data source, including visibility analysis (determining what can be seen from a particular location), slope and aspect analysis, and terrain routing, planning, and path calculations. DEMs are sometimes also referred to as digital terrain models (DTMs), although a DTM would normally represent other topographic features in addition to elevation. Burrough et al. (2015) contains a useful survey of DEM principles, techniques, and applications.

region of the Earth's surface (see Box 4.5).

It is worth emphasizing that the conceptual models under discussion are independent of any implementation or any physical representation of the data, although certain types of spatial framework may be suited to certain types of data structure. For example, if the spatial framework is a regular tessellation, especially a regular square grid, it may be natural to represent this using raster-based data structures. In contrast, irregular tessellations, especially triangulations, often lend themselves more readily to vector-based data structures.

4.2.1 *Properties of fields*

As already highlighted, a spatial framework is a partition that may be based on any one of a range of different mathematical models of space, such as those discussed in the previous chapter. From the perspective of the field-based model, an important characteristic of a spatial framework is whether it is regular or irregular. In addition to the spatial framework, the attribute domains and the fields themselves have several important characteristics, described in this section.

Properties of the attribute domain The attribute domain may contain values which are commonly classified into four levels of measurement, originally devised by Stevens (1946):

- An attribute domain that consists of simple labels is termed a *nominal* attribute. A land cover classification that consists of woodland, grassland, agricultural land, and urban land is an example of a nominal attribute. By definition, nominal attribute domains are qualitative, cannot be ordered, and arithmetic operators are not permissible for nominal attributes. nominal

- An attribute domain that consists of ordered labels is termed an *ordinal* attribute. For example, the results of a site suitability analysis to determine the best place to locate a new superstore might be expressed as values from ordinal the ordinal attribute domain: "highly suitable," "suitable," "moderately suitable," and "unsuitable." Like nominal attributes, ordinal attribute domains are qualitative and cannot be subjected to arithmetic operators, apart from ordering.

interval

- An attribute domain that consists of quantities on a scale without any fixed point is termed an *interval* attribute. The Celsius and Fahrenheit temperature scales are examples of interval attributes. Interval attributes can be compared for size, with the magnitude of the difference being a meaningful notion. However, the ratio of two interval attribute values is not a meaningful notion. For example, if Madrid is 20°C and Edinburgh is 10°C, it makes sense to say that Madrid is 10°C hotter than Edinburgh, but not that Madrid is twice as hot as Edinburgh.

ratio

- An attribute domain that consists of quantities on a scale with respect to a fixed point is termed a *ratio* attribute. The Kelvin temperature scale is an example of a ratio attribute, because zero Kelvin is absolute zero, the lowest physically possible temperature. Ratio measurements are capable of supporting a wide range of arithmetical operations, including addition, subtraction, multiplication, and division. There are many geographic examples of ratio measurement domains, including annual rainfall and topographical altitude above sea level. So, it makes perfect sense to say that Boston had twice as much rain in March as Washington, or that one place is three times higher than another with respect to sea level.

While Stevens' four levels of measurement are a widely used guide, it is important to note that they do not provide a complete description of number scales. Probabilities, with two fixed points at 0 and 1, do not fit comfortably into any of Stevens' levels. An example of a common geographic measurement that also does not fit Stevens' four levels is population density. Population density has one fixed point (it is not possible to have less than zero population density) and can be compared like a ratio (e.g., a population density of 200 people per square kilometer is twice as dense as 100 people per square kilometer). However, population density does not behave quite like a ratio. For example, it is not in general possible to add population densities like ratios. The combined population density of two regions containing 100 and 200 people per square kilometer will lie somewhere between 100 and 200 people per square kilometer dependent on the relative sizes of the two regions. Further information on the nuances of measurement levels in geographic information can be found in Chrisman (2006).

continuous

Continuous, differentiable, and discrete fields A spatial field is said to be *continuous* if small changes in location lead to small changes in the corresponding attribute value (recall that a spatial field is a *function* from a spatial framework to some attribute). Continuity is only appropriate if the notion of a "small change" is well defined in both the spatial framework and attribute domain, that is, if these domains are themselves continuous. A spatial field is *differentiable* if its rate of change (slope) is defined everywhere. As with continuity, differentiability only makes sense for a continuous spatial framework and continuous attribute domain. Every differentiable field must also be continuous; but not every continuous field is differentiable.

differentiable

Examples of various continuous and differentiable functions are shown in Figure 4.18. For simplicity, assume that the spatial framework is one-dimensional, ranged along a horizontal line. Then the field may be plotted as a graph of attribute value against (one-dimensional) spatial framework. In Figure 4.18a the variation is represented as a continuous smooth curve, clearly differentiable because the slope of the curve can be defined at every point. In Figure 4.18b, although the field is continuous (the graph is connected), it is not everywhere differentiable. There is an ambiguity in the slope, with two choices at the articulation point between the two straight-line segments. In Figure 4.18c, the graph is not connected and so the field is not continuous and *a fortiori* not differentiable. Figure 4.18d shows a curve known as the *blancmange function*. The blancmange function is another example of a fractal, first introduced in Section 3.4.3. Recall that fractals have the property that they exhibit statistically similar levels of detail at all scales, termed self-similarity, no matter how far one "zooms in." As a result, fractals such as the blancmange function are *nowhere differentiable*.

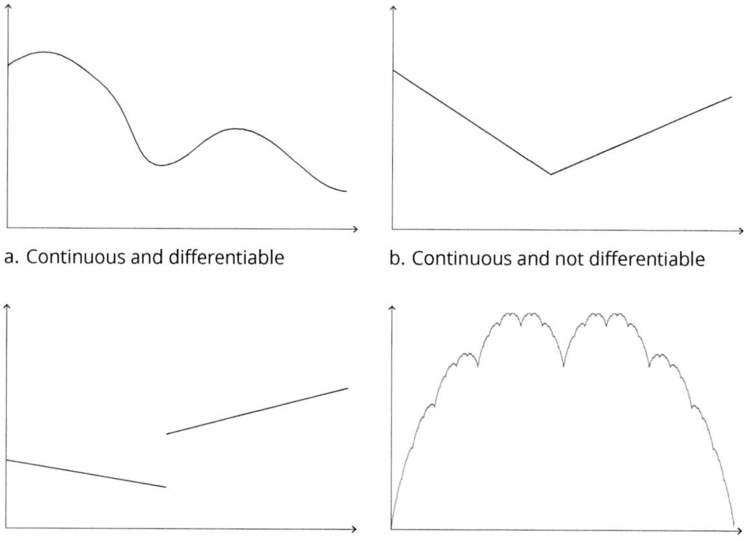

a. Continuous and differentiable

b. Continuous and not differentiable

c. Not continuous and not differentiable

d. Continuous and nowhere differentiable

Figure 4.18: Examples of continuity and differentiability for fields plotted as 1D spatial framework (x-axis) against attribute value (y-axis)

In the case where the spatial framework has dimension two (or higher), the slope is dependent not only on the particular location but also on bearing at that location (see Figure 4.19).

Isotropic and anisotropic fields A characteristic feature of a spatial field is whether its properties vary with direction. A field whose properties are independent of direction is called an *isotropic* field. Consider travel time

isotropic

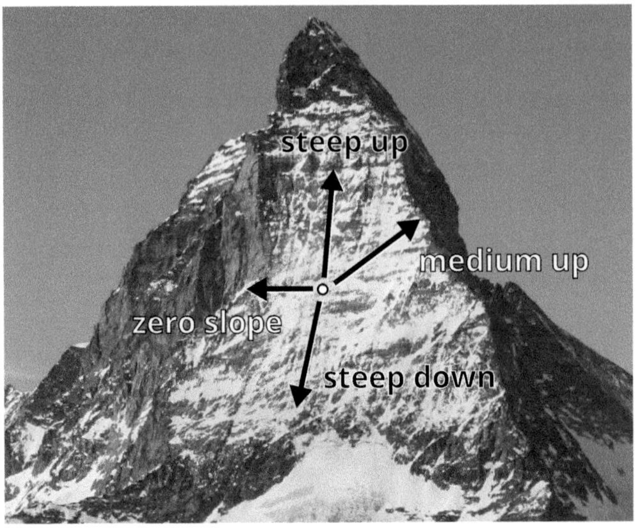

Figure 4.19: Elevation field where the slope at a point depends upon the bearing

in a spatial framework. Let us start with the simplest possible assumption that the time taken to travel between locations is in direct proportion to the Euclidean distance between them. Then, the locus of all points that are a constant time from an arbitrary point *X* (an isochrone) is a circle, shown in Figure 4.20a. This field is isotropic, because the time from *X* to any point *Y* is dependent only upon the distance between *X* and *Y* and independent of the bearing of *Y* from *X*.

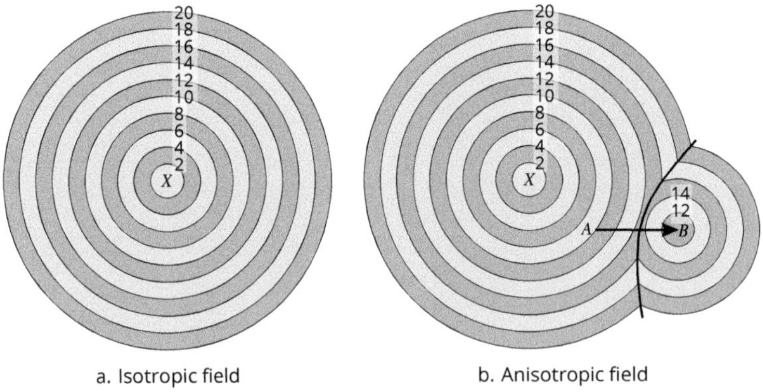

Figure 4.20: Travel time from *X* in isotropic and anisotropic fields

Now, let us make life more interesting. Suppose that there is a high-speed link *AB* in the spatial framework (Figure 4.20b). For simplicity, assume that the link is so fast as to make the travel time from *A* to *B* negligible compared with the travel times anywhere else in the field. Then, when traveling between two points, there is a choice of whether or not to use the high-speed link. Consider points close to *X* (say, within 14 time units) in Figure 4.20b. For many of the points near *X*, the minimum travel time is unchanged by the presence of the link. However, for points near *B* it would be better for the traveler from *X* to travel to *A*, take the link (zero-time), and continue on from *B* to the destination. The hyperbola in Figure 4.20b marks the boundary between destinations where it is better/worse to use the link. In the second case, to determine the travel time, it clearly matters in which direction the destination location is from X. The field is said to be *anisotropic*.

anisotropic

Anisotropic fields are common in real-world situations. They are often linked to networks (in our example, the high-speed link was a very simple network). Other examples that introduce anisotropic conditions include natural and artificial barriers to direct accessibility.

Spatial autocorrelation Spatial autocorrelation measures the degree of clustering of values in a spatial field. Spatial autocorrelation is a quantitative expression of Waldo Tobler's famous proposition, often referred to as the first law of geography (Tobler, 1970), that "everything is related to everything else, but near things are more related than distant things." If a spatial field has the property that like values tend to cluster together, then the field exhibits high positive spatial autocorrelation. If there is no apparent relationship between attribute value and location neighborhoods, then there is zero spatial autocorrelation. If there is the propensity of like values to be located away from each other, then there is negative spatial autocorrelation. Spatial autocorrelation therefore measures the relationship between attribute values at a location and attribute values in the location's neighborhood. Figure 4.21 gives examples of a high positive (left-hand bull's eye), zero (random noise), and high negative spatial autocorrelation (right-hand checker).

spatial autocorrelation

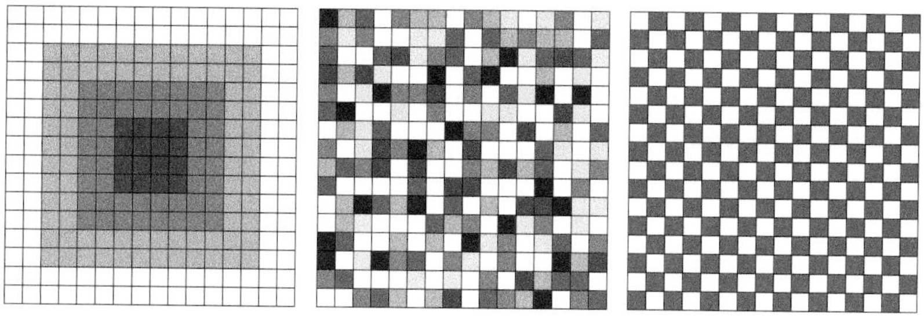

Figure 4.21: Patterns with high positive, zero, and high negative spatial autocorrelation (left to right)

4.2.2 *Operations on fields*

A field operation takes as input one or more fields and returns a resultant field; thus, fields form a closed structure under field operations. This section describes some of the typical operations that field-based models allow, dividing them into three main classes: *local*, *focal*, and *zonal* operations. The system of possible operations on fields in a field-based model is referred to as *map algebra*, first classified and formalized by Tomlin (1983). Before the different classes of map algebra operations are described, the definition of a neighborhood function is needed.

map algebra

- Given a spatial framework F, a *neighborhood function* $n : F \rightarrow \mathcal{P}(F)$ is a function that associates with each location x a set of locations that are "near" to x.

neighborhood function

Section 3.2 defined the power set of a set F, $\mathcal{P}(F)$, as the set of all subsets of F. Every member of $\mathcal{P}(F)$ is therefore a subset of F. For each location x in F, $n(x)$ will then be a subset of F, called the neighborhood of x (cf. neighborhood topologies in Chapter 3). If the space is metric, then the neighborhood of x might be the set of points within a specified distance of x. If the space is topological, then the neighborhood could be defined in terms of topological neighborhood. The neighborhood of x will usually contain x itself, although this is not a requirement of a neighborhood.

local operation

Local operations A local operation acts upon one or more spatial fields to produce a new field. The distinguishing feature of a local operation is that the value of the new field at any location is dependent only on the values of the input field functions at that location. Local operations may be unary (transforming a single field), binary (transforming two fields), or n-ary (transforming any number of fields). We give a more formal definition in the binary case. The general case may easily be extrapolated from this.

Formally, suppose we are given a framework F and spatial field functions f and g. Suppose further that $\bullet$ is a binary operation that acts on values in the attribute domains of f and g, respectively, to produce a value in another attribute domain. Then, we may pointwise construct a new spatial field function h defined as follows:

1. For each location x, $h(x) = f(x) \bullet g(x)$.

This binary combination of the two fields f and g with spatial framework F is shown in Figure 4.22. The figure highlights one element of the spatial framework, x. Note that although the spatial framework in Figure 4.22 (and following figures) is a regular square tessellation, this is not a requirement of operations on fields: any spatial framework (partition) is admissible.

Figure 4.22: Local operation, $\bullet$, on two fields, f and g

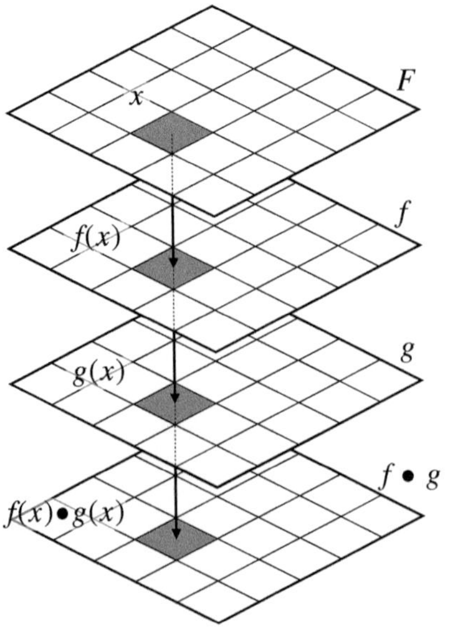

For example, suppose the attribute domains of the two fields f and g are real numbers describing population and incidence of cancer, respectively. For locations with non-zero population, cancer incidence may be divided by population. Then in this case, the • operation is division, and $f \bullet g$ is the rate of cancer incidence per unit population. The derivation of a value at any location depends only upon the values of cancer incidence and population at that location. Therefore, this operation is local.

focal operation

Focal operations For a focal operation, the attribute value derived at a location x may depend not only on the attributes of the input spatial field functions at x, but also on the attributes of these functions in the neighborhood $n(x)$ of x. Thus, the value of the derived field at a location may be influenced by the values of the input field nearby that location. In the unary case, suppose F is a spatial framework, n is a neighborhood function, f is a spatial field function, and φ (phi) is a unary focal operation. For each location x:

1. Compute $n(x)$ as the set of neighborhood points of x (usually including x itself).
2. Compute the values of the field function f applied to appropriate points in $n(x)$.
3. Derive a single value $\varphi(x)$ of the derived field from the values computed in step 2, possibly taking special account of the value of the field at x.

A unary focal operation φ on the field f with neighborhood function n is shown in Figure 4.23.

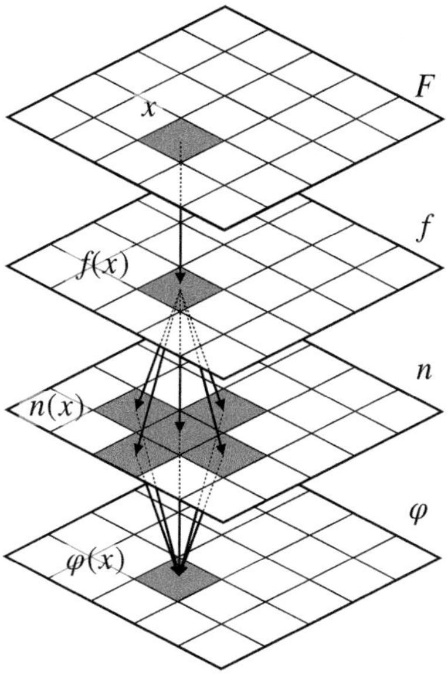

Figure 4.23: Focal operation φ on a field

A good example of a focal operation is computing the slope of a topographical altitude field. The height of the terrain surface is an important type of field in its own right—a *digital elevation model* (DEM, see Box 4.5 on page 143). In this case, f is the topographic field that maps locations x in the spatial framework to heights of the terrain surface at those locations; $n(x)$ will be close neighbors of x; and φ will compute the differences between values of the neighbors of x in different directions to produce a slope field.

zonal operation

Zonal operations A zonal operation aggregates values of a field over each of a set of zones (arising in general from another field function) in the spatial framework. A zonal operation ζ (zeta) derives a new field based on a spatial framework F, a spatial field f, and set of k zones $\{Z_1, ..., Z_k\}$ that partition F. For each location x:

1. Find the zone Z_i in which x is contained.
2. Compute the values of the field function f applied to each point in Z_i.
3. Derive a single value $\zeta(x)$ of the new field from the values computed in step 2.

A unary zonal operation ζ on the field f with zones $\{Z_1, ..., Z_k\}$ is shown in Figure 4.24.

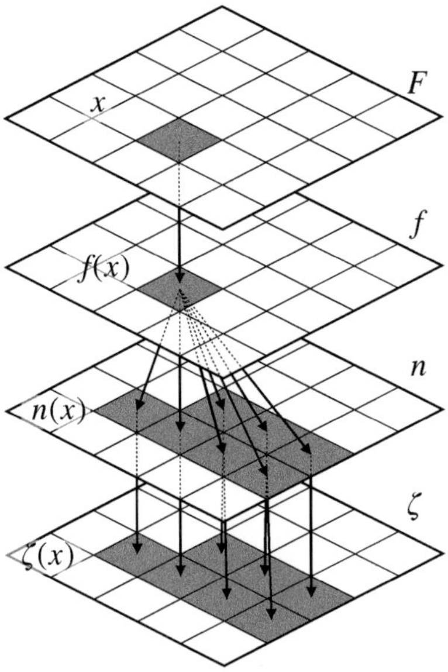

Figure 4.24: Zonal operation ζ on a field with zones $\{Z_1, ..., Z_k\}$

For example, given a layer of precipitation and a zoning into administrative regions, a zonal operation might derive a new layer of average precipitation for each region. In this case f is the precipitation field, $\{Z_1, ..., Z_k\}$ will be the different administrative regions, and ζ will calculate average precipitation over all locations within each zone.

Summary of field operations By way of summary, Table 4.1 contains brief notes on a sample of general field operations. For some of these, where the spatial framework is assumed to be a Euclidean plane and the field values are real numbers, it is useful to view the field as a surface. The local operations in Table 4.1 compute local arithmetical combinations of fields. The focal operations in Table 4.1 are all unary and compute a value at each location dependent on that location's neighborhood. The zonal operations in Table 4.1 are also unary and at each location they compute a value dependent upon the field values in the zone containing that location.

Type	Name	Degree	Description
Local	**lsum, ldiff, lprod, lquot**	binary	Pointwise sums, differences, products, and quotients
	lmax, lmin	*n*-ary	Pointwise maximums and minimums
	lmean	*n*-ary	Pointwise means
Focal	**slope**	unary	Maximum gradient at locations
	aspect	unary	Bearing of steepest slope at each location
	fmean	unary	Weighted average based on neighborhood
	fsum, fprod	unary	Sum and product of values in the neighborhood of each location
Zonal	**zmin, zmax**	unary	Minimum and maximum values in each zone
	zsum, zprod	unary	Sum and product of values in each zone
	zmean	unary	Mean values of field in each zone

Table 4.1: Sample of field operations

A fundamental GIS operation is the overlay of one layer upon another, introduced in Chapter 1. For a field, overlay produces a composite function upon the underlying spatial framework. Without using the language of overlay explicitly, it should be clear that overlays are implicit in many of the field operations summarized above. Indeed, the example of a local operation involved the overlay of population and cancer incidence, in order to perform an arithmetic operation (division) upon the individual pairs of elements in the layers.

4.3 *Time, events, and processes*

The discussion so far in this chapter, indeed in this book, has been focused primarily on static spatial information. This section introduces the modeling of the essential companion to space: time. As we have seen in the introduction to this chapter, at a fundamental level entities in the world may be divided into *continuants*, which include objects and fields, and *occurrents*—things that "happen." Adding temporality allows the possibility of managing information about the changing states of objects and fields at different times and places. Temporality will enable us to answer "what–where–when" questions—for example, "Who owned the land at location *l* and at time *t*?"—as well to start to ask *how* change occurred, and perhaps even *why*.

To set the scene, it is worth considering the following stages in a "brief history of time" in GIS.

4.3.1 *Stage zero: Static representations*

Traditional spatial and geographic information systems hold only a single state of the "real world." This state is almost always the most recent in time for which the data was captured. Interactions with the system are "timeless," in that only information contained in the single state can be retrieved. Most of the systems and models we have met up to now in this book have been concerned with static representations.

4.3.2 *Stage one: The snapshot metaphor*

snapshot
timestamp

Stage one uses the *snapshot* metaphor to show dynamic phenomena as a collection of *timestamped* states. Snapshots can be a powerful mechanism for understanding change. For example, Figure 4.25 shows the growth of the main firefront from our running example, the wildfire in Figure 4.3.

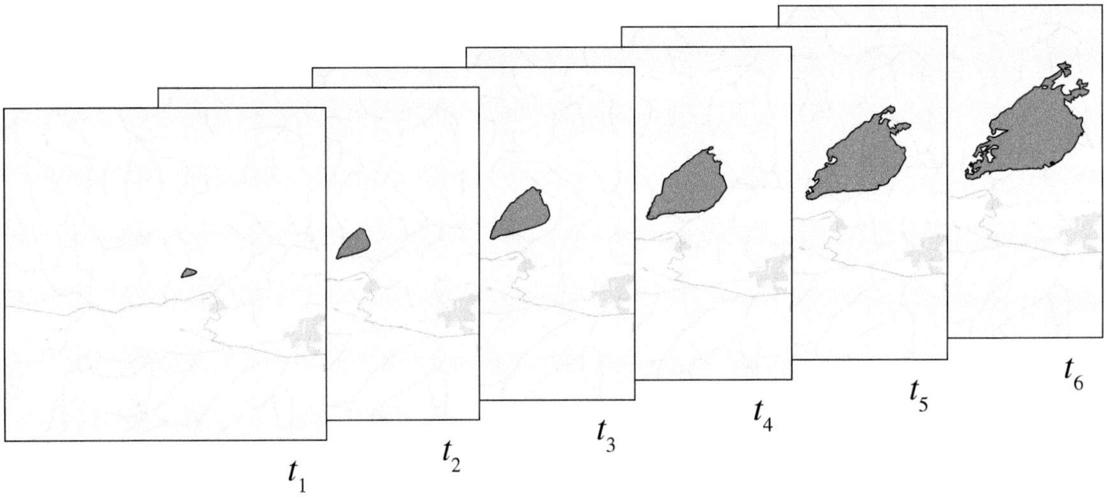

Figure 4.25: Stage one: Sequence of temporal snapshots of the development of the main firefront from Figure 4.2

branching time
cyclic time

Underlying such stage-one representations are models of time. For example, if time is viewed as a linear dimension, shown on a time-line (i.e., timestamps are linearly ordered), then the snapshot collection becomes a sequence. In addition to linear time, other more elaborate models are sometimes used to conceptualize time, including *branching time*, in which past and future possibilities may be represented, and *cyclic time*, in which periodic phenomena such as seasons or times of the day may be represented. However, analysis using these complex variants can usually be reduced to aggregates of linear timelines.

4.3.3 *Stage two: Object histories*

The difficulty with stage one is that the static nature of the individual timestamped states dictates the generally static nature of the representation. Thus, it is a problem in stage one representations to identify dynamic phenomena, such as birth, change, and death. Figure 4.26 shows a detail from two historic topographic maps of Marsh Island, near the University of Maine. Examining the maps clearly shows that an airport was constructed and part of the railroad discontinued in the north between 1902 and 1946. However, this important information is only available implicitly through comparison of the two states. The snapshot metaphor offers no mechanism for *explicitly* representing information about the time or the occurrence of events such as the construction of an airport.

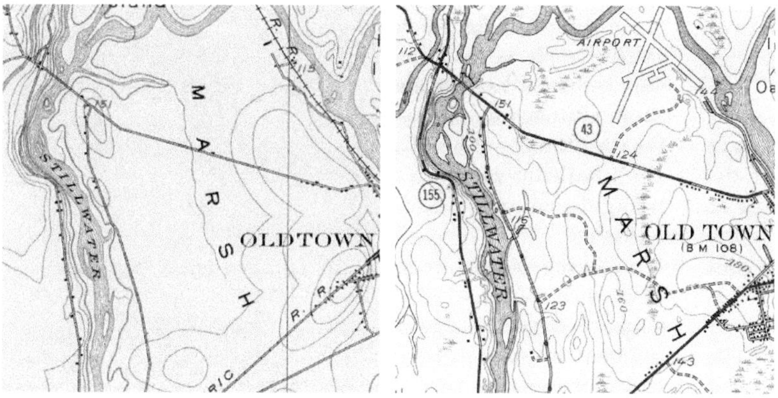

a. 1902

b. 1946

Figure 4.26: Implicit events in stage one: Detail from 1902 (left) and 1946 (right) topographic maps of north Marsh Island, near Orono, Maine (Source: USGS)

Stage two begins to address these issues. Object *histories* are designed explicitly to represent changes of state in a single object, and interactions between different objects. For example, there are seven basic types of changes to objects defined in Renolen's 2000 *history graph model*: creation, destruction, disappearance and reappearance, split and deduction, merge and annexation, reallocation, and alteration, discussed in more detail below with reference to some recent changes to English administrative regions (Figure 4.27).

history graph model

- *Creation* occurs when an object is first generated. For example, in 1972 the English administrative county of Cleveland was *created* as part of a major redesign of local government in England and Wales.

creation

- *Destruction* occurs when an object is permanently eradicated. The English county of Cleveland was subsequently *abolished* (destroyed) by another local government reorganization in 1996.

destruction

- *Disappearance* and *reappearance* are distinct from deletion and creation, because changes may not be permanent. For example, the smallest county in England, Rutland, disappeared in 1974 when as a result of administrative restructures the region was incorporated into a larger county, Leicestershire. Rutland later reappeared as a county in 1997 following further administrative changes.

disappearance

reappearance

- A *split* occurs when two or more now objects are created from a subdivision of an object. For example, in 2009 the English county of Bedfordshire was replaced with two new areas, Central Bedfordshire and the Borough of Bedford. *Deduction* is a special case of a split, which occurs when one or more new objects are deducted from an existing object.

split

deduction

merge

- A *merge* creates a new object from the combination of two or more objects. For example, in 1964 the old English county of Cambridgeshire was merged with the area known as Isle of Ely to form a new county called—perhaps unsuprisingly—Cambridgeshire and Isle of Ely. Similar to deduction, *annexation* is a special case of merging where one or more objects become incorporated into an existing object.

annexation

reallocation

- *Reallocation* can be seen as a complex combination of merging and splitting, where two or more objects are recombined into two or more different resulting objects. For example, in 1974 the English county of Yorkshire, previously divided into the three regions of West Riding, North Riding, and East Riding, was reallocated into four new counties of North Yorkshire, South Yorkshire, West Yorkshire, and East Riding of Yorkshire.

alteration

- An *alteration* occurs when an object is changed or modified in some other way. For example, in 1988 the English district of The Wreakin became the borough of Telford and Wreakin.

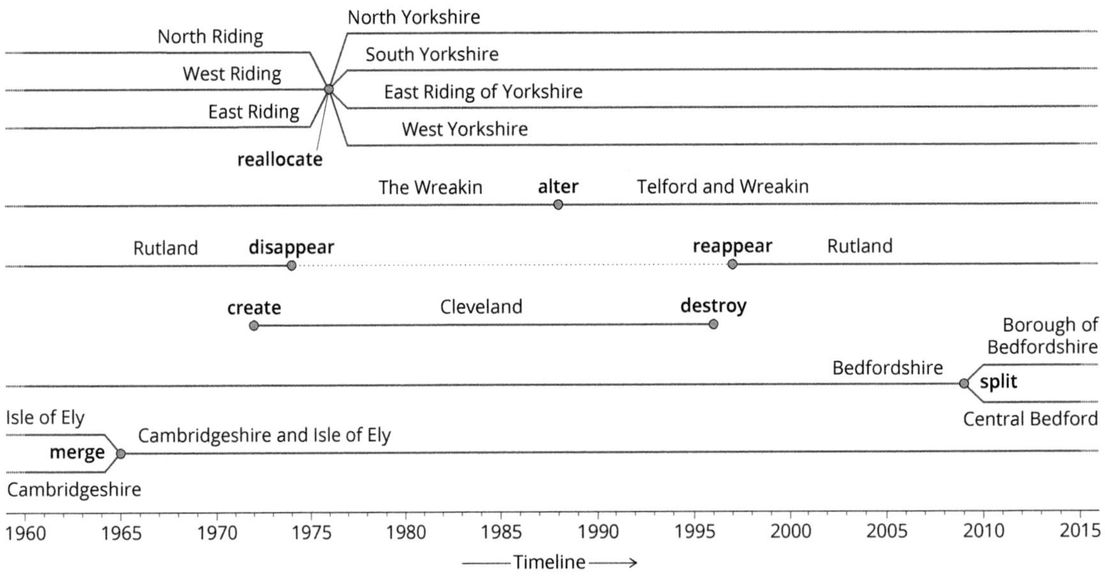

Figure 4.27: Recent histories of some local government counties in England

Figure 4.28 depicts some further spatial examples of selected changes, in terms of the evolution of plots of land within a town. The figure highlights the spatial aspects of destruction (or disappearance if the plot subsequently reappears), creation (or reappearance if the plot previously had disappeared), split (or deduction, if one of the areas is viewed as persisting), merge (or annexation if similarly one of the areas persists), and reallocation.

In terms of alterations to an object, we can further distinguish several finer-grained types of change, including:

- spatial changes such as transformations in the shape, size, and position of an object;

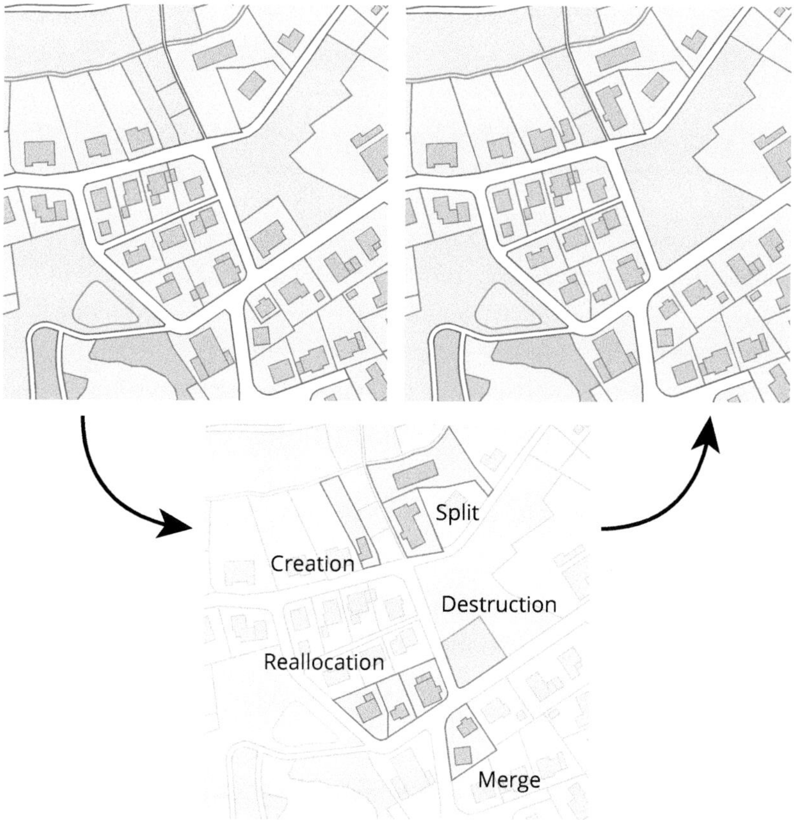

Figure 4.28: Evolution of a neighborhood, highlighting creation, destruction, split, merge, and reallocation

- aspatial changes, involving transformations in the name, color, or other attributes of an object;
- mereological changes to "part of" relationships (*mereology* is the study of parts and wholes); and
- typological changes to the underlying classification of objects.

mereology

4.3.4 *Stage three: Events and processes*

Stage three in this evolution is a full-blooded treatment of change, in terms of complexes of events and processes. Events, actions, and processes are allowed to become explicit entities in stage three representations. From an ontological perspective, stage three involves modeling *occurrents*—things which happen or occur (lectures, people's lives, boat races)—as first-class entities, and not only continuants, such as tables, houses, and people, and changes to continuants, such as creation, alteration, and disappearance.

An initial difficulty arises concerning the meaning of stage three terms. Many accounts use different definitions, even for the most fundamental terms: *event* and *process*. One common definition of an event is an *instantaneous* (punctual or point-based) occurrence, such as the *point* at which someone enters a room or that the 21st century began. In this account, events are often thought of as instantaneous transitions between one state of the world

event
process

achievements

accomplishments

and another, sometimes also called *achievements*. Accordingly, a *process* can be defined as a *durative* (interval-based) occurrence, sometimes also termed *accomplishments*. Processes in the sense of durative occurrences may involve continuous change, such as the erosion of a beach, or structured sequences of events, such as the construction of a house.

activities

The "instantaneous versus durative" (or equally "achievements versus accomplishments") definition of events and processes has the advantage of simplicity. However, it does not leave room for another major ontological class of entity that must be accounted for, termed *activities*. Activities are durative occurrences, akin to accomplishments. Unlike accomplishments, however, activities are open-ended and internally homogeneous. The activity of "erosion," for example, is an unbounded occurrent where each part can likewise be called "erosion". In contrast, the "erosion of a beach" occurs over some time period, for example, ending in a cliff's collapse or all the sand being gone.

count noun

mass noun

A second common and incompatible definition of events versus processes, then, is as achievements/accomplishments (punctual/durative events) versus activities (processes). This different distinction comes to life when we consider the analogy to the distinction between *count nouns* and *mass nouns*, summarized in Table 4.2. As the name suggests, a *count noun* refers to something that can be counted, whereas a *mass noun* refers to something that cannot. In the case of continuants, this distinction manifests itself as a distinction between *things* and *stuff*. For example, you can count lakes (things) but not water (stuff). In the case of occurrents, the same distinction may be used to distinguish events (both punctual achievements and durative accomplishments) and processes (activities). There may be many athletics races, but "running" cannot be counted. Similarly, "volcanos" (continuant count noun) and "volcanic eruptions" (events) can be counted; "magma" (continuant mass noun) and "vulcanism" (the process of volcanic activity) cannot.

Table 4.2: Classification of events and processes through count and mass nouns, and continuants and occurrents

	continuant	occurrent
count noun	thing (e.g., a "lake" or "volcano")	event (e.g., a "flood" or "volcanic eruption")
mass noun	stuff, (e.g., "water" or "magma")	process (e.g., "flooding" or "vulcanism")

situation

state

This second account of the difference between events and processes is after Mourelatos (1978) and summarized in Figure 4.29. The top-level entity in the taxonomy in Figure 4.29 is the general concept *situation*. A situation involves some combination of *states* and *occurrences*. A *state* is a static attribute of a situation, such as the temperature at a location or the speed of a vehicle. Events and processes may result in changes in states, but states themselves are not dynamic. Occurrences may be events or processes, as we have seen. Events may be further subdivided into point-based achievements or durative accomplishments.

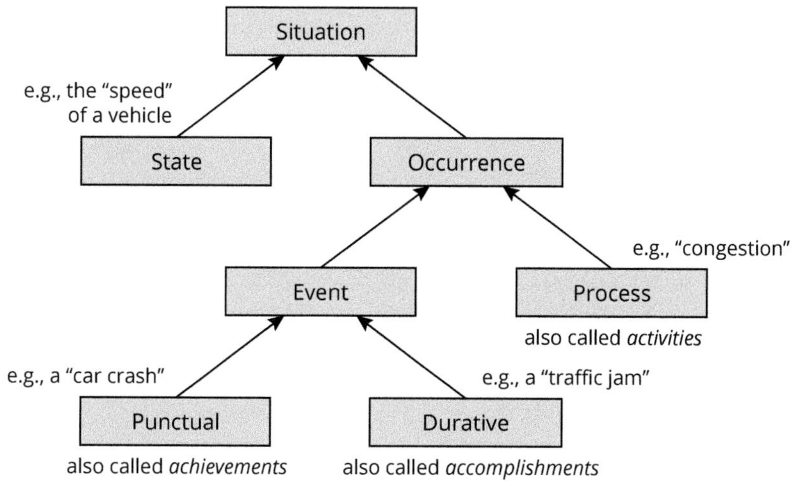

Figure 4.29: A taxonomy of situations after Mourelatos (1978)

In general, this latter Mourelatos account of events (as achievements/accomplishments) and processes (as activities) is preferred, in order to avoid the potential ambiguity of distinguishing between punctual and durative occurrences that can arise when using the former definitions. For example, is a car crash a punctual occurrence or only a very short durative occurrence? Often such distinctions reduce to a matter of application granularity, with arguably the only truly punctual occurrences being fiat events (such as the creation by fiat of the administrative region of Cleveland, Figure 4.27). Whichever terminology in preferred, though, it is essential to remain clear about the underlying ontological distinctions above for avoidance of costly and confusing errors.

4.3.5 Stage four: Causes

The discussion above leads to one final stage in our brief history of time: *causes*. The nature of causality has been debated by philosophers and scientists for centuries, and the notion of a cause is at one and the same time both elusive and essential (see also Box 4.6 on the following page). Causes remain at the cutting-edge of research into GIS, with no practical algorithms, functions, or toolkits available today to the GIS user to analyze "causes" directly. Yet, paradoxically, most applications of GIS arguably have "causes" at their root. Wildfires and floods, traffic jams and migration, epidemics and environmental change are all examples of critical applications of GIS that are fundamentally connected with causes, whether directly or indirectly, through the need to make predictions and explain observations.

In more precisely defining causes, it is helpful to start with some examples of two types of things that do not count as causes. First, whilst important in any discussion of causes, *states* cannot themselves be *causes*. We may informally use statements such as: "The slippery road caused the accident." But the *state* of the road (in this case "slippery") cannot itself be the cause, or is at best

Box 4.6: Temporal metaphors

Metaphors for time and temporal relationships are often spatial. Future events may be viewed metaphorically as coming toward us, as in the traditional Christmas song "Christmas is coming / The goose is getting fat / Please put a penny in the old man's hat." Possible future events may be viewed metaphorically as having a spatial relationship to us, as in "Love is just around the corner," or "Harry's career can take one of two possible paths from here." The last example shows a movement through time. There are many examples of this, such as "We have just left the 20th century," or "Joan is approaching retirement." Basic notions of time, like timelines, branching time, and cyclic time also have their roots in physical space. Human cognition about events and processes also has a rich metaphorical structure, with a spatial component. Events may be locations, as in "My mother was in the war" (event as container) or "The country is going to war" (event as place). Events may be objects, as in "I have a headache" (event as something to be possessed). Finally, events may be causes as we have seen, as in "The war changed the boundary of our country" (event changing the state of an object). This *spatialization* is an example of an important metaphor for HCI, discussed further in Chapter 8.

an incomplete cause. As the thought experiment goes, presumably the road was slippery before the accident, and remained slippery after the accident. If the slippery state of the road was the cause of the accident, it still leaves unanswered the question of why, then, did not the accident occur earlier, or later? A similar logic can be applied to processes (in the sense of activities above, i.e., processes as unbounded occurrences, such as "speeding", distinct from bounded events), meaning processes too cannot strictly be said to be a cause.

Following Galton (2012), Figure 4.30 summarizes an account of causes that is also compatible with the taxonomy of situations in Figure 4.29. In this account, a *cause* is strictly a relationship between two events, where a first event leads directly to the occurrence of a subsequent event. Note that *causes* may only directly relate *events*; states and processes can neither cause nor be caused. Events, however, may *change* a state (such as a rain storm that *terminates* the "dry" state of the road) or *modify* a process (such as the driver depressing the brake that *initiates* the process of decelerating). Closing the loop, states are involved in *allowing* causes. Although a slippery road (state) does not directly cause the accident, it is pivotal in allowing the braking event to cause the accident.

cause

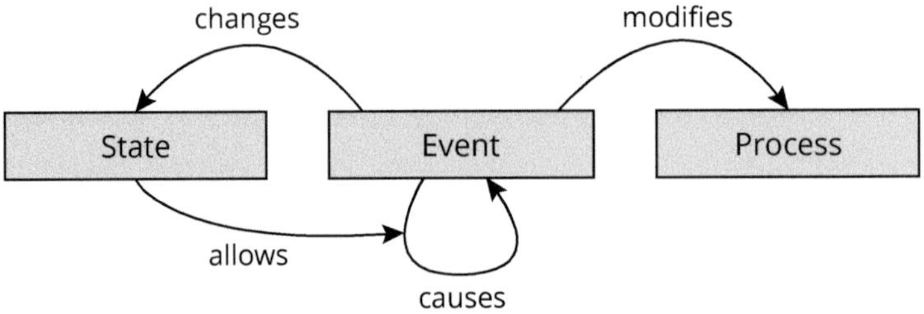

Figure 4.30: A taxonomy of causes after Galton (2012)

Further interrelationships, such as between processes and states, are required for a more complete account. We may also wish to further classify different types of states, processes, events, and causes. Intentions, for example, are important in understanding the causes of situations involving

Box 4.7: Movement models

The archetypal example of an intrinsically spatial and temporal phenomenon is *movement*. Consequently, it is not possible to formulate an atemporal, Stage 0 model of movement. GNSS trajectories are the most familiar example of a Stage 1 model, where an object's movement is represented using a sequence of timestamped coordinate locations, called "fixes." In Stage 2 the focus shifts to a history of each object's movements. For example, *geospatial lifelines* are a history of the movements of an object in terms of timestamped records of the significant places that object has been located (Hornsby & Egenhofer, 2002). Stage 3 shifts further towards a view of events and processes as first-class objects, such as explicitly representing a meeting (an event) or flocking (a process). Finally, Stage 4 involves descriptions of the events and associated states that cause movement, such as high water temperature as a driver of fish migration or the intention of illegal fishing vessel captains to evade detection. To illustrate, the figure right captures the geospatial lifelines (Stage 2, history of movement) of two vehicles around The Potteries. A (Stage 3) meeting occurred in Stoke between the pink and purple vehicles. The pink vehicle's (Stage 1) trajectory is annotated with fixes on the leg from Stoke to Longton, with the vehicle slowing as it moves through Fenton.

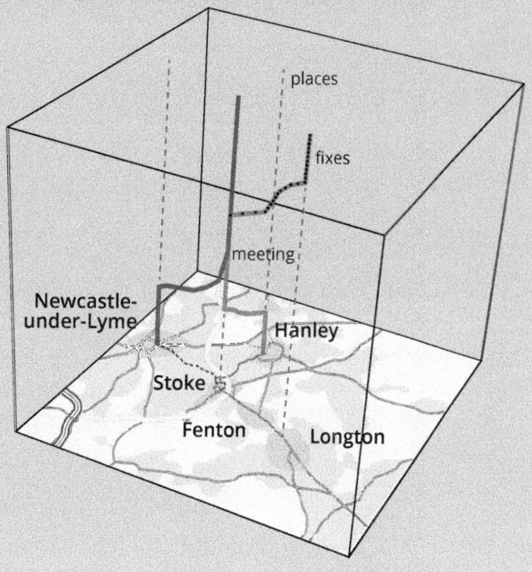

cognitive agents, such as people. An *intent* is a state of mind of an agent that potentially will enable certain causes to occur, such as the intention to drive safely inhibiting the chain of causal links between events that led up to the accident. However, the central observation—that only events can *be* caused, and only caused *by* other events—can help to bring clarity to the muddy waters that often surround causal reasoning.

intent

4.3.6 *Temporal information systems*

The stages above capture increasingly sophisticated representations of time, moving from static scenes (Stage zero) and timestamped objects and fields (Stage one) through to explicit descriptions of events and processes (Stage three) and, in the future, perhaps of causes and intents (Stage four) too. These different stages are best viewed as complementary models, each with their own strengths and weaknesses, rather than an inexorable progression towards a single, overarching model. There will always be a need for static, atemporal (Stage zero) representations of the world and for simple, timestamped (Stage one) representations of objects and fields, even though those representations cannot hope to capture (Stage three) events and processes explicitly (see Box 4.7).

Irrespective of the specific Stage used, a number of further crosscutting issues with time face any temporal information system, beginning with the need for a clear distinction between *revision* and *update*.

update

revision

Revision and update An information system undergoes an *update* when new information arrives that indicates a change in the application domain. An information system undergoes a *revision* when new information arrives that indicates a change in our knowledge about the application domain, but the application domain itself may not have changed.

For example, if new information from satellite imagery indicates that the land cover in region A is different from that recorded in the database, then two possibilities exist. The land cover for region A may have actually changed from that recorded in the database to that detected by the remote sensing satellite. This might occur when a farmer plants a new crop in a field, for example. In this case, the database must be updated to reflect this change. For a temporal database, it will be necessary to record the timing and history of any changes.

However, if there is no reason to suspect any actual change, then it will be necessary to revise our stored information about the land cover in region A. If we are able to invoke a preference between the information from two sources, this will inform the decision as to which information to retract. For example, if as a result of new satellite sensor technology the new land cover information is expected to be more reliable than the old information stored in the database, we may decide to prefer the new information.

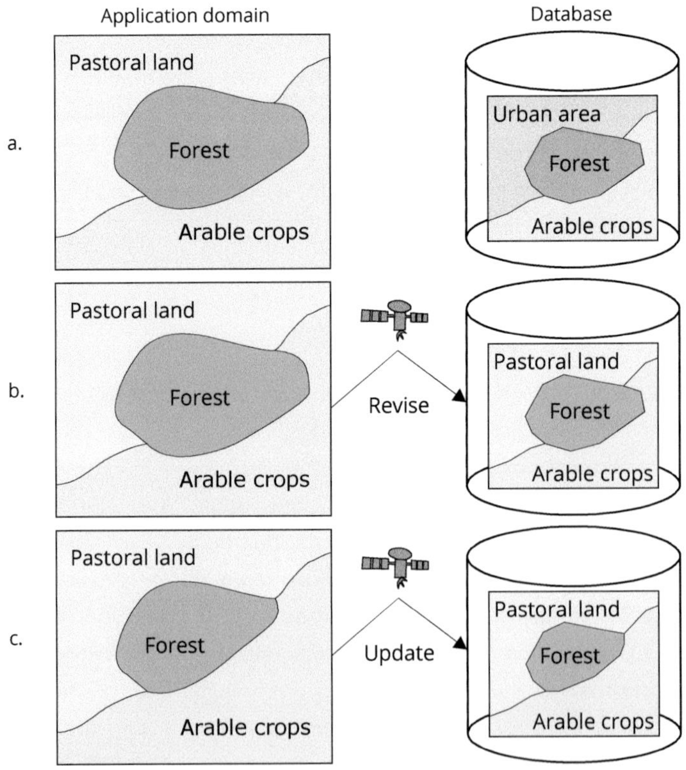

Figure 4.31: Contrasting update (c) and revision (b) to a land cover database (a)

Figure 4.31 illustrates the distinction between update and revision. Figure 4.31a shows the initial state of the application domain on the left and the database on the right. In Figure 4.31b new information from satellite imagery indicates that the region with stored land cover type "Urban area" is in fact a region of land cover type "Pastoral land." Since the application domain has not changed, a revision operation is needed. A subsequent change in the application domain, shown in Figure 4.31c where part of the forested region has become cropland, requires an update of the database. At both stages in Figure 4.31 the database is changed, but at only one stage does the application domain change.

transaction time

valid time

Valid and transaction time Acknowledging the distinction between revision and update, two types of temporal references are important to temporal systems: *transaction time* references and *valid time* references. Transaction time is the time when data was entered into the database. Valid time is the time when the event relating to that data actually occurred in the world. These two "temporal dimensions" are orthogonal. Which times need to be represented in the database will depend on the application domain. If the key item of information is the situation of real-world objects in real-world time, then valid time is required; if the issue of importance is tracking the evolution of our knowledge in the system, then transaction time is required. If both times are needed, then we require a so-called *bitemporal* information system, discussed further below.

Valid and transaction times have different properties and need to be handled in different ways. A fundamental distinction is that transaction time monotonically increases with the life of the information system, whereas valid time may not. So, if transaction T_1 comes before transaction T_2, then the transaction time associated with the data entered by T_1 must be before the transaction time associated with the data entered by T_2. This is not the case for valid time. It is perfectly possible that the real-world event to which T_1 refers is *later* than that described by the data in T_2. It follows from this that transaction times cannot be changed, whereas valid times can be changed retroactively.

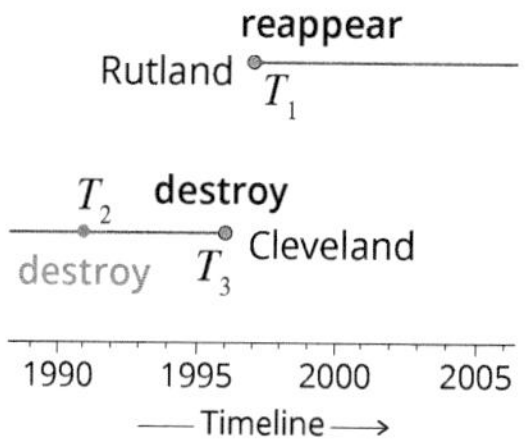

Figure 4.32: Some transactions in a temporal database of English local government counties, cf. Figure 4.27

For example, referring back to Figure 4.27, a temporal database of English local government counties created in 2010 might involve storing the appearance of Rutland in 1997 as transaction T_1 (see Figure 4.32). A subsequent update to the database in 2011 might involve inaccurately storing the destruction of Cleveland in 1991 as transaction T_2. The transaction time for T_1 is *before* that for T_2, but the valid time for T_1 is *after* T_2. If we subsequently uncover this error in 2012, noting that the valid time for transaction T_2 is actually 1996, not 1986, then a third transaction T_3 would be needed to alter the valid time for T_2. The transaction times for T_1, T_2, and T_3 can never be altered.

Hence, transaction times maintain a history of system activity, and the capability to roll back to previous states of the system. Each state of the system is usually called a *version*, and the business of organizing this activity is called *versioning* or *version management*. At any time, a temporal information system with version management has access to the current state and all previous versions. Valid times maintain the history of real-world occurrences. A temporal

versioning

information system with valid time capabilities allows querying of current, past, and possibly future states of the real-world objects in its database.

bitemporal model

Bitemporal models Valid time and transaction time can be maintained together in a database using the *bitemporal model*. A bitemporal timestamp is a sequence of ordered pairs (r, v), where $r = [r_s, r_e]$ is a transaction time interval and $v = [v_s, v_e]$ is a valid time interval. For example, imagine recording in 2022 the fact that a house H was owned by person P up until 2021. We might represent this bitemporal timestamp, $([2020, 2021], [2022, 2023])$ graphically as in Figure 4.33.

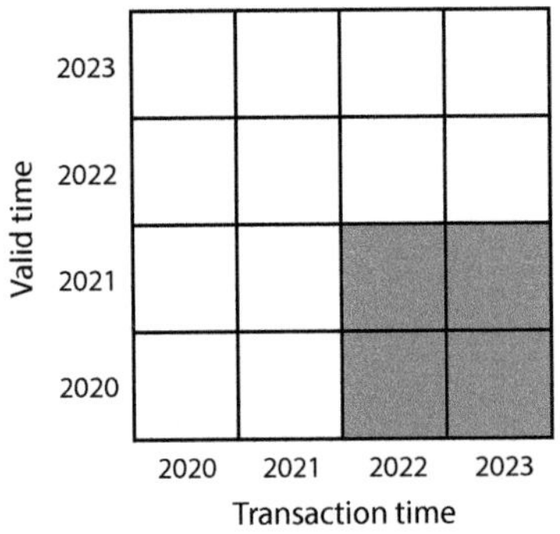

Figure 4.33: A graphical representation of the bitemporal timestamp $([2020, 2021], [2022, 2023])$

Now consider the following more complex sequence of events recorded in a bitemporal database of properties and owners, established in 2021.

- The initial 2021 database includes houses H_1 and H_2, already owned by persons P_1 and P_2, respectively. It also includes the information that house H_3, not yet built, is planned for construction in 2022.
- In 2022, person P_4 purchases house H_1 from P_1, and the purchase is reflected in a prompt update to the database.
- In 2023, person P_3 takes ownership of house H_3, when it also becomes clear that house H_3 was in fact completed in 2023, not 2022, necessitating a further revision to the database.

Figure 4.34 depicts this situation, using our graphic representation of each bitemporal timestamp. The four owners, P_1 to P_4 are shown in the left-hand side of Figure 4.34. The arrows connecting owners with their houses are annotated with a bitemporal icon to show the valid and transaction times stored in the database. The arrows and bitemporal icons associated with P_1 and P_4, for example, indicate that P_1 was the owner of H_1 in 2020 and 2021 (as indicated in the database in 2021). The change of ownership to P_4 in 2022 appears as a database update with transaction time 2022. The construction of the house H_3 planned in 2021 to occur in 2022, but updated in 2023 to have been built in 2023 is shown by the bitemporal timestamp icon adjacent to the new house H_3.

Note that this simple example does not enforce any consistency constraints upon bitemporal timestamps (such as, for example, that a house must always

Figure 4.34: A bitemporal database of properties and owners, annotated using the bitemporal timestamp icon design from Figure 4.33

be owned by exactly one person). Nevertheless, the database contains the necessary information for such consistency rules to be defined and added to the database.

Reflections

If databases lie at the *technical* core of GIS, then models—of objects, fields, events, and processes—lie at its *conceptual* core. Models in GIS provide the vital link between the structure of spatial data in the database, and the structure of the geographic world we wish to describe using that data. When mistakes occur—for example, when one confuses a raster for a field, a forest for forested, a timestamp for an event—the resulting confusion cannot usually be repaired by mere technical wizardry. As a consequence, the most valuable skills of a geographic information scientist are not technical; they lie in the analytical ability to identify the underlying models of space and time needed to support an application of GIS.

Few modeling mistakes are more costly—or more easy to overlook—than ontological mistakes. As a result, time invested in developing a more sophisticated ontological grasp of geographic information is usually time rewarded. In addition to the core ontological work already cited in connection with objects and fields (Couclelis, 1992), fiat and bona fide boundaries (Smith & Varzi, 2000), events and processes (Mourelatos, 1978), and causes (Galton, 2012), there is a wealth of other important and interesting ontological work to be uncovered in the field of GI science. This body of work includes, for example, investigations into the ontology of geographic continuants and occurrents (Bittner, Donnelly, & Smith, 2004; Grenon & Smith, 2004); moving objects (Dodge, Weibel, & Lautenschütz, 2008; Parent et al., 2013); collectives of, and interactions between moving objects (Orellana & Renso, 2010; Wood & Galton, 2009); wayfinding and navigation (Timpf, 2002; Wang, Duckham, & Worboys, 2016); amongst many other phenomena.

Finally, whilst this chapter is given over to models of objects, of fields, and of time in roughly equal proportion, in practice the capabilities of GIS today remain strongly biased towards serving the former two—(static) objects and fields. Despite a long history of research into time in GI science (e.g., Claramunt, Parent, & Thériault, 1998; Hornsby & Egenhofer, 2000; Langran, 1992; Peuquet, 1994; Yuan, 1999), practical spatiotemporal capabilities in GIS today remain an exception rather than the rule. The scale of the effort that has already been invested in bridging this gap is underlined by a 2018 study, which cataloged more than 150 distinct spatiotemporal models across almost 1500 research articles in the field (Siabato, Claramunt, Ilarri, & Manso-Callejo, 2018). As the authors of that study conclude, the challenge faced is not technical; it is instead a modeling challenge: to date "no general consensus for the modeling of dynamic geographic phenomena in the temporal dimension has been achieved" (Siabato et al., 2018).

Representation and Algorithms

5

Sections

The manner in which spatial data is represented in an information system is key to the efficiency of the computational processes that will act upon it. Computing with spatial data requires specialized representations, and in turn specialized algorithms to process those representations. Through the topics covered in this chapter, you will discover how to:

- contrast the key challenges in computing with one- and **two-dimensional data**;
- discuss the major computational representations of spatial objects, spatial fields, and networks found in a GIS, including the **spaghetti** model, **node-arc-area** data, **Delaunay triangulations**, and **adjacency matrices**;
- demonstrate the use of different representations spatial in efficient data processing, through spatial algorithms such as **point-in-polygon**, **overlay**, and **shortest path**.

Efficient computing with spatial data requires that the *representation* of data and the *algorithms* for processing that data go hand-in-hand. This chapter is concerned with how spatial data is represented in a GIS, and how those representations enable efficient spatial operations to be performed on that data.

The focus in the previous chapter was on high-level conceptual models for computing with spatial data. In this chapter we start to consider questions that are closer to lower-level computation. The first section sets the scene with a discussion of the general computational issues facing any GIS, including efficiency. Sections 5.2 and 5.3 introduce the most important computational representations of spatial objects and fields. If databases are the engine of GIS, then algorithms are the drivetrain, converting stored data into useful outputs, analysis, and applications. Fundamental geometric algorithms and network algorithms are covered in Sections 5.4 and 5.5, respectively.

5.1 *Computing with spatial data*

Much traditional computing is based upon 1D data representations. The following example is intended to give a feel for the jump from one to two dimensions. Figure 5.1 shows some of the different possible relationships that can exist between two straight-line segments, each embedded in the same 1D Euclidean space (the real line). To categorize the relationships, we distinguish only those that are topologically different. Thus two segments may be disjoint, touching internally or externally, overlapping, nested, or equal. These six relationships exhaust all the possibilities (and cf. the set-oriented and topological operators of Chapter 3).

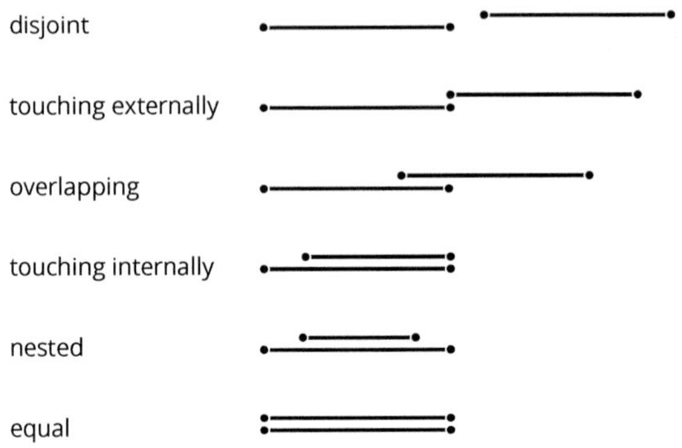

Figure 5.1: Relationships between two line segments embedded in a 1D Euclidean space

Let us now see what happens when we extend the investigation to two dimensions. In a sense, the simplest finite 1D object is the line segment (1-simplex), introduced in Section 3.3.5. In two dimensions, assuming discreteness, the corresponding object is the triangle (2-simplex). The extended problem is to analyze the topologically distinct relationships that can exist between two triangles embedded in the same plane. Figure 5.2 shows a handful of the many possibilities.[1]

[1] There are 32 possibilities in total (Li & Liu, 2010). It is an interesting exercise to try to enumerate all of them for yourself.

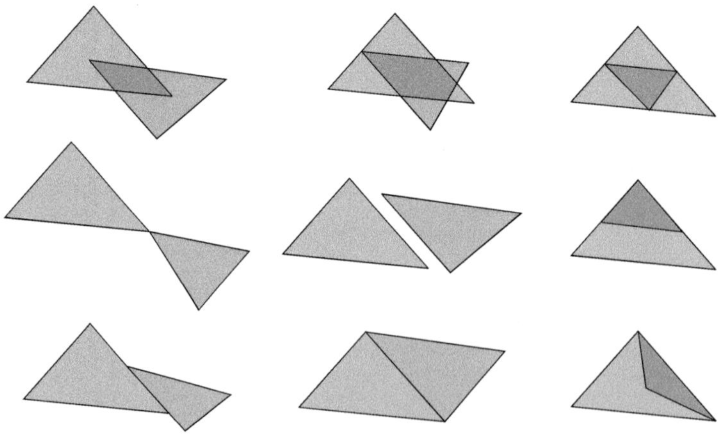

Figure 5.2: Nine of the 32 topologically distinct relationships between two triangles embedded in the Euclidean plane

As triangles are in a sense the simplest two-dimensional objects, it should be clear that the step from one to two dimensions can lead to an explosion of possibilities. It is then natural to infer that the operations underlying two-dimensional spatial data models will likewise require more elaborate algorithmic support. The further jump to three dimensions introduces even greater complexities.

5.1.1 Geometric algorithms and computational geometry

algorithm

An *algorithm* is a specification of the computational processes required to perform an operation. Thus, an algorithm for the addition of two numbers in a particular base will specify precisely the steps required to compute their sum. Algorithms may be straightforwardly translated into machine instructions (usually by means of a high-level programming language) that a computer is able to process.

geometric algorithm

A *geometric algorithm* is an algorithm that operates upon geometric (or spatial) objects. Geometric algorithms have some special features. Consider the problem of determining whether a point is inside a closed loop or not. Unless the loop is extremely convoluted (Figure 5.3, right), the human eye and brain are able to decide this instantly and without much discernible effort (Figure 5.3, left). But what instructions would you need to specify in order to make a machine perform this task? We shall see that this classical point-in-polygon decision procedure, while not being all that difficult, is nevertheless not trivial. This property of geometric algorithms is in stark contrast to arithmetic algorithms where many complex calculations, requiring paper, pencil, and much head-scratching from a person, can be solved with ease by a computer.

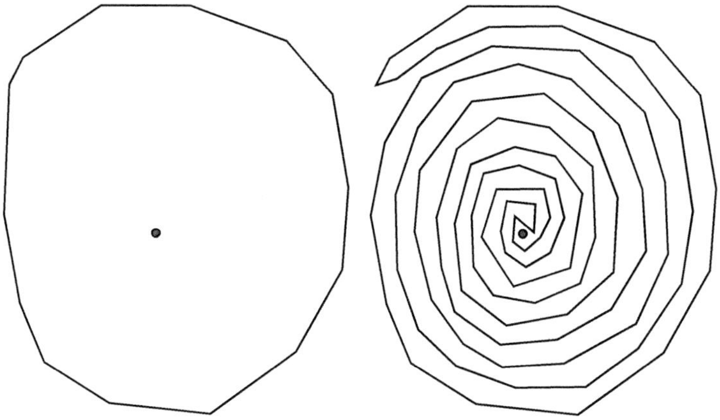

Figure 5.3: Is the point in the polygon?

Another characteristic of geometric algorithms is that they are frequently plagued by special cases. What seems at first a simple problem often becomes more complex as special cases are considered. This feature has been vividly described by Douglas (1990), where what initially appears to be a trivial algorithm to determine whether two straight lines intersect becomes beset by a multitude of special cases.

computational geometry

Computational geometry is the study of the properties of algorithms for solving geometrical problems. Typical questions in computational geometry include:

1. Is it possible to find *any* algorithm to solve a particular geometrical problem?
2. What is the most efficient algorithm to solve a geometrical problem?

Box 5.1: Geometric algorithms and computation

The notion of a geometric algorithm goes back at least as far as the geometer Euclid, living and teaching in the city of Alexandria circa 300 BC, in whose works geometric constructions are important features. In Euclid's geometry, the constructions are executed using a specified collection of instruments—straight-edge and compass—and proceeding according to a given set of basic moves. At that time, the issue was not the efficiency of the constructions but the class of objects resulting from them. A typical problem for investigation was the possibility of trisecting an angle using straight-edge and compass alone. It is only in the past century that efficiency of computation has become important. Work by Czech-born mathematician Kurt Gödel showed that a large class of logical systems, including arithmetic, must be *incomplete*, in the formal sense that these logics must contain true statements that cannot be proved. British mathematician Alan Turing was able to set this result within the framework of computation and show that the same limitations apply to any computational system (of which, we shall consider further in Chapter 9).

3. What is the best way of structuring the geometric data to make the search for a solution to a problem most efficient?

The first question is about computability, where a function or problem is *computable* if it is possible to find some algorithm to compute the function or solve the problem. Interest in geometric algorithms and computability dates back to ancient Greece (see Box 5.1).

computable

Finding an algorithm to solve a problem in theory is quite different from solving it in practice. This is the issue raised by questions 2 and 3, above. The efficiency of an algorithm (*algorithmic complexity*) may be measured in terms of the time that an algorithm consumes (*time complexity*) or storage space used (*space complexity*). The time taken for an algorithm to execute is the sum of the times of all the constituent operations within the algorithm.

algorithmic complexity
time complexity
space complexity

It is almost always the case that the time and space taken to solve a problem increase with the size of the input data. For example, we would expect the time taken to compute the area of a simple polygon to increase as the number of edges of the polygon increases. The question is, "What is the relationship between the input size and time or space required by the algorithm?" This relationship may be expressed as a function, called the *complexity function*, from input size to amount of time (or space). In the remainder of this chapter we are concerned primarily with time complexity.

complexity function

In the case of the area of a simple polygon, it is not too hard to show that the time taken to compute the area is directly proportional to the number of its edges, and so the complexity function is $f : n \rightarrow kn$, where k is a constant of proportionality and n is the number of edges. If we are interested in the general way that the computation time is related to input size, then the value of the constant k does not matter; in any case it will depend upon details of the speed of the computer system and other factors. Also, the input size is not n (the number of edges) but is proportional to n. It is the linearity of the relationship that is important.

These considerations have led to the notation $O(n)$ being used for the set of functions that are at most as large as the product of some constant and n. In general, the *order* $O(f(n))$ of a function $f(n)$ is the set of functions that have a

order

time complexity that is at most as large as the product of some constant and $f(n)$. The notation itself is termed *"big-oh" notation*. Big-oh notation allows us to describe the relationship between computational time (or space) and input size, without needing to be precise about the scalar multipliers.

"big-oh" notation

The most important functions that are used as yardsticks to measure the complexity of algorithms are, in increasing order of size, logarithms, fractional powers (e.g., square roots), polynomials, and exponentials. To give an idea about the rate of increase of these functions, Table 5.1 gives the approximate values for inputs between 1 and 100. Note that it is not just the static value of the function that is important, but the way that it changes as the input size n increases. It is often of interest to consider the *asymptotic behavior* nature of the function: that is, how it behaves for very large input sizes.

asymptotic behavior

n	$\log_e n$	$\sqrt{n}$	n^2	2^n
1	0.0	1.0	1	2.0
25	3.1	5.0	625	3.3×10^{7}
50	3.9	7.1	2500	1.1×10^{15}
75	4.3	8.6	5625	3.8×10^{22}
100	4.6	10.0	10000	1.3×10^{30}

Table 5.1: Approximate values of some common functions

A function order may then be associated with an algorithm to give some indication of how the algorithm will perform relative to the size of input. Of course, a change in the input to an algorithm will usually result in a change in performance of the algorithm, even if the size of the input remains constant. It is usual to take a pessimistic view and give functions that measure the *worst-case* performance of the algorithm: that is, the performance on the input of each size that results in the most time or space consumption.

worst-case performance

Order	Name	Speed
$O(1)$	Constant time	Very fast, independent of input
$O(\log n)$	Logarithmic time	Fast (e.g., binary search)
$O(n)$	Linear time	Moderate (e.g., linear search)
$O(n \log n)$	Sublinear time	Moderate (e.g., optimal sorting algorithms)
$O(n^k)$	Polynomial time	Moderate or slow (e.g., shortest path algorithms)
$O(k^n)$	Exponential time	Intractable (e.g., traveling salesperson algorithms)

Table 5.2: Rough guide to time complexity

Table 5.2 gives a rough idea about the performance of algorithms with different complexity functions. Algorithms that are of order $O(k^n)$ (i.e., exponential time complexity) and greater are often called *intractable*, because they can only be used for very small input sizes.

intractable

For each geometric operation, we may now compare the performance of algorithms that implement that operation. The comparison may be experimental, generating sets of inputs and measuring the performance of a particular computer system on the input sets for each of the algorithms. Drawbacks with this approach are that:

- the results depend on the particular computer system being used as the platform for the experiments, and
- generating a truly representative sample of inputs can be difficult or impossible.

Alternatively, we may take the theoretical approach discussed above, finding and comparing the function class of each algorithm. Drawbacks here are that:

- calculating the complexity of the algorithms may be difficult;
- worst-case analysis may not be a true reflection of experience (for example, an algorithm may perform very badly in only a few cases that hardly ever occur); and
- comparing the function class may be misleading if the scalar multipliers are very different.

Overall, the theoretical approach is more general and is independent of any particular computational platform.

5.1.2 *The discrete Euclidean plane*

Discretization emerged in the previous chapter as an important theme when computing with spatial data. The issues arising from the discretization of field-based and object-based data are quite distinct. For field-based data, the field functions are already discretized to the locations in a fixed spatial framework. On the other hand, with object-based data discretization has wider ramifications, because there is no explicit spatial framework. Spatially referenced objects in a GIS are usually based on Cartesian coordinate pairs within the Euclidean plane. However, Euclidean geometry assumes an underlying continuum that can only be finitely approximated using a digital computational model. As we shall see, this approximation can lead to a large and uncontrolled accumulation of errors, which may alter the geometrical embedding of objects in the Euclidean plane and even distort topology.

Geometric domains Assume that the number pairs chosen to represent points in a discretized Euclidean space are integers. Next, set up a coordinate frame consisting of a fixed, distinguished point (origin) and a pair of orthogonal lines (axes) intersecting in the origin. A point in the plane has associated with it a unique pair of integers (x, y) measuring its distance from the origin in the direction of each axis. The collection of all such points is the *discrete Euclidean plane*, $\mathbb{Z}^2$. Straight-line segments may be defined by giving a quadruple of integers, the two pairs of which represent the end-points of the line segment.

discrete Euclidean plane

We are now ready to define more precisely the notion of a *geometric domain* as a triple (G, P, S), where:

geometric domain

domain grid

- G, the *domain grid*, is a finite connected portion of the discrete Euclidean plane, $\mathbb{Z}^2$

- P is a set of points in $\mathbb{Z}^2$
- S is a set of line segments in $\mathbb{Z}^2$

subject to the following closure conditions:

- Each point of P is a point in the domain grid G.
- Any line segment in S must have its end-points as members of P.
- Any point in P that is incident with a line segment in S must be one of its end-points.
- If any two line segments in S intersect at a point, then that point must be a member of P.

The idea is that higher-level spatial objects may be constructed from objects in the geometric domain. Figure 5.4a shows a small and simple example of a geometric domain. Here the grid G is a 10-by-10 square; the set $P = \{a, b, c, d, e, f, g, h, i\}$; and the set $S = \{ab, ac, dh, eh, fh, gh\}$. Note that only points on the grid are allowed. Line segments must intersect at a point in S, so the configuration in Figure 5.4b is not a geometric domain.

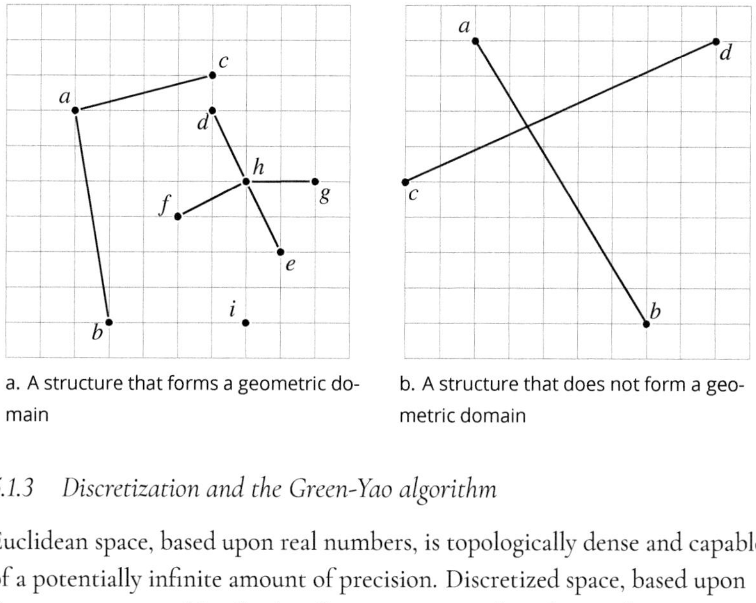

a. A structure that forms a geometric domain

b. A structure that does not form a geometric domain

Figure 5.4: Grid structures

5.1.3 Discretization and the Green-Yao algorithm

Euclidean space, based upon real numbers, is topologically dense and capable of a potentially infinite amount of precision. Discretized space, based upon the integers, is capable of only a finite precision. Thus, during the process of *discretization* (moving data from a continuous to a discrete domain) some precision will inevitably be lost. Loss of precision is unfortunate, as it can lead to errors. Most people who have used a computer or calculator are aware of the rounding errors that can occur during discrete numerical applications. Similar types of errors, described below, can also occur in the geometric domain.

discretization

Figure 5.5a shows a configuration that is not classed as a geometric domain. The problem is that the line segments ab and cd intersect at a point that is not a member of the domain's point set. Indeed, it is not possible to add this point of intersection to the point set, because it is not a domain grid point. So what should we do? One solution is to introduce a grid point x to

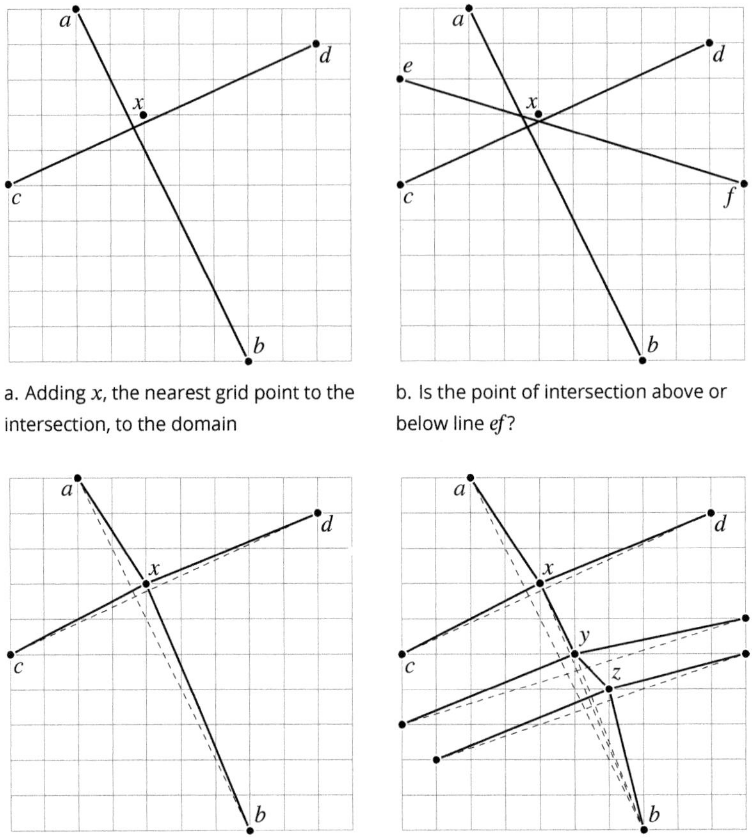

Figure 5.5: Shifting points and drifting lines

a. Adding *x*, the nearest grid point to the intersection, to the domain

b. Is the point of intersection above or below line *ef*?

c. New split lines meet at intersection point *x*

d. Line *ab* drifts away from its original position with further intersections

the domain that is the nearest to the point of intersection (or use a convention to choose if several points are equally near). This is the 2D equivalent of numerical rounding and is illustrated in Figure 5.5a. Of course, we still do not have a configuration that satisfies the domain axioms, but we could extend our notion of geometric domain to include such structures.

However, this solution on its own is unsatisfactory, because topological constraints are broken. For example, the incidence relationship—that the point of intersection of two lines lies on both the lines—does not hold here. Also, as Figure 5.5b shows, it is possible for a point to "shift" from one side of a line to the other. The "real" point of intersection of lines *ab* and *cd* lies below line *ef*, but the rounded point of intersection *x* lies above line *ef*.

A first attempt at a modification that avoids these problems is to split the line segments so as to join at the rounded intersection point, as shown in Figure 5.5c. Thus lines *ab* and *cd* have been split into lines *ax*, *bx*, *cx*, and *dx*. The configuration is now a geometric domain. The problem with this solution is that we lose control of the shift of the line segments as more and more intersections occur in a dynamic situation. Figure 5.5d shows some further line segments that have been added. Notice that the chain *axyzb* has strayed well away from the original line segment *ab*.

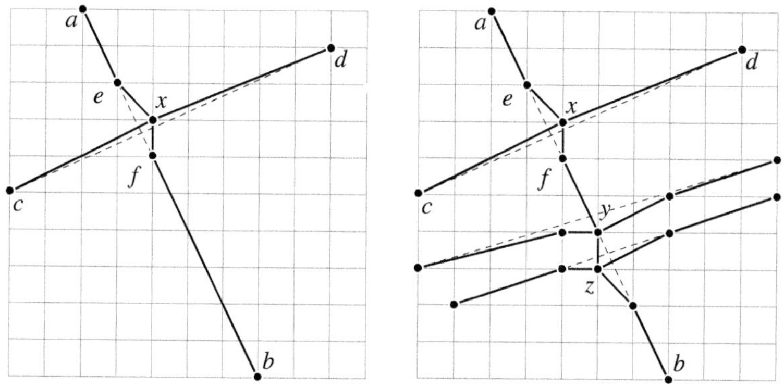

a. Greene-Yao algorithm applied to the *ab* and *cd*

b. Further intersections using the Greene-Yao algorithm

Figure 5.6: Greene-Yao handling of segment intersections

Greene-Yao algorithm

The problem of drifting lines can be solved using the *Greene-Yao algorithm*. Imagine that the grid points are pegs on a pegboard and that the lines ab and cd are elastic bands stretching between their end-points, shown in Figure 5.6a. When we move their point of intersection to the nearest point on the grid, we do so in such a way that the bands cannot pass over the pegs but may come to rest against some of them. In our example, points e and f form such a barrier to the movement of segment ax and xb, respectively. Thus segment ab is no longer split into two segments, but in this case four, namely, ae, ex, xf, and fb.

We have tried to give an appreciation of the algorithm by means of an example. In fact, the formal expression of the algorithm requires some knowledge of the mathematics of continued fractions and is omitted from this account. The important result is that this algorithm places a limit on the drift of a line segment to its neighboring points (the so-called "envelope" of the segment), no matter how many line segment intersections are involved. In the example above, further intersections (Figure 5.6b) do not lead to the same line drift as for the earlier example (Figure 5.5d). Therefore, we now have a discretization process that is both well defined and results in a bounded error accumulation.

5.1.4 Discretizing arcs

The previous section concentrated upon problems caused by points and straight-line segments in a discretized plane. However, many spatial objects are composed of smooth curves, so here we describe approaches to the approximation of curvilinear arcs in a discrete space. The usual technique is to approximate the curve by a polyline.

There are of course an unlimited number of ways that a curve may be approximated by a polyline, due to the number and positioning of the segments. In general, the more segments used, the better the approximation, providing that the segments have been chosen appropriately. The problem of approximating a curve with a polyline is closely related to the problem of *line*

line simplification

simplification. Line simplification is the task of reducing the level of detail in the representation of a polyline, while still retaining its essential geometric character. Line simplification is commonly used in *cartographic generalization*, where maps need to be displayed at different scales or levels of detail (explored further in Section 8.2.4).

Douglas-Peucker algorithm

A technique based on the *Douglas-Peucker algorithm* for arc simplification is shown in Figure 5.7. Consider the curve in Figure 5.7a. First, a point is assigned to each end of the arc. A straight-line segment ***ab*** is then constructed between these points. The point *c* on the arc farthest from the line ***ab*** (in terms of length of a line perpendicular to ***ab***) is added to the point set. The arc is split into two parts, ***ac*** and ***cb***, and the algorithm applied recursively to each of the pieces of arc. In Figure 5.7b, the arc has been split into ***ad***, ***dc***, ***ce***, and ***eb***. When the distance from a straight-line segment to the farthest point on the corresponding arc segment is below a predefined threshold, then that iteration stops. In Figure 5.7c ***ce*** and ***eb*** have this property, indicated with a faint line. The smaller the threshold, the more points will be added and the better will be the approximation. The algorithm continues (Figure 5.7d) until there are no more arcs to split. In our example, this stage is reached in Figure 5.7e with arcs ***ah***, ***hf***, ***fd***, ***dg***, ***gj***, ***ji***, ***ik***, ***kc***, ***ce***, ***eb***. The approximating polyline is shown in Figure 5.7f.

Figure 5.7: Douglas-Peucker algorithm for discretizing a curve

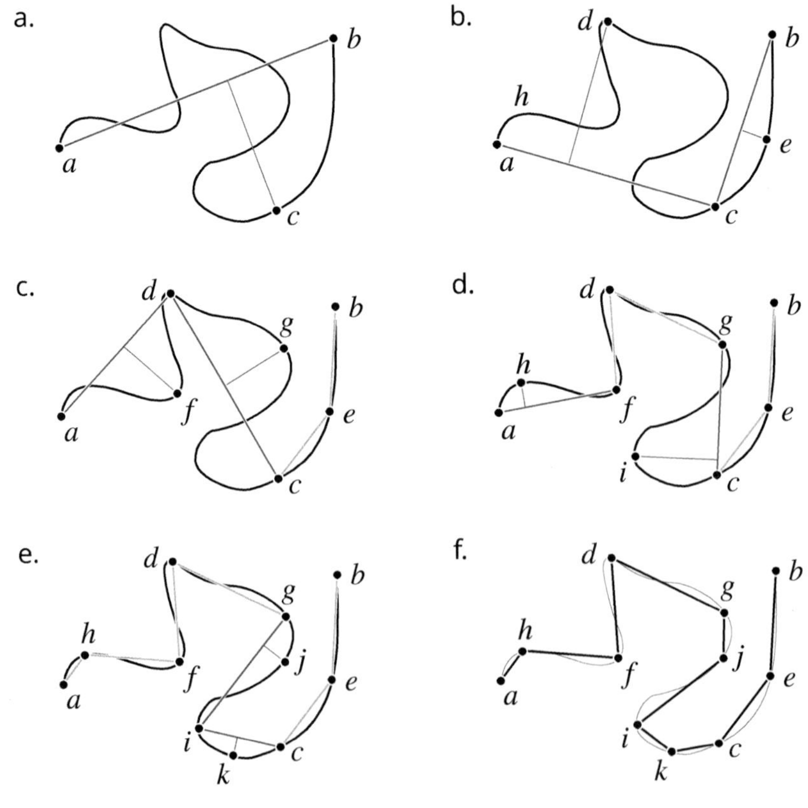

5.2 Representing spatial objects

This section considers some different approaches to representing the structure of spatial objects, often called "capturing the topology." The first representation—spaghetti—provides only a minimal level of topological information. We then progress to representations that handle topology more explicitly. Some of these representations mix topology with the spatial embedding.

5.2.1 Spaghetti

The *spaghetti* data structure represents a planar configuration of points, arcs, and areas. Geometry is represented as a set of lists of straight-line segments. Each such list is the discretization of an arc that might exist independently, or as part of the boundary of an area. There is no explicit representation of the topological interrelationships of the configuration, such as adjacency relationships between constituent areas. The term "spaghetti" is an expressive metaphor for this representation: lists of straight line segments forming arcs are akin to strands of spaghetti on a plate. spaghetti

An example is shown below, where the planar configuration of areas shown in Figure 5.8 is represented as spaghetti. Each of the polygonal areas is represented by its boundary loop. Each loop is discretized as a closed polyline. Each such polyline is represented as a list of vertices (points), each vertex being an extreme of a line segment in the polyline. For the example in Figure 5.8, the vertex lists are:

A: $[(3,10),(3.5,14),(3,15),(6,15.5),(12,15),(11,13),(11,9),(12,7),(9,8),(2,14)]$
B: $[(6,15.5),(12,15),(16,17.5),(17.5,17),(16,19),(10,20),(5,19),(3,15)]$
C: $[(12,7),(12,3),(15,1),(19,2),(20,5),(18,11),(17.5,17),(16,17.5),(12,15),(11,13),(11,9)]$
D: $[(9,8),(2,14),(0,4),(1,1),(5,0),(12,3),(12,7)]$

The spatial structure has thus been translated into a set of lists. Any geometrical configuration based upon points and lines embedded in the Euclidean plane may be represented in this way. The model is conceptually simple and useful for very basic geometric operations, such a graphical plotting on a screen or printer. However, the spaghetti model has a number of serious drawbacks:

1. The representation includes redundant duplication of data (e.g., the point $(12,7)$ is stored three times above), which is inefficient and can lead to inconsistencies (for example, if the location of the vertex $(12,7)$ needs to be changed, but one or two of the three copies are omitted from the update).
2. The representation only implicitly captures the closure of polygon loops, unless the first vertex is also repeated as the last vertex (further increasing duplication).

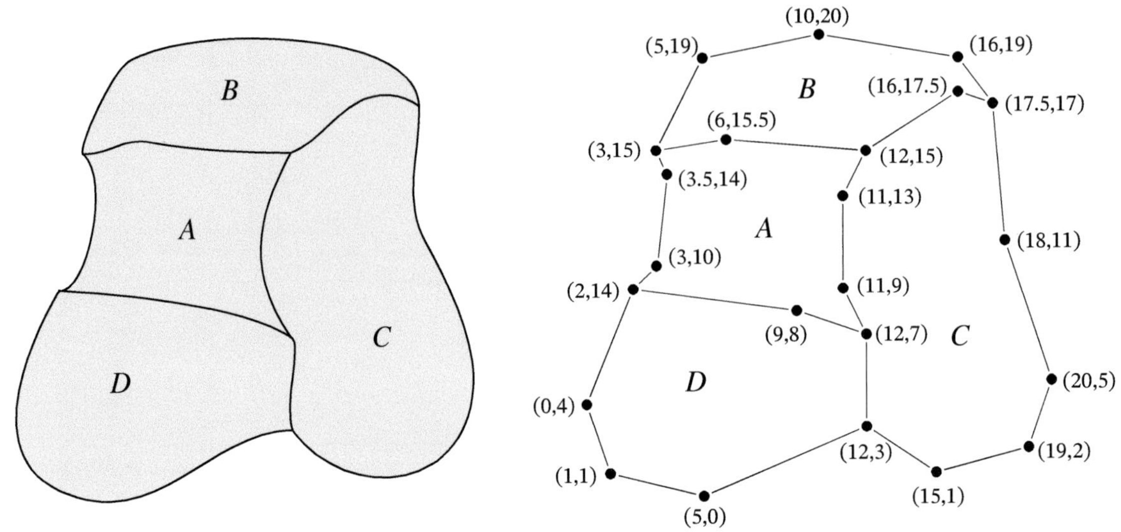

Figure 5.8: The spaghetti model discretization of four continuous regions, *A*, *B*, *C*, and *D*

3. The sequences are structured as cycles, so, for example, $[p_1, p_2, p_3, p_4] = [p_4, p_1, p_2, p_3] = [p_3, p_4, p_1, p_2] = [p_2, p_3, p_4, p_1]$. As a result there are many different representations of the same polygon, complicating some operations (such as testing for equality).
4. While spaghetti provides basic connectivity between points, few other spatial relationships are explicitly represented. For example, the fact that areas *A* and *B* are adjacent must be calculated by noting a common point sequence in their point lists.

Extensions of the spaghetti model can address some of these issues. For example, the vertex directory shown in Figure 5.9 augments the basic spaghetti data structure with two tables, one to list the raw coordinates of the vertices (Figure 5.9, left-hand side) and another to relate the vertices to the polygons (Figure 5.9, right-hand side). The approach removes the duplication in vertices (1 above) as well as offers the potential for conventions that yield a unique representation of each polygon (beginning to address 3 above). Indeed, the idea of such auxiliary data structures is an important one in GIS, and the focus of the next chapter on indexes and access methods.

However, the lack of topology (4 above) means such representations remain inefficient for whole classes of spatial analyses where spatial relationships are important. This leads us to the one of the most important ideas at the genesis of GIS: the node-arc-area model.

5.2.2 *Representing more topology*

The remainder of this section introduces two representations that address the key shortcoming of the spaghetti model, discussed above, and capture a richer variety of spatial relationships. Such representations are often called

Vertex	x	y
1	5	19
2	10	20
3	16	19
4	17.5	17
5	18	11
6	20	5
7	19	2
8	15	1
...	...	...

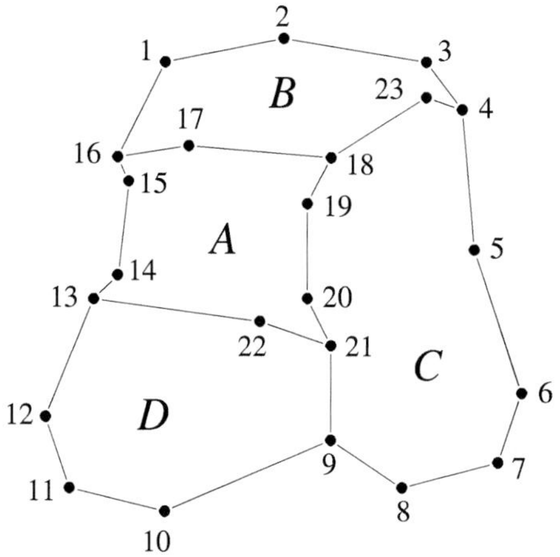

Polygon	Vertices
A	13, 14, 15, 16, 17, 18, 19, 20, 21, 22
B	1, 2, 3, 4, 23, 18, 17, 16
C	4, 5, 6, 7, 8, 9, 21, 20, 19, 18, 23
D	9, 10, 11, 12, 13, 22, 21

Figure 5.9: Vertex directory for the spaghetti model

"topological," to show that they are able to represent topological relationships, such as adjacency. However, this terminology is not very helpful, as even spaghetti is capable of expressing the topological connectivity relationship between adjacent vertices.

NAA

The next structure, called NAA for *node-arc-area*, explicitly represents the adjacency relationships between areas in a subdivision of a surface. The NAA model was one of the central innovations at the origins of GIS more than half a century ago. Using ideas drawn from the combinatorial map in Section 3.3.5, the primary constituent entities are (directed) **arc**, **node**, and **area**. As for the combinatorial map, an **area** is a face enclosed by arcs and homeomorphic to a cell (except for the external area). The system rules for NAA are:

1. Each directed arc has exactly one start and one end node.
2. Each node must be the start node or end node (maybe both) of at least one directed arc.
3. Each area is bounded by one or more directed arcs.
4. Directed arcs may intersect only at their end nodes.
5. Each directed arc has exactly one area on its right and one area on its left.
6. Each area must be the left area or right area (maybe both) of at least one directed arc.

Figure 5.10 captures the essential features of the NAA model in a UML E-R diagram. NAA shows clearly in its symmetry the duality between nodes and areas in a surficial subdivision. Each instance of **area** enters into a many-to-one relationship with the directed arcs that bound it on the left and right. An **arc** must bound exactly one **area** on its left and one on its right; and **area** must be bounded by at least one **arc**, but possibly many. Similarly, each **node** enters into a many-to-one relationship with the arcs for which it is the begin or end node. An **arc** must have exactly one begin and one end **node**; a **node** must begin or end at least one **arc**, but possibly many.

Further, the E-R model in Figure 5.10 captures the geometric embedding of nodes, arcs, and areas as points, polylines, and polygons (connecting up

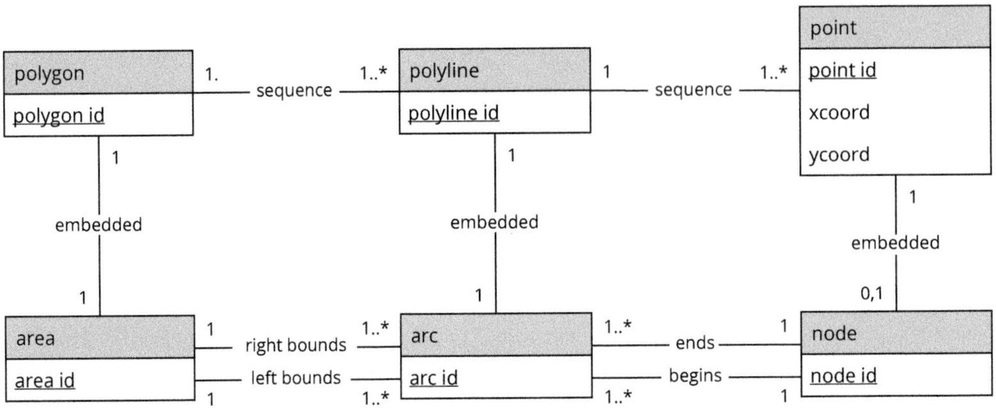

Figure 5.10: E-R diagram for the NAA representation

with the entities in the simple feature model in the previous chapter). Every **polygon** is an embedding of an **area** and is constituted as a sequence (cycle) of **polyline** entities. Each **polyline** is an embedding of a (directed) **arc** and is constituted as a sequence of **point** entities. Every **node** is a **point**, but not every **point** is a **node**, because points trace the embedding of arcs between nodes. We also require an x, y coordinate to pin down point positions to the surface.

Note that the E-R diagram in Figure 5.10 is not quite expressive enough to capture the nuance of system rules 2 and 6 above. An **area** might not have *both* a right arc **arc** *and* a left **arc** (e.g., consider the external area), but must have at minimum either one right arc **arc** *or* one left **arc**. The cardinality constraints in Figure 5.10 do not exactly capture this condition, as the E-R model lacks the ability to aggregate constraints over multiple different relationships. Similarly, the **begins** and **ends** relationships in Figure 5.10 do not exactly capture that a node must engage in a minimum of two relationships: either beginning and ending one or more arcs or beginning or ending at least two arcs (as the model disallows "dangling" nodes that end or begin a sole arc).

Nevertheless, the E-R diagram does capture the main features of the NAA model, and translates conveniently into the following five relation schemes for a relational database.

ARC (ARC ID, BEGIN, END, LEFT, RIGHT)
NODE (NODE ID, POINT_ID)
POLYGON (AREA ID, ARC ID, SEQUENCE_NO)
POLYLINE (ARC ID, POINT ID, SEQUENCE_NO)
POINT (POINT ID, X_COORD,Y_COORD)

The relation ARC is at the core of this relation scheme. Foreign keys are posted to the ARC relation from the (begin and end) NODE relation. The one-to-one relationship between **polygon** and **area** in Figure 5.10 means a potential AREA relation is not required. Instead, the POLYGON relation captures both the polyline sequence (posting the foreign key from POLYLINE to POLYGON) and the left and right bounding arcs (posting the foreign key from POLYGON to ARC). A POINT relation, distinct from NODE is required, however, in order to capture the optional participation of points in the embedding relationship with nodes as well as the coordinate attributes of the points themselves.

Figure 5.11 shows a decomposition of the configuration in Figure 5.8 into nodes, directed arcs, and areas ready for NAA representation. The constituent areas are labeled A, B, C, D. The external area, X, is necessary so that the system rules are satisfied. Arcs are labeled $a, \ldots, i$ and nodes are labeled $1, \ldots, 6$. Area A is bounded by four arcs a, e, h, and g. Arcs a and e have A as a left area, while arcs g and h have A as a right area. Node 1 has incident arcs a, b, and e. Arcs e and b have node 1 as an end node, while arc a has node 1 as a begin node.

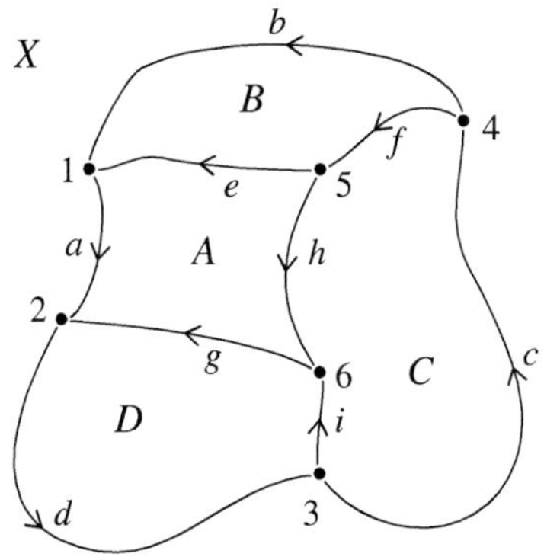

Figure 5.11: Decomposition of the planar configuration in Figure 5.8 into nodes, directed arcs, and areas

The corresponding relation for the example in Figure 5.11 is shown in Table 5.3. The further embeddings captured through the POINT, NODE, POLYLINE, and POLYGON relations are not given here, but they are constructed along similar lines to the vertex directory in Figure 5.9.

ARC ID	BEGIN	END	LEFT	RIGHT
a	1	2	*A*	*X*
b	4	1	*B*	*X*
c	3	4	*C*	*X*
d	2	3	*D*	*X*
e	5	1	*A*	*B*
f	4	5	*C*	*B*
g	6	2	*D*	*A*
h	5	6	*C*	*A*
i	3	6	*D*	*C*

Table 5.3: Relation corresponding to areal configuration in Figure 5.11

The duality between node and area leads to a pleasing symmetry in the NAA representation. The symmetry is only marred by the existence of the specially distinguished area that is the external (X in Figure 5.11): there is no such specially distinguished node. However, if we change our embedding space to be the surface of the *sphere* (the perfect 2-space?), then the need for an external area no longer exists and we have perfect symmetry! So much for mysticism, back to computation.

The NAA representation gives a more complete topological description of a planar object than the spaghetti model. However, the NAA representation falls short of a full topological representation in at least two ways.

1. The more detailed connectivity of the object is not explicitly given. Thus there is no explicit representation of weak, strong, or simple connectedness.
2. The representation is not *faithful*, in the sense that two different topological configurations may have the same representation.

For example, consider the weakly connected cellular arrangement of shape *A* in Figure 5.12. Is it a disk, *D*, with an ellipse, *E*, removed or the union of two lunes, *B* and *C*?

Figure 5.12: Ambiguity in representing a planar object

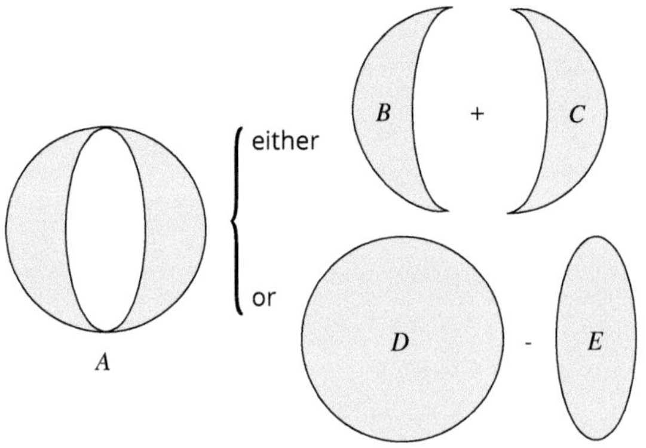

The NAA model is unable to distinguish these two topologically distinct possibilities. For such further refinement, we will need a fuller implementation of the combinatorial map as the DCEL.

5.2.3 *Doubly connected edge list*

DCEL

The *doubly connected edge list* (DCEL) allows further capture of topology in surficial configurations, providing a complete representation of the topology of a connected planar graph. The DCEL provides in an extended arc table the information necessary to construct:

- the sequence (cycle) of arcs around the node for each node in the configuration; and
- the sequence (cycle) of arcs around the area for each node in the configuration.

The word "sequence" is important here. The earlier NAA representation will give the sets of arcs that bound an area or meet at a vertex, but the sequencing cannot be determined unless the embedding is also given. Further, even with the embedding, NAA cannot distinguish certain topological details, such as the ambiguous configuration in Figure 5.12.

The E-R diagram for the DCEL, shown in Figure 5.13, is an extension of the NAA representation (Figure 5.10). Two new relationships have been added: every arc has a unique next arc and a unique previous arc. Given an arc a, to find the previous arc to a, select the begin node (say, 1) of a and starting at a proceed around 1 in an counterclockwise direction until the preceding arc is found (e). To find the next arc to a, proceed around the end node of a, 2, in an counterclockwise direction starting at a, until the next arc is found (d).

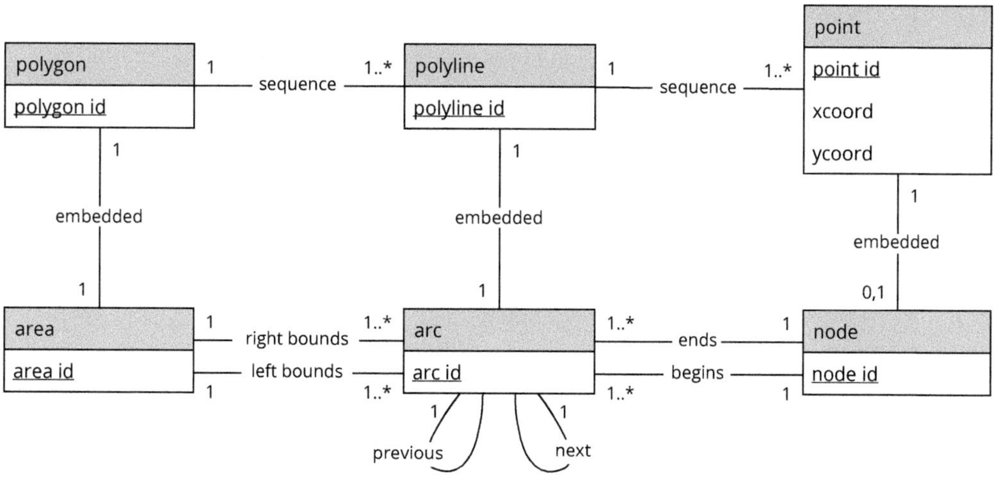

Figure 5.13: E-R diagram showing the data structure of the DCEL, extending Figure 5.10

The relationships **previous** and **next** are shown in Figure 5.14. An example of a DCEL representation is given in Table 5.4, showing the configuration from Figure 5.11. Here we omit the details of the actual embedding (i.e., polygons, polylines, and coordinates of points) and focus solely upon the topological relationships embodied in the entities node, arc, and area captured in the DCEL relation, extending the previous ARC relation.

Algorithm 5.1 gives the procedure for computing the counterclockwise sequence (cycle) of arcs surrounding node n for a DCEL. The input to the algorithm is the node_id n. The algorithm then arbitrarily selects an arc x that is incident with n (line 1). For the relation DCEL in Table 5.4, we might specify this step as an expression in the relational algebra. For example:

$$\pi_{\text{ARC\_ID}}(\sigma_{\text{BEGIN}=n \text{ OR END}=n}(\text{DCEL}))$$

returns the relation containing all arcs incident with node n (see Chapter 2 for an explanation of relational algebra). The algorithm stores this initial arc

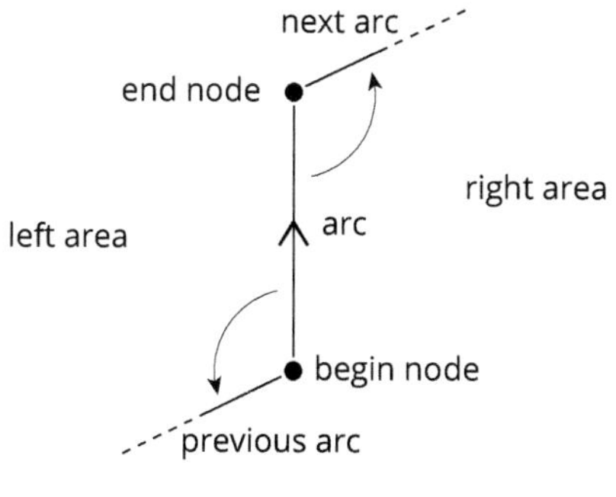

Figure 5.14: Relationships to a single arc in the DCEL

x as the first in a sequence (indicated with the assignment operator $\leftarrow$ in line 3 of Algorithm 5.1), then finds the arc that is counterclockwise from x (lines 6–9) using relational algebra to retrieve the relevant data items in lines 6, 7, and 9. The process reiterates, storing each newly discovered arc in sequence (line 5), until we return to the original arc x (line 10).

Algorithm 5.1: Computing the sequence of arcs surrounding node n

Input Node n

1: find some arc x which is incident with n
2: initialize empty queue Q
3: arc $\leftarrow x$
4: **repeat**
5: add arc to queue Q
6: **if** $n \in \pi_{\text{BEGIN}}(\sigma_{\text{ARC\_ID}=\text{arc}}(\text{DCEL}))$ **then**
7: arc $\leftarrow \pi_{\text{PREV}}(\sigma_{\text{ARC\_ID}=\text{arc}}(\text{DCEL}))$
8: **else**
9: arc $\leftarrow \pi_{\text{NEXT}}(\sigma_{\text{ARC\_ID}=\text{arc}}(\text{DCEL}))$
10: **until** arc $= x$

Output Counterclockwise sequence of arcs Q around n

To illustrate, consider applying Algorithm 5.1 to node 6 in Figure 5.11 (Table 5.4). The algorithm might select arc i as the initial arc (line 1). The first iteration of the algorithm would then store i in queue Q (line 5). Node 6 is the end node for i (Table 5.4). Hence, line 6 will evaluate to false and the variable arc will be updated in line 9 with arc h. After storing h the algorithm iterates to find the next arc g. Finally, g is stored, and the algorithm terminates because the previous arc for g is i, which completes the sequence.

In a similar way, the clockwise sequence (cycle) of arcs around an area X may be computed by modifying Algorithm 5.1 as in Algorithm 5.2. For example, consider the clockwise cycle of arcs around area C in Figure 5.11 (Table 5.4). If the left area for arc c is arbitrarily selected to initialize variable arc (line 1), then on the first iteration of Algorithm 5.2, arc c is stored in Q (line 5). Using SQL this time, instead of relational algebra, the condition on line 6 will evaluate to true, resulting in the selection arc i as the previous arc from c (line 7). The next iteration stores i and selects h as the next arc, which in turn is stored and used to select the arc f. Finally, f is stored. The algorithm

ARC ID	BEGIN	END	LEFT	RIGHT	PREV	NEXT
a	1	2	*A*	*X*	*e*	*d*
b	4	1	*B*	*X*	*f*	*a*
c	3	4	*C*	*X*	*i*	*b*
d	2	3	*D*	*X*	*g*	*c*
e	5	1	*A*	*B*	*h*	*b*
f	4	5	*C*	*B*	*c*	*e*
g	6	2	*D*	*A*	*i*	*a*
h	5	6	*C*	*A*	*f*	*g*
i	3	6	*D*	*C*	*d*	*h*

Table 5.4: DCEL relation representing the configuration in Figure 5.11

then terminates because *c* is the previous arc from *f*, which completes the sequence *cihf*.

Algorithm 5.2: Computing the sequence of arcs surrounding area X

```
Input Area X
find some arc x which bounds X
initialize empty queue Q
arc ← x
repeat
    add arc to queue Q
    if X ∈ SELECT LEFT FROM DCEL WHERE ARC_ID= arc then
        arc ← SELECT PREV FROM DCEL WHERE ARC_ID= arc
    else
        arc ← SELECT NEXT FROM DCEL WHERE ARC_ID= arc
until arc = x
Output Clockwise sequence of arcs Q around X
```

The DCEL representation provides a complete and faithful description of the topology of strongly connected planar configurations. Dealing additionally with weakly connected planar objects, such as that in Figure 5.12, requires the addition of one further table to identify the connected *components* in the configuration. The additional table is used to specify which of the two possible interpretations in Figure 5.12 is the required one, explicitly identifying as components either B and C or D and E. There exist a number of further related representations of planar areal objects, two of which are mentioned briefly in Box 5.2 on the next page.

5.3 *Representing fields*

The overall development of the last section steered a path in the direction of ever more comprehensive representations of spatial objects embedded in the plane. These representations explicitly hold information about topological relationships between constituent parts of spatial objects. We now change course in order to describe an important class of representations based upon tessellated structures, for which topological relationships are often implicit.

tessellation

A *tessellation*, also called tiling, is a partition of the plane, or a portion of the plane, as the union of a set of disjoint areal objects. If these constituent objects are all exact copies of the same regular polygon and each of the vertices is also regular (in a sense shortly to be explained), then the tessellation is *regular*; otherwise, the tessellation is *irregular*.

Tessellations provide an important class of *spatial framework* for field-based models, introduced in the previous chapter. The most important class of regular tessellated representations is the grid or *raster* representation, based upon a tessellation of squares. The irregular tessellated representations of most interest are TINs (triangulated irregular networks). Triangles are the

Box 5.2: Alternative spatial object representations

Spaghetti, NAA, and DCEL are three of the most common representations of planar spatial objects. However, several other representations exist, particularly tailored to specialized domains. The *winged-edge* representation (Baumgart, 1975) is a variant of the DCEL. It may be used to represent subdivisions of orientable surfaces (i.e., surfaces for which the notion of "clockwise" is well defined). Unlike the DCEL, where edge orientation is explicitly stored as begin and end nodes, in a winged-edge representation orientation is implicit within the data structure. For a winged-edge representation, each edge is associated with four other edges: the next edge clockwise, the previous edge clockwise, the next edge counterclockwise, and the previous edge counterclockwise. The basic winged-edge representation does not allow loops and isthmuses, but it can be extended to handle such features. The *quad-edge* representation (Guibas & Stolfi, 1985) allows orientable or non-orientable surfaces to be represented (i.e., the notion of "clockwise," invariant over the surface, is not required). The quad-edge representation is generally more complicated than the winged-edge representation. The main advantage of quad-edge over winged-edge is that the *dual graph* (see Chapter 3) is automatically embedded within the quad-edge representation.

simplest discrete areal objects that populate the Euclidean plane, and so it is interesting to use them as the primitive areal elements in representations.

5.3.1 *Regular tessellated representations*

regular polygon

vertex figure

regular tessellation

A tessellation of a surface is a covering of the surface with an arrangement of non-overlapping polygons. A *regular polygon* is a polygon for which all edges have the same length, and all internal angles are equal. At each vertex of a tessellation, the *vertex figure* is the polygon formed by joining in order the mid-points of all edges incident with the vertex. A tessellation of a surface for which all the participating polygons and vertex figures are regular and equal is termed a *regular tessellation*. For example, a regular tessellation of equilateral triangles has a regular hexagon for its vertex figure.

Of the regular tessellations (see Box 5.3 on page 186), by far the most commonly used for spatial representations is the square grid. This provides the raster representation of spatial data, whereby planar spatial configurations are decomposed into a pattern of squares (*pixels*) on a grid. Regular triangular and hexagonal tessellations are less frequently used for planar data representation, because the regular tessellation representation based upon squares fits well with standard and well-supported programming data types, such as 2D arrays.[2] Nested regular triangular tessellations have been used for representing spherical data (see Chapter 6). However, despite their simplicity and elegance, regular tessellations do not necessarily accord with the way that the data has been collected, often with measurement locations arranged irregularly through the region. In this case irregular tessellations are more appropriate, and such representations are now considered.

[2] Note, though, that the regular hexagonal tessellation is becoming more widely supported by GIS, and offers a number of modeling advantages, including a more regular neighborhood (i.e., 6 neighbors each sharing a single edge, in contrast to 4 edge and 4 vertex neighbors in a square raster) and a shorter perimeter than a square of the same area (which can reduce edge effects in models).

5.3.2 *Irregular tessellated representations*

irregular tessellation

TIN

An *irregular tessellation* is a tessellation for which the participating polygons are not all regular and equal. The most commonly used irregular tessellation is the TIN—triangulated irregular network. A TIN may represent the variation of a field function over a spatial framework. Its irregularity allows the resolution to vary over the surface, capturing finer details where required. A useful concept in what follows is again the notion of duality of planar graphs, discussed in Section 3.4.1, where faces become nodes, and nodes become faces. If we label each triangular face of the TIN with a node and join with edges those faces that are adjacent in the TIN, then the planar mesh that results is the *dual* of the original mesh. If T^* is the dual of the TIN T, then the degree of each node (the number of edges incident with that node) in T^* must be three.

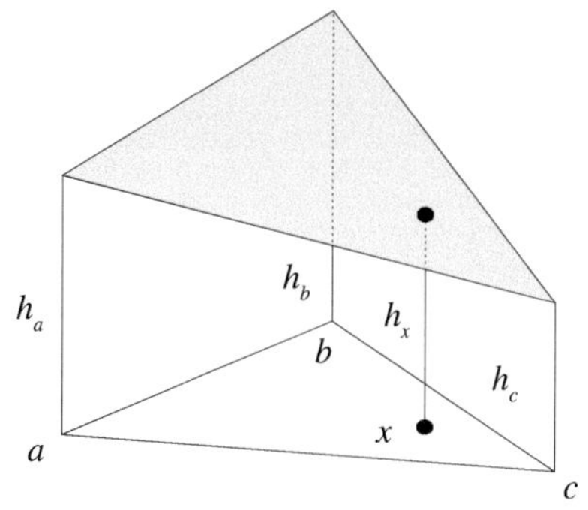

Figure 5.15: Interpolation is required to find the height of point X

The height at any point on a TIN can be determined using linear interpolation. Figure 5.15 shows the linear interpolation process working for a single general triangular facet of the surface. The triangular facet abc is part of a TIN. The coordinates (in vector form) $\vec{a}$, $\vec{b}$, and $\vec{c}$ of the vertices are assumed known, as are the heights h_a, h_b, and h_c, at a, b, and c, respectively. The problem is to estimate the height h_x of the surface at point x, the coordinates of which are known to be $\vec{x}$. Point x is inside or on the boundary of the triangle abc, so we have the following relationship:

$$x = \alpha\vec{a} + \beta\vec{b} + \gamma\vec{c}$$

where α, β, and γ are scalar coefficients that can be uniquely determined, such that:

$$\alpha + \beta + \gamma = 1$$

Assuming discretization into planar facets (that is, first-order linear surfaces as in Figure 5.15), the height h_x can now be found by using the equation:

$$h_x = \alpha h_a + \beta h_b + \gamma h_c$$

5.3.3 Delaunay triangulations and Voronoi diagrams

Given a set of irregularly spaced points within a region, there remains the problem of constructing the tessellation of irregular triangles, the vertices of which are the given set. In general, there will be many candidate tessellations, and some may be judged better than others. Figure 5.16 shows two possible

Box 5.3: Regular tesselations

Tessellations have been used for hundreds of years in the decorative tiling of the floors, walls, and ceilings of buildings. In fact, *tiling* is an alternative term for tessellation. In the 17th century, the mathematician and astronomer Johannes Kepler was the first person to systematically study the geometry of tessellations. Kepler observed that in the case of the Euclidean plane, only three regular tessellations are possible, based upon equilateral triangles, squares, and regular hexagons (illustrated below). We can show that no other regular tessellations are possible by noticing that in order to tessellate, the interior angles of the regular polygon at each vertex of a tessellation must be a factor of 360°. Triangles (interior angle 60°), squares (interior angle 90°), and hexagons (interior angle 120°) are the only regular polygons that satisfy this criterion.

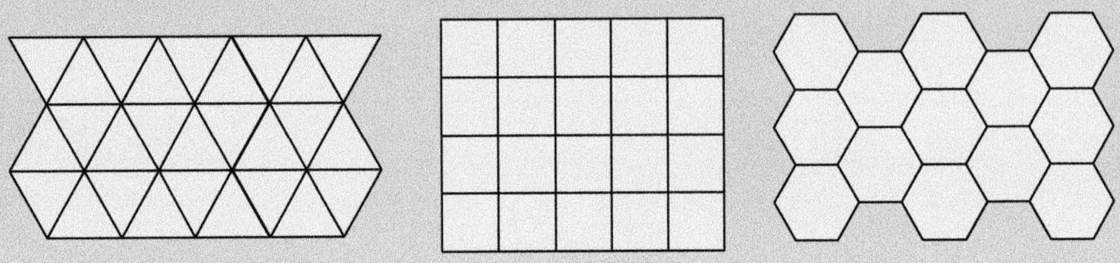

triangulations based upon the same set of vertices. The right-hand triangulation has a higher proportion of long, thin triangles and so may be judged inferior for some purposes.

Figure 5.16: Two triangulations based upon the same vertex set

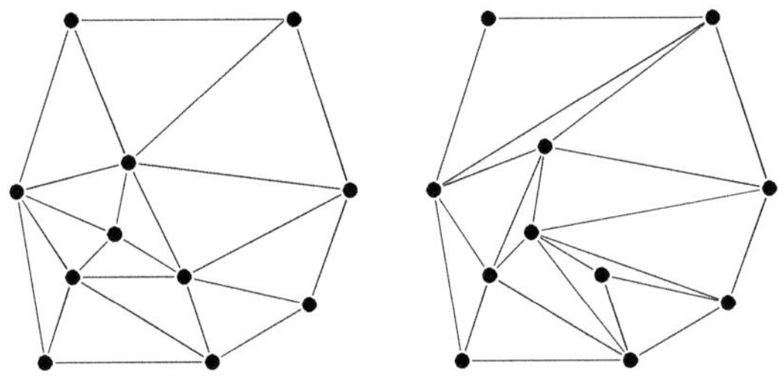

Delaunay triangulation

A triangulation that has many desirable properties, both for itself and for its dual, is the *Delaunay triangulation*. Loosely, it can be said that the constituent triangles in a Delaunay triangulation are "as nearly equilateral as possible." The dual of a Delaunay triangulation is a *Voronoi diagram* (or *Thiessen polygons*). Delaunay triangulations can be introduced by considering their dual, Voronoi diagrams.

Voronoi diagram

Thiessen polygon

Imagine that we are given the point locations of fire emergency response units in an area. Suppose that we wish to surround each response point p by a region R_p which has the property that every location in R_p is nearer to p than to any other response unit. Suppose for simplicity that distance is measured "as the crow flies" using the usual Euclidean metric. Then it turns out that these areas of closest proximity, R_p, are polygons, termed *proximal polygons*. The set of proximal polygons constitutes a Voronoi diagram. Given a set of points with the property that no three are collinear (to avoid degenerate cases), a Voronoi diagram surrounds each point with an area that contains all the locations that have that "seed" point as their nearest.

proximal polygon

The Delaunay triangulation is now simply formed as the dual of the Voronoi diagram. Points in the Voronoi diagram become vertices in the Delaunay triangulation. Edges in the Delaunay triangulation connect adjacent proximal polygons in the Voronoi diagram (Figure 5.17). A Delaunay triangulation has the property that each *circumcircle* of a constituent triangle does not include any other triangulation point within it. This property is illustrated in Figure 5.18.

circumcircle

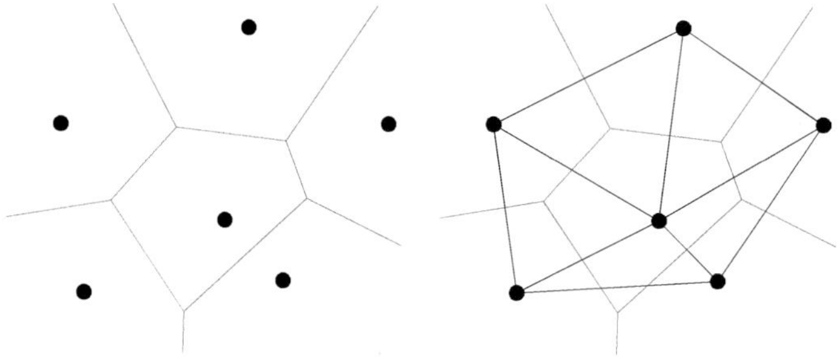

a. Voronoi diagram   b. Delaunay triangulation

Figure 5.17: Voronoi diagram of proximal polygons and its dual Delaunay triangulation

The list below gives some of the more striking properties of the remarkable Delaunay construction. Given an initial point set P for which no sets of three points are collinear (again to avoid degenerate cases):

1. The Delaunay triangulation is unique.
2. The external edges of the triangulation form the convex hull of P (i.e., the smallest convex set containing P).
3. The circumcircles of the triangles contain no members of P in their interior. Note that this is a defining property, in that if it holds, then the triangulation must be a Delaunay triangulation.
4. The triangles in a Delaunay triangulation are best-possible with respect to regularity (closest to equilateral).

Property 4 is vague and needs some explanation. Suppose that we have a triangulation (not necessarily Delaunay) and make a list of the minimum angles in each of the constituent triangles in such a way that the list starts with the smallest angle and moves along in sequence to the largest angle. Call this list the *ordered minimum angle vector* for the triangulation. We may com-

ordered minimum angle vector

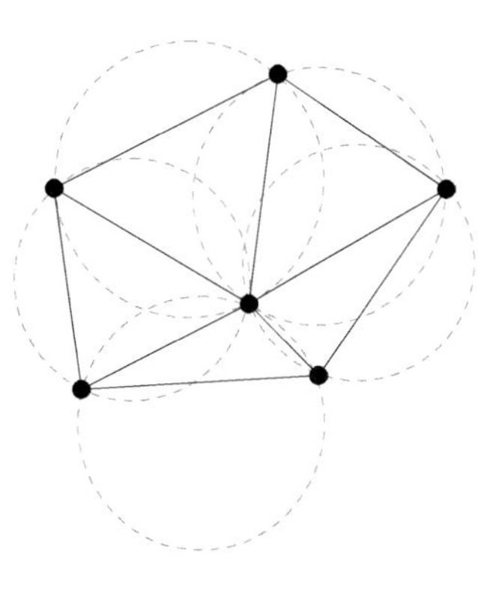

Figure 5.18: Circumcircles of a Delaunay triangulation

pare two such vectors *lexicographically* as if they were ordered in a telephone directory. Now Property 4 can be made more precise as follows:

4′. The triangles in a Delaunay triangulation have the greatest ordered minimum angle vector of any triangulation of P.

A Delaunay triangulation has the greatest possible ordered minimum angle vector (property 4′), so it will contain the smallest collection of long, thin triangles. Informally, this implies that the Delaunay triangulation is "as regular as possible" (i.e., property 4).

5.3.4 *Triangulations of polygons*

In general, a triangulation tessellates an unbounded region with triangles based upon a prescribed set of vertices. This section describes several approaches to the triangulation of a bounded polygonal region. It would be nice to modify the Delaunay triangulation for an unbounded region, but if we take the collection of points comprising the vertices of a polygon and form the Delaunay triangulation based upon them, this will not necessarily be the triangulation of the polygon, because some of the edges of the polygon may not be edges in the triangulation. However, it is possible to use the *constrained Delaunay triangulation*, which is constrained to follow a given set of edges.

constrained Delaunay triangulation

Figure 5.19 shows a simple example. The Delaunay triangulation is as in Figure 5.19a. The triangulation in Figure 5.19b has been constrained by a polygon (boundary highlighted). The constrained triangulation includes the edge ab, which is not part of the Delaunay triangulation in Figure 5.19a.

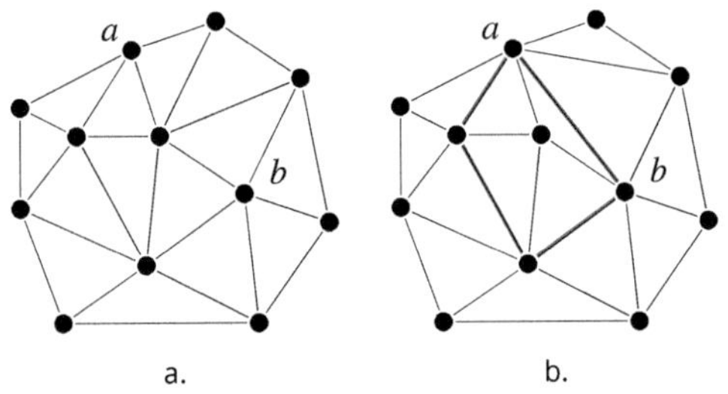

Figure 5.19: Unconstrained and constrained Delaunay triangulations

Apart from Delaunay methods, there are many other methods for triangulations of sets of points and polygons (see bibliographic notes). For example, the *greedy triangulation* has the objective of minimizing the total edge length in the triangulation. This is achieved by introducing the shortest possible internal diagonal at each stage. Figure 5.20 shows the greedy and Delaunay triangulations of a simple quadrilateral. Algorithms for triangulations are considered later in the chapter.

5.3.5 *Medial axis of a polygon*

medial axis

The *medial axis* of a polygon is the Voronoi diagram computed for the line segments that make up the boundary of that polygon. For example, Figure 5.21 shows the medial axis (thick lines) of the boundary of the polygon (thin lines).

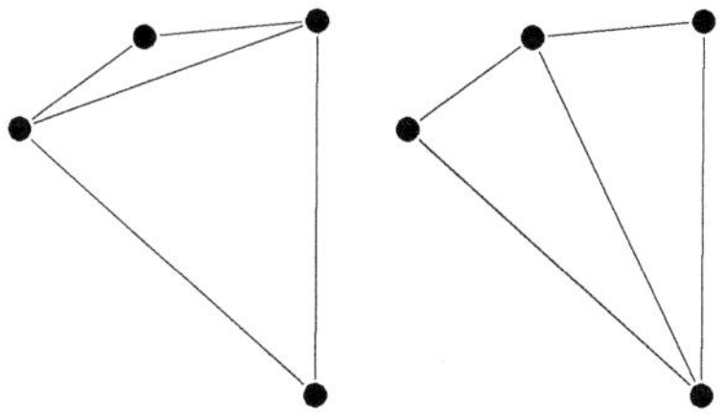

Figure 5.20: "Greedy" (left) and Delaunay (right) triangulations of a simple polygon

The medial axis is used within a variety of algorithms, some of which are mentioned later on, and is sometimes referred to as the *skeleton* of a polygon. The medial axis for a convex polygon is composed of straight-line segments, but in general it may comprise arcs of a parabola, as in Figure 5.21. The medial axis of a polygon can be computed in linear time using a *medial axis transform* (MAT), based on an extension of the algorithm for computing the Voronoi diagram.

skeleton

medial axis transform

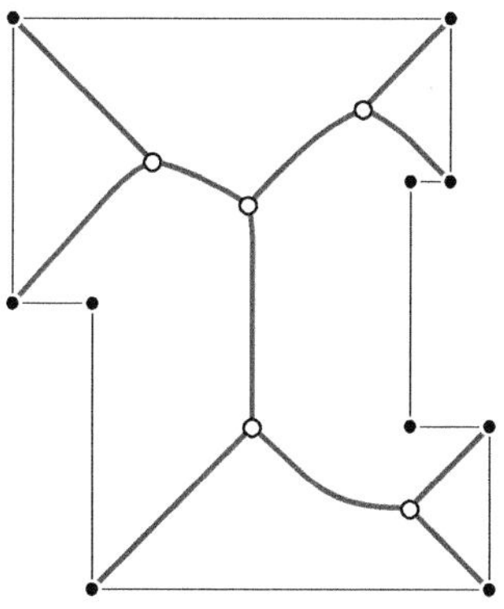

Figure 5.21: Medial axis of a polygon

5.3.6 *Tessellations of the sphere*

The representations described so far have all been referenced to planar objects. Of course, the Earth is topologically a sphere, and so it is natural to consider representations appropriate for spherically referenced data. The reasons that so much emphasis has been placed upon the plane are:

- Planar data is computationally simpler than the sphere.
- Small areas of the Earth's surface may be approximated by planes.

Regular tessellations of the sphere correspond to the five Platonic solids of antiquity: tetrahedron, cube, octahedron, dodecahedron, and icosahedron. For example, the tetrahedral tessellation comprises four spherical triangles (called *triangular facets*) bounded by parts of great circles. Each spherical triangle has internal angle of 120°. Three triangles meet at each vertex. In the case of the octahedral tessellation, there are initially eight triangular facets, each having an internal angle of 90°. Four triangles meet at each vertex. Figure 5.22 shows an octahedral tessellation on the sphere, having triangular facets ***nab***, ***nbc***, ***ncd***, ***nda***, ***sab***, ***sbc***, ***scd***, ***sda***. Unfortunately, unlike the plane, regular tessellations of arbitrary fine resolution are not possible on the sphere.

triangular facets

Box 5.4: Nested tessellations

A *nested tessellation* is a tessellation the cells of which are themselves partitioned using a further and finer tessellation. This process can be recursive, providing a graded granularity to which spatial phenomena may be referenced. *Regular nested tessellations* occur where the regular constituents are themselves partitioned using the same regular figure. Examples of planar regular and irregular nested tessellations are shown below. The first two (triangle and square) are regular, but the third (hexagon) is not, because hexagons cannot fit together to make a larger hexagon (see Peuquet, 1984). The nested square tessellation of the plane leads to the quadtree data structure, discussed in Chapter 6. The nested triangular tessellation may be used to provide the nested octahedral tessellation of the sphere, also considered in Chapter 6. Returning to the plane, there is a distinction between *hierarchical* models, based upon recursive regular or irregular tessellations and *stratified* models in which each stratum is defined to have particular properties. A survey of hierarchical models is given by De Floriani, Marzano, & Puppo (1994a) with particular emphasis on hierarchical TIN in De Floriani & Puppo (1992). A stratified Delaunay triangulation, where the circumcircle property is satisfied at each level, is defined in the *Delaunay pyramid* of De Floriani (1989). Stratified models are discussed in general by Bertolotto, De Floriani, & Marzano (1994).

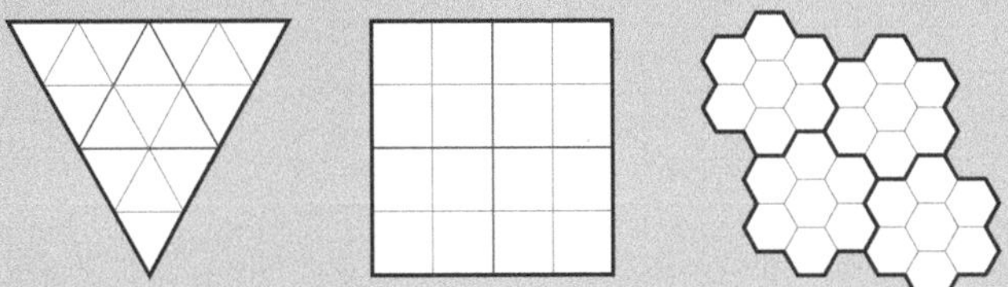

nested tessellation

However, it is possible to recursively *nest* regular polygons within the facets of the Platonic solids to produce the resolutions required (see Box 5.4).

5.3.7 Vectorization

rasterization

scan conversion

vectorization

Having introduced object and field representations, we briefly explore the conversion between the two. The conversion from vector to raster, termed *rasterization*, is typically straightforward. Indeed, displaying vector data on a computer screen, or taking a photo of a vector data set, is effectively a form of rasterization. This topic is known in the graphics literature as *scan conversion*.

However, conversion from raster to vector format, termed *vectorization* is more challenging because the structure of spatial objects is not explicitly represented in the raster data format. Consequently, most vectorization systems are semi-automatic, requiring human input either to guide an automatic system or more often to help train a machine learning algorithm (returned to in Chapter 9 on AI and GIS).

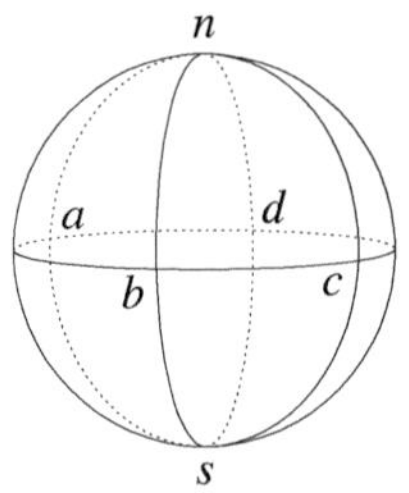

Figure 5.22: Octahedral tessellation of the sphere

A simple semi-automated vectorization process may be summarized as:

1. Thin lines so that they are one pixel in width. An example *thinning* algorithm is given below. Other thinning algorithms make use of the medial axis of a region (Section 5.3.5). thinning
2. Transform the thinned raster image into a collection of chains of pixels each representing an arc, termed *chain coding* (see also Box 6.2 on page 235). chain coding
3. Transform each chain of pixels into a sequence of vectors. In this final stage there is a balance to be struck between line smoothness and conversion accuracy. Vectors that follow the arc chains accurately will often be jagged, exhibiting a "staircase" effect. Smoothing vectors will lead to greater deviations from the original raster.

A commonly used thinning algorithm originated with Zhang & Suen (1984). The approach progressively erodes pixels at the boundary of a raster region, while attempting to maintain shape and connectivity. Looking at each pixel p in the region, p is marked for erosion if:

- between two and six neighbors of p are also in the region;
- cycling around p there is just one transition from in to out of the region;
- on every even iteration, at least one of the south and east neighbors or both the north and west neighbors are outside the region;
- on every odd iteration, at least one of the north and west neighbors or both the south and east neighbors are outside the region.

The thinning algorithm terminates when there are no more pixels marked for erosion. The top line of Figure 5.23 shows an example of three successive stages of erosion. Chain-coding then converts the thinned pixels into arcs. For example, *Freeman chain coding* uses the numbers 0 to 7, arranged clockwise around the eight directions from N = 0 through to NW = 7. Starting at a one end of a strand (i.e., a terminating pixel with only one neighbor in the region) and then following the pixels along the line, arcs can be created that aggregate chains of adjacent neighbors with the same direction encoding. Figure 5.23d shows the result of chain encoding our example as a vector. Freeman chain coding

Finally, vector reduction simplifies the geometry according to the desired level of accuracy. A long, "wiggly" line will of course need more vectors than a straighter line. The Douglas-Peucker algorithm (Section 5.1.4) is often used for the vector-reduction process.

5.4 Geometric algorithms

It would be impossible to describe in this single section all the spatial algorithms that have been devised. The aim of this section is to provide a representative sample, giving preference to commonly used algorithms and those that show an example of a generic approach. The representations introduced in earlier sections will be used in the algorithms.

Figure 5.23: Rasterization example, using Zhang-Suen thinning (a–c), chain coding (d), and smoothing (e)

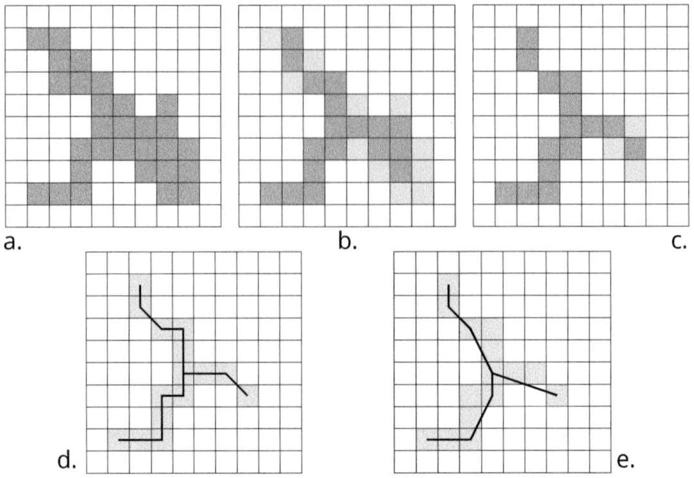

All of the algorithms described below assume an embedding in the Euclidean plane. Thus point objects have a position that may be given by a coordinate pair (x, y). The set of points on a straight line in the Euclidean plane may be constrained by a linear equation to give $\{(x, y) \mid ax + by + c = 0\}$, where a, b, c, are constants. The set of points on a straight-line segment between two distinct points p and q may also be represented in parametric form as $\{\lambda\vec{p} + (1 - \lambda)\vec{q} \mid \lambda \in [0, 1]\}$ (see Chapter 3).

5.4.1 *Metric and Euclidean algorithms*

Distance and angle between points For ease of reference, we restate these formulas from Chapter 3. The (Euclidean) distance $|pq|$ between points $p = (x_p, y_p)$ and $q = (x_q, y_q)$ is given by the formula:

$$|pq| = \sqrt{(x_q - x_p)^2 + (y_q - y_p)^2}$$

The bearing, θ, of q from p is given by the unique solution in the interval $[0, 360[$ of the simultaneous equations:

$$\sin\theta = \frac{x_q - x_p}{|pq|}$$
$$\cos\theta = \frac{y_q - y_p}{|pq|}$$

Distance from point to line Distance is meant here as the *minimum* distance between spatial objects. To compute the distance between a point and a straight line, the most compact formula arises when the straight-line l is given in the form $\{(x, y) \mid ax + by + c = 0\}$. Suppose that the point p is given by the coordinate pair (x_p, y_p). Then the distance from p to l, measured by the length of the line segment through p and orthogonal to l, is given by the formula:

$$\text{distance}(p, l) = \frac{|ax_p + by_p + c|}{\sqrt{a^2 + b^2}}$$

For the distance between a point p and a straight-line *segment* l, the calculation is no longer quite so straightforward. Figure 5.24 shows that the distance will be measured as the distance from p to one of the end-points of l or from p to the line incident with the end-points of l, depending on the relative positions of p and l. The line segment l defines a partition of the plane into two point sets, a connected set that we have called **middle** (l) and a disconnected set called **end** (l). The calculation to determine the distance of p from l depends upon whether p is in **middle** (l) or **end** (l).

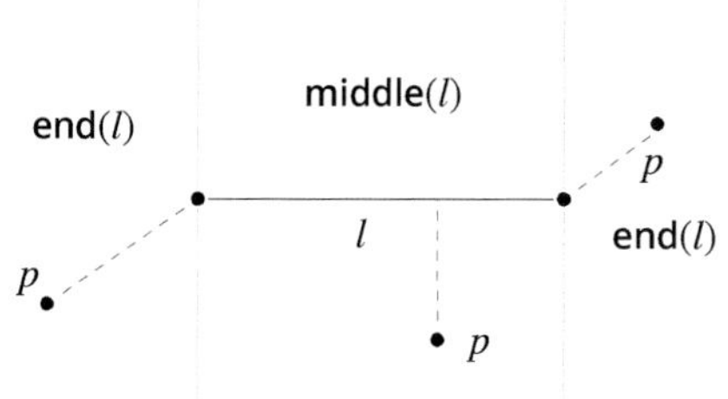

Figure 5.24: Distance between a point p and line segment l

The problem becomes more complex for a polyline, as can be seen from Figure 5.25. First, the distance to each line segment in the polyline must be calculated, as above. The distance from p to the polyline will then be the minimum of all these distances. If the number of segments in the chain is n, then the time complexity of the computation is $O(n)$. An approximation to the distance may be achieved by considering only the distances to vertices of the chain. This approximation will be good in general if the segment lengths are small compared with the distance of the point from the polyline.

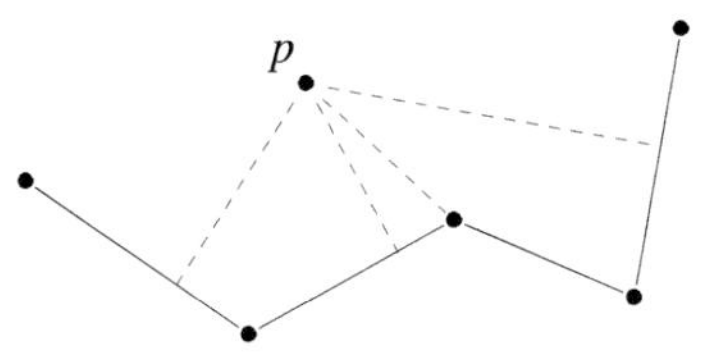

Figure 5.25: Distance between a point and a polyline

Distances from a point to a polygon or between polygons may be required in terms of their boundaries. Thus the distance between two polygons may be interpreted as the distance between their closest points. In that case, a calculation involving boundary polylines may be necessary. More usually, the distance between two polygons is interpreted as the distance between their centroids (discussed in a later section).

Area of a simple polygon Let P be a simple polygon (no boundary self-intersections) with vertex vectors $(x_1, y_1), (x_2, y_2), ..., (x_n, y_n)$ where $(x_1, y_1) = (x_n, y_n)$ (i.e., the polygon is closed and has the same start and end vertices). Then a formula for the area is:

$$\textbf{area}(P) = \frac{1}{2} \sum_{i=1}^{n-1} x_i y_{i+1} - x_{i+1} y_i$$

In the case of a triangle *pqr*, where $p = (x_p, y_p)$, $q = (x_q, y_q)$, $r = (x_r, y_r)$, the formula gives:

$$\textsf{area}(pqr) = \frac{x_p y_q - x_q y_p + x_q y_r - x_r y_q + x_r y_p - x_p y_r}{2}$$

Notice that the area calculated has a sign (positive or negative). Also, according to this formula $\textsf{area}(pqr) = -\textsf{area}(qpr)$. If point p is to the left-hand side of the directed segment qr, then $\textsf{area}(pqr)$ will be positive; if point p is to the right-hand side of the directed segment qr, then $\textsf{area}(pqr)$ will be negative; and if p, q, r are collinear, then $\textsf{area}(pqr)$ will be zero. Thus a useful spin-off from our work on area is an algorithm for the **side** operation (Figure 5.26):

$$\textsf{side}(p,q,r) = \begin{cases} 1 & \text{if } \textsf{area}(pqr) > 0 \quad (p \text{ is left of } qr) \\ 0 & \text{if } \textsf{area}(pqr) = 0 \quad (pqr \text{ are collinear}) \\ -1 & \text{if } \textsf{area}(pqr) < 0 \quad (p \text{ is right of } qr) \end{cases}$$

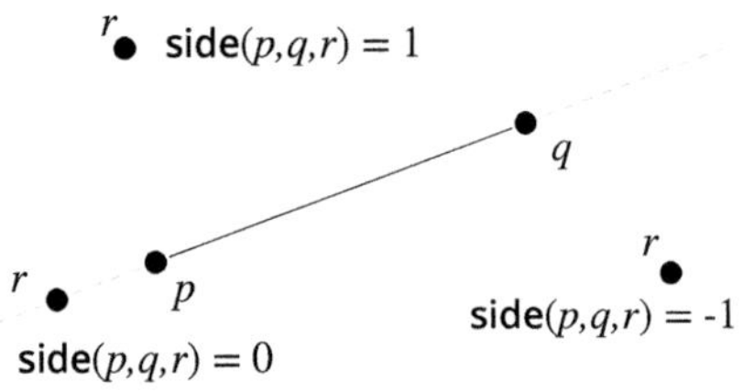

Figure 5.26: The **side** operation

centroid

Centroid of a polygon The *centroid* or *center of gravity* of an areal object is the point at which it would balance if it were cut out of a sheet of material of uniform density. For a regular polygon or a set of points the centroid is simply the mean of the vertex vectors. Computing the centroid of an irregular polygon requires a more complex calculation, based on the weighted sum of vertex vectors. For a (possibly irregular) polygon P with n vertex vectors $(x_1, y_1), (x_2, y_2), ..., (x_n, y_n)$ where again $(x_1, y_1) = (x_n, y_n)$ the x and y coordinates of the centroid are given by:

$$\textsf{centroid}_x(P) = \frac{1}{6.\textsf{area}(P)} \sum_{i=1}^{n-1} (x_i + x_{i+1})(x_i y_{i+1} - x_{i+1} y_i)$$

$$\textsf{centroid}_y(P) = \frac{1}{6.\textsf{area}(P)} \sum_{i=1}^{n-1} (y_i + y_{i+1})(x_i y_{i+1} - x_{i+1} y_i)$$

Notice that the center of gravity of a polygon is not guaranteed to lie within its boundary (consider the center of gravity of a "horseshoe" shape, for example).

5.4.2 *Topological algorithms*

point-in-polygon

Point-in-polygon Point-in-polygon is one of the most common operations in

GIS. Given a point p and a polygon P, **point_in_polygon**(p, P) = **true** if and only if p is in the interior of P.

If the polygon is convex, then the operation **side** discussed in the previous section may be used. Assume that the vertices of the polygon are $p_1, p_2, ..., p_n$, ordered counterclockwise around the polygon (again $p_1 = p_n$). Then **point_in_polygon**(p, P) = **true** if and only if:

$$\mathbf{side}(p, p_1, p_2) = \mathbf{side}(p, p_2, p_3) = ... = \mathbf{side}(p, p_{n-1}, p_n) = 1$$

The more usual and interesting case is where the polygon is not necessarily convex. Two algorithms are considered here: the *semi-line algorithm* and the *winding number algorithm*.

The underlying principle behind the *semi-line algorithm* is simple. Given a point p and polygon P, assume that p is not on the boundary of P (to avoid degenerate cases). Draw an infinite semi-line (or *ray*) out from p in any direction. Count the number of times that the ray intersects with the boundary of P. If the number is even, then p is outside P. If the number is odd, then p is inside P (see Figure 5.27).

semi-line algorithm

ray

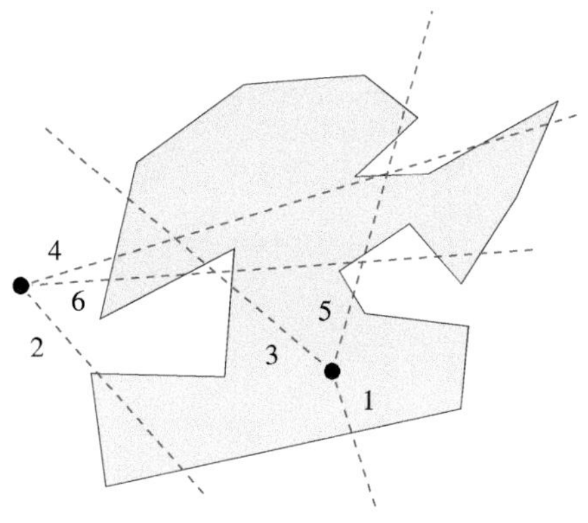

Figure 5.27: Semi-line algorithm for determining whether a point is inside a polygon

However, as with many of these algorithms, several special cases can arise. For example, if a ray intersects with a vertex of the polygon, how should it be counted? Figure 5.28 shows two such cases. At vertex a, the intersection should not be counted, while at vertex b, the intersection should be counted. It is important that the polygon actually crosses the ray for the intersection to be counted, and this does not happen at a. Figure 5.28 also shows the intersection of the entire segment cd with the ray. The same principle may be applied: it is only the *crossing* that is counted.

With regard to implementation, it simplifies matters to choose the semi-line to be horizontal and directed to the right, then to translate the polygon so that the semi-line is the positive x-axis with the point p as the origin. The polygon is then traversed and, for each edge, a simple calculation with coordinates determines whether the edge crosses the positive x-axis. The algorithm has time complexity $O(n)$, where n is the number of edges of the polygon.

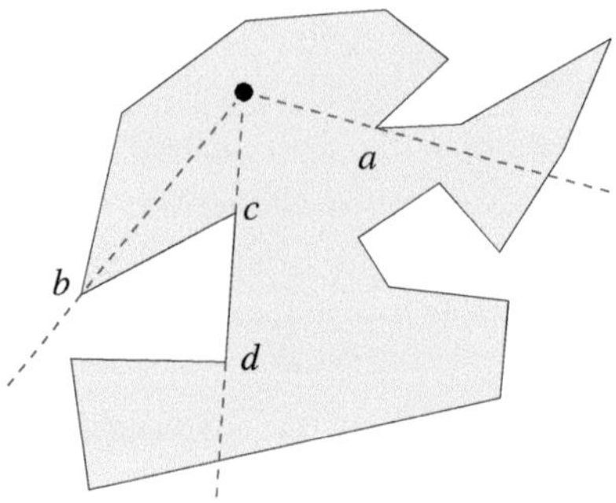

Figure 5.28: Special cases in the semi-line algorithm

winding number algorithm

The *winding number algorithm* provides an alternative and quite distinct approach. It has the same complexity as the semi-line algorithm, but in practice is much less efficient due to the trigonometrical calculations. Imagine that an observer is moving counterclockwise along the boundary of P, always facing the point p. If, after having completed the journey round P, the observer has turned through one complete circle, then p is within P. Otherwise, p is outside P. Figure 5.29 shows examples of this. In each case a path is walked visiting the vertices in the order 1-2-3-4-5-6-1. In the left-hand configuration, where the point is inside the polygon, the observer will rotate through a full turn. In the right-hand configuration, where the point is outside the polygon, the net turning effect will be zero.

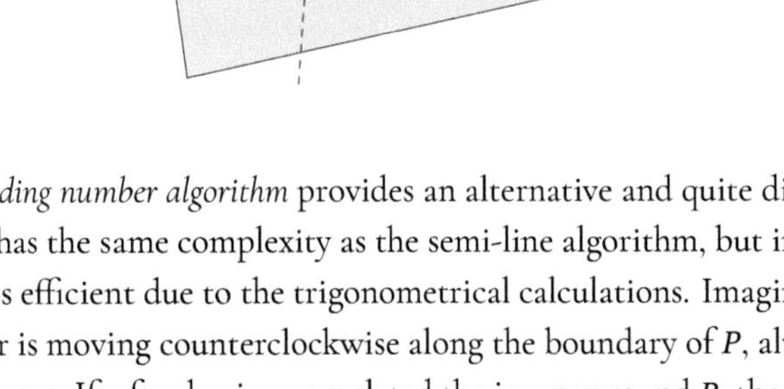

Figure 5.29: Traversing the boundary of the polygon 123456 always facing point p

Given the coordinates of the polygon vertices, the angle formed by each pair of consecutive vertices and point p may be calculated using elementary trigonometry, as in Chapter 3. The angles are then summed over the entire walk round the polygon to find the total turn.

5.4.3 *Set-based algorithms*

Throughout this section spatial objects are treated as sets of points. It is assumed, unless the contrary is indicated, that all the sets are topologically closed; that is, they contain their boundary points and edges.

Collinearity Given three points a, b, and c, the operation **collinear**(a, b, c) determines whether a, b, and c are collinear (lie on the same straight line). The algorithm for this operation needs no new ideas. From Section 5.4.1:

$$\textbf{collinear}(a, b, c) = \textbf{true} \text{ if and only if } \textbf{side}(a, b, c) = 0$$

Point on segment Let straight-line segment l have end-points $q = (x_q, y_q)$ and $r = (x_r, y_r)$. For any point $p = (x_p, y_p)$, the operation **point_on_segment**(p, l) returns the Boolean value **true** if $p \in l$.

The first step in calculating **point_on_segment**(p, l) is to use the previous algorithm to decide whether p, q, and r are collinear. If not, the determination is negative and there is no need to proceed further. Otherwise, consider the *minimum bounding box* (MBB) of l, **mbb**(l), which is the smallest rectangle with sides parallel to the coordinate axes that contains l. Then, $p \in l$ if and only if $p \in$ **mbb**(l). Finally $p \in$ **mbb**(l) if and only if both the following inequations hold:

minimum bounding box

$$\min(x_q, x_r) \leq x_p \leq \max(x_q, x_r)$$

$$\min(y_q, y_r) \leq y_p \leq \max(y_q, y_r)$$

where **min** and **max** return the minimum and maximum of the arguments, respectively.

Segment intersection There are two related operations for straight-line segment intersection: one Boolean, testing whether the two lines have at least one point in common; the other returning the point of intersection if it exists. Clearly, the result of the latter operation immediately gives a result for the former. Assume, as usual, that the segments are closed and therefore contain their end-points.

The intersection detection operation is a decision method, returning **true** if and only if the segments have at least one point in common. The operation may be defined in terms of previous operations, in particular the **side** operation. Figure 5.30 shows an example of how the **side** operation is used. Segments ab and cd will intersect if points a and b are on opposite sides of the line through cd and points c and d are on opposite sides of the line through ab. More formally, ab and cd intersect if both the inequations below hold:

$$\textsf{side}(a, b, c) \neq \textsf{side}(a, b, d)$$

$$\textsf{side}(c, d, a) \neq \textsf{side}(c, d, b)$$

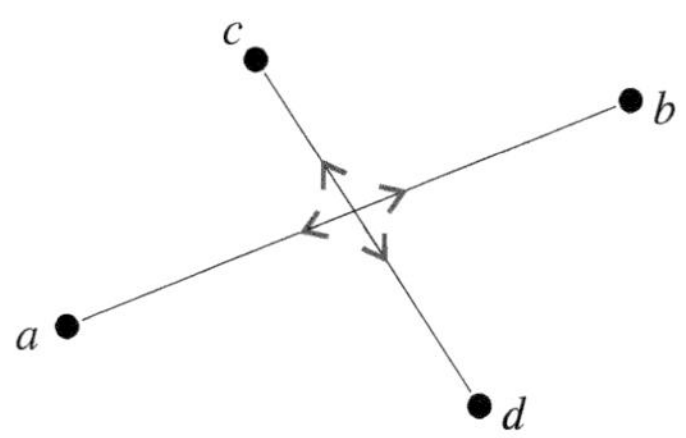

Figure 5.30: Using the **side** operation for intersection detection

The converse of this statement is not true, however. It is possible for ab, cd to intersect when one or both of these inequations is false. This might occur when one of segments partially lies on the other. Thus, a robust intersection detection algorithm must also take account of the special cases where **side** takes the value zero. This may be accomplished by also checking the following possibilities:

$$\textbf{point\_on\_segment}(a, cd) = \textbf{true}$$
$$\textbf{point\_on\_segment}(b, cd) = \textbf{true}$$
$$\textbf{point\_on\_segment}(c, ab) = \textbf{true}$$
$$\textbf{point\_on\_segment}(d, ab) = \textbf{true}$$

Our work is not yet done. Even if two segments are shown to intersect by the above decision procedure, it still remains to determine their point of intersection. The calculation of a point of intersection of two line segments, if it exists, is at first sight a trivial exercise in coordinate geometry, reducing to the simultaneous solution of two simple linear equations. However, as with many geometric algorithms, special cases (e.g., vertical segments) and errors due to the discretization process provide several degenerate cases for which a robust algorithm must allow (see Douglas, 1990).

Suppose that we are given segments l and l', with respective parametric forms for their point sets:

$$l = \{\lambda\vec{p} + (1-\lambda)\vec{q} \mid \lambda \in [0,1]\}$$
$$l' = \{\lambda'\vec{p}\,' + (1-\lambda')\vec{q}\,' \mid \lambda' \in [0,1]\}$$

Let r be a point of intersection of these sets. Then there are $\alpha, \beta \in [0,1]$ such that:

$$\alpha\vec{p} + (1-\alpha)\vec{q} = \vec{r} = \beta\vec{p}\,' + (1-\beta)\vec{q}\,'$$

which may be rearranged to give $\alpha(\vec{p} - \vec{q}) + \beta(\vec{q}\,' - \vec{p}\,') = \vec{q}\,' - \vec{q}$. This vector equation may be decomposed into two simultaneous equations for the two unknowns α and β, based on the x and y coordinates of the end-points of l and l'. Solving for α and β yields:

$$\alpha = \frac{(x_q - x_{q'})(y_{p'} - y_{q'}) - (x_{p'} - x_{q'})(y_q - y_{q'})}{(x_{p'} - x_{q'})(y_p - y_q) - (x_p - x_q)(y_{p'} - y_{q'})}$$
$$\beta = \frac{(x_p - x_q)(y_{q'} - y_q) - (x_{q'} - x_q)(y_p - y_q)}{(x_{p'} - x_{q'})(y_p - y_q) - (x_p - x_q)(y_{p'} - y_{q'})}$$

Provided that l and l' are not parallel, the solutions for these two equations must then be checked to ensure that the line segments do intersect (i.e., $\alpha, \beta \in [0,1]$). If so, the coordinate of the intersection point can be recovered by substituting α or β into the corresponding original parametric line equation.

Intersection, union, and overlay of polygons Algorithms for the intersection, union, and overlay of polygons are based on iterative application of the line segment intersection procedure, described in the previous section. For two simple polygons $P = [l_1, \ldots, l_m]$ and $P' = [l'_1, \ldots, l'_n]$, each given as a cycle of edge segments, the key is to find the points of intersection of line segments from

P with line segments from P'. In the general case, we must examine each pair (l_i, l'_j) where $1 \leq i \leq m, 1 \leq j \leq n$ and determine the point of intersection, if it exists. (If the segments overlap or meet at their end-points, then special cases must also be constructed, not considered here.) There are mn such pairs to examine and so the time complexity of intersection, union, and overlay algorithms is usually $O(mn)$.

Examples of individual algorithms are not described in detail here. Figure 5.31 illustrates the core features of intersection, union, and overlay for a pair of intersecting line segments ab and cd within two polygons, P and P'. Note that the topology of the polygons is important in deciding which line segments to discard and which to keep in the case of intersection and union. For intersection, the points a and d along with the point of intersection x form part of the resulting structure, while xc and xb are discarded. For union, the points c, b, and x are part of the resulting polygon, and ax and dx are discarded. For overlay, all the points are retained.

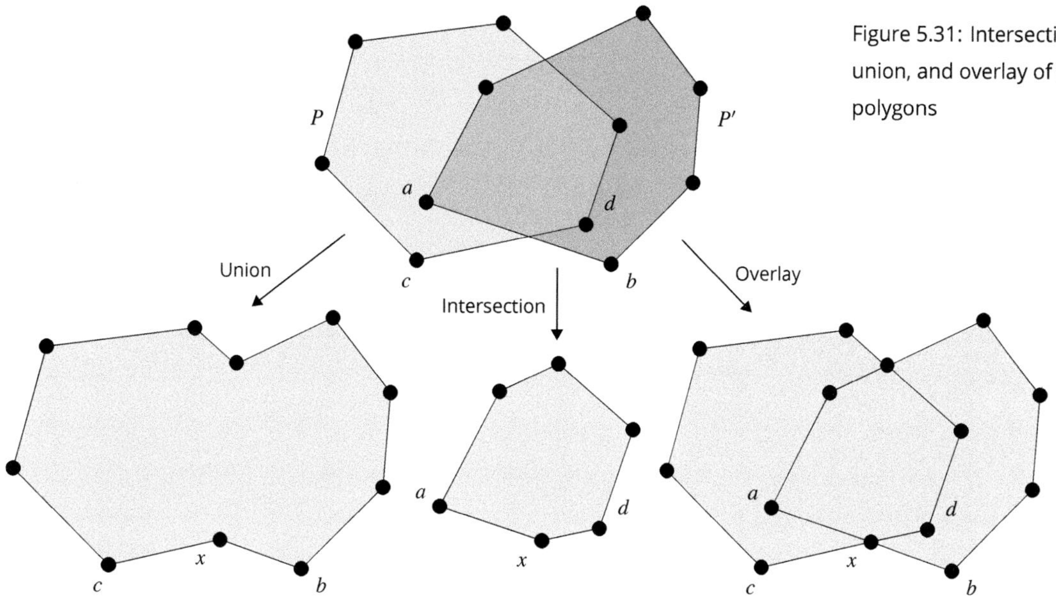

Figure 5.31: Intersection, union, and overlay of two polygons

5.4.4 *Polygon triangulation algorithms*

This section describes two of the simpler non-Delaunay algorithms for triangulation, taking triangulation of a polygon as an exemplar. The triangulations will not introduce any extra interior vertices (Steiner points). The objective is to arrive at reasonably regular triangulations: that is, with few long, thin triangles.

Greedy triangulation algorithm *Greedy triangulation* has been briefly considered in Section 5.3.4 and illustrated in Figure 5.20. In general, a *greedy algorithm* is a localized processing procedure that never revisits data that has already been processed. Algorithm 5.3 provides a procedure for the greedy triangulation

greedy triangulation

greedy algorithm

of a simple polygon. The algorithm first sorts the diagonals of the polygon vertices into ascending order of length (line 2). For each diagonal d_i, beginning with the shortest, the algorithm tests whether d_i properly intersects any diagonals already within the triangulation T (i.e., intersects at a point in the interior of a diagonal, line 5) and whether d_i is an internal diagonal (line 6). If not, d_i is added to the triangulation T (line 7).

Algorithm 5.3: Greedy polygon triangulation algorithm

```
Input Simple polygon P with n vertices
1: form the set D of m = n(n−3)/2 diagonals of P
2: sort D into ascending order of length d_1, ..., d_m
3: triangulation T ← P
4: for i ← 1 to m do
5:     if d_i does not properly intersect any line segment in T then
6:         if d_i is an internal diagonal of P then
7:             T ← T ∪ d_i
Output Triangulation T of P
```

The algorithm is termed "greedy" because it processes at each step the next available diagonal, never revisiting diagonals that have already been processed. However, it is also "greedy" for computational time. In Algorithm 5.3 all diagonals $m = n(n-3)/2$ will be considered. Each of these edges must be tested for intersection with the other edges in the triangulation. The triangulation will contain n edges. Assuming a naive test for intersection of two edges, running in constant time, the algorithm is of worst-case complexity $O(n^3)$.

We can make some immediate improvements to the efficiency of this algorithm by keeping a count of the edges added to the triangulation. The number of such edges must be $n - 3$, so when we arrive at this point the algorithm may terminate. This change will not improve the worst-case complexity of $O(n^3)$, although it will improve the average-case complexity. Consequently, while the greedy algorithm is useful for illustration purposes, as it is simple to understand, it is not a good choice for practical implementations due to its high algorithmic complexity.

The triangulation algorithm of Garey, Johnson, Preparata, & Tarjan The algorithm of Garey et al. (1978) is more suitable for the triangulation of simple polygons. With respect to time complexity, it improves on the greedy method above. The algorithm proceeds by first decomposing the polygon into monotone polygons and then triangulating each of the monotone polygons. Monotonicity was introduced in Section 3.1.3. Monotone polygons have the property that their vertices are sorted with respect to a particular orientation, the *line of monotonicity*. As a result of this property, monotone polygons can be triangulated in linear time, $O(n)$, where n is the number of vertices.

line of monotonicity

A procedure for triangulating a monotone polygon is outlined in Algorithm 5.4. Let P be a monotone polygon with respect to the y-axis. Merging the two monotone chains of P to produce a list $V = [v_1, ..., v_n]$ can be

Algorithm 5.4: Triangulation of a monotone polygon

Input Monotone polygon P with n vertices

1: merge the two monotone chains of P into a list of vertices $V = [v_1, ..., v_n]$, sorted according to descending y-coordinate
2: define triangulation $T \leftarrow P$ and list $X = [x_1, ..., x_m]$ of nodes to be processed
3: initialize list $X \leftarrow [v_2, v_1]$
4: for $i \leftarrow 3$ to $n - 1$ do
5: if v_i and x_1 are in different monotone chains then
6: add diagonals $v_i x_1, ..., v_i x_{m-1}$ to triangulation T
7: update $X \leftarrow [v_i, v_{i-1}]$
8: else
9: for $j = 2$ to m do
10: if diagonal $v_i x_j$ lies entirely within P then
11: add diagonal $v_i x_j$ to T
12: remove x_{j-1} from X
13: add v_i to the beginning of X
14: add diagonals $v_n x_k$ to T for $x_k \in X$ where $1 < k < m$

Output Triangulation T of monotone polygon P

achieved in linear time, because the two chains are already ordered. Processing each element of V in turn, diagonals are added to the triangulation when either a pair of vertices are on opposite monotone chains or a pair of vertices are intervisible, in the sense that their diagonal lies entirely within the polygon P.

To illustrate Algorithm 5.4, consider the monotone polygon shown in Figure 5.32. The two monotone chains are $[a, b, c, d, e, f, g]$ and $[a, m, l, j, k, i, h, g]$. Sort-merging the chains gives the vertex list $V = [a, m, b, c, d, l, k, j, e, i, f, h, g]$. List X is initialized to $[m, a]$. The first vertex to be processed is b. Since b and m are members of different monotone chains in P, we add the edge mb to the triangulation. The following two vertices to be processed, c and d, do not result in any new edges in the triangulation. Subsequently processing vertex l results in three new edges bl, cl, and dl. Processing vertex k results in no new edges, but subsequently processing j leads to the edge jl being added to the triangulation. Although j and l are members of the same monotone chain, they satisfy the condition that edge jl lies completely within P (line 10). The algorithm continues in this way until eventually 10 new edges are added and the algorithm terminates.

Having established that it is possible to triangulate a monotone polygon in linear time, the next question is how to partition a simple polygon into monotone polygons. The details of this procedure are beyond the scope of this text. In brief, let us assume that monotonicity is with respect to the y-axis and that no two different vertices of the polygon have the same y-coordinate. Then the initial polygon P must be *regularized* to ensure that:

1. Edges intersect only at polygon vertices.
2. Apart from the highest vertex (with respect to the y-axis), every vertex is joined by at least one edge to a higher vertex.

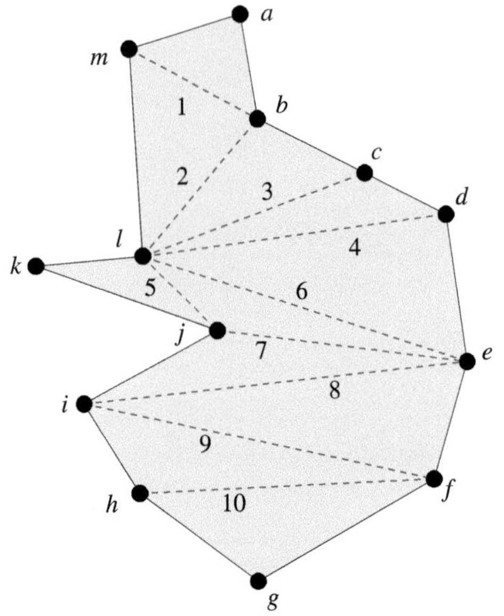

Figure 5.32: Triangulation of monotone polygon (edges numbered in order of processing)

3. Apart from the lowest vertex (with respect to the y-axis), every vertex is joined by at least one edge to a lower vertex.

It can be proved that this regularization[3] process results in the polygon P being partitioned into regions, each of which is monotone. After regularization, any new edges that lie outside the polygon must be removed. The time complexity of the regularization stage is $O(n \log n)$, where n is the number of vertices of the polygon. As a result, the time complexity of this procedure dominates the algorithm, and the entire triangulation algorithm can be completed in $O(n \log n)$ time. Figure 5.33 illustrates the different stages of the Garey et al. triangulation algorithm.

[3] Note that this is a different sense of the term "regularization" from that used in Chapter 3.

5.4.5 *Delaunay triangulation algorithms*

Algorithms for the Delaunay triangulation and its dual Voronoi diagram are abundant in the literature. One particularly elegant example uses a technique common in many algorithms termed the *divide-and-conquer* strategy. Using divide-and-conquer, a complex problem is recursively broken up into smaller and simpler problems, which are solved individually and then combined to give a complete solution.

divide-and-conquer

Algorithm 5.5 summarizes the divide-and-conquer method for Delaunay triangulation. The algorithm begins by sorting points into ascending order of x coordinate (line 1). As long as the cardinality of the set of points is greater than 3 (line 6), then the point set is divided into two halves, a left and a right half (line 7), and then recursively subdivided (lines 8–9). When a subdivision reaches a cardinality of 3, a corresponding triangle is added to the interim results (line 5). Similarly, a subdivision of cardinality 2 results in a line added to the interim results (line 4). Figure 5.34a illustrates the recursive divide stage of the divide-and-conquer Delaunay algorithm for 14 ordered points.

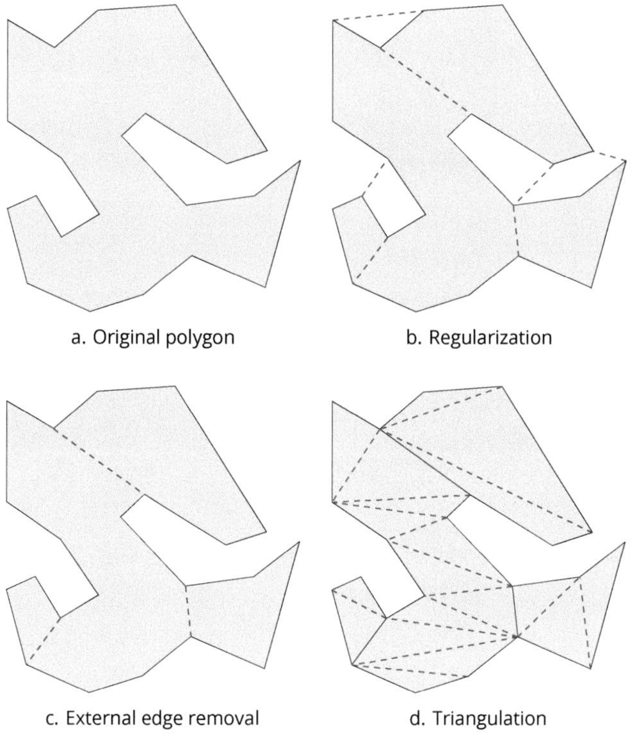

Figure 5.33: Stages of Garey, Johnson, Preparata, & Tarjan algorithm for $O(n \log n)$ simple polygon triangulation

The subdivision requires three levels: a top-level division between points 7 and 8; second-level subdivisions between points 3 and 4, and 10 and 11; and the final subdivisions between points 5 and 6, and 12 and 13.

Algorithm 5.5: Divide-and-conquer Delaunay triangulation

```
Input Set P of points p_1, ..., p_n in the Euclidean plane
1: sort P into ascending order of x-coordinates (ties sorted by y-coordinate)
2: T ← divide(P)
3: function divide(P)
4:     if |P| = 2 (i.e., P = {p_1, p_2}) then return line p_1 p_2
5:     if |P| = 3 (i.e., P = {p_1, p_2, p_3}) then return triangle p_1 p_2 p_3
6:     if |P| > 3 then
7:         divide the points into two (roughly) equal halves L and R
8:         T_l ← divide(L)
9:         T_r ← divide(R)
10:        create single triangulation T by merging T_1 and T_2
11:        return T
12: end function
Output Delaunay triangulation T of points p_1, ..., p_n
```

Algorithm 5.5 glosses over the crucial details of how the merger is conducted (line 10). The key to understanding the merge is the observation that any new edges created by the merge must connect points in the left and right triangulations. While it may be necessary to delete some edges already in the left or the right triangulations, it will never be necessary to create new edges except across the boundary between the two halves. The merge itself begins by adding to the combined triangulation the lower common tangent—an edge in the convex hull of the two triangulations that must necessarily be in the merged triangulation. The merge then proceeds upwards from this edge stitching together points across the boundary subject to the Delaunay circumcircle property: that the circumcircle of a constituent triangle does not include any other triangulation point within it (Figure 5.18).

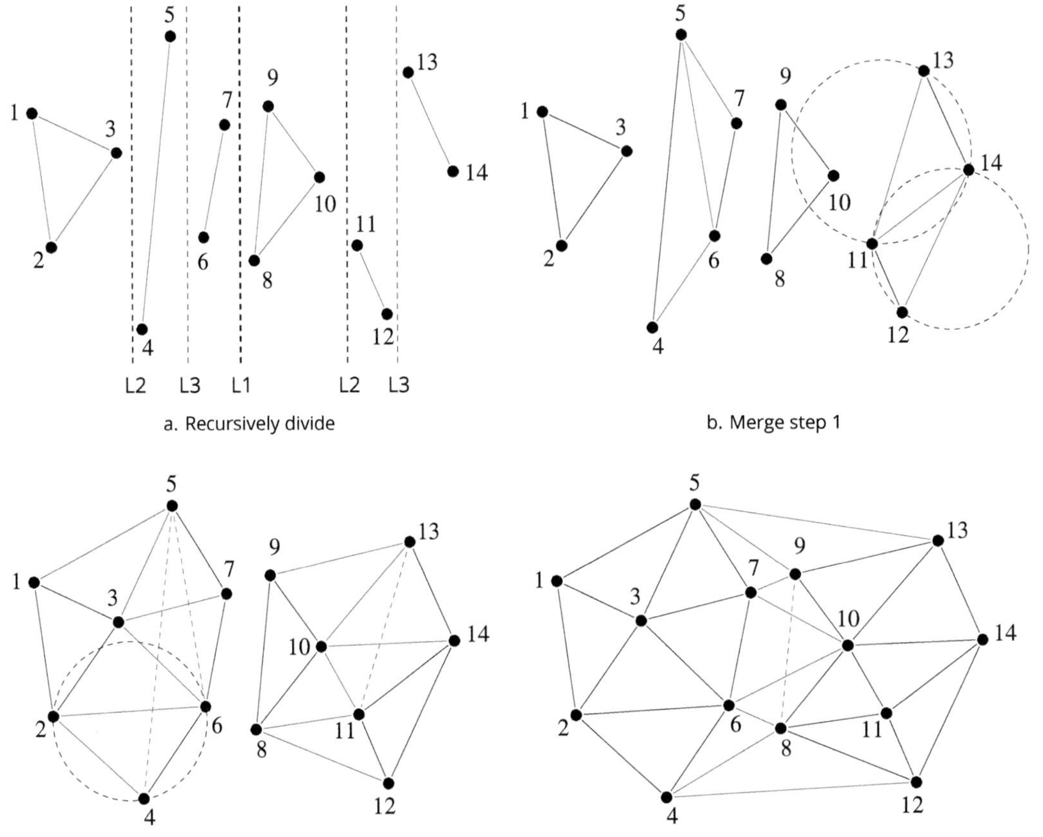

a. Recursively divide

b. Merge step 1

c. Merge step 2

d. Merge final step

Figure 5.34: Stages of $O(n \log n)$ divide and conquer algorithm for Delaunay triangulation

Figure 5.34b illustrates the first step in the merge, as well as the circumcircle properties for the merged point sets {11, 12} and {13, 14}. Note that point 10 appears in violation of the circumcircle property for the newly added edge between 11 and 13. However, this is a *global* violation, which will be ironed out in subsequent merge steps. *Locally*, the new edges added between {11, 12} and {13, 14} do satisfy the circumcircle property as required.

The specific computation involved in determining which edges to add when stitching up the merge is beyond the scope of this text. However, the intuition behind the merge can be grasped by imagining a growing circumcircle "bubbling" up from each newly added edge. As the bubble grows, the next point it meets will uncover the next edge to add to the triangulation. For example, in the next merge step in Figure 5.34c, a circumcircle "bubble" growing up from the lower common tangent edge between 2 and 4 meets point 6 first,

Box 5.5: Faster triangulation algorithms

The history of triangulations is interesting. We have seen that a greedy triangulation has a time complexity of $O(n^3)$. Triangulation algorithms of time complexity $O(n^2)$ date back to Lennes (1911). Garey et al. published their $O(n \log n)$-time algorithm in 1978, an achievement of the newborn field of computational geometry. Although a lower bound on the time complexity of a general planar triangulation was known to be $O(n \log n)$-time, it was not known whether this still held for triangulations of simple polygons or whether an even lower bound was possible. It was not until nearly 10 years later that Tarjan & Wyk (1988) published an $O(n \log\log n)$-time algorithm. Of course, the best that can be hoped for is a linear time algorithm, and this was actually achieved by Chazelle (1991), thus providing a climax to years of intense research effort. A concise introduction to Chazelle's remarkable algorithm can be found in O'Rourke (1997, p56). The worst-case Delaunay triangulation algorithm is limited by the vertex sorting algorithm, which requires at least $O(n \log n)$-time. In fact, optimal $O(n \log n)$ Delaunay algorithms have been achieved (Lee & Schachter, 1980). Despite a worst-case time complexity of $O(n^2)$, an incremental method enhanced by an efficient data structure (Ohya, Iri, & Murota, 1984) is often more useful in practice because it is conceptually simple and has linear average time complexity. Dwyer (1987) published a variant of the *divide-and-conquer* approach that led to an algorithm with optimal worst-case time complexity and average time complexity of $O(n \log\log n)$.

before growing in to any other point. As a result, the edge between 2 and 6 must be added to the merge triangulation, in turn requiring the deletion of the edge between points 4 and 5. Figure 5.34d shows the final merge step and the completed Delaunay triangulation.

The time complexity of the resulting merger operation is linear in the number of nodes, $O(n)$. The basic partial triangulations are constructed only from sets of two or three edges (Algorithm 5.5, lines 4–5) so this stage can be completed in constant time, $O(1)$. As a result, the sorting stage (line 1) of the algorithm dominates the time complexity, so the whole algorithm runs in $O(n \log n)$ time. Indeed, the worst-case time complexity of any Delaunay triangulation algorithm is at least $O(n \log n)$, so the divide-and-conquer algorithm is optimal in terms of time complexity. Nevertheless, a number of other ingenious algorithms to compute Delaunay triangulations have been devised, also yielding some important theoretical insights into the fundamentals of computational geometry (see Box 5.5). Indeed, computational geometry has traditionally been concerned with measures of efficiency derived from theory—especially complexity theory. However, theory and practice are not always in accord. Methods that are theoretically efficient may not be practical to implement. However, computational geometry provides the basis from which practical approaches may be constructed and evaluated.

As noted in Section 5.3.3, the Voronoi diagram is the dual of the Delaunay triangulation. Consequently, algorithms for computing the Delaunay triangulation can also be used to compute Voronoi diagrams, and vice versa. Figure 5.35 illustrates one method for computing a Voronoi diagram *incrementally*. The general idea is to construct the Voronoi diagram by modifying it as new vertices from the vertex set are considered. At each stage, the newly introduced vertex is highlighted in Figure 5.35. The polygon in which the new vertex lies is calculated, and a trail (highlighted) is made around the new vertex, creating a new proximal polygon. The other edges are then modified appropriately and the process continues until no more new vertices are available. We return to incremental algorithms (including an incremental Delaunay triangulation algorithm in Algorithm 7.1) in Chapter 7.

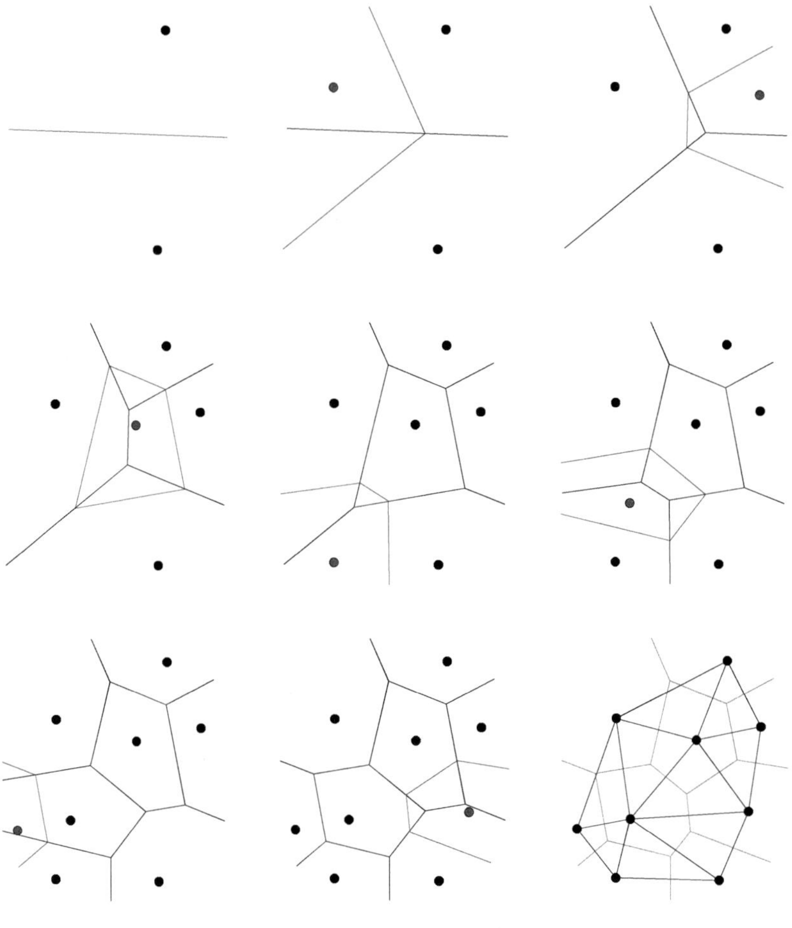

Figure 5.35: Some steps in an incremental method for computing the Voronoi diagram

5.5 *Network representation and algorithms*

As introduced in Chapter 2 and elaborated on in Chapter 3, the fundamental structure for modeling a network is the graph. This section explores the representation of graphs in a computer and goes on to consider algorithms for some of the most common operations on networks.

5.5.1 *Network representation*

The most basic representation in a computer of a graph is as a set of nodes and a set of edges (pairs of nodes). Additional information associated with nodes or edges, such as node labels or edge weights, can be added as extra fields or associated attributes. Node and edge sets are highly efficient in terms of storage, but much less efficient for computing. Most graph operations require searching through the set of nodes or edges to determine basic connectivity.

An alternative to support more efficient computation is a square n by n *adjacency matrix*. In an adjacency matrix, each row and column is labeled by a node. A "1" is placed in a matrix cell if and only if the nodes corresponding to the appropriate row and column are connected by an edge, otherwise a "0" is placed in the cell.[4]

adjacency matrix

[4] The closely related *incidence matrix* similarly stores nodes as rows, edges as columns, and places a "1" in cells to indicate the corresponding node is incident with that edge; "0" otherwise.

If the network is directed, then the matrix may not be symmetric. If the graph is weighted, then non-zero numbers can be interpreted as weights. Edge labels may similarly be accommodated placing the label in the matrix. Figure 5.36 shows a hypothetical example of a network of average travel times (in minutes) for tram routes between places of interest in The Potteries. The adjacency matrix representing this weighted graph is given in Table 5.5. A non-zero entry in the matrix indicates the existence of an edge between the appropriate row and column nodes and gives the average tram travel time. A zero entry indicates no edge between the appropriate row and column nodes.

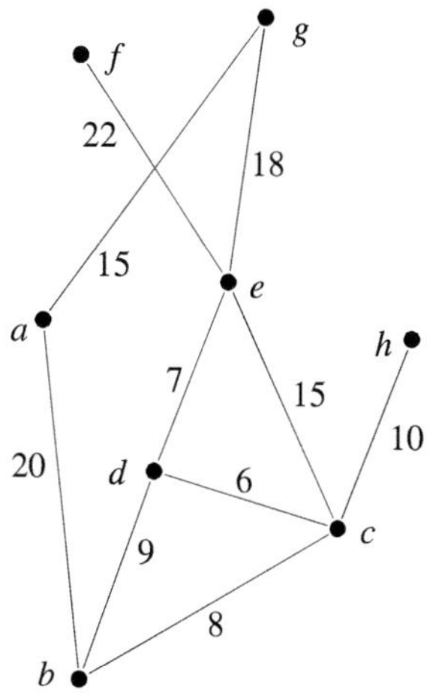

a. Newcastle Museum
b. Trentham Gardens
c. Beswick Pottery
d. Minton Pottery
e. City Museum
f. Westport Lake
g. Ford Green Hall
h. Park Hall Country Park

Figure 5.36: Network of hypothetical tram routes between some Potteries locations, labeled by average travel times

	a	*b*	*c*	*d*	*e*	*f*	*g*	*h*
a	0	20	0	0	0	0	15	0
b	20	0	8	9	0	0	0	0
c	0	8	0	6	15	0	0	10
d	0	9	6	0	7	0	0	0
e	0	0	15	7	0	22	18	0
f	0	0	0	0	22	0	0	0
g	15	0	0	0	18	0	0	0
h	0	0	10	0	0	0	0	0

Table 5.5: Adjacency matrix for the network in Figure 5.36

The adjacency matrix is particularly efficient for deciding whether a particular edge exists within the graph. Checking for a non-zero entry in the cell corresponding to a particular pair of nodes can be computed in constant time. However, since geographic networks tend to have large numbers of nodes, but relatively few connections at each node, the adjacency matrix will sparse matrix

usually be *sparse*, containing mainly redundant 0s in most cells. Consequently, the adjacency matrix tends to be much less efficient than node-edge sets in terms of storage space.

adjacency list

As a result, computational geometry algorithms often use a close relation to the adjacency matrix, the *adjacency list*. For each vertex, the adjacency list stores only those edges that are incident with that vertex, effectively eliminating all the 0s from the adjacency matrix. Consequently, the adjacency list achieves a good balance between storage efficiency and computational efficiency, especially for relatively sparse graphs. Table 5.6 shows the adjacency list for the network in Figure 5.36.

Table 5.6: Adjacency list for the network in Figure 5.36

a	$(b, 20), (g, 15)$
b	$(a, 20), (c, 8), (d, 9)$
c	$(b, 8), (d, 6), (e, 15), (h, 10)$
d	$(b, 9), (c, 6), (e, 7)$
e	$(c, 15), (d, 7), (f, 22), (g, 18)$
f	$(e, 22)$
g	$(a, 15), (e, 18)$
h	$(c, 10)$

We will encounter one further structure—the linked list—that is especially well adapted for storing and indexing data structured as a network (e.g., in a graph database) in the next chapter, Section 6.1.

5.5.2 *Depth-first and breadth-first traversals*

A fundamental operation on any connected network is the systematic traversal of the nodes within it. Such traversals have many applications, for example, searching for disconnections in a drainage network. There are two main alternative approaches: *depth-first* or *breadth-first* traversal. We may either traverse the network by probing ever deeper and only coming up when all alternatives below are exhausted (*depth-first traversal*), or traverse the network by examining all nodes connected to a given node before going deeper (*breadth-first traversal*). A single breadth- or depth-first traversal will visit every node in the network if and only if the underlying graph is connected (there exists a path from every node to every other node in the network).

depth-first traversal

breadth-first traversal

Algorithm 5.6 gives the procedure for a breadth-first traversal of a graph. To illustrate the breadth-first traversal, Figure 5.37 gives the series of iterations for a breadth-first traversal of the tram network in Figure 5.36. Starting with node b, each of the adjacent nodes a, c, d is visited before the search proceeds to the next level. The set V in Algorithm 5.6 is used to keep track of the nodes that have been visited by the algorithm. The tree on the right-hand side of Figure 5.37 shows the traversal order diagrammatically.

queue

FIFO

Algorithm 5.6 uses a special type of list to store the active nodes waiting to be processed, called a *queue* (Q, line 1, also indirectly encountered in earlier algorithms in this chapter). In a queue, the first elements to be added to the list are also first to be removed. A queue is often referred to as a *FIFO*

Algorithm 5.6: Breadth-first network traversal

```
Input Adjacency matrix M, starting node s
1: queue Q ← [s], visited set V ← ∅
2: while Q is not empty do
3:     remove the first node x from Q
4:     add x to V
5:     for each node y ∈ M adjacent to x do
6:         if y ∉ V and y ∉ Q then add y to the end of Q
7: end while
```

structure, meaning *first-in-first-out*. The complementary structure used in the depth-first traversal is the *stack*, in which the last elements to be added are removed first, termed *LIFO* or *last-in-first-out*.

stack

LIFO

n	Q	V
1	$[b]$	$\{\}$
2	$[a, c, d]$	$\{b\}$
3	$[c, d, g]$	$\{b, a\}$
4	$[d, g, e, h]$	$\{b, a, c\}$
5	$[g, e, h]$	$\{b, a, c, d\}$
6	$[e, h]$	$\{b, a, c, d, g\}$
7	$[h, f]$	$\{b, a, c, d, g, e\}$
8	$[f]$	$\{b, a, c, d, g, e, h\}$
9	$[]$	$\{b, a, c, d, g, e, h, f\}$

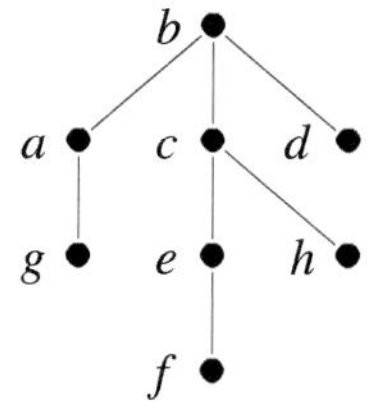

Figure 5.37: Iterations for breadth-first traversal of the tram network in Figure 5.36

Algorithm 5.7 gives the procedure for a depth-first traversal. Algorithm 5.7 is almost identical to Algorithm 5.6. The key difference is that Algorithm 5.7 uses a *stack* where Algorithm 5.6 uses a *queue*. As a consequence, in line 6 of Algorithm 5.7 the adjacent nodes are added to the top of the stack, ensuring the last nodes added to the list are the first to be processed.

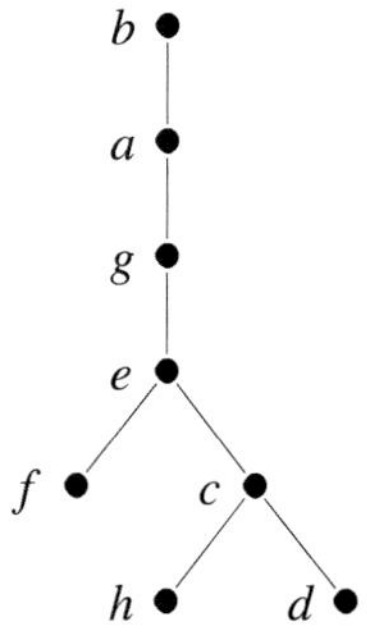

n	Q	V
1	$[b]$	$\{\}$
2	$[a, c, d]$	$\{b\}$
3	$[g, c, d]$	$\{b, a\}$
4	$[e, c, d]$	$\{b, a, g\}$
5	$[f, c, d]$	$\{b, a, g, e\}$
6	$[c, d]$	$\{b, a, g, e, f\}$
7	$[h, d]$	$\{b, a, g, e, f, c\}$
8	$[d]$	$\{b, a, g, e, f, c, h\}$
9	$[]$	$\{b, a, g, e, f, c, h, d\}$

Figure 5.38: Iterations for depth-first traversal of the tram network in Figure 5.36

Figure 5.38 illustrates the series of iterations for the depth-first traversal of the same tram network in Figure 5.36, again starting with node *b*. The tree on the left-hand side of Figure 5.38 illustrates the way the depth-first traversal follows each path until exhausted, before backtracking to search the next path.

Algorithm 5.7: Depth-first network traversal

Input Adjacency matrix M, starting node s

stack $S \leftarrow [s]$, visited set $V \leftarrow \emptyset$
while S is not empty do
    remove the first node x from S
    add x to V
    for each node $y \in M$ adjacent to x do
        if $y \notin V$ and $y \notin S$ then add y to the *top* of stack S
end while

5.5.3 *Shortest path*

shortest path

Perhaps the most fundamental and important of all network operations is the computation of the *shortest path* between nodes in a network. Shortest path algorithms are needed, for example, by navigation systems to determine the most direct route, in terms of Euclidean distance, through a street network to some destination. However, the shortest path algorithm is behind a host of other network analyses, including finding the shortest route in terms of travel time.

Dijkstra's algorithm

The most well-known shortest path algorithm is *Dijkstra's algorithm* (Dijkstra, 1959), devised by Edsger Dijkstra, an influential pioneer in computer science. Dijkstra's algorithm, outlined in Algorithm 5.8, requires a weighted graph in which the edge weights are non-negative. This restriction is not normally problematic because negative distances or travel times are not sensible in most applications (cf. the discussion of metric spaces in Section 3.4.2). In Algorithm 5.8, distances are represented as a weighting function w from the set of edges E to the positive reals $\mathbb{R}^+$. An additional target weighting function $t : N \to \mathbb{R}^+$ is used to store the minimum distances from the start node s to each node in the graph.

Algorithm 5.8: Dijkstra's shortest path algorithm

Input Undirected simple connected graph $G = (N, E)$, starting node $s \in N$, weighting function $w : E \to \mathbb{R}^+$, target weighting function $t : N \to \mathbb{R}^+$

1: initialize $t(n) \leftarrow \infty$ for all $n \in N$, visited node set $V \leftarrow \{s\}$
2: set $t(s) \leftarrow 0$
3: for all $n \in N$ such that edge $sn \in E$ do
4:     set $t(n) \leftarrow w(sn)$
5: while $N \neq V$ do
6:     find, by sorting, $n \in N \backslash V$ such that $t(n)$ is minimized
7:     add n to V
8:     for all $m \in N \backslash V$ such that edge $nm \in E$ do
9:         $t(m) \leftarrow \min(t(m), t(n) + w(nm))$
10: end while

Output Graph weights $t : N \to \mathbb{R}^+$

Dijkstra's algorithm initializes the target weights t to infinity (line 1), except for the starting node ($t(s) \leftarrow 0$, line 2) and the nodes adjacent to s

(lines 3–4). The algorithm then proceeds to traverse the entire graph from the start node, at each step sorting the unvisited nodes $N \backslash V$ into ascending order of target weight (line 6) and recalculating the minimum target weights t (line 9).

The progress of Dijkstra's shortest path algorithm for the tram network in Figure 5.36 is shown in Figure 5.39. At the initial iteration (Figure 5.39a), all the target weights are set to infinity, except for the start node, b, and the nodes adjacent to b: a, c, and d. Since c has the lowest weight of the unvisited nodes, $t(c) = 8$, c is the next node to be visited (Figure 5.39b). The target weights for the adjacent nodes to c (h and e) are updated, $t(h) = 18$ and $t(e) = 23$. At the third iteration (Figure 5.39c) node d is visited. At this iteration, the distance from b to e via d is found to be less than b to e via c. As a result $t(e)$ is updated to 16. The next iteration (Figure 5.39d) visits e, generating new target weights for f and g. The algorithm continues until all nodes are visited, although in this case no further changes to the target weights would occur.

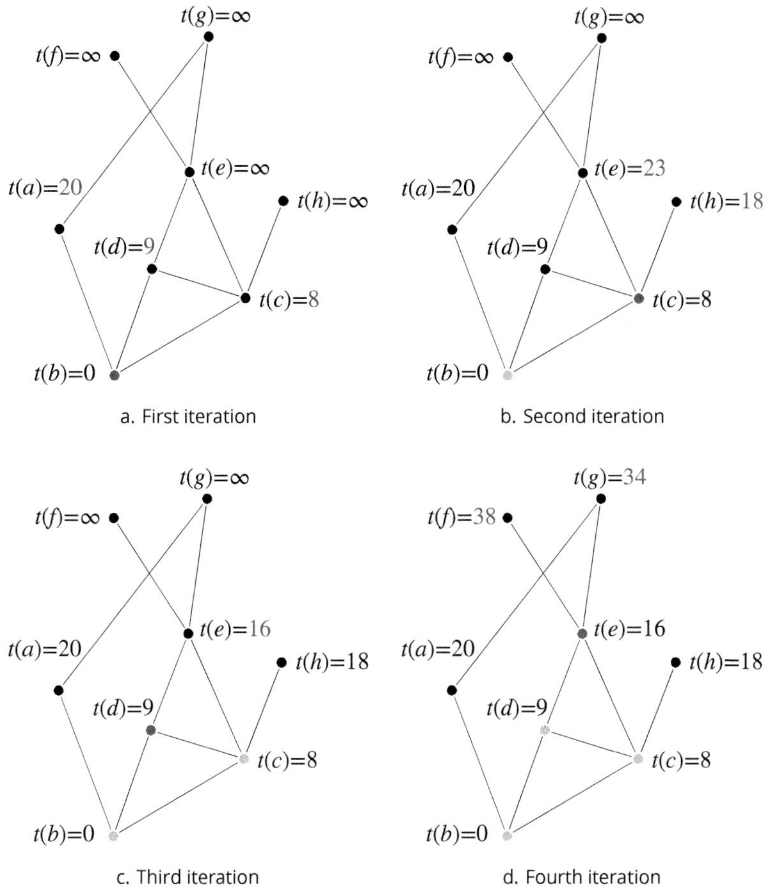

Figure 5.39: First four iterations of Dijkstra's algorithm from node b through the tram network in Figure 5.36. Red indicates the current active node being processed, and the associated neighbor weight updates; pale blue indicates visited nodes

Algorithm 5.8 only stores target weights t, which give the length of the shortest path, but does not explicitly store the actual paths themselves. It requires only a simple modification of Algorithm 5.8 to store the path itself

[5] This is usually achieved by storing at each node its neighboring previous node in the shortest path, updated at the same time as node weights.

single-source shortest path

too.[5] Note also that Dijkstra's algorithm is able to calculate the shortest path from a single node to *all* the other nodes in the network, termed a *single-source shortest path* algorithm. Dijkstra's algorithm is able to achieve this with a time complexity of $O(n^2)$, where n is the number of nodes (because it visits each node once, and must recalculate target weights each time it visits a node). If we are interested in the shortest path from a single starting node to a single destination node, we could improve the average-case time complexity by halting the algorithm when the destination node is added to the visited list. However, this change does not affect the worst-case time complexity, as the destination node might still be the very last node to be visited.

A* algorithm

goal-directed

evaluation function

A close relative of Dijkstra's algorithm is the A* (pronounced "ay-star") algorithm. The A* algorithm is a *goal-directed* version of Dijkstra's algorithm, meaning that it aims to find the shortest path from the starting node to a particular destination (goal). A* uses a *heuristic* to achieve this: at each iteration, it preferentially visits those nodes that are closest to the destination node. To do this requires an *evaluation function* that gives a consistent underestimate of the remaining distance from each node to the destination. A* can be particularly useful for geospatial information, since the Euclidean distance between two points often forms a convenient evaluation function. For example, when traveling through a road network from node n_1 to n_2, the shortest possible distance is the Euclidean straight-line distance between n_1 and n_2 (assuming the Earth is locally planar). The actual distance we need to travel will always be at least as far as the Euclidean distance, because the twists and turns of a road network usually prevent us from traveling directly to our destination. In the example of our travel time network, in Figure 5.36, there is no such obvious evaluation function, so A* could not be used.

all-pairs shortest path

The A* algorithm is essentially the same as Dijkstra's algorithm, so the worst-case time complexity for A* is still $O(n^2)$. However, in practice A* offers considerable improvements on Dijkstra's algorithm for average-case time complexity when computing goal-directed shortest paths where a suitable evaluation function exists. There are occasions when the shortest path from *all* possible starting locations to *all* possible destinations is required, termed an *all-pairs shortest path* algorithm. Clearly, a simple way of achieving this is to iterate Dijkstra's algorithm using every node in the network in turn as the start node. As a result, the time complexity of an all-pairs shortest path algorithm is usually $O(n^3)$, although some algorithms exist that are able to improve on this slightly.

5.5.4 *Other network operations*

subgraph

spanning tree

In addition to the operations described so far in this section, there are many other operations on graphs. Figure 5.40 illustrates five of the most common, based on a simple weighted undirected graph (Figure 5.40a). A graph that has a subset of nodes and edges of another graph G is called a *subgraph* of G. A *spanning tree* is a subgraph of a graph G that contains all the nodes of G and has no cycles (i.e., is a tree). The *minimum spanning tree* (see Figure 5.40b) is the spanning tree of a weighted graph with the minimum total edge weight. Figure 5.40b gives the minimum spanning tree of the graph in Figure 5.40a. The minimum spanning tree provides a unique summary of the essential structure of a graph, used for example in transportation network analysis and planning.

minimum spanning tree

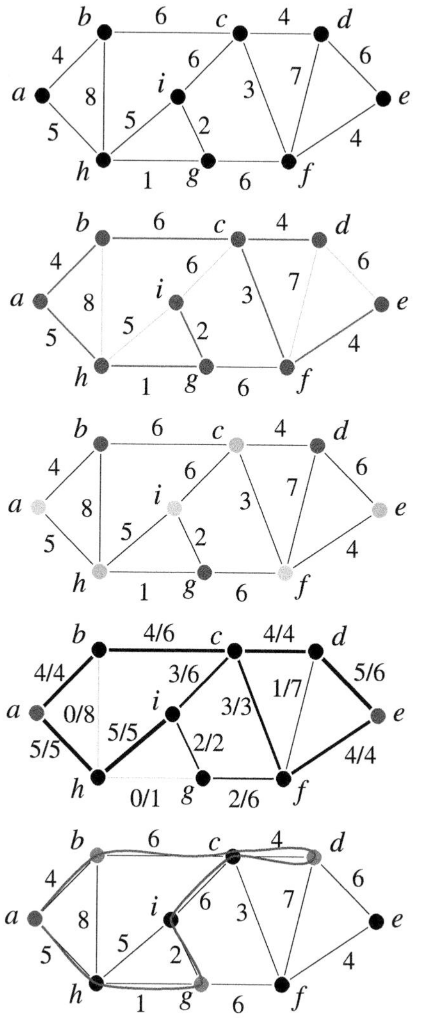

Figure 5.40: Gallery of graph algorithms

a. Weighted undirected graph with 9 nodes and 14 edges.

b. Minimum spanning tree with total weight 29.

c. Graph (node) coloring with chromatic number 3.

d. Maximum flow between *a* and *e* of 9.

e. Traveling salesperson path from *a* visiting *b*, *d*, and *g*.

Graph coloring is a type of graph labeling such that no two adjacent nodes have the same label (color). The *chromatic number* of a graph is the smallest number of colors (labels) that can be used to color the graph. The chromatic number of the graph in Figure 5.40c is 3: fewer colors would leave it impossible to avoid two neighboring nodes having the same color. Graph coloring has been famously applied in connection with maps through the four-color theorem. The theorem states that four colors are enough to color any map such that no two adjacent regions have the same color. Proving the four-color theorem required the translation of the map into a planar graph, with

graph coloring

chromatic number

Box 5.6: NP-completeness

Many common search problems, including the traveling salesperson problem, belong to a class of problems known as *NP-complete*. NP-complete problems have the property that it is relatively easy to check whether a solution is correct, but very difficult to actually find the solution. As an example, the NP-complete *subset sum problem* aims to find whether a subset of a finite set of integers sums to a given number. For example, even for the small set $S = \{-15, -3, 7, 11, 29, 34, 38\}$ deciding whether any subset of S sums to 55 is not immediately obvious. It is, however, trivial to check that the subset $\{-15, 7, 29, 34\}$ does indeed sum to 55. Similarly, we may rephrase the traveling salesperson problem as a *decision problem* (a problem with a "yes" or "no" answer) by asking "For an edge-weighted graph is there a Hamiltonian circuit with total weight less than k?" A *Hamiltonian circuit* is a cycle that visits each node only once. In general, the answer to this question requires exponential time to compute; checking whether a specific route is indeed a Hamiltonian circuit with total weight less that k can be computed very rapidly. All known NP-complete problems can be solved in exponential time (they are intractable). However, no one has ever managed to *prove* that NP-complete problems cannot be solved in polynomial time. What makes NP-complete problems even more interesting is that all NP-complete problems are interchangeable, in the sense that if an algorithm were ever found for solving one NP-complete problem in polynomial time, the same solution could be applied to every other NP-complete problem. As a result, NP-completeness has been the subject of extensive research by the theoretical computer science community (e.g., Fellows et al., 2011).

nodes representing regions and edges representing adjacency between regions (termed an *adjacency graph*). A manual proof eluded mathematicians since the mid-1800s and was only finally proven with the assistance of computers in the 1970s, one of the first long-standing mathematical theorems to be proven in this way.

adjacency graph

maximum flow problem

The *maximum flow problem* treats the weights on a graph's edges as maximum capacities and searches for the maximum flow through the graph from a source node to a sink node. For example, in Figure 5.40d the capacity of the two edges leaving node a is 4 (edge ab) and 5 (edge ac), giving an upper limit to the flow leaving a of 9. In fact, propagating maximum flows through the graph, the maximum flow between a and e is able to reach this total, giving a maximum flow of 9. Figure 5.40d shows the individual loads on edges required to achieve this maximum flow, with edges ab, ah, hi, ig, cd, cf, and fe all at capacity. Maximum flow has clear applications to network analysis, such as transport planning.

traveling salesperson algorithm

Finally, the *traveling salesperson algorithm* computes the round-trip traversal of an edge-weighted network, visiting all the nodes in such a way that the total weight for the traversal is minimized. A naive solution, calculating all possible routes and choosing the best, has exponential time complexity and so is impractical for any but the smallest examples. Heuristic methods, such as at each stage visiting the nearest unvisited node, allow good approximations in reasonable time. In fact, the traveling salesperson operation is a member of a large class of problems termed *NP-complete* (see Box 5.6).

NP-complete

Reflections

Most GIS today will offer all of the representations and algorithms encountered in this chapter, and many more besides. GIS are typically designed for those algorithms to be accessible to these without requiring the user to unpack how the algorithms actually work, as we have done in this chapter. Taking the time to "look under the hood" and understand some of the ingenious algorithms behind GIS can be enjoyable and rewarding; we hope you will agree. But beyond intellectual curiosity, a grounding in spatial representations and algorithms can furnish us with a deeper of sense of what is possible. This "sense of the possible" is what enables a skilled geographic information scientist to quickly deduce what functions will be useful for an application using any GIS, without needing to be familiar with the particular software package or even needing to consult the manual.

If from amongst all these representations and algorithms we were to select just one that every geographic information scientist should know by heart, it would be Dijkstra's shortest path algorithm. Versions of Dijkstra's shortest path algorithm power the ability to compute millions of routes through dynamic networks that people and businesses around the world rely on every day (Delling & Wagner, 2007). The algorithm underpins a wide variety of other vital GIS applications, including ride sharing, multimodal and public transport trip planning, ambulance and emergency response, traffic simulation, and even robot path planning (Bast et al., 2016). Dijkstra's algorithm has blossomed into dozens of fascinating routing variants too, including finding routes that are simple to explain (Duckham & Kulik, 2003); routes based on landmarks (Nuhn & Timpf, 2022); routes that avoid turns (Haque, Kulik, & Klippel, 2006); and even routes to minimize the potential impacts of wrong turns (Amores, Tanin, & Vasardani, 2021).

For those who find their appetite for spatial algorithms grows beyond this chapter, there lies a feast of further work in the field of computational geometry (De Floriani, Magillo, & Puppo, 2000). Preparata & Shamos (1985) and de Berg, Cheong, van Kreveld, & Overmars (2008) are two classic texts for broader background on computational geometry. The many applications of computational geometry to GIS include terrain analysis and TINs (De Floriani, Marzano, & Puppo, 1994b; Mitasova, Mitas, & Harmon, 2005); cartography and map design applications (e.g., Buchin, Speckmann, & Verbeek, 2011; Van Kreveld & Speckmann, 2007); moving object analysis and trajectory segmentation (e.g., Buchin, Driemel, Kreveld, & Sacristán Adinolfi, 2011; Dodge, Weibel, Ahearn, Buchin, & Miller, 2016); and similarity metrics and map matching algorithms (e.g., Alt, Efrat, Rote, & Wenk, 2003; Driemel, 2013) amongst many other examples.

Structures and Access Methods

6

Sections

The organization of data in computer storage plays a vital role in enabling efficient data retrieval. The special characteristics of spatial data make storing and retrieving spatial data a particularly challenging prospect. Through the chapter you will learn to:

- justify the need for **indexes** as the foundation of efficient database access;
- discuss the specialized techniques for storage and retrieval of 2D spatial data required by a **spatial database**; and
- construct and apply a range of indexes to spatial and non-spatial data, including **B-trees, quadtrees, R-trees**, and **z-ordering trees.**

Structures and access methods take us a step further on the path from high-level conceptual models toward machine-level considerations for GIS. In the previous chapter, the emphasis was upon computationally appropriate ways of representing and manipulating different kinds of spatial data. We now move on to consider questions of storage, and in particular storage structures that allow acceptable *performance*. In this chapter, the performance of an information system is taken to be measured in terms of database size and response times to queries. Other aspects of performance, such as the suitability and usability of the interface, are considered in Chapter 8. The discussion in this chapter begins with an introduction to the main issues of performance for general-purpose databases, before taking up these issues in the special context of spatial data. We will see that while the fundamental principles of good database indexing practice still apply, spatial data storage and access presents its own special problems.

6.1 *Physical storage and access*

We have already seen that a fundamental requirement for any database is the persistent storage of the data it contains (Section 2.1). Ultimately, persistence requires that any database must store its data on physical storage devices directly connected to the computer, such as hard disk drives (HDDs) and solid state "flash" drives (SSDs, see Section 1.4 for a refresher on computing technologies and hardware issues). Recall that such persistent, connected storage is referred to as *secondary storage*, in contrast to volatile memory directly accessible to the computer CPU, termed *primary storage* (Section 1.4.3).

block

Irrespective of the specific secondary storage device, however, the atomic unit of data held on secondary storage is known as a *block*. A block is the smallest amount of data that can be accessed in a single read or write operation. Different types of secondary storage media entail different block sizes. Typical HDDs have a block size of 512 bytes (a byte is 8 *bits*, binary digits), for example. For reference, 512 bytes equates to the storage of about 500 characters of plain text. Each block has a unique address that enables the physical storage device to access the data in that block.

byte

bit

file organization

The next level of data storage is termed *file organization*. File organization structures stored data into fields, records, and files.

field

record

file

- A (named) place for a data item in a record is termed a field. Note that this sense of "field" is unrelated to "field" in the sense of field-based models.
- A sequence of fields related to a single logical entity is termed a *record*.
- A sequence of records, usually all of the same type, is organized into a *file*.

Collections of related files form a database.[1] Files are physically placed upon a disk by assigning blocks to hold records. If the disk block is smaller than the record size, then records will be spread across multiple blocks. Otherwise (and more usually) each block contains several records. A field that serves to identify each record within a file unambiguously is called a *key field*. Figure 6.1 summarizes the file organization in secondary storage, with example text, integer, and identity (key) fields in each record, and up to two records in a block.

[1] Note that although the terminology used is more general, there is a clear relationship between the concepts defined in connection with file organization and those used in databases (cf. "field" and "attribute," "record" and "tuple," "file" and "relation" or "table").

key field

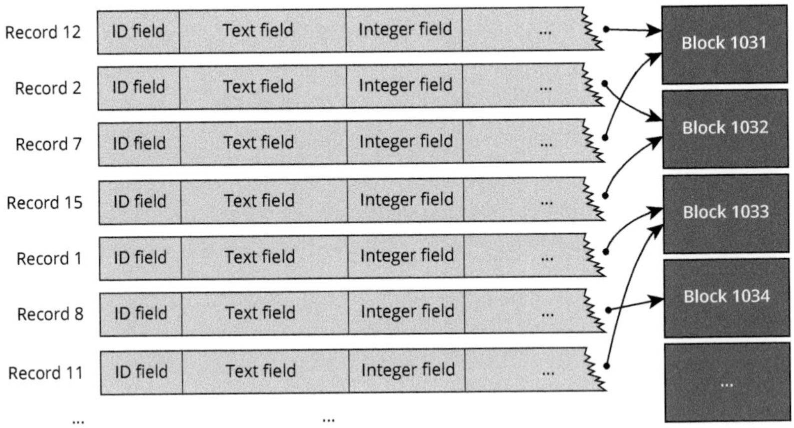

Figure 6.1: Fields and records in a file, and the physical storage in blocks

Understanding how data is physically organized in secondary storage is important because it has implications for database performance. Persistent data must be retrieved from storage before it can be used in computations, such as the algorithms in the previous chapter. The results of those computations may themselves need to be returned to persistent storage. The physical characteristics of storage devices and file organization will affect the performance of these database read and write operations. For example, an important factor in disk-based storage, such as HDDs, is the time taken to rotate the disk and move the read head to the physical location on disk containing the required block, termed *seek time*. SSDs, in contrast, have no moving parts (see Section 1.4) and require no seek time. The manner in which blocks of records are organized, linked, and the way new records are inserted into storage, though, always have an impact on database performance regardless of storage device.

seek time

6.1.1 *File organization*

Irrespective of the type of DBMS, the data in a database must be stored persistently as a collection of files, each containing a collection of records and fields. Just as the conceptual data model provided the database with *logical data independence* from the underlying data structures (i.e., the logical data model), so the logical data model helps ensure *physical data independence* from the underlying data storage (physical data model). Nevertheless, efficient physical file organization of persistent data in secondary storage is an important determinant of database performance.

Heap file organization The most basic file organization structure is called *heap file organization*. Heap file organization simply inserts new records at the end of files, either in the last used block or in a new block if the last used block is full. The order in which records will be created is entirely dependent on the specific application instance, and cannot be controlled or known in advance by the database. Consequently, using heap file organization is *unordered* with no relationship between a record's attributes and its physical storage location.

heap file organization

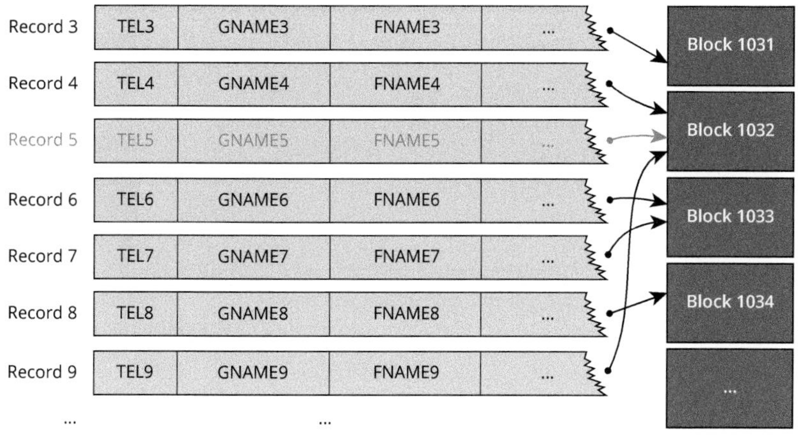

Figure 6.2: Heap file organization, showing insertion in order of creation, with a hole created by deletion of record 5 subsequently filled by record 9

Heap file organization makes insertion of a new record highly efficient. If an operation only requires that a whole file is accessed sequentially—for example, to print out all the records in file or all the tuples in a table—then retrieval using heap file organization may also perform acceptably. However, because the unordered file has no structure beyond entry order, most retrievals must search through each record of the file in sequence.

As a result, heap file organization makes retrieving specific records, or records in a specific order, highly inefficient. Searching for a specific record, for example, requires each record to be read and examined in turn. Such a search is called a *linear search*, or a *brute-force* approach, and is to be avoided if at all possible. For n records, a linear search may need to retrieve every record. A linear search therefore has linear time complexity, $O(n)$. We will see shortly how indexes can help to solve this problem.

linear search

Another difficulty arises for highly dynamic files. As records are deleted, "holes" are created. It is possible to modify heap file organization to mark holes and insert new records into these positions, but this will slow the insertion operation.

sequential file organization

ordering field

ordering key

Sequential file organization In *sequential file organization*, each record is inserted into the file in the order of values in one or more of its fields. The values in the chosen field, called the *ordering field* must be capable of being totally ordered (e.g., integers with the usual ordering, character strings with lexicographic ordering, dates with temporal ordering). For example, in the Nutty Nuggets database in Appendix A we might decide to use the contact phone TEL as the ordering field in the file that contains the data for the CUSTOMER relation. In cases such as this where the ordering field is also the key field we can also refer to this field as the *ordering key*. Figure 6.3 summarizes the approach graphically.

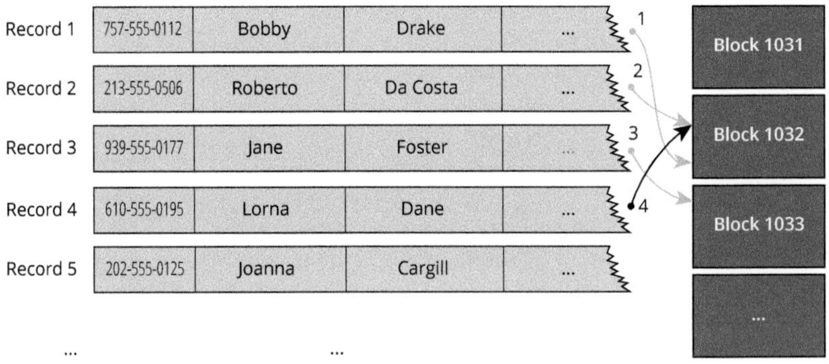

Figure 6.3: Sequential file organization, with two records to each block. The insertion of the fourth record will cause an overflow in block 1032, necessitating shunting of other records to accommodate

Compared to heap file organization, creation of records using sequential file organization will be slower because the correct insertion location must be found before inserting the new record. However, the great advantage of an ordered file organization is that *binary searches* on ordered key fields become possible.

binary search

The *binary search* algorithm is given in Algorithm 6.1. Binary search takes advantage of the ordering in the file to successively "chop" the file in half until the targeted records are found and retrieved or no matching records are found. Note that the operation **div** in Algorithm 6.1 forms the integer part of the quotient of two integers (thus 15 **div** 2 = 7). With a file utilizing n disk blocks, the number of chops can be at most $\log_2(n)$. Thus the time complexity of the binary search method is logarithmic for retrieval of a single record—a huge improvement on linear search.

Algorithm 6.1: Binary search

Input An ordered file with an ordering field, placed on n disk blocks (labeled 1 to n), and a search value v

```
low ← 1;
high ← n;
while high ≥ low do
    mid ← (low + high) div 2
    read block mid into memory
    if v < value of ordering field in first record of block mid then
        high ← mid − 1
    else
        if v > value of ordering field in last record of block mid then
            low ← mid + 1
        else
            linear search block mid for records with value v in ordering field
            store search hits in new table result
            break
end while
```

Output Table *result* containing records with value v in their ordering field

For our previous example, retrieving a specific customer's record, suppose that the file is placed upon 1000 blocks. Then linear search would require on average 500 block accesses. Ordering by the key field TEL, binary search would require approximately $\log_2(1000) = 10$ accesses. Note that if multiple records match the search condition, then the time complexity may no longer be logarithmic. In the worst case, where all records match the search condition, the time complexity is clearly linear. However, in all those cases where the number of records retrieved is small compared with the size of the data file, the complexity is approximately logarithmic.

Many retrievals are based on range search conditions, rather than equality conditions. For example, the search might involve finding all the customers with family names beginning with "D". Assuming family name has been used as the ordering field, then records close in the range will be close in the ordering, so a binary search will also effect significant performance improvement for range queries. It is straightforward to modify Algorithm 6.1 to handle range searching.

For a regularly updated file, insertion of a record can be expensive, as all succeeding records must be shunted along, as illustrated with the insertion of record 4 in Figure 6.3. The same problem arises with deletion, although we could just mark records that are deleted and clean the file up periodically. Nevertheless, ordering a file makes a significant improvement to performance with searches on the ordering fields. Retrievals on non-ordering fields, however, are once again reduced to linear search. In the example above (of a range search for family name beginning with "D") if the ordering field is contact telephone (as in Figure 6.3), then sequential file organization provides no advantages to the search. This limitation is central to the rationale of a secondary index, tackled shortly in Section 6.1.2.

linked list file organization

pointer

Linked list file organization Sequential file organization provides a mechanism to order records into physically contiguous blocks in secondary storage using the logical ordering of records by ordering field. *Linked list file organization* goes a step further and embeds the logical ordering in the data items themselves by means of *pointers*. A *pointer* or *pointer field* is a data item that contains the addresses of another record in storage.

list head

singly linked list

doubly linked list

Pointers allow the records to be arbitrarily ordered, or sorted according to any field. For example, Figure 6.4 shows six customer records connected together in a linked list sorted by the TITLE field. The first record (for Dr Jane Foster, contact number '939-555-0177') is termed the head of the list. Starting with the head, the last field in each record contains a pointer to the address of the beginning of each successive record in the list. Figure 6.4 shows a *singly linked list*, which enables all the records to be traversed in order from the head. To aid in traversal of the list in either direction, a second pointer to the prior record is often added to each record, termed a *doubly linked list*.

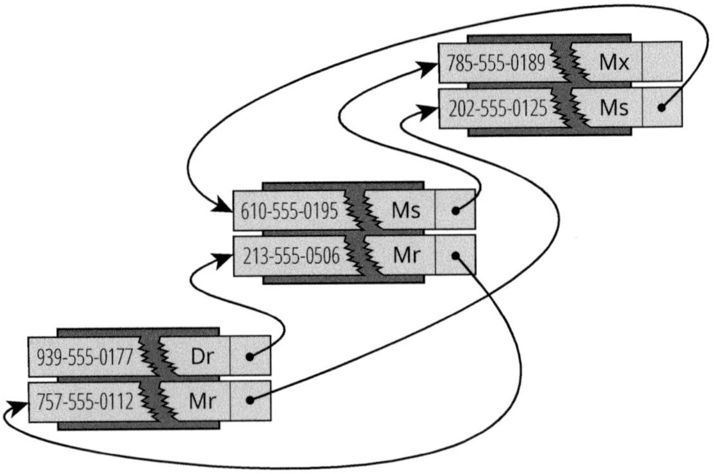

Figure 6.4: Singly linked list file organization. Beginning with the record with key field '939-555-0177', the entire list can be traversed in title order

Linked list file organization makes insertion end deletion of records much simpler than sequential file organization. To insert a new record in a doubly linked list, for example, one only needs to update the inserted, next, and prior records' pointers. Deletion likewise requires only the next and prior records' pointers be updated. To avoid the entire list being traversed for every query, a separate directory file can list all the pointers in order. Combining the pointer and the index field value in the directory file enables as binary search on that field, as for sequential file organization.

Linked lists are particularly useful for dynamic data sets and where queries frequently require all records in a set to be retrieved. As a result, linked

Box 6.1: Hashing

Hashing is a form of file organization that has much in common with a simple index. A *hash file* is organized using a *hash function*. For each record, the hash function transforms the values of a field into the addresses of the disk blocks where each record is stored. The field upon which a hash function operates is referred to as a *hash field*. To search for records on a hash field, the hash function is applied to the search value, and the address of the disk block is calculated. As a result, only one disk block access is required for each read or write operation. Taking a highly simplified example, suppose that we have 100 blocks to store a file containing one year's worth of weather reports. We could hash on the date field by using a key field for each block based on the day, month, and final two digits of the year. To ensure an even distribution of records across all blocks, a popular hash function generates a block address using the formula $k \bmod p$, where k is the key field value, p is a selected prime number lower than the number of available blocks, and $\bmod$ is the modulo operation (the remainder of k/p). For example, taking $p = 59$ the weather report for 21 August 2023 would be placed in a block with address $210823 \bmod 59 = 16$. To retrieve the record for 21 August 2023 then simply involves using the hash function to find the block address 16 from key value 210823. Similarly, the record for 22 August 2023 can be found at block address 45. Hashing is in principle a simple technique for file organization. However, complexities arise when we consider what happens when disk blocks *overflow*. Also, in order to choose the most suitable hash function, it helps to know in advance the number of disk blocks to be allocated.

lists are a common file organization scheme in graph databases, where the highly flexible and evolving structure of data requires storage that is similarly flexible. In a graph database, the record representing a node might be the head of a linked list that traverses, in turn, each of the properties of that node. Indeed, linked lists need not be limited to singly or doubly linked linear orders. A record representing an edge in a graph database might be the head of a linked list of edge properties, in addition to containing pointer fields to the records for the two nodes it connects.

From file organization to indexes Databases, by their very nature, are designed to be flexible and to respond well to a wide variety of queries. In general, queries are unknown in advance, and database content can change considerably in the course of its lifetime. File organization techniques, such as those above and other schemes such as *hashing* (see Box 6.1), aim to ensure the placement of data on secondary storage supports efficient retrieval. However, physical file organization can only hope to provide a partial solution to performance issues. The idea of the directory file used in the linked list leads us directly to the next logical step in achieving good performance: secondary indexes.

6.1.2 *Indexes*

An *index* is an auxiliary data structure that is specifically designed to speed the retrieval of records. Most databases rely on indexes as the primary mechanism for efficient search and retrieval. At root, indexes trade space for speed: an index speeds retrieval at the expense of the extra space required for the index itself. index

The concept of an index to a file is similar to the index to a book. The index for this book, for example, contains an ordered sequence of important terms used to reference pieces of text. Without an index, the book would need to be searched from cover to cover every time you needed to refer to a particular topic not listed in the contents. An index is ordered, so it can be searched quickly (perhaps with a binary search). To find a topic in the main text, you may search for the item in the index (for example, "database," listed alphabetically) and retrieve a reference to the positions (addresses) in the main text where the item occurs (for example, page 2).

single-level index

A *single-level index* acts in just the same fashion as a book index or the directory file for a linked list, described above. A single-level index is an ordered file, each of whose records contain two fields:

index field

- an *index field* that contains the ordered values of the field to be indexed in the data file (i.e., the *indexed* field); and
- a *pointer field* that contains a pointer to the indexed record.

block pointer

record pointer

The pointer field may contain a pointer to the address of the block containing the record with that specified index value (termed a *block pointer*); or it may contain a combination of a block address and a record identifier or offset to the specific record within that block (termed a *record pointer*). In the case that the indexing field is a key field, then the pointer field of each index record will contain just a single pointer; otherwise, an index field may require multiple pointers within a pointer field.

Figure 6.5 shows a schematic for two indexes to the ADDRESS table in the Nutty Nuggets database in Appendix A, physically ordered (sequential file organization) by address number. The index on the right of Figure 6.5 uses STATE as its index field with block pointers to the relevant block or blocks. The index on the left of Figure 6.5 uses CITY as its index field with single record pointers to each individual record.

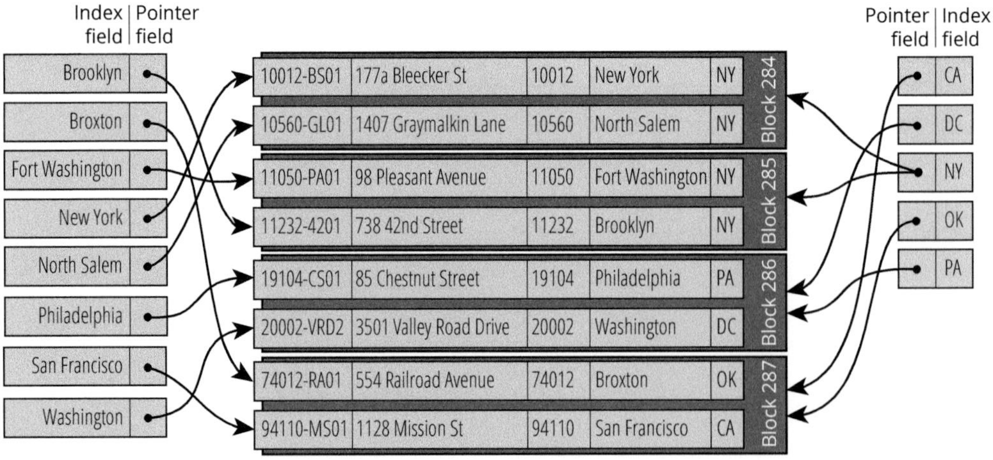

Figure 6.5: The ADDRESS table with a block pointer (right) and record pointer (left) indexes

To retrieve a record from the file based upon a search condition on the indexed field, the index is searched. The index is ordered, so the search can be binary. Having located the relevant record in the index, the pointer indicates the location of the indexed block or record in the data file. In the case of a block pointer, an additional linear search of the few records within the block may be needed to identify exact matches. Just as for a sequential file ordering, the ability to perform binary search ensures the index provides an improvement from linear to logarithmic time complexity for retrieval. Unlike sequential file ordering, multiple fields may be indexed, as in Figure 6.5.

One factor so far neglected in the discussion is the blocking of the index file. For a large data file, the index may itself form a substantial file and may need to be stored across multiple blocks. The search time on the index will be proportional to the logarithm of the number of these blocks. We can cut down the search time even further by allowing the index to be itself indexed, and so on recursively. This is the principle of the *multilevel index*. Alternatively, we can choose only to index a subset of the values in the indexed field, termed a *sparse index*. In a *dense index*, every value in the indexed field also appears in the index itself; both indexes in Figure 6.5 are dense.

multilevel index

sparse index

dense index

B-trees As index structures become more complex, maintaining the index through changes and updates becomes more difficult. When a record is deleted or added to a data file, the index must also be modified. If the index is multilevel, then a modification may have ramifications throughout its levels. A *B-tree* is a fundamental index structure that handles modifications in a particularly elegant fashion. B-trees remain *balanced* in a dynamic setting, in the sense that the various branches of a balanced tree remain of equal length as the tree grows and shrinks.

B-tree

Suppose that we have a set of data records, indexed by an index field (any type that is capable of being linearly ordered can be used as a B-tree index domain).

Each node in a B-tree can store up to a fixed maximum number of indexed values—at least two, but often more. Within each node, the indexed values are in strictly increasing order. Each indexed value is accompanied by a pointer to the stored data associated with that value, much like the basic indexes we have already seen. Figure 6.6 shows a single B-tree node holding up to three indexed values and associated pointers. For clarity, the data pointers in Figure 6.6 are not shown in the subsequent B-tree figures.

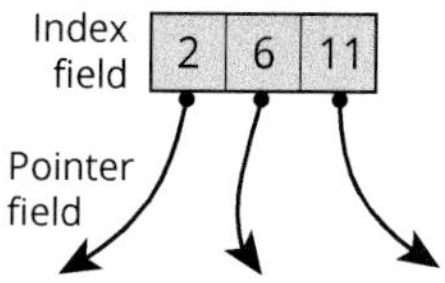

Figure 6.6: A B-tree node with three indexed values (fan-out ratio 4) and associated data pointers

Like any tree, a B-tree contains two sorts of nodes: leaf nodes and internal (non-leaf) nodes. Each internal node additionally has internal *tree* pointers to immediate descendant nodes in the tree. Internal nodes always hold one more tree pointer than indexed values. The (fixed) maximum number of tree pointers in a tree is called the *fan-out ratio*. The B-tree node in Figure 6.6 has a fan-out ratio of 4 (i.e., maximum of three indexed values at each node),

fan-out ratio

for example. That in Figure 6.7, below, has a fan-out ratio of 3 (two indexed values at each node).

The basic property of a B-tree is that the value of the index field for all descendants is within the range set by the index fields of the ancestor node. In Figure 6.7, for example, the left-most tree pointer will point to nodes containing index values less than 6; the second tree pointer will point to nodes containing index values between 6 and 14; and the right-most pointer will point to nodes greater than 14. Remember: data pointers are omitted from Figure 6.7, with only internal, tree pointers shown.

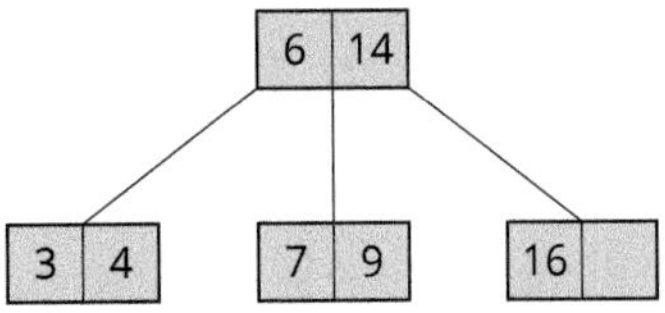

Figure 6.7: A two-level B-tree with fan-out ratio 3

The B-tree operations of search, insert, and delete are described below. Searching a B-tree, based upon a value of the indexing field, is the most straightforward operation, because no restructuring of the tree is required. The other operations restructure the tree to keep it balanced.

Search The search begins at the root node. Depending upon the value of the indexing field, a route will be traced to one of the root's descendants, and so on down the tree until an exact match is found or the bottom of the tree is reached. Upon exact match, the value of the pointer field enables the required records to be retrieved from the data file.

overflow

Insert When a record is inserted into the data file, the B-tree is searched to determine where its index record is placed. If the record index can be inserted into an empty space in the B-tree, no restructuring is required. For example, a new record with index value 19 can be inserted into the B-tree in Figure 6.7 simply by adding 19 to the empty space in the existing right-hand leaf node. If inserting the record index leads to an *overflow* for a non-root node, then that node is split into two equal parts, with the middle value promoted to the parent level. If there is an overflow in the root node, that node is split into two equal parts, each demoted one level, with the middle value retained as the new root node. For example, in Figure 6.8 inserting index value 10 into the B-tree leads to the 7–9 node being split, with 9 promoted to the next level. This in turn leads to the root node being split, with 6 and 14 being demoted one level.

underflow

Delete A similar algorithm to insertion ensures a balanced tree on restructuring the tree following a deletion. If the deletion of the appropriate index value from a node *b* leaves that node less than half-full (an *underflow*), then a process of merging nodes will take place, propagated from *b*. This process is inverse to the insertion restructuring and will not be described in detail. Of course, if *b* is still at least half-full after deletion, then no merging is required.

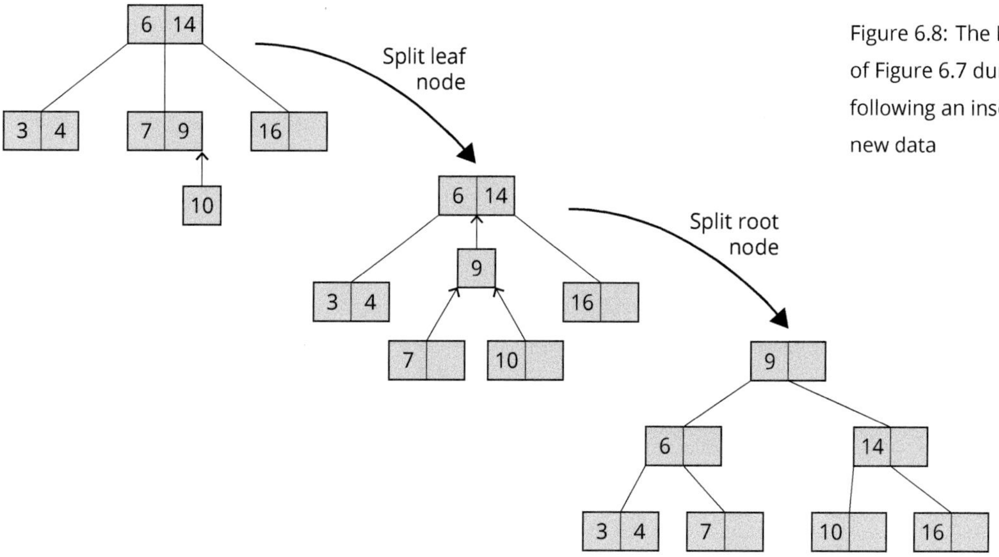

Figure 6.8: The B-tree of Figure 6.7 during and following an insertion of new data

The B-tree structure has the following properties, which are maintained as the structure is dynamically modified with insertion and deletion of records from the data file. Let n be the number of records in the data file to be indexed.

- A B-tree is completely balanced: at any stage in its evolution, the path from root to leaf is of constant depth regardless of the leaf. Search time is bounded by the length of this path and therefore has time complexity $O(\log n)$.
- Insertion and deletion of a record from the file (and possible restructuring of the index) have time complexity $O(\log n)$.
- Each node is guaranteed to be at least half-full (or almost half-full for B-trees with an even fan-out ratio) at all stages of the tree's evolution.

The B-tree provides a balanced multilevel index in a dynamic setting and constitutes one of the first substantive achievements for research into general-purpose data structures. B-trees allow indexes to be placed in a dynamic environment and achieve significant performance improvements. B-trees, or their close relations, are at the core of most modern database management systems. In fact, the *B^+-tree* is the refinement most often used in actual implementations. In a B^+-tree, pointers to records are stored only at the leaf nodes. Thus the leaf nodes must contain every index value and will be more numerous than for the B-tree; on the other hand, the non-leaf nodes will have a simpler structure and will occupy less storage space.

B^+-tree

6.2 *From one to two dimensions*

The discussion so far has applied to any general-purpose database. Files in such a database are multidimensional, each entity having several independent attributes. The customer file used as an example has dimensions corresponding to at least customer contact, given name, family name, and title. While there may be dependencies between the attributes in some cases, usually attributes are independent. When dependencies do occur, they are ideally eliminated by normalization or are of a more statistical nature (e.g., correlations between job descriptions and salaries in an employee database).

With spatial data, the spatial dimensions are orthogonal, but there is a dependency between them expressed in Euclidean space by the Euclidean metric. To illustrate the point, consider the data on the positions of key points in The Potteries, as shown in tabular form in Table 6.1 and on a grid in Figure 6.9.

Table 6.1: Potteries places of interest

ID	Site	East	North
1	Newcastle Museum	14	58
2	Waterworld	31	65
3	Gladstone Pottery Museum	74	23
4	Trentham Gardens	20	00
5	New Victoria Theater	18	55
6	Beswick Pottery	66	25
7	Coalport Pottery	54	36
8	Spode Pottery	37	43
9	Minton Pottery	36	39
10	Royal Doulton Pottery	31	87
11	City Museum	41	62
12	Westport Lake	17	92
13	Ford Green Hall	53	99
14	Park Hall Country Park	86	44

Consider now the effect of retrievals on this data. We distinguish two types of queries on the planar points: point and range queries.

point query

- A *point query* retrieves all records with spatial references located at a given point coordinate.

range query

- A *range query* retrieves all records with spatial references located within a given range. The range may be of any shape, although typically a rectangular area will be specified by the coordinates of two opposite vertices or a disk identified by center and radius.

Consider the following three example queries:

Query One (non-spatial query) Retrieve the point location of Trentham Gardens.

Query Two (spatial point query) Retrieve any site at location $(37, 43)$.

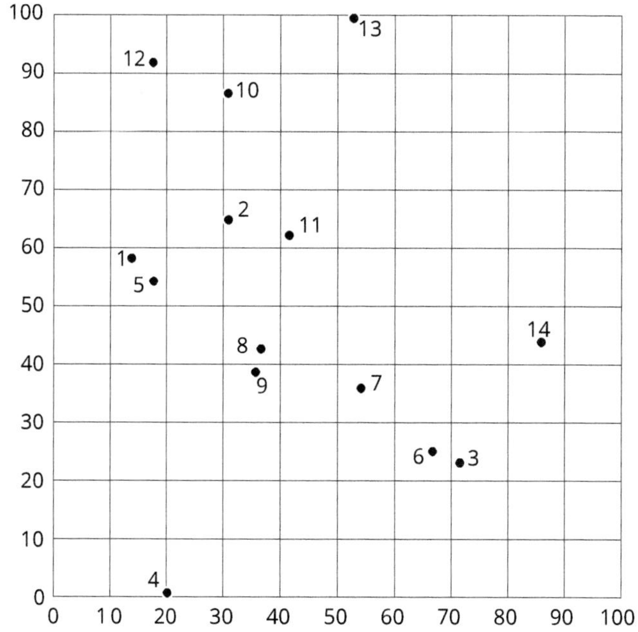

Figure 6.9: Location of some places of interest in The Potteries (labels refer to Table 6.1)

Query Three (spatial range query) Retrieve any site in the rectangle with southwest and northeast vertices (20, 20), (40, 50), respectively.

The first query, although involving spatial location in the result, is concerned only with searching the file using a non-spatial field. Consequently, the general-purpose access methods discussed above would be appropriate for satisfying this query. The second query is a retrieval using a spatial search condition. Without any index or file ordering, a linear time search would be necessary, such as in Algorithm 6.2.

Algorithm 6.2: Point query linear search

```
open The Potteries file
while there are records to examine do
    get the next record
    if the value of the first coordinate field is 37 then
        if the value of the second coordinate field is 43 then
            retrieve site name from record
end while
```

For the third range query, a similar linear-time search may be adopted, as in Algorithm 6.3.

To improve on these linear time queries, we might build two indexes for the two spatial coordinate fields, using the general-purpose database approach. The indexes are shown (with the site names indicating the pointer fields) in Table 6.2, and the two sequences of points are shown as paths in Figure 6.10. For the second query, we conduct a binary search of the east index to locate records that have a first coordinate value of 37. We then go to the full records to check if the second coordinates have value 43, and retrieve

Algorithm 6.3: Range query linear search

```
1: open The Potteries file
2: while there are records to examine do
3:     get the next record
4:     if the value of the first coordinate field is in the range [20, 40] then
5:         if the value of the second coordinate field is in the range [20, 50] then
6:             retrieve site name from record
7: end while
```

those records for which this is the case. For the third query, we could do a range [20, 40] search on the east index, resulting in a list of pointers to the data file. For each pointer in the list, the data file record may be accessed and the north value checked to be in the range [20, 50], in which case the site name is retrieved.

Table 6.2: East and North indexes to The Potteries places of interest

East	Site	North	Site
14	Newcastle Museum	00	Trentham Gardens
17	Westport Lake	23	Gladstone Pottery Msm
18	New Victoria Theater	25	Beswick Pottery
20	Trentham Gardens	36	Coalport Pottery
31	Waterworld	39	Minton Pottery
31	Royal Doulton Pottery	43	Spode Pottery
36	Minton Pottery	44	Park Hall Country Park
37	Spode Pottery	55	New Victoria Theater
41	City Museum	58	Newcastle Museum
53	Ford Green Hall	62	City Museum
54	Coalport Pottery	65	Waterworld
66	Beswick Pottery	87	Royal Doulton Pottery
74	Gladstone Pottery Msm	92	Westport Lake
86	Park Hall Country Park	99	Ford Green Hall

The key observation in this discussion is that only *one* of the indexes is used in these retrievals. What we would like to do is to construct an index that takes advantage of ordering in *two* dimensions. Particularly for range searches, we would like to order the records in such a way that records that are near each other in space are near each other in the index.

6.2.1 2D orderings

The principal problem confronting designers of data structures for 2D and higher-dimensional Euclidean space is that computer storage is essentially one-dimensional. Even though storage devices of course physically exists in 3D space, the computational model of secondary storage is one-dimensional (as can be seen by noting its 1D addressing scheme). Thus we have to unravel the higher-dimensional space so as to minimize the distortion, in a similar way to unraveling a piece of 2D knitting into its one-dimensional yarn. This section examines some of the ingenious techniques that have been proposed for performing the unraveling process in the case of two dimensions.

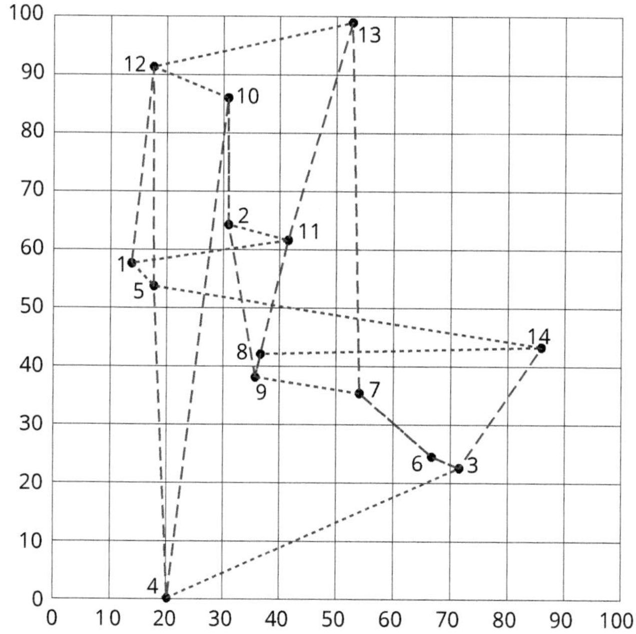

Figure 6.10: The two indexes shown as paths through the places of interest in The Potteries (labels refer to Table 6.1)

More formally, let us take a finite portion of the Euclidean plane P. Without loss of generality, let S be the real interval $[0, 1]$ where the ordering is the usual ordering of real numbers, and let P be the unit square $[0, 1] \times [0, 1]$. We are looking for a bijective function $f : P \rightarrow S$ that has the property that, for all $x, y \in P$, x is near y in the region if and only if $f(x)$ is near $f(y)$ in the interval $[0, 1]$.

In the discrete case, assume that S is regularly tessellated into squares. Then the function f can be represented as a path through the grid. Suitable functions, called *tile indexes*, will have the property that points close in P will be close on the path, and vice versa. In practice, it is not possible to find tile indexes that have this nearness property for all pairs of points, but functions have been found that work well in most cases.

tile index

Figure 6.11 shows six common tile indexes. The *row ordering* shown in the top-left of the figure is the first and most obvious possibility. It corresponds to the scan of a raster image across the rows of pixels. Although simple, it has the unfortunate property that close points at row ends are often quite distant on the path, particularly for tessellations with a large number of cells. The *row prime* order provides a simple modification that allows some improvement, but there is still a problem with half of the end-points. Somewhat better at preserving nearness are the *Cantor-diagonal* order (row-prime rotated, top-right) and the *spiral* order (bottom-left).

row order

row-prime order

Cantor-diagonal order

spiral order

The *Morton* (bottom-center) and *Peano-Hilbert* (bottom-right) orderings are rather more ingenious. The Morton ordering can be seen as a recursive "Z," with the letter "Z" at various scales, or if rotated through a right-angle, recursive "N." Indeed, Morton ordering is often referred to as *z-ordering*. The *Peano-Hilbert* ordering relates to the Morton ordering as the row-prime to the row

Morton order

z-order

Peano-Hilbert order

ordering. Both Morton and Peano-Hilbert orderings are discretized fractals, generated using a finite number of recursive iterations of a fractal generator (see Chapter 3). Morton and Peano-Hilbert orderings are therefore space-filling and can be constructed at arbitrary levels of detail (Section 3.4.3). They are also better at preserving nearness than the other orderings in Figure 6.11.

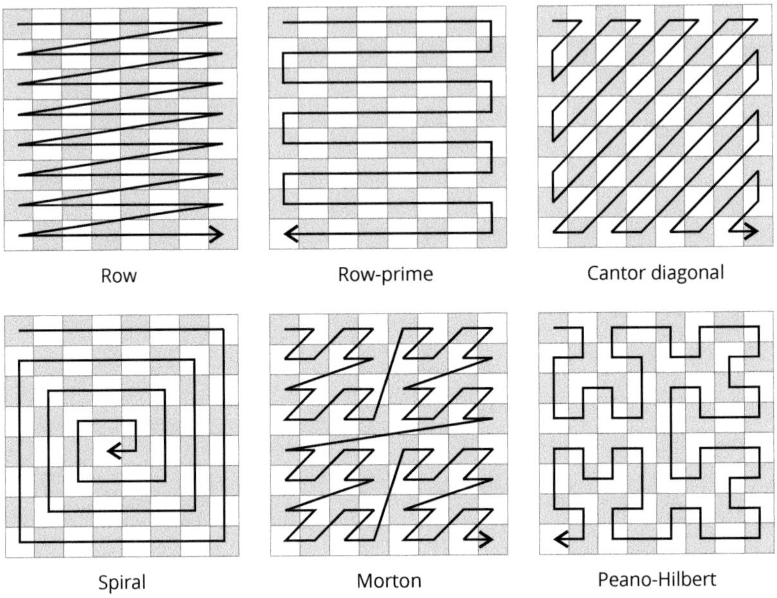

Figure 6.11: Six common tile indexes

6.3 Raster structures

This section reviews some of the basic methods for storing and compressing raster data. Each cell in a raster is addressed by its position in the array (row and column number). For example, in Figure 6.12 (reproduced from Figure 1.16), the raster may be represented as a 32×32 array, each cell occupied by 0 or 1. Rasters are able to represent a large range of computable spatial objects: a point may be represented by a single cell; a strand by a sequence of neighboring cells; and a connected area by a collection of contiguous cells. Rasters are managed naturally in computers because most commonly used programming languages support array handling. However, a raster when stored in a raw state with no compression can be extremely inefficient in terms of computer storage. For example, a large uniform area with no special characteristics must be stored as a large collection of cells, each holding the same value. In this section, we examine ways of improving raster space efficiency.

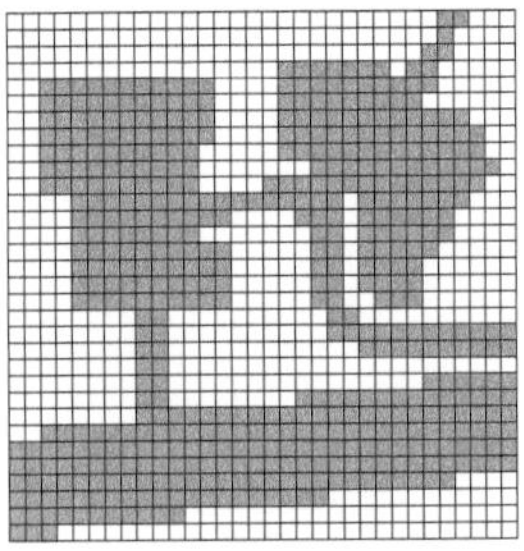

Figure 6.12: An example raster structure after Figure 1.16

6.3.1 Region quadtrees

quadtree

A widely used structure for holding planar areal raster data is the *region quadtree*. The term *quadtree* arises because the underlying data structure is a tree where all non-leaf nodes have exactly four descendants. The principle of the region quadtree is a recursive subdivision of a non-homogeneous square array of cells into four equal-sized quadrants. The decomposition is applied to each subarray until all the subarrays bound homogeneous regions. This process has been applied to the raster in Figure 6.12, and the result is shown in Figure 6.13. To see the process of subdivision in detail, Figure 6.14 shows successive subdivisions of part of the northwest quadrant of the raster.

region quadtree

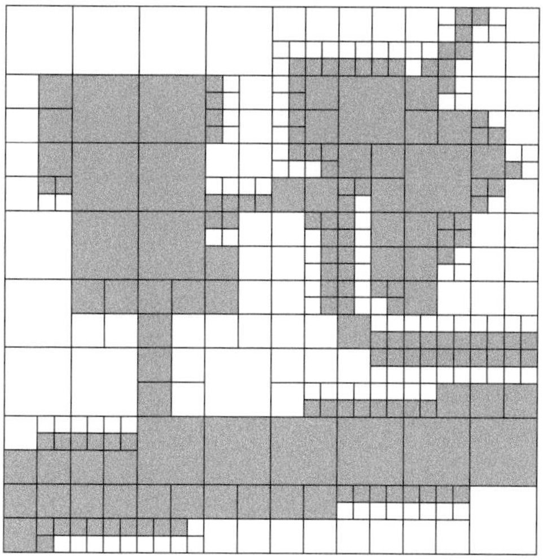

Figure 6.13: Quadtree subdivision for the raster in Figure 6.12

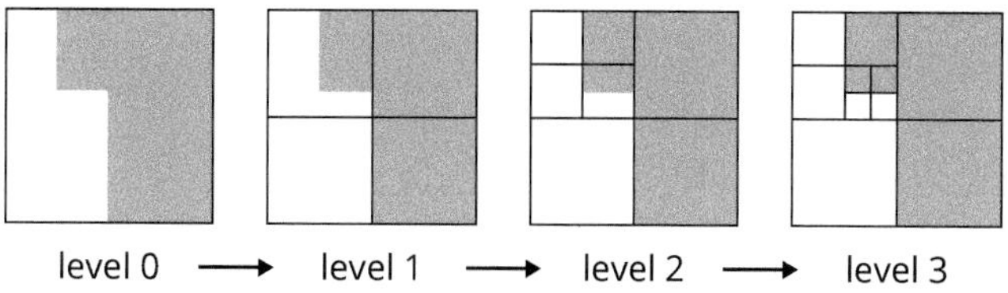

Figure 6.14: Successive subdivisions of a small portion of the raster

Quadtrees are stored in a leveled tree data structure, with the root at the top level (level 0). For each non-leaf node, its four constituent quadrants are represented by four descendant nodes. A homogeneous quadrant, in which no further subdivision is required, is stored as a leaf node. Leaf nodes may have attributes associated with them, such as a color code, or they may point to records of more detailed information in a data file. We need a convention for the order of the descendants at each level of the tree—usually, NW, NE, SW, and SE. Figure 6.15 shows the tree corresponding to the small portion of the raster in Figure 6.14.

The region quadtree structure takes full advantage of the 2D nature of the data; this is not the case for some other simple representations, such as chain encoding (see Box 6.2 on page 235). Further, region quadtrees have variable adaptive resolution, devoting more depth of the tree in places as needed to capture more detail. There is a body of results on space complexity of the quadtree structure, for example:

Figure 6.15: The quadtree for part of the raster in Figure 6.12

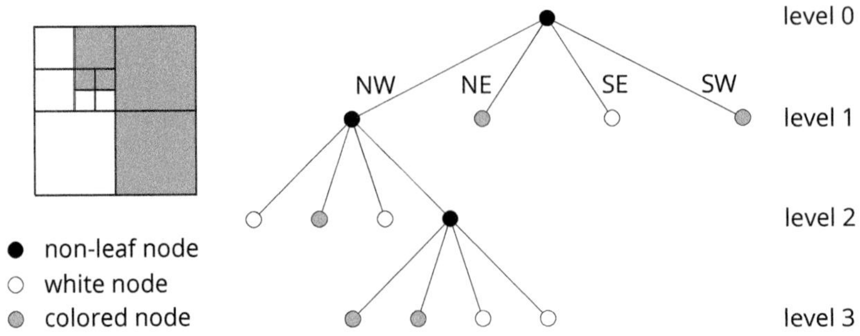

> The quadtree representing a polygon with perimeter m embedded in a [raster] image of size $2^n \times 2^n$ contains $O(m+n)$ nodes in the worst case. (Rosenfeld et al. quoted by Samet, 1990).

Algorithms for Boolean operations such as union, intersection, and difference of rasters based on the region quadtree structure can be elegantly coded and perform well. These algorithms are able to take advantage of the recursion and variable resolution inherent in the quadtree structure. Algorithms for other operations, such as labeling and counting the number of connected components in an areal spatial object, also perform well. Figure 6.16 shows the complement, intersection, union, and difference operations applied to two planar spatial regions Q and R.

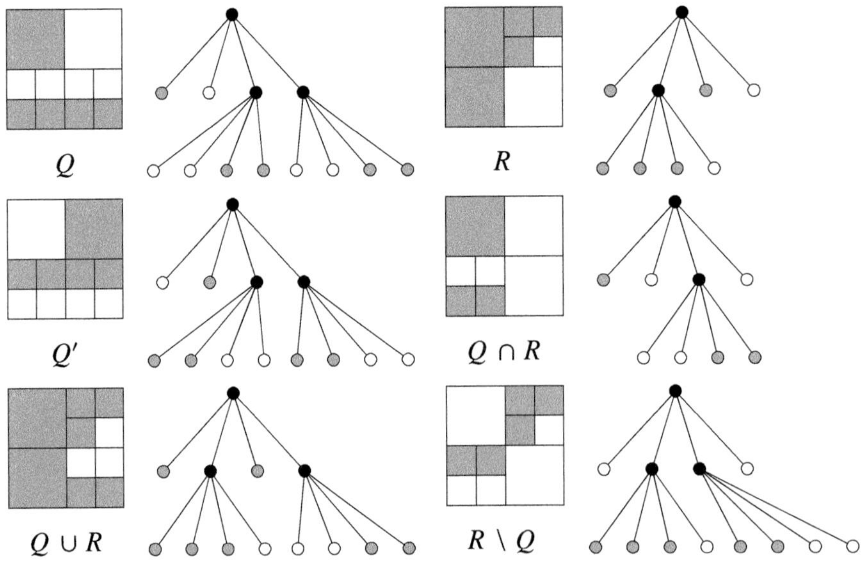

Figure 6.16: Quadtree complement, intersection, union, and difference

For example, the algorithm for quadtree complement involves a simple recursive traversal of the tree. Algorithm 6.4 shows the complement algorithm, based on a breadth-first traversal. Algorithm 6.4 is simpler than the breadth-first traversal algorithm given in the last chapter because the quadtree structure means there is no need to keep a list of visited nodes (see Section 5.5.2).

Box 6.2: Chain encoding

Chain codes, introduced in the previous chapter, may be used to represent a raster region. *Run-length encoding* (RLE) counts the length of "runs" of consecutive cells with the same value, 1 or 0. For example, counting in row order the figure below can be run-length encoded as: $[18, 11, 5, 11, 5, 11, 5, 11, 5, 10, 6, 10, 6, 10, 8, 8, 8, 8, 8, 8, 8, 10, 6, 10, 6, 10, 6, 10, 18]$.

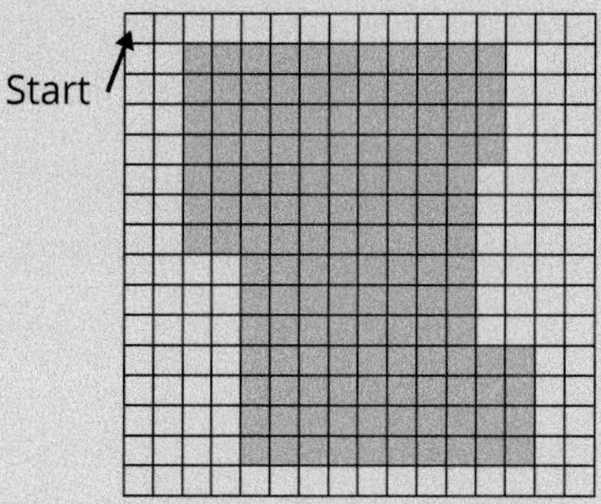

Using a different tile index results in a different run-length encoding. For example, using Morton ordering the same figure may be encoded as: $[6, 2, 4, 4, 2, 2, 2, 10, 4, 4, 4, 20, 2, 2, 2, 10, 2, 1, 5, 1, 1, 1, 5, 17, 31, 16, 16, 10, 2, 2, 2, 16, 10, 2, 4, 10, 2, 2, 2, 4, 4, 2, 2, 4]$. *Freeman chain coding* can encode the boundary of a region, such as already encountered for vectorization (Section 5.3.7). Freeman chain encoding uses the numbers 0 to 7, arranged clockwise around the 8 directions N=0, NE=1, E=2, ..., NW=7. Starting at the northeast corner and proceeding clockwise, the Freeman chain code for the figure below would be: $[2, 2, 2, 2, 2, 2, 2, 2, 2, 2, 4, 4, 4, 6, 4, 4, 4, 4, 4, 4, 4, 2, 2, 4, 4, 4, 6, 6, 6, 6, 6, 6, 6, 6, 0, 0, 0, 0, 0, 0, 0, 6, 6, 0, 0, 0, 0, 0, 0]$.

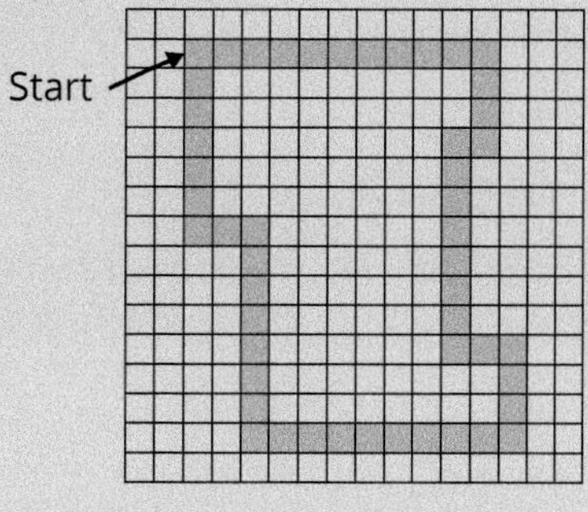

Freeman chain coding and RLE can be used together to produce a compact boundary representation. For example, the Freeman chain code representation above could also be run-length encoded as: $[2, 10, 4, 3, 6, 1, 4, 7, 2, 2, 4, 3, 6, 9, 0, 7, 6, 2, 0, 6]$. Direction codes and the number of consecutive occurrences of each direction code are interlaced in the encoding above (i.e., $[2, 10, \ldots]$ means 10 consecutive occurrences of the Freeman code for "east," 2).

An algorithm for constructing a quadtree S from the Boolean intersection of two binary quadtrees Q and R is given in Algorithm 6.5, similar to the breadth-first complement algorithm given in Algorithm 6.4. When constructing a new node in S, the Algorithm 6.5 must deal with four cases:

Case 1 If q or r (or both) is a leaf node representing a white area, then S is constructed with a white leaf node in the position of q and r (lines 5–6 in Algorithm 6.5).

Case 2 If q is a leaf node representing a gray area, then S is formed using a copy of the subtree with r as its root (lines 7–8).

Case 3 If r is a leaf node representing a gray area, then S is formed using a copy of the subtree with q as its root (lines 9–10).

Case 4 If both nodes q and r are non-leaf nodes, the algorithm is applied recursively to the subnodes of q and r after creating a new non-leaf node in S (lines 11–14).

Other Boolean operations can be composed from the complement and intersection operations, using equational relationships, such as De Morgan's laws. For example, $Q \cup R = (Q' \cap R')'$ and $Q \backslash R = Q \cap R'$.

Algorithm 6.4: Quadtree complement algorithm (breadth-first)

```
Input A non-null binary quadtree Q
r ← root of Q
queue L ← [r]
while L is not empty do
    remove the first node x from L
    if x is a leaf node then
        invert the value of x (0 to 1 or 1 to 0)
    else
        for each descendant y of x do
            add y to the end of L
end while
Output A binary quadtree that represents Q′, the complement of Q
```

A link with the previous section on 2D orderings becomes clear from Figure 6.17. If the checkered pattern is structured as a quadtree and the leaf nodes of the quadtree are read from left to right, then the pattern that the trail follows in the plane corresponds to the Morton ordering of the plane (see also Box 6.3 on page 238).

Algorithm 6.5: Quadtree intersection algorithm (breadth-first)

```
Input Binary quadtrees Q, R
q ← root of Q, r ← root of R
queue L ← [(q, r)]
while L is not empty do
    remove the first node pair (x, y) from L
    if x or y is a white leaf then
        add white leaf to output quadtree S
    if x is a non-white leaf then
        add y and all subnodes to output quadtree S
    if y is a non-white leaf then
        add x and all subnodes to output quadtree S
    if x and y are non-leaf nodes then
        add a new non-leaf node to output quadtree S
        for pairwise descendants x′ of x and y′ of y do
            add (x′, y′) to the end of L
end while
Output A binary quadtree S that represents the intersection Q ∩ R
```

octree

The quadtree is simply generalized to higher dimensions. Thus, in three dimensions we have the *octree*, in which each non-leaf node has eight (2^3) descendants.

Despite its clear strengths, the region quadtree structure has some associated problems. The structure is highly dependent on the embedding of the spatial objects in the raster space: a small translation or rotation of the object will usually result in a quite different quadtree structure. Similarly, dynamic raster data sets that are subject to regular updates are inefficient if stored

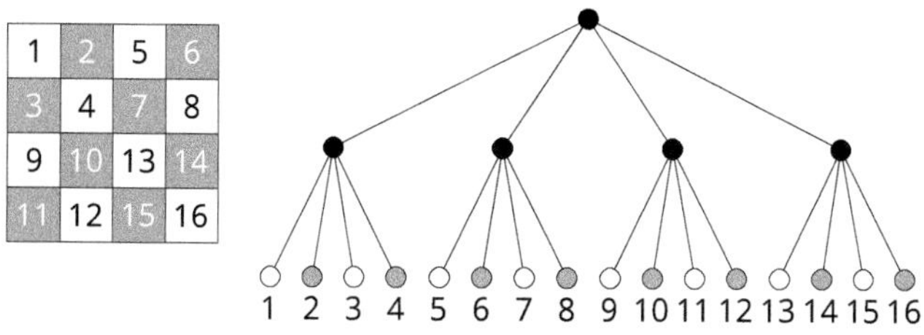

Figure 6.17: A region quadtree that is a Morton order of its leaf nodes

as a quadtree because each update may require major reorganization of the quadtree structure. Highly inhomogeneous rasters, such as the alternating checkered pattern in Figure 6.17, are not efficient to store in a quadtree because there are no homogeneous regions to be amalgamated. Rasters largely composed of point and linear features also have low storage efficiency.

6.4 *Point object structures*

This section considers data structures that are designed specifically with point objects in mind. The three structures that are considered are grid files, point quadtrees, and 2D-trees. As usual, assume that the point objects are located in the plane by means of two coordinates and may have other attributes.

6.4.1 *Grid structures*

As an introduction to grid structures, and to the bucket methods that underlie them, we begin with the fixed grid structure, which allows binary search methods to be applied in more than one dimension. We then introduce the variable grid file structure, which overcomes some of the disadvantages of the fixed grid. We will use the example data set from The Potteries given in Table 6.1 and Figure 6.9.

Fixed grid structure The *fixed grid* or *fixed cell* structure is a partition of the planar region into equal-sized cells; squares in our example. Each packet of data with point locations sharing the same cell of the grid is stored in a contiguous area of secondary storage. The usual term to describe a cell of this structure is a *bucket*. Each bucket represents a physical location in storage where the complete records are held. Thus the grid cell structure is a way of partitioning objects that have planar point references so that neighboring objects are more likely to have their attribute data stored in the same or nearby areas of storage. The performance of range queries is improved as a consequence.

fixed grid

bucket

An example of a fixed grid structure for our Potteries point data set is given in Figure 6.18. The region has been partitioned into 16 squares, each containing between zero and two points. Boundary ambiguities are resolved by making the convention that each square "owns" the points on its south and

Box 6.3: Bit interleaving and geohashing

Using the Morton order has the handy property that translating between row-column and Morton-order cell references is straightforward, using the process of *bit interleaving*. To understand bit interleaving, consider the 16 by 16 grid and its Morton order pictured. If we wish to convert to Morton ordering, we first represent our chosen cell's row and column numbers as binary digits. For example, the highlighted cell at row 1 column 6 (counting from 0 as the first row and column) translates into binary digits 001 (1) and 110 (6). Next we interleave the two binary numbers bitwise, leading with the row. In our example, interleaving gives us the binary number 010110 = 22 (i.e., 22nd cell in the Morton order). The process works in reverse too. For example, cell 50 in the Morton order can be represented as binary digit 110010. Decomposing 110010 into two interleaved binary digits gives us row 101 = 5 and column 100 = 4, also highlighted in the figure. The technique of *geohashing* makes direct use of Morton-order bit-interleaving to provide a compact multi-resolution system for location referencing. Geohashing labels the first 32 5-bit strings with the numbers 0–9 (00000–01001) and 22 letters of the alphabet (omitting four potentially ambiguous letters "a," "i," "l," and "o," so that 01010 = b, 10001 = j, 10011 = m, 10101 = p, and so forth). Geohashed codes have the desirable property that progressively greater spatial precision in location references can be achieved simply by appending additional characters. A geohashing location reference system based on bit-interleaving of latitude and longitude coordinates requires strings of 8 characters (8 × 32 bits = 256 bits) to resolve location at the equator to within about 20 m. A geohash string of fixed length references progressively larger areas further from the equator, as a consequence of the changing great circle distances between spherical coordinates at different latitudes (see Box 3.7 on page 117).

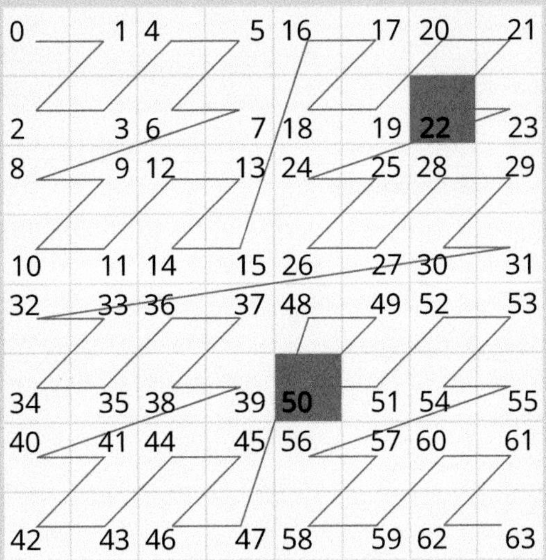

west boundaries, but not points on the other two boundaries. Thus the point labeled "6" belongs to the square containing the point labeled "7."

The ideal size for the partition is dependent upon at least two considerations. First, the larger the total number of points, the more cells will be required (because the buckets correspond to blocks of secondary storage that can hold only a limited number of records). Second, the size of a cell is dependent on the magnitude of the range in the average range query supported by the system. There is no purpose to be served by making the cell size much smaller than the average range: in such a situation, range queries degenerate to near-linear searches.

Looking at Figure 6.18 reveals a major disadvantage of the fixed grid method. The population of cells with points varies, and in some cases there are no points in cells. This problem becomes more acute for less uniformly distributed points. If the distribution is too far from uniform, there will be many empty cells, while other cells will be full to overflowing. Instead, we would like a partition that takes account of and adapts to the density of points in particular local areas. This leads us to the next structure, the grid file (and later point quadtrees).

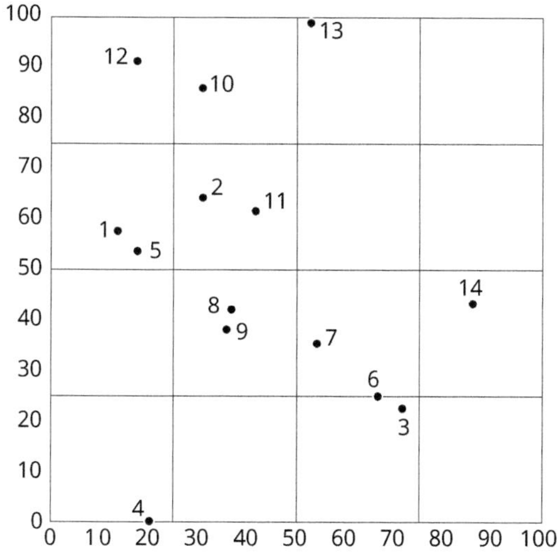

Figure 6.18: Fixed grid partition of Potteries point data (labels refer to Table 6.1)

If the points are uniformly distributed, then the fixed grid structure provides a simple and easy-to-implement solution. The fixed grid is applicable to an arbitrary number of dimensions. It may be appropriate to use a tile planar ordering (e.g., Morton order) for the cells of the grid.

Grid file The *grid file* is an extension of the fixed grid structure that allows the vertical and horizontal subdividing lines to be at arbitrary positions, taking into account the distribution of the points. This structure is designed for dynamic data, entering and leaving the system as time passes. A new level of indirection, called the *grid directory*, is introduced. Grid cells may have their data sharing the same bucket only if their union is a rectangle. Figure 6.19 shows an example, in which a bucket size of two records is assumed (much smaller than a real application). The total area is divided as a three-by-three grid. The two grid cells in the southwest region are amalgamated into a single bucket, because there is spare bucket capacity at that location, and the amalgamation results in a rectangle. The grid directory shows the relationships between cells and buckets. The two linear scales, dynamically updated, show the positions of the partitions.

grid file

grid directory

The grid cell is designed to expand and contract as new data is inserted and deleted. A rectangle may be divided if it becomes too full, and cells may be amalgamated if the space becomes too empty. Note that in our example, we have reduced from 16 buckets in the fixed grid structure to eight buckets with the grid file structure. Like the fixed grid, the grid file is designed to be applicable to an arbitrary number of dimensions.

6.4.2 *Point quadtree*

The *point quadtree* combines the grid approach with a multidimensional generalization of the binary search tree. It has many characteristics in common with the region quadtree, described earlier in this chapter. Assuming as usual

point quadtree

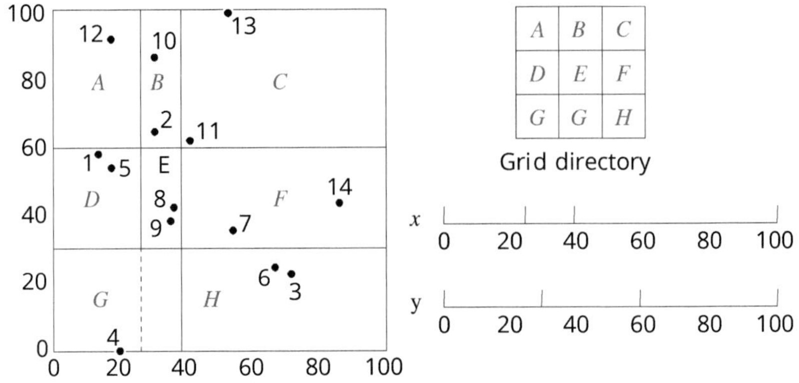

Figure 6.19: Grid file structure for Potteries point data (labels refer to Table 6.1)

that the data is planar, each non-leaf node is associated with a data record for a point location and has four descendants (NW, NE, SW, SE). Thus, each data record comprises two fields to hold the coordinates, four fields pointing to the four descendants, and further fields to hold the data associated with the point, such as the name of the city at that point location (see Figure 6.20).

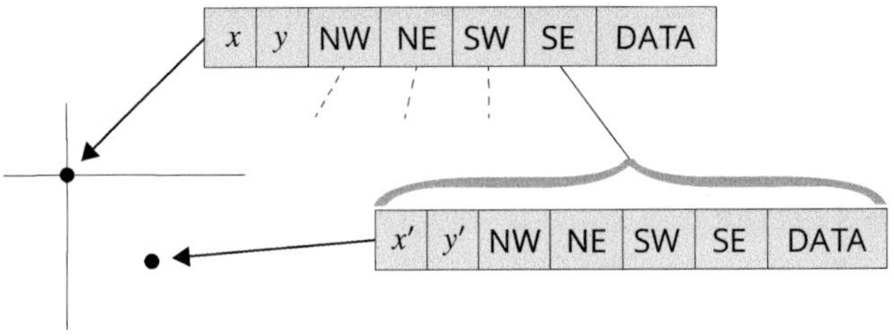

Figure 6.20: Point quadtree records and the points to which they refer

The point quadtree has the property that the position of each partition into quadrants is centered on a data point, unlike the region quadtree in which a quadrant is always partitioned into four equal-sized subquadrants. A data structure in which the positions of subdivisions are independent of the data points is called a *trie* structure, in opposition to a *tree* structure in which the positions of subdivisions are dependent. Thus, a more strict term for the region quadtree is the *region quadtrie*. The point quadtree is a true tree structure, because the center of a subdivision into four is always a data point.

trie

region quadtrie

The point quadtree is described with reference to an example. Consider once again the point data from The Potteries given in Table 6.1 and Figure 6.9. Assume that the points are to be entered into a quadtree index in order of numerical identifier. The first point (Newcastle Museum) is placed at the root of the tree, and the plane is divided into four quadrants as shown in Figure 6.21a. The second point (Waterworld) is compared with the root. We assume the usual ordering of the cardinal directions, NW, NE, SW, SE. Point 2 is northeast of point 1, and so it is placed as the NE (second) descendant of the root.

The northeast quadrant of the plane is divided into four quadrants, as shown in Figure 6.21b. Point 3 (Gladstone Pottery Museum) must now be placed in the tree. It is first compared with the root (SE), and since there are no further nodes along the SE branch, it is placed as the SE descendant of point 1. The southeast quadrant of the plane is divided into four quadrants, as shown in Figure 6.21c. Point 4 is processed next: first, it is compared with the root (SE), and then with point 3, the SE descendant of the root. There are no further nodes for comparison, so point 4 is placed as the SW descendant of point 3. The appropriate quadrant of the plane is divided into four quadrants. Figure 6.21d shows the insertion of points 4 and 5 into the tree and the resulting planar subdivisions. Figure 6.22 shows the full quadtree and planar subdivisions for this data set.

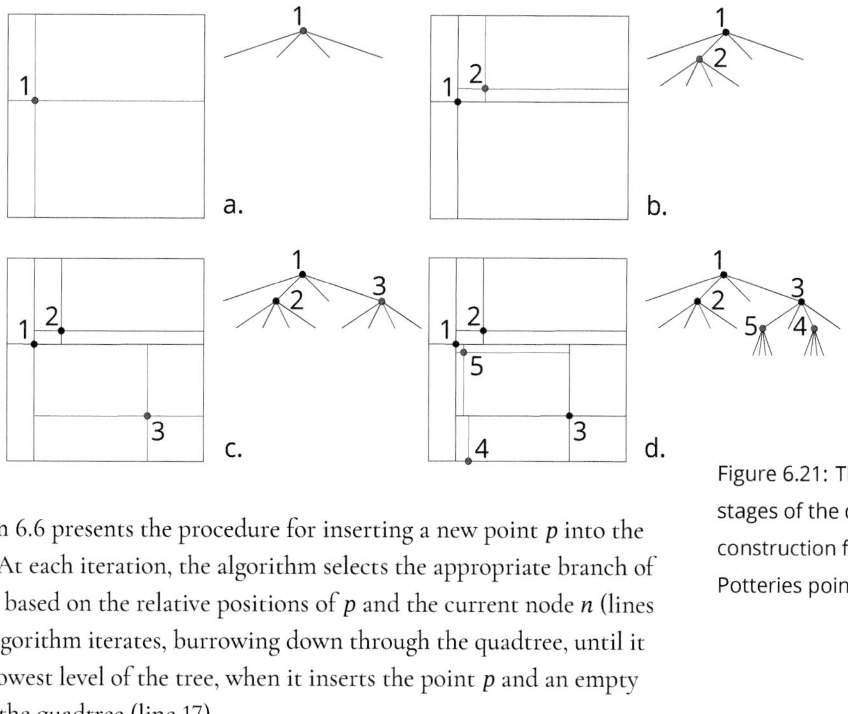

Figure 6.21: The first stages of the quadtree construction for The Potteries point data

Algorithm 6.6 presents the procedure for inserting a new point p into the quadtree Q. At each iteration, the algorithm selects the appropriate branch of the quadtree based on the relative positions of p and the current node n (lines 5–14). The algorithm iterates, burrowing down through the quadtree, until it reaches the lowest level of the tree, when it inserts the point p and an empty subtree into the quadtree (line 17).

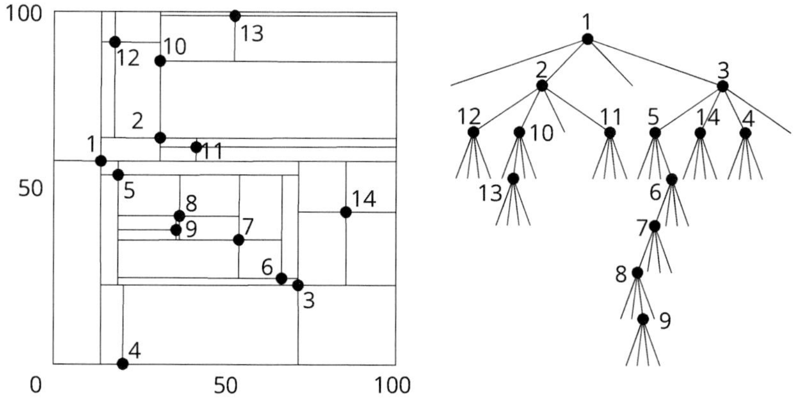

Figure 6.22: Point quadtree for The Potteries point data (labels refer to Table 6.1)

Algorithm 6.6: Quadtree insert algorithm

```
Input Point p and quadtree Q
if Q is an empty quadtree then
    Q ← a new quadtree with root p and four null tree descendants
node n ← the root of Q
repeat
    if x-coordinate of p < x-coordinate of n then
        if y-coordinate of p < y-coordinate of n then
            n' ← the SW node from n
        else
            n' ← the NW node from n
    else
        if y-coordinate of p < y-coordinate of n then
            n' ← the SE node from n
        else
            n' ← the NE node from n
    n ← n'
until n = null tree
insert new node p at position of n with four null tree descendants
Output The quadtree Q updated by the insertion of the point p
```

The shape of a quadtree is highly dependent upon the order in which the points are inserted into it. Figure 6.23 shows the quadtree associated with an insertion of The Potteries data in inverted numerical order: 14–1. This dependence has implications for dynamic data. If a point near the top of the tree is deleted, the resulting tree may be substantially changed. Therefore, point quadtrees are not well suited to dynamic geospatial data.

Some worst-case performance measures can be given for the point quadtree. Let n be the number of points structured by the quadtree. Then, the quadtree build time is proportional to the total path length $O(n \log n)$, and the point query search time is $O(\log n)$. We have not presented the algorithms for point and range query for this structure: these operations will be given in detail for the next structure, the 2D-tree, and can be modified for the point quadtree.

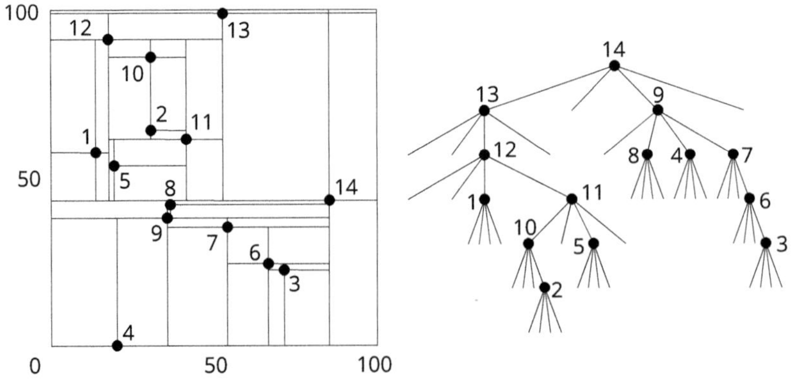

Figure 6.23: Point quadtree for a reversed insertion of The Potteries point data (labels refer to Table 6.1)

6.4.3 2D-tree

The point quadtree takes full advantage of the embedding of the points in a Euclidean plane. One problem facing the point quadtree is the exponential increase in the number of descendants of non-leaf nodes as the dimension of the embedding space increases: for k dimensions, each node has 2^k descendants. The *2D-tree* (in general, kD-tree) solves these problems at the expense of a deeper tree structure. The kD-tree is a binary tree (each non-leaf node has two descendants) regardless of the dimension k of the embedding space.

2D-tree

kD-tree

For planar embeddings, the 2D-tree does not compare points with respect to both dimensions at all depths, but compares x-coordinates at even depths and y-coordinates at odd depths (assume the root is at depth 0). Each record with a point location has two fields storing the coordinates of the point, two fields pointing to the descendant records, and any subsequent fields holding additional data. For the binary tree structure, we make the convention that the left descendant is less than the right descendant, when compared with respect to the appropriate coordinate (see Figure 6.24).

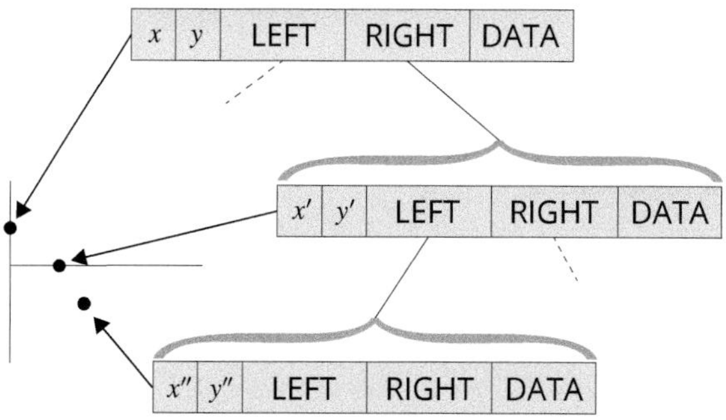

Figure 6.24: 2D-tree records and the points to which they refer

The 2D-tree and planar decomposition for our Potteries point data inserted in numerical order is given in Figure 6.25. Point 1 is inserted at the root. Point 2 has a greater x-coordinate than point 1 and so is inserted as the right descendant of the root. Point 3 has a greater x-coordinate than point 1 and a lesser y-coordinate than point 2 and so is inserted as the left descendant of point 2. The process continues with each of the 14 points in turn. As for point quadtrees, dynamic data leads to problems restructuring the tree. However, the 2D-tree is a simpler structure and leaves fewer "dangling" nodes (about half as many).

Algorithm 6.7 gives a modified version of Algorithm 6.6 to insert a new point p into a 2D-tree T. The algorithm must keep track of the level l of the current node n to ensure the correct test is applied (lines 6–11).

Algorithm 6.8 performs a range search of a 2D-tree T using a rectangular search range specified by diagonal points p_1 (southwest extreme) and p_2 (northeast extreme). The range search in Algorithm 6.8 is based on a breadth-first traversal (see Algorithm 5.6): if the range rectangle contains every point in the 2D-tree, then Algorithm 6.8 simply performs a breadth-first traversal of the entire 2D-tree.

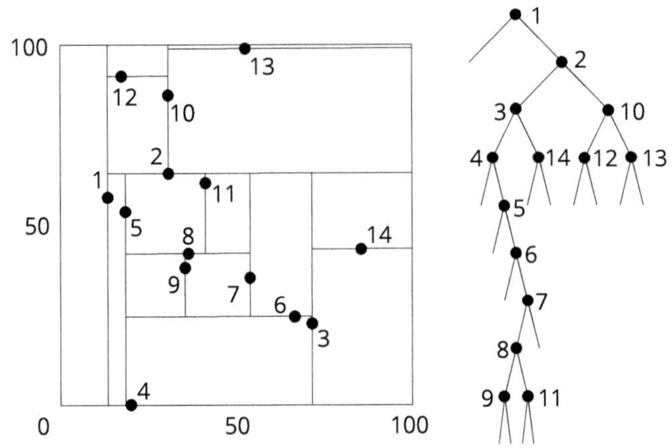

Figure 6.25: 2D-tree for The Potteries point data (labels refer to Table 6.1)

Algorithm 6.7: 2D-tree insert algorithm

```
Input Point p and 2D-tree T
if T is an empty 2D-tree then
    T ← a new 2D-tree with root p and two null tree descendants
node n ← the root of T
tree level l ← 0
repeat
    if l is even then
        a ← x-coordinate of p
        b ← x-coordinate of n
    else
        a ← y-coordinate of p
        b ← y-coordinate of n
    if a < b then n′ ← the LEFT node from n
    elsen′ ← the RIGHT node from n
    n ← n′
    l ← l + 1
until n = null tree
insert new node p at position of n with two null tree descendants
Output The 2D-tree T updated by the insertion of the point p
```

Figure 6.26 shows an example of the algorithm in action with the usual Potteries point data in a 2D-tree. The rectangular range is shown with a dashed boundary. The search begins with point 1 at the root of the tree. This point is not in the rectangle, so it is not retrieved (Algorithm 6.8, lines 6–7). Point 1 occupies an even level in the tree, so the x-coordinates of point 1 and the extremes of the rectangular search range, p_1 and p_2, are compared (lines 9–10). The x-coordinate of p_1 is larger than the x-coordinate of point 1, so the left subtree of point 1 is not examined (lines 13–14). In any case, this subtree is null. However, the x-coordinate of point 1 is smaller than the x-coordinate of p_2, so the right subtree of point 1 is examined (lines 15–16). The algorithm is applied to the root of this subtree, point 2. Point 2 is within the search rectangle, so its record is retrieved (lines 6–8) and its left and right subtrees are searched according to the algorithm, which continues recursively to the

Algorithm 6.8: 2D-tree range query algorithm

Input A non-null 2D-tree T and a rectangular range specified by diagonal points p_1 (southwest extreme) and p_2 (northeast extreme)

node $r \leftarrow$ the root node in T
queue $Q \leftarrow [r]$
initialize set of nodes R as the empty set
while Q is not empty **do**
    remove the first node n from Q
    if x-coord of $p_1 < x$-coord of $n < x$-coord of p_2 **then**
        if y-coord of $p_1 < y$-coord of $n < y$-coord of p_2 **then**
            add node n to R
    if node n occupies an even level in T **then**
        $a \leftarrow x$-coord of p_1, $b \leftarrow x$-coord of n, $c \leftarrow x$-coord of p_2
    else
        $a \leftarrow y$-coord of p_1, $b \leftarrow y$-coord of n, $c \leftarrow y$-coord of p_2
    if $a < b$ and LEFT node of n is non-null **then**
        add LEFT node of n end of Q
    if $b < c$ and RIGHT node of n is non-null **then**
        add RIGHT node of n end of Q
end while

Output The set of points R within the specified range

bottom of the tree. The figure shows only the part of tree directly examined by the algorithm. Points 2, 7, 8, 9, and 11, shown highlighted in Figure 6.26, are all retrieved.

As with the point quadtree, the 2D-tree suffers from the problem of its structure being dependent upon the order in which points are inserted. In the worst case, the insertion of each point will result in a further level of the tree and a finished tree with the same number of levels as points.

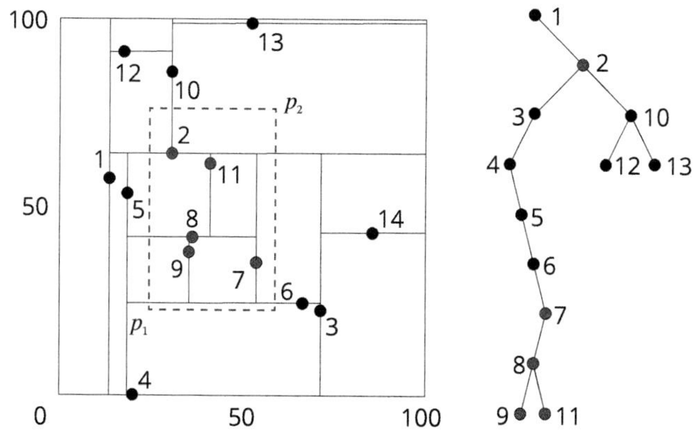

Figure 6.26: Range search in a 2D-tree (labels refer to Table 6.1)

6.5 *Linear object structures*

The spatial data structures described so far have handled points and raster areas. In this section we introduce an index used for linear vector structures, like polylines, linear networks, or the boundary of an areal object. For these linear objects, a first thought might be to use the region quadtree structure, considering linear objects as very thin regions. Unfortunately, the unmodified region quadtree is not usually appropriate, using too much space and resulting in unnecessarily deep tree structures. Instead the *PM quadtree* adapts the region quadtree idea to storing linear objects.

6.5.1 *PM quadtrees*

PM quadtree

The *PM quadtree* is a variant of the region quadtree specially designed for indexing polygonal objects. There are several types of PM quadtree, including PM_1, PM_2, and PM_3 quadtrees. We will describe the PM_1 quadtree, which can be formulated in the general setting of a planar network of vertices and edges, all of which are straight-line segments. Assume such a network of vertices and edges is enclosed in a square region of the plane. The region is divided into quadrants as for the region quadtree (i.e., the trie subdivision). The subdivision is such that vertices and edges are separated into distinct leaf nodes. To be precise, the region is divided into the minimum number of quadrants, subquadrants, etc., such that the quadtree satisfies the following constraints:

1. Each leaf node of the quadtree represents a region that contains at most one vertex of the network.
2. If the leaf node of the quadtree represents a region that contains one vertex of the network, then it can contain no part of an edge of the network unless the edge is incident with that vertex.
3. If the leaf node of the quadtree represents a region that contains no vertex of the network, then it can contain only one part of an edge of the network.

Figure 6.27 shows the partial construction of a PM_1 quadtree for the boundary of the raster region in Figure 6.12 after the first two subdivisions have been made. In the first stage shown (Figure 6.27a), the highlighted quadrant is an example violating condition 1 and so must be further subdivided. In the next stage, in Figure 6.27b, the highlighted quadrant violates condition 2. To make this quadrant visible, it has been exaggerated in size by a factor of two. In Figure 6.27c, the highlighted quadrant (exaggerated in size by a factor of four) violates condition 3, and so further subdivision is still required. Figure 6.28 shows the final PM quadtree, in which all conditions are satisfied for all cells.

The structuring of records in the PM_1 quadtree allows different field structures for vertices and edges. In fact, entire edges are not stored in a

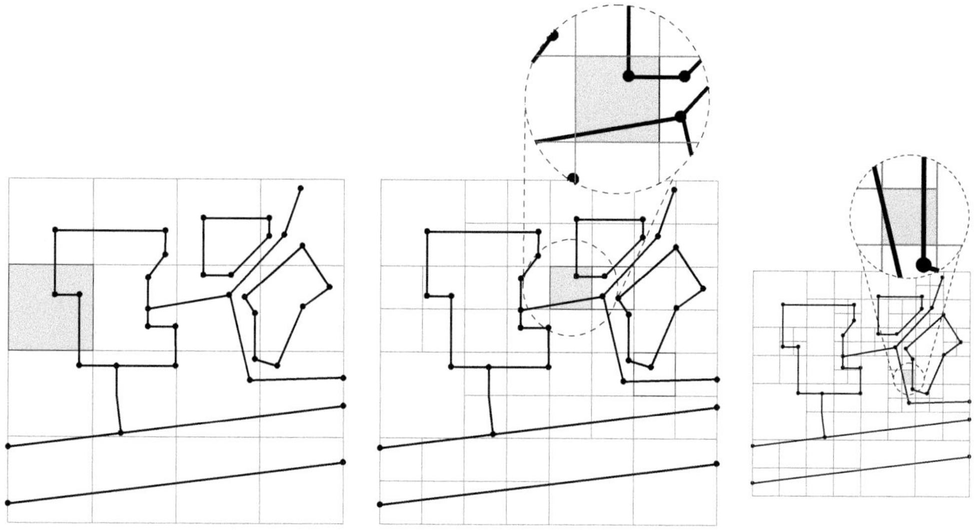

Figure 6.27: Stages in the construction of the PM_1 quadtree

q-edge

quadtree node, only parts of edges clipped by the appropriate subquadrants, termed *q-edges*. When an edge is inserted into the PM_1 quadtree, the tree is searched for the appropriate places to insert parts of the edge. Moving down the levels of the tree, the edge is successively clipped by the boundaries of the subquadrants. Deletion is a similar but inverse process.

The PM_2 and PM_3 quadtrees are variants of the PM_1 quadtree, where the constraints are altered to provide advantages under some conditions. The PM_2 quadtree has a structure that is more stable under translations and rotations of the planar graph. Both PM_2 and PM_3 quadtrees have a smaller tree structure but correspondingly more complex associated records.

6.6 *Collections of objects*

A very common requirement in spatial databases is to structure a large collection of objects so as to facilitate good performance on point and range queries (retrieving all those objects located at a given point or within a given range). We have already seen examples of such structures when the constituent objects in the collection are all points (point quadtree and 2D-tree). This section introduces some of the structures designed to index collections containing more general classes of constituent objects, such as rectangles, polygons, and complex spatial objects.

6.6.1 *Rectangles and minimum bounding boxes*

Efficient indexes for rectangles are important because rectangles can be used to approximate bounded planar spatial objects. Each geometric object is enclosed in its MBB (minimum bounding box), the smallest bounding rectangle with sides parallel to the axes of the Cartesian frame, introduced in

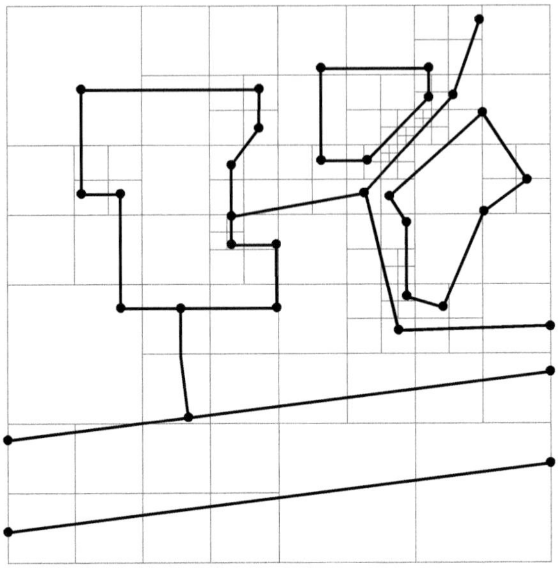

Figure 6.28: Final stage of construction of PM_1 quadtree

the previous chapter). Figure 6.29 shows a simple polygon with its MBB. An efficient indexing of MBBs facilitates queries on the objects themselves.

Normally, the MBB will be stored separately from the detailed geometry of the bounded object, for example, in the form of a quadtree. The geometry of the bounded object can then be referenced by the record associated with its bounding box. The advantage of using an MBB is that some queries may be answered simply by processing the MBB of an object, rather than retrieving the entire bounded object. For example, consider the range query "Find all the objects which lie in their entirety in a specified disk," applied to the objects shown in Figure 6.30. In order to answer this, the following steps are required:

1. Identify all the MBBs that lie wholly *inside* the given circular range. The objects inside these bounding boxes must lie in their entirety in the disk and therefore are to be retrieved. Figure 6.30 shows this case applying to object *A*. Similarly, identify all the MBBs that lie wholly *outside* the given circular range. These objects must lie entirely outside the disk and need

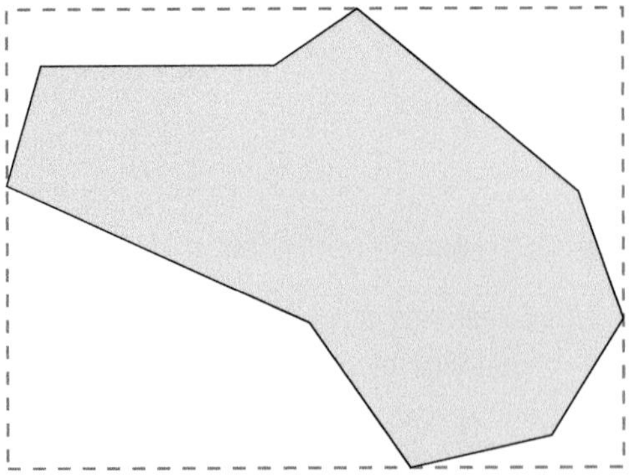

Figure 6.29: A minimum bounding box (dashed line) for a simple polygon

not be considered further. The MBB of object E lies entirely outside the range in Figure 6.30.

2. The remaining MBBs intersect but do not lie wholly inside or outside the given circular range. The objects inside these MBBs may or may not lie in their entirety in the disk. In these cases, further computation on the detailed geometry of each object is required to determine whether the object lies in its entirety within the given range. In Figure 6.30, B (entirely in the range) should be retrieved, while C (partially in the range) and D (wholly outside the range) should not be retrieved.

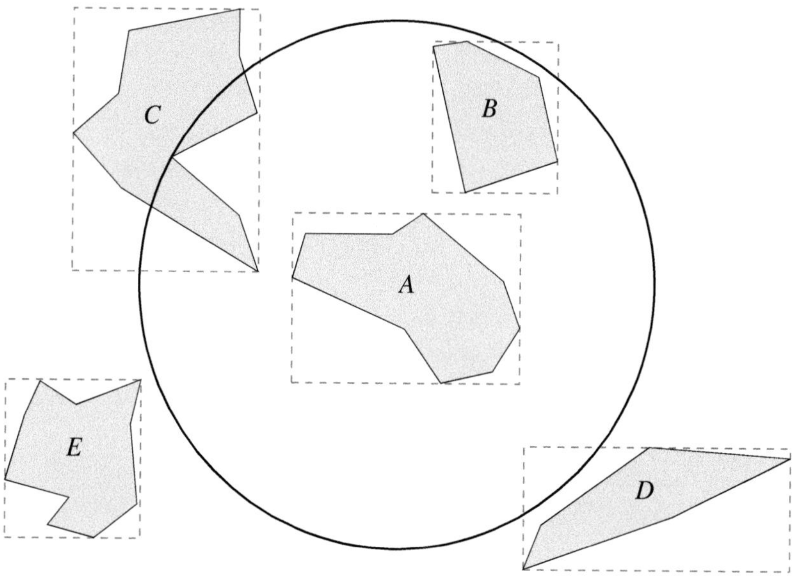

Figure 6.30: Using minimum bounding boxes for a range query

The two steps above illustrate an important principle used for many efficient queries using spatial indexes, termed the *filter/refine strategy*. In many cases, a spatial index cannot completely satisfy a query in a single step, because the index only holds an approximation of the spatial objects. For example, in Figure 6.30 the MBB approximation can only definitively identify one of the two polygons that satisfies the query, object A. Hence, the approximation is first used to *filter* out any objects that definitely do not satisfy the query criteria (i.e., object E in Figure 6.30). Then in the subsequent *refine* step, exact geometric computations with each individual polygon are required to determine the correct response for the objects remaining in the filtered set.

filter/refine

Geometric and spatial algorithms are often computationally intensive, as we have seen in previous chapters. Consequently, there is usually a significant advantage to filtering out as many objects as possible before resorting to refinement of the exact results. The most efficient spatial indexes, therefore, are those that are able to filter out as many objects as possible for the widest range of queries.

6.6.2 R-trees

Having identified the MBB as a useful approximation of a geometric extent, the next step is to find ways of indexing rectangles. The *R-tree* is a rooted tree in which each node represents a rectangle. The leaf nodes represent containers for the actual rectangles to be indexed. Each higher-level node represents the smallest axes-parallel rectangle that contains the rectangles represented by its descendants. It should be noted that the containing rectangles at any given level may overlap one another.

R-tree

The R-tree is essentially a multidimensional extension of the B-tree (see Section 6.1.2), where in the 2D case point pairs define axes-parallel rectangles. The R-tree is a dynamic structure, constructed in a similar way to a B-tree. As rectangles are inserted into the structure, and leaf nodes become full, the effects are back-propagated through the tree nodes, so the tree grows. Conversely, the tree contracts as rectangles are dropped from the collection. The overall tree structure remains balanced and is always at least half-full.

The primary question then is how to divide the space of rectangles into groups (represented by higher-level nodes) as the rectangle collection expands; and conversely, how to coalesce groups as the collection contracts. Point and range searches will be less efficient when the search for an object must traverse multiple different branches of the tree. Larger overlaps between indexed rectangles tend to lead to weaker filters that are not as good at excluding objects that definitely do not satisfy the query.

As a result, good R-tree subdivisions typically aim to minimize the:

- area of non-leaf, containing rectangles;
- area of overlap between non-leaf, containing rectangles; and
- perimeter length of non-leaf, containing rectangles, also termed the "margin" of the rectangles.

Figure 6.31 shows an example of a subdivision and simple R-tree for a set of MBBs for the towns in The Potteries (see Figure 1.4). We assume a fan-out ratio of two, such that each node may contain at most two rectangles and thus have at most two descendants.

Queries on the index then search the R-tree to filter out objects that definitely do not satisfy query criteria before refining the filtered set using geometric computations. For example, Algorithm 6.9 gives a filter/refine strategy for a *spatial join* of two R-trees. Whereas a regular relational join identifies tuples from two tables based on a matching attribute, a spatial join likewise identifies all objects from two spatial data sets that satisfy some spatial criterion. Algorithm 6.9 identifies all pairs of objects from spatial data sets M and N that intersect.

spatial join

Clearly, a brute force strategy to performing such spatial join will be very expensive to perform, requiring $|M| \times |N|$ detailed geometric checks for intersecting geometries. Algorithm 6.9 gives one of several possible procedures (a depth-first search) for using the R-tree index to more efficiently satisfy the query using a filter/refine strategy. The filter step, initiated on line 4 of

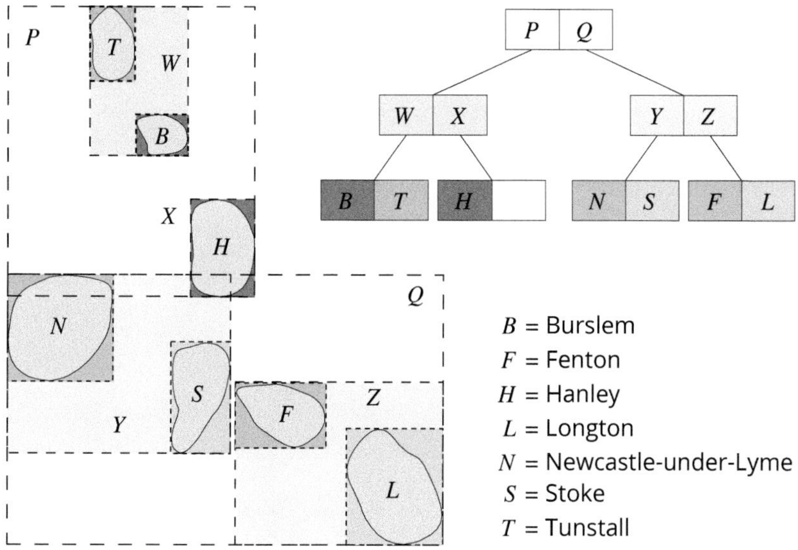

Figure 6.31: R-tree structure for minimum bounding boxes of The Potteries towns

Algorithm 6.9, recursively applies the RTJ-DFS (R-tree join depth-first search) function, lines 9–18. The function searches the two trees for nodes with MBBs that intersect, adding intersecting pairs of leaves to the filtered object set. Filtering is fast because only rectangle intersection is checked, and the R-tree structure means many branches will likely be pruned from the search. Lines 5–8 then perform the exact check for intersections between the geometries that are indexed by the filtered nodes.

6.6.3 R⁺-trees

The problem with the R-tree structure identified above arises from the overlap of containing rectangles, particularly for rectangles that are large compared with the total space. A refinement of the R-tree, more frequently used in practice, is the *R*-tree* (Beckmann, Kriegel, Schneider, & Seeger, 1990). The R*-tree improves on standard R-tree performance by extending the insertion operation with routines to minimize the area, overlap, and margins of indexed rectangles.

R*-tree

The *R⁺-tree* is another refinement of the R-tree that does not permit overlapping rectangles associated with non-leaf nodes, although it of course cannot prevent overlapping of leaf rectangles. The R⁺-tree achieves no overlaps by partitioning rectangles and storing parts in different nodes of the tree. Studies have shown that it admits more efficient searches for large objects. However, while the R⁺-tree improves the efficiency of point and range queries, more complex insertion and deletion algorithms are needed to achieve efficient use of disk space, and to ensure the tree is always at least half full.

R⁺-tree

Algorithm 6.9: R-tree spatial join (depth-first search)

Input Two R-trees R_A and R_B indexing spatial data sets A, B

```
initialize filter as an empty set
initialize refine as an empty set
retrieve root nodes r_a and r_b from R_A and R_B
filter ← RTJ-DFS(r_a,r_b,filter)
for each (n_a,n_b) ∈ filter do
    retrieve from A, B the geometry a, b indexed by nodes n_a, n_b
    if geometries a and b intersect then
        add (a,b) to refine
function RTJ-DFS(x,y,r)  # Function R-tree join depth-first search
    if MBBs x and y intersect then
        if x and y are leafs then
            add (x,y) to r
        else
            for each child node x' of x and y' of y do
                if MBBs x' and y' intersect then
                    r ← r ∪ RTJ-DFS(x',y',r)
    return r
end function
```

Output The set refine of pairs of geometries from A, B that intersect

The Potteries example is structured using an R$^+$-tree in Figure 6.32. When compared with Figure 6.31, X has been extended to contain part of N (the MBB for Newcastle-under-Lyme) that before was contained in Y.

6.6.4 *z-ordering trees*

The R-tree is the most famous of all spatial indexes. However, one further spatial index for object collections, the z-ordering tree, deserves a special mention. The z-ordering tree combines (once again!) the B-tree (Section 6.1.2) with an important concept derived from the region quadtree (Section 6.3.1): *quadtree addressing*. Recall from Figure 6.14, the region quadtree recursively decomposes space into quadrants at successive levels using the Morton ("z") order. If we assign each quadrant a number in the z-order, say NW=0, NE=1, SW=2, SE=3, then we can uniquely address any cell in a quadtree using its leveled sequence of quadrant numbers. For example, if the NE quadrant has the address 1, then the SW quadrant of that NE quadrant is assigned the address 12. The NW quadrant of that quadrant has address 120 and so on. Figure 6.33 summarizes the quadtree addressing scheme, highlighting the different address levels.

quadtree addressing

The z-ordering tree imposes a grid over the spatial data set with some predefined finest level of granularity. Each spatial object can then be approximated as the set of quadtree addresses that intersect that object. Quadtree addresses themselves can be linearly ordered lexographically, with 012 coming before 12, coming before 120, for example. Using this fact, the z-ordering tree simply stores quadtree addresses, along with pointers to indexed spatial

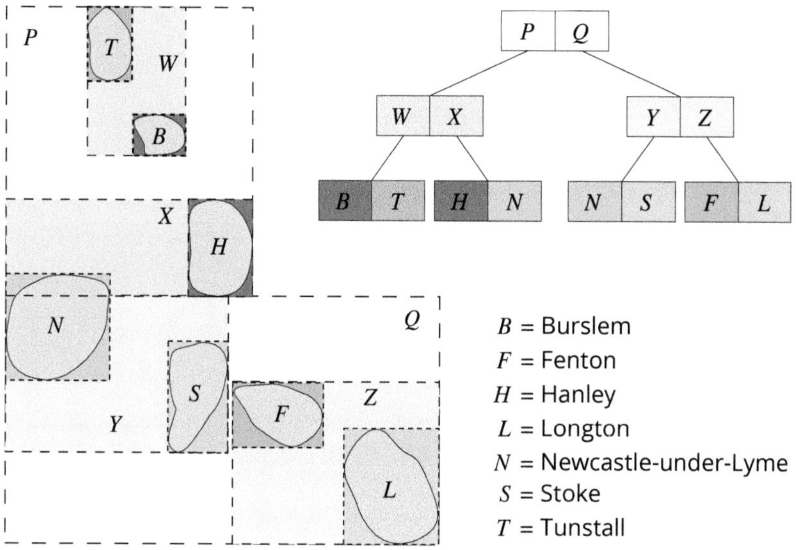

Figure 6.32: R^{+}-tree structure for minimum bounding boxes of The Potteries towns

objects, in a B-tree. Figure 6.34 summarizes a z-ordering tree for The Potteries example data.

On first inspection, the z-ordering tree appears to have a number of drawbacks as an index, including:

- The same object may appear multiple times in the z-ordering tree, including in different branches and at different levels of the tree, such as polygon *H* composed of quadtree addresses 122, 123, 300, 301 in Figure 6.34.
- The same quadtree address may refer to multiple objects, such as address 300 indexing both polygon *S* and polygon *H* in Figure 6.34.
- The level of approximation depends on the size of the finest grid cell, requiring up to 4^l stored addresses, where l is the level of quadtree subdivision (3 levels equates to up to $4^3 = 64$ addresses in Figure 6.34).

Despite these apparent limitations, the z-ordering does have two important advantages. First, quadtree addressing with appropriate levels of subdi-

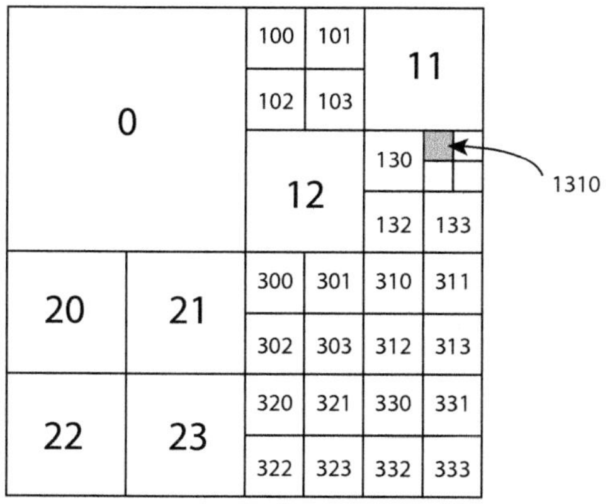

Figure 6.33: Quadtree addressing, using leveled sequence of z-ordered quadrant numbers, NW=0, NE=1, SW=2, SE=3

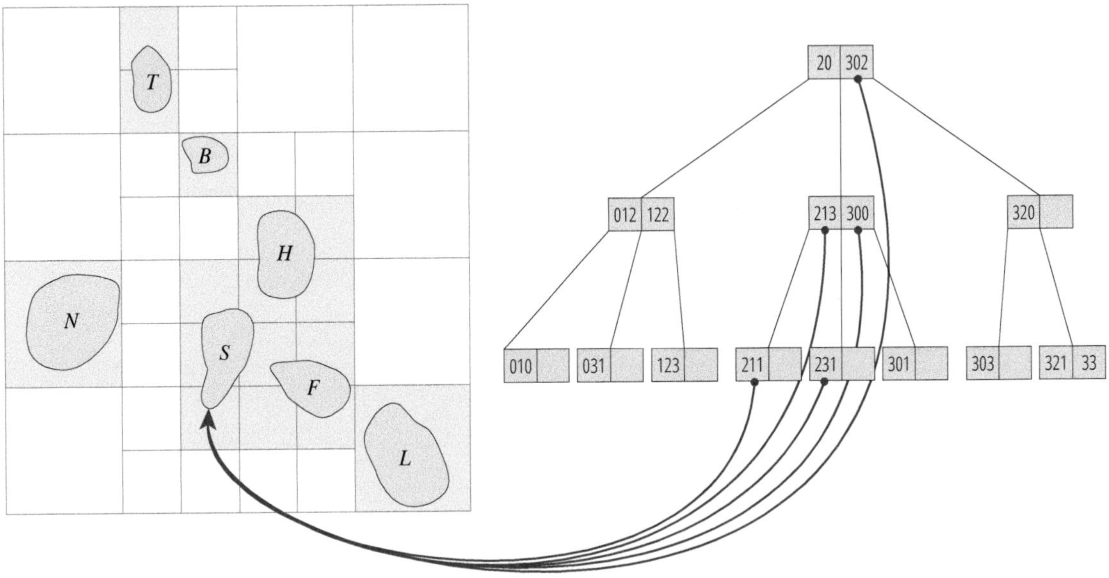

Figure 6.34: Z-ordering tree structure for Potteries towns, highlighting pointers to region *S*, Stoke

vision provides a better approximation of the shape of spatial objects than the MBB used by the R-tree. A better approximation to shape leads to a more accurate filter, reducing the number of expensive refinement operations for a query, resulting in more efficient query satisfaction. Second, where R-trees are an *extension* of the B-tree, the z-ordering tree is a *combination* of quadtree addressing with an unmodified B-tree index. While the leading spatial DBMS today have implemented R-tree indexes, every DBMS implements the B-tree. As a result, z-ordering trees are more easily adopted when using DBMS without such well-developed spatial capabilities as the leading spatial DBMS, such as some newer graph databases.

6.7 *Further indexes*

We end this chapter by looking at some further structures for indexing data beyond two planar dimensions, as well as indexes for temporal data. In general, such structures all follow the now-familiar principles of 2D indexes: mapping to 1D orderings from two or more interdependent dimensions. Hence, the aim in this section is only to give a flavor of the diversity and ingenuity of multidimensional indexes available.

6.7.1 *Spherical data structures*

Spherical tessellations provide better approximations to the surface of the Earth than planar models over large areas. Chapter 5 introduced some regular

Box 6.4: Triangular point quadtree

The QTM construction leads to triangular versions of other standard quadtree structures. For example, it is possible to imagine a triangular version of the point quadtree, as shown below left. A triangular frame *abc* surrounds the data points. These points are inserted into the triangular frame in the order 1, 2, 3, ... as labeled. The first insertion induces a partition of the triangle into three subtriangles and a consistent labeling of these subtriangles as shown; further insertions induce further subdivisions. The corresponding tree is shown below, with color-coded edges for clarity. This structure has the advantage over the standard point quadtree that each node has only three descendants rather than four.

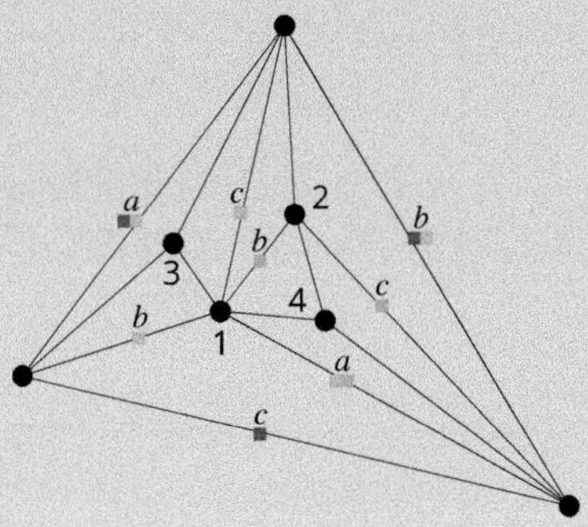

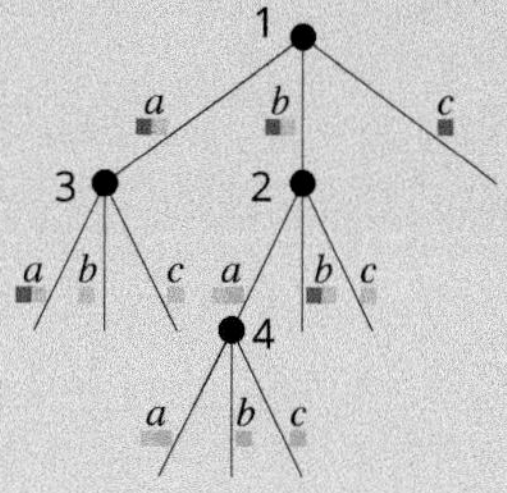

spherical tessellations. The octahedral tessellation in Section 5.3.6 is the most convenient of these, as it is the only regular spherical tessellation that can be oriented with two vertices at the poles and a set of edges at the equator.

quaternary triangular mesh

The *quaternary triangular mesh* (QTM) recursively approximates locations on the surface of a sphere by a nested collection of equilateral triangles. A particular case of a QTM arises from the central projection of the edges of an octahedron onto the surface of the globe, termed an *inscribed octahedron* (see Figure 6.35a, similar to Figure 5.22, and see the triangular point quadtree in Box 6.4). The initial eight faces can be further subdivided, illustrated in Figure 6.35b and c where the vertices of each triangle are shown joined by the arcs of great circles.

inscribed octahedron

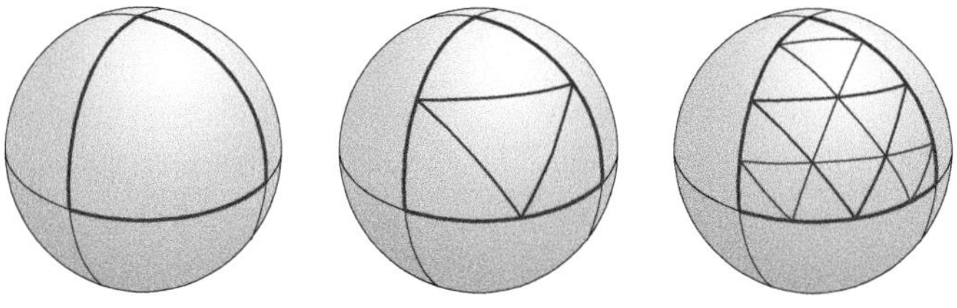

a. b. c.

Figure 6.35: Initial levels of the QTM based on the inscribed octahedron

For simplicity, each face of the inscribed octahedron may be represented as a planar equilateral triangle, which may be recursively subdivided as in Figure 6.36. At the first level (Figure 6.36a), the nodes of the triangle are labeled 1, 2, and 3. The triangle is partitioned as shown into four equilateral triangles. Each subtriangle is labeled according to the label of the node nearest to it, except that the inner subtriangle is labeled 0.

At the second level (Figure 6.36b), the unlabeled nodes of the three subtriangles of Figure 6.36a are labeled so that each edge contains the labels 1,

2, and 3 in some conventional order. Each subtriangle is now further divided into four. Every sub-subtriangle is then labeled with two digits: the left digit is the label of the level one triangle to which it belongs; the right digit is the label of the node nearest to it. Again, the innermost triangle has right digit 0. This process continues for subsequent levels, as far as desired.

Figure 6.36: Two levels of QTM cell numbering (bold numbers outside triangle indicate labeling scheme)

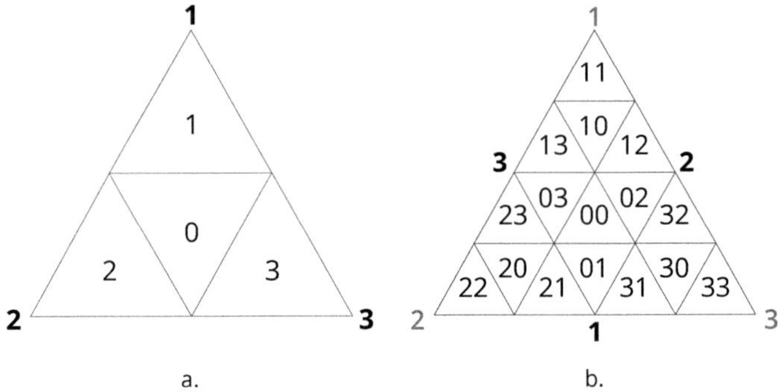

Figure 6.37: QTM region quadtree

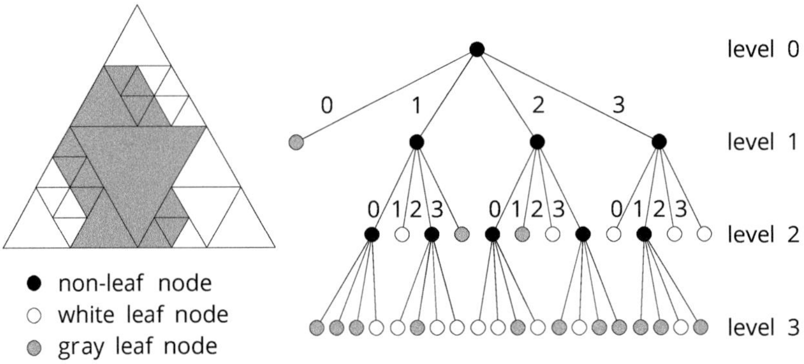

The QTM may be used as a triangular version of the region quadtree. An example of this use is shown in Figure 6.37, where a region is decomposed into its maximal triangular pieces. In the traditional rectangular region quadtree, each node has four descendants, corresponding to the northwest, northeast, southwest, and southeast subregions represented by its descendant nodes. Similarly with this "QTM region quadtree," each node has four subnodes, corresponding to the subtriangles labeled 0, 1, 2, and 3. As with the traditional region quadtree, the QTM region quadtree is a trie structure. Triangular versions of other standard quadtree structures are also possible, such as a triangular point quadtree (see Box 6.4 on the preceding page).

6.7.2 *BSP-tree*

BSP-tree

The *binary space partitioning* (BSP) tree is a binary tree that hierarchically subdivides a space into two regions. The BSP-tree has been widely used in 3D gaming engines and first-person perspective computer games because it is a fast index for searching and sorting the polygons of the spatial environments through which gamers guide their characters.

The BSP-tree may be used for indexing n-dimensional space, partitioning using planes in 3D, and hyperplanes in higher dimensional space. In two dimensions, the BSP tree hierarchically structures sets of directed line segments. Given a sequence of directed line segments $s_1, s_2, ..., s_n$ in the plane, the construction of the 2D BSP-tree takes place as follows:

1. Place segment s_1 at the root of the tree.
2. Extend segment s_1 in both directions to form the infinite directed line l_1. If any other segment in the sequence is cut by l_1, then replace that segment with two segments, taking account of the cut.
3. Examine the next segment, say s_2' in the (possibly revised) sequence. Determine whether s_2' is to the left or right of l_1 and place s_2' in the tree as the left or right descendant of s_1.
4. Extend segment s_2' in both directions either indefinitely or up to its intersection with l_1 to form the directed line l_2. If any other segment in the sequence is cut by l_2, then replace that segment with two segments taking account of the cut.
5. Continue in this way until every line segment is added to the tree.

An example of this construction is given in Figure 6.38. The segment sequence is initially a, b, c, d, e, f, g, as shown in Figure 6.38a. Segment a becomes the root of the BSP-tree, and its associated line (shown as a dashed line) cuts segment e into parts e_1 and e_2. Segment b is processed next: it falls to the left of a and thus becomes the left descendant of a in the tree. The process continues, eventually giving the planar subdivision, shown as Figure 6.38b, and tree, shown as Figure 6.38c.

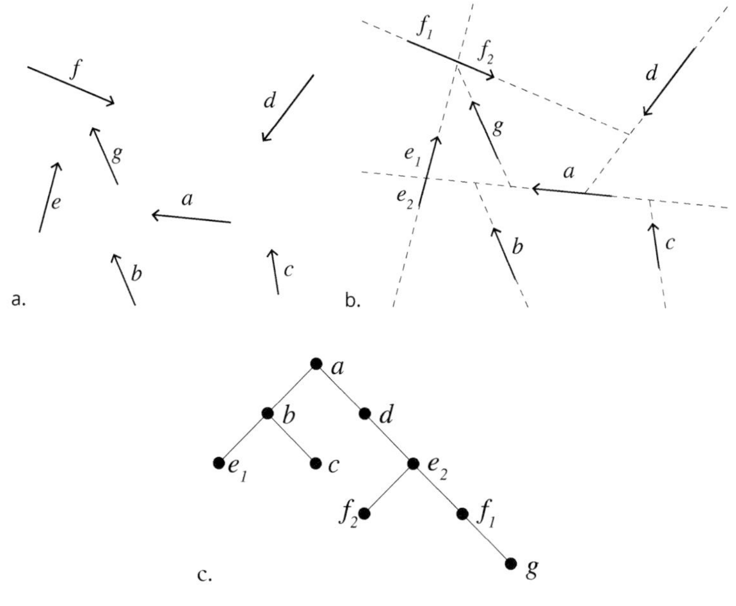

Figure 6.38: Construction of a BSP-tree in the plane from a collection of directed line segments

A balanced tree structure implies minimal depth for the leaves and efficient tree searching. As the figure shows, the BSP-tree may be unbalanced.

The structure of the tree is dependent upon the order of insertion of the segments. It is possible to find cases, such as edges of a convex polygon with all segments oriented in the same direction, where there exists no ordering of the nodes that results in a balanced tree. Another factor influencing efficiency of searching and retrieval is the number of edge cuts that are made. In a BSP-tree, the more edge cuts, the more nodes in the tree.

6.7.3 *Transaction time indexes*

The discussion of different models of time, events, and processes in Chapter 4 introduced the concepts of *valid time* (the time when a stored event occurred in the world) and *transaction time* (the time when data was entered into the database, Section 4.3.6). Thus, a transaction time database must store current and past states of the database and provide a structure for creation, deletion, and modification transactions. The index for a transaction time database then needs to allow efficient access to objects in both past and current states. There have been several solutions to the problem of indexing data in transaction time databases.

A first solution to indexing transactions would be to simply create a new B-tree index for every change in the database. Figure 6.39 illustrates the idea (cf. Figure 6.8), in which after each update a complete new version of the B-tree is stored (in this case, the insertion of the value 10). Changes (as opposed to insertions or deletions) may be achieved using compound deletion and insertion operations (e.g., changing the value 9 to 10 in Figure 6.39 can be accomplished by deleting 9 and inserting 10 as a single transaction). While this solution does enable efficient query processing for certain types of query, it should immediately be clear that using multiple B-trees for transaction time databases is prohibitively inefficient in terms of storage space.

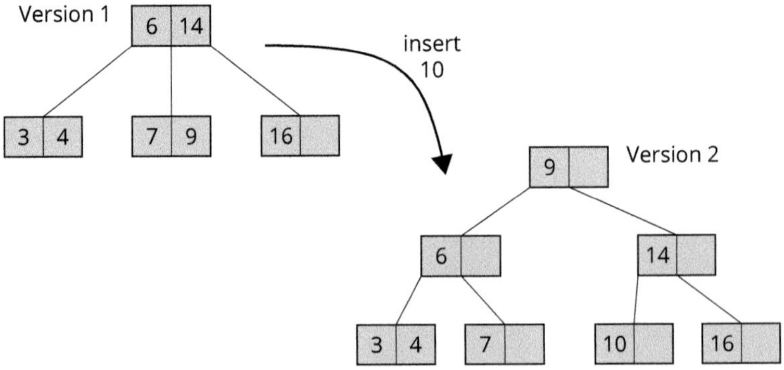

Figure 6.39: Multiple B-trees used to index a transaction time database

overlapping B-tree

One simple solution that extends this naive idea is the *overlapping B-tree*. Instead of creating a completely new tree for each update, the overlapping B-tree duplicates only those nodes where some value has changed. Different transactions share any nodes that are unchanged. Figure 6.40 shows the same insert operation as that in Figure 6.39, but now using an overlapping B-tree structure.

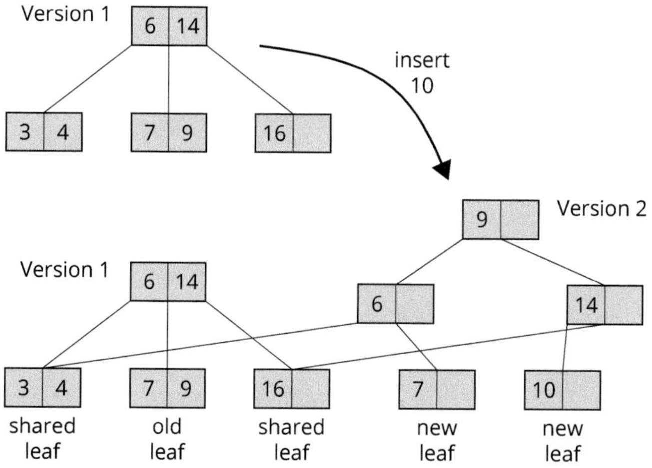

Figure 6.40: Overlapping B-tree representation of insert operation in Figure 6.39

In this way, the overlapping B-tree is able to store changes without large amounts of redundant duplicated data. Note that the overlapping B-tree only stores transaction times at the roots. Overlapping B-trees require a further index for the root nodes themselves.

Overlapping B-trees have the advantage that they are simple to understand and implement. Other more complex temporal indexes provide more advanced capabilities. The *multiversion B-tree*, for example, explicitly represents the insertion and deletion of objects into the database. Deletion time is only implicitly stored in the overlapping B-tree. Attribute changes are modeled as the deletion and creation of a new object. Each object, when it is created, is timestamped with the temporal reference interval $[t_c, \mathsf{UC}]$, where t_c is the time the creation transaction is made, and UC always indicates the transaction is valid "until changed" (i.e., currently valid). If the object is later deleted, its timestamp is modified to $[t_c, t_d]$, where t_d is the time the deletion transaction is made.

multiversion B-tree

By explicitly storing the creation and deletion time for objects, the multiversion B-tree is able to cluster data within the tree according to whether it is *alive* (referenced by a timestamp of the form $[t_x, \mathsf{UC}]$, and more likely to be involved in future active transactions) or *dead* (referenced by a timestamp of the form $[t_x, t_y]$, much less likely to be involved in future transactions), potentially improving query efficiency.

6.7.4 *Valid-time indexes*

A valid-time database may need to store changing attribute histories as a collection of interval timestamped values. A valid-time database needs to store only a single current state, detailing the evolution of its constituent real-world objects. In this case the database evolution is not stored, so an update has a permanent effect in that the past database state is lost.

A *segment tree* is one example of a useful non-spatial index for such valid-time databases. A segment tree structures a collection of linear (time) in-

segment tree

tervals so that it becomes efficient to retrieve all intervals enclosing a given point. For example, a segment tree allows efficient queries of the type:

- Given a time t, retrieve all objects that were valid at time t.

The segment tree structure can be described with reference to an example. Suppose that we have a finite set of time intervals. The first step is to sequence the boundary points in ascending order. For example, if the intervals are:

$$A[1,10], B[3,15], C[4,7], D[4,9], E[6,15], F[8,12], G[10,14], H[16,22], I[18,20], J[10,19]$$

then the sequence is:

$$1,3,4,6,7,8,9,10,12,14,15,16,18,19,20,22$$

In a segment tree, these points become labels for the leaves of a binary tree, as shown in Figure 6.41. Each segment is now split into maximal chunks and distributed as labels of nodes in the tree. For example, segment A labels a single high-level node $[1,10]$, whereas B labels leaf nodes $[3]$ and $[15]$ and non-leaf nodes $[4,6]$, $[7,10]$, and $[12,14]$. Each node therefore represents an interval (or point) and is labeled by zero or more intervals.

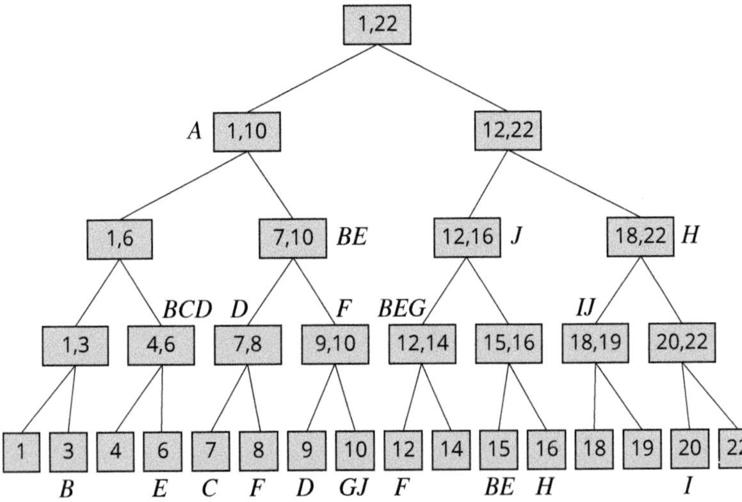

Figure 6.41: Example of a segment tree

Given a time t, the process of searching the tree for intervals containing that point is straightforward. We begin at the root. If the root is labeled with any intervals, insert them into the result list (assuming the root interval contains the search point). Next, examine the immediate descendants of the root. If t does not belong to the interval represented by either of the nodes, then halt the procedure. Otherwise, go to the node representing the interval containing t, add any intervals labeling the node to the result list, and continue this procedure until the leaf is reached or the procedure halts. The result list will hold all the intervals containing t. For example, to search for intervals containing the point 5, start at the root. Proceed to the node $[1, 10]$ and add interval A to the result list. Proceed to node $[1, 6]$, then to node $[4, 6]$ and add intervals B, C, and D to the result list. Then halt the procedure. The intervals found are A, B, C, and D.

The segment tree is a static structure, because the set of interval end-points is given in advance. However, new intervals whose end-points are in the given set may be dynamically inserted into the tree structure by labeling appropriate notes. Also, intervals may be deleted from the structure.

6.7.5 *Bitemporal and spatiotemporal indexes*

Indexing valid and transaction time together—a bitemporal index, see Section 4.3.6—can be achieved by treating valid and transaction time as independent dimensions of an MBB. Standard spatial indexes for MBBs, such as the R-tree, can then be used to index each bitemporal MBB. A further extension is to use two R-tree indexes—one to store objects that are "alive" (from a transaction-time perspective) and another to store those that are "dead" (again, from a transaction-time perspective). Using two R-tree indexes in this way tends to reduce the area of overlap of indexed regions because it is only necessary to store the first element of the transaction timestamp in the "alive" R-tree. By definition, the second element will be UC (until changed) for all objects in the "alive" R-tree.

The approach of using R-trees for bitemporal indexes has several advantages, not least simplicity. It can also be extended directly to spatiotemporal database indexing, where bitemporal spatial regions can be indexed as 4D hyper-rectangles (i.e., two spatial and two temporal dimensions). However, in practice, such an approach suffers from considerable query inefficiencies, for two main reasons:

- bitemporal time regions (comprising a transaction time interval and a valid-time interval) tend to overlap much more frequently than spatial regions' MBBs, since many spatially dispersed objects with non-overlapping MBBs may still share similar transaction and valid times; and
- "alive" bitemporal time regions with transaction time upper bound UC—and the valid-time equivalent of NOW—will continue to grow as the current time advances, changing the size and potentially the shape of their indexed rectangle.

Consequently, optimizing the R-tree process for subdividing the 4D hyper-space into indexed hyper-rectangles requires special attention in a spatiotemporal database. Several different options have been proposed and explored. The *R^{ST}-tree*, for example, approaches the problem by indexing bounding regions with raked (trapezoidal or "stair-shaped") profiles that can efficiently accommodate growth in bitemporal time regions, in addition to regular bounding rectangles (see Figure 6.42 below). The R^{ST}-tree also attempts to R^{ST}-tree account for anticipated future growth in the area and overlap of bounding regions in selecting where to insert of new indexed regions into the tree.

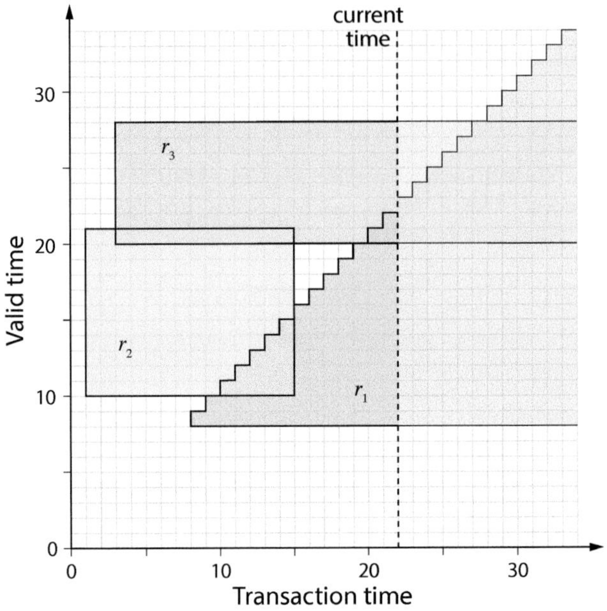

Figure 6.42: Indexing in the R^{ST}-tree can allow for growth in indexed regions

For example, three bitemporal R^{ST}-tree regions, r_1, r_2, and r_3 are shown in Figure 6.42. Only the (valid and transaction) temporal extents are pictured in Figure 6.42 because of the difficulty in drawing 4D hyper-rectangles in two dimensions. Region r_1 has bitemporal timestamp ([8, UC], [8, NOW]).[2] In the R^{ST}-tree, such a region can have a raked, "staircase" shape, which grows as time proceeds. The raked shape can reduce unnecessary overlap (and so increase search efficiency) with other indexed regions, such as r_2 with bitemporal index ([1, 15], [10, 21]). A new indexed region with bitemporal index ([3, UC], [20, 28]) has a larger overlap with r_2 than r_1 at current time 22. However, despite the larger overlap at the current time, the R^{ST}-tree insert operation might still elect to group r_3 with r_2 rather than r_1 because of the much larger overlap between r_3 and r_1 if both regions grow as expected into the future (right of the "current time" line in Figure 6.42).

[2] Bitemporal timestamps comprise a transaction time interval and a valid time interval, see Section 4.3.6.

Reflections

Indexes represent the last step in our journey at the core of this book, from conceptual models in Chapter 4, through algorithms and representations in Chapter 5, to indexes here in Chapter 6. In the next chapter, we take a step sideways into the system architecture of GIS before moving on to interfaces and AI. So this is the last point in the book at which the structure of spatial data itself has the primary focus.

Happily, for those who want to uncover more of the computational structure of spatial data—argued in Section 1.5 to be one of the five key elements that make "spatial special"—Samet (2006) is a veritable encyclopedia of indexes, including details on every spatial index imaginable. Rigaux, Scholl, & Voisard (2002) is also highly recommended for its treatment of spatial indexes, and its coverage of spatial databases more generally.

While a central message of this chapter is connected with the diversity and ingenuity of spatial indexes, a few indexes are undoubtedly more widely used, and consequently more important for every geographic information scientist to be familiar with. Arguably, the four most important indexes in GI science are: the quadtree, explored in classic papers in the field such as Gahegan (1989) and Holroyd & Bell (1992); the R-tree and its family of related indexes, with applications not only in 2D but further spatial and temporal dimensions, such as in Zlatanova & Gruber (1998) and Bertino & Ooi (1999); the z-ordering tree, as explored in Belussi, Bertino, & Catania (2002) and Rigaux et al. (2002); together with the B-tree family—the single most important database index, both in its own right and as the foundation for many other spatial and temporal indexes (Bertino et al., 1997).

ARCHITECTURES

7

The **architecture** of an information system is the structure and organization of the system's functional components. The architecture of a GIS, like most information systems today, is typically made up of distinct functional units connected together by a digital communication network. In this chapter you will learn how to:

- disaggregate a GIS into a **distributed system**, composed of different **interoperable** components with distinct and coherent functions, called **modules**;
- contrast different mechanisms for **coordination** of distributed GIS modules, such as the **client-server model** and **event-driven architectures**;
- discuss the role of **transparency** in GIS architecture, for instance, contrasting **web mapping** applications that shield users from components' details versus **location-based services** which depend on exposing **location-awareness** in certain components; and
- understand the importance of **autonomy** in a modular GIS, distinguishing between a distributed and a **decentralized system**, for example, in developing decentralized algorithms for **geosensor networks**.

SECTIONS

STRUCTURING and organizing the different components of an information system—termed the *system architecture*—plays a critical role in the overall efficiency and efficacy of that system. Chapter 1 introduced several different functional components of a GIS, including:

system architecture

- data capture, storage, and retrieval functions, sometimes also called the *data layer*;
- data manipulation, analysis, and processing, sometimes also called the *application layer*; and
- data modeling, interaction, and usage, sometimes also called the *presentation layer*.

data layer

application layer

presentation layer

Figure 7.1 summarizes these three layers with reference to the functional components of a GIS, introduced in Chapter 1. The essential storage and retrieval functions of a GIS in the "data layer" have been key focus of earlier chapters, in particular Chapters 2, 3, and 6. The processing and algorithms that make up the "application layer" have also been introduced in the context of Chapters 4, 5, and 6, as well as being an important focus of Chapter 9. The role of data models as a bridge between the system and its users—introduced in Chapter 4—is central to the idea of a "presentation layer" (together with user interaction covered in depth in the next chapter).[1]

Instead of the database at the core of the functions of a GIS in Figure 1.1, Figure 7.1 highlights the sharing and communication functions as key enablers of a GIS. This view of GIS—with communication and sharing at its foundations—aligns more closely with the system architecture perspective

[1] The terms "application layer" and "presentation layer" can be confusing, since *applications* are often taken to be synonymous with the practical *uses* of GIS, and *presentation* is only one element of user interaction. Consequently, these terms are used sparingly in this book.

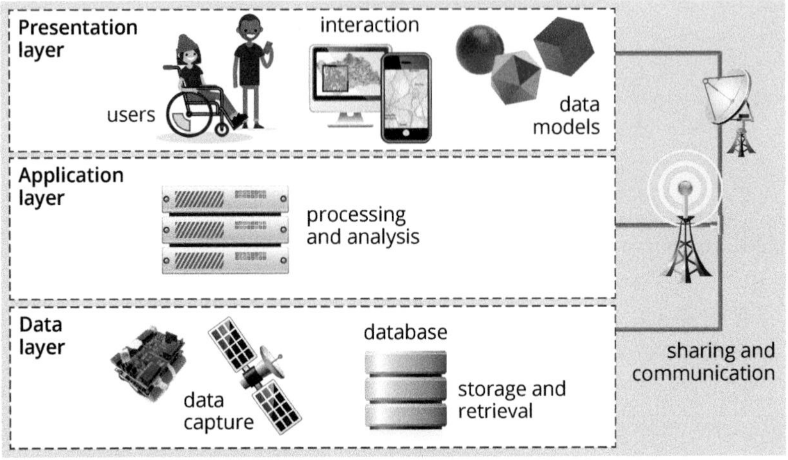

Figure 7.1: Data, application, and presentation layers in GIS, after Figure 1.1

taken in this chapter. Sharing of data, communication between functional elements, and networking of system components underpin the architecture of a GIS. In this view of a GIS, "the network is the computer[2]."

[2] John Gage, Sun Microsystems chief scientist, originally coined this phrase in 1984.

Adopting this perspective, Section 7.1 introduces two cross-cutting concepts important to any system architecture: *interoperability* and *modularity*. Section 7.2 then steps through a foundation of modern system architecture—the distributed system—and explores the different ways that the functional components of a GIS can be decomposed and reconnected through a digital communications network. Section 7.3 then explores the issue of *coordination* in system architectures, in particular through the examples of *cloud computing* and *stream processing*. Section 7.4 then introduces another key ingredient in GIS today: location-based computing, where the location of some system components and devices—usually hidden from users of distributed systems, termed *transparency*—becomes vital to the computing itself. Finally, Section 7.5 explores the emergence of highly distributed computing environments, termed *decentralized systems*, where components enjoy high levels of *autonomy* such as the Internet of things (IoT) and sensor networks.

7.1 Interoperability and modularity

The origins of GIS lie in the development of unconnected, unitary software packages (with names such as SYMAP, Odyssey, and GRASS) each of which had to integrate all the required functions of the three layers in Figure 7.1. However, even within individual software packages, it was still necessary to separate components within the package into distinct functional components. For example, perhaps most successful of all such software packages called ARC/INFO was released in the early 1980s by ESRI (Environmental Systems Research Institute). As the name suggests, ARC/INFO was constructed from the combination of two distinct functional components: one for storing and manipulating vector-based spatial data (ARC) and another for storing and manipulating table-based attribute data (INFO).[3]

The primary motivation for choosing a hybrid approach was its *modularity*. The special models and structures needed for storing and processing spatial data demanded special software and spatial database management technologies (covered, for example, in Chapters 5 and 6). At the same time, tabular attribute data could be handled by regular non-spatial relational database technologies (introduced in Chapter 2). Enabling these technologies to work together as cooperating modules helped achieve the "best of both worlds." This modular, so-called "hybrid" GIS approach is illustrated in Figure 7.2.

Modularity, then, is the extent to which an information system can be constructed from independent units with standardized or clearly defined functions. Modularity makes complex software easier to develop, maintain, and adapt to meet the requirements of particular users or specific application areas.

[3] Although more recent ESRI GIS software draws on a variety of sophisticated spatial and non-spatial technology, the historical importance of the hybrid GIS model is still evident in the ESRI Shapefile format, used for storing and exchanging geospatial data. The Shapefile format consists of three separate files—one to store geometry (a .shp file), one to store the non-spatial data tables (a .dbf file), and one to store an index to relate the two (a .shx file).

modularity

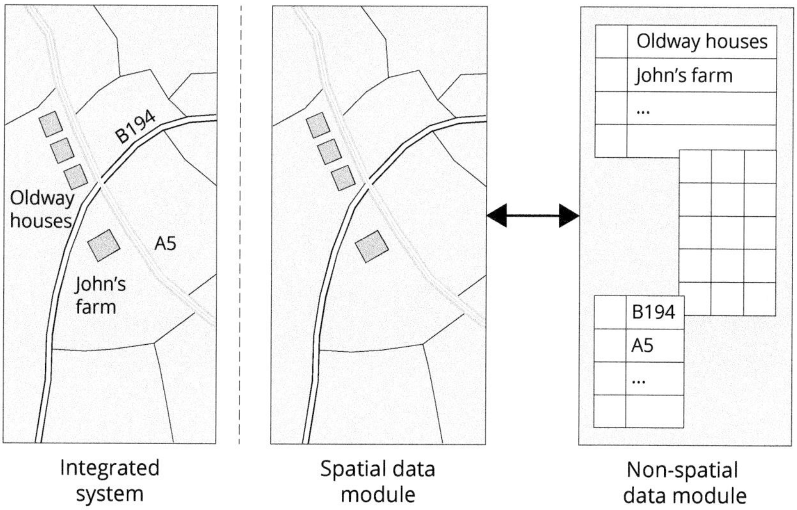

Figure 7.2: Modularity and the hybrid GIS approach

Modularity requires that there exists some method to communicate between different modules. In the hybrid ARC/INFO GIS example above, illustrated in Figure 7.2, the different modules ARC and INFO were tightly coupled, such that the communication between the two was deeply embedded in the software itself, and it was not possible to operate one without the other. Sharing between less tightly coupled modules, such as between different software packages, requires the ability to communicate in a more transparent, loosely coupled way. For example, *file transfer* involves exporting data from a source software package to a file with a particular predefined format, and then importing that file into the next, target software package.

file transfer

Interoperability, then, is the capability of two or more modules or information systems to share and exchange data. Interoperability has always been especially important to GIS because spatial data arises from so many different sources, and it is used in so many different types of analyses. There are two

interoperability

main barriers to data sharing that any interoperable system must overcome: *syntactic heterogeneity* and *semantic heterogeneity*.

7.1.1 Syntactic heterogenity

syntactic heterogeneity

Syntactic heterogeneity occurs when two or more information systems use incompatible encoding or formats for information. Syntactic heterogeneity may arise as a result of differences in data file formats, for example. Whatever the source of syntactic heterogeneity, its consequence is that data must be converted into a compatible format before the two systems can interoperate.

Converting directly between pairs of data formats is inefficient because it produces many different *conversion paths*. Each conversion path requires that computer programs need to be written and maintained to perform the data conversion. Translating data between the wide variety of GIS software packages and spatial data file formats that exist, for example, has historically presented a real headache for GIS technicians. Rather than convert directly between different heterogeneous formats, syntactic heterogeneity is usually overcome by adopting a standard intermediate format, called an *exchange format*.

exchange format

In Figure 7.3a, converting between six data formats demands 30 different conversion paths (the figure has 15 double-headed arrows). In general, n data formats will require $n(n-1)$ different conversion paths. Instead of converting directly between pairs of data formats, an exchange format ensures that in general only $2n$ conversion paths are needed. In Figure 7.3b, using a exchange format reduces the number of conversion paths from 30 to 12. The greater the number of data formats, the greater the efficiency gains from using an exchange format.

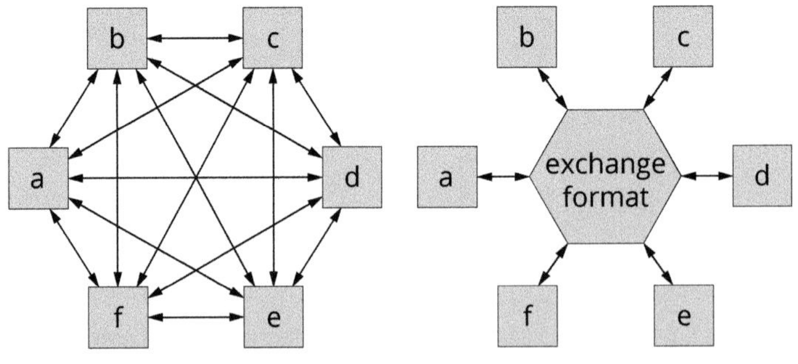

Figure 7.3: Exchange formats reduce the number of required conversion paths

Exchange formats have a long history in GIS. The costs of spatial data transfer are sufficiently high that most countries have developed national strategies for sharing and communicating geospatial data, with international coordination also happening through bodies such as the UN-GGIM (United Nations Committee of Experts on Global Geospatial Information Management), OGC (the Open Geospatial Consortium), and ISO (International Standards Organization). A diversity of standard spatial data exchange formats exists today, both official and unofficial (see Box 7.1 on the next page and Box 7.2 on page 273).

Box 7.1: De facto and de jure standards

As illustrated in Figure 7.3, exchange formats address syntactic heterogeneity by providing a standard intermediate format for data conversion. A *de jure* standard is administered by an official standards organization. For example, the simple feature model and associated WKT (well-known text) data format (introduced in Section 4.1.3) are official specifications developed by the OGC (Open Geospatial Consortium) and integrated into standards by the ISO (International Standards Organization). A *de facto* standard is a data format that is widely used as a standard, even though it has not been approved by an official standards organization. For example, the Shapefile format developed by ESRI (see above) became a *de facto* exchange standard. However, the boundary between de facto and de jure standards is often blurred. GeoJSON, for example, is a widely used specification for exchanging geospatial data. This open standard is maintained by a working group of the IETF (Internet Engineering Task Force), which while not strictly an official standards organization in practice performs a similar function.

7.1.2 Semantic heterogenity

Distinct from syntactic heterogeneity, *semantic heterogeneity* occurs when two or more information systems use different or in some way incompatible meanings for data. For example, some different words have the same or very similar meaning, such "road" and "street," termed *synonymy*. Other words, such as "bank," may have multiple different meanings (e.g., a river bank and a savings bank), termed *homonymy*. Unfortunately, deciding on exactly what is meant by a piece of information can be difficult, as meaning is highly dependent on context. The statement "Let's go to the bank" can assume rather different meanings depending on whether it is followed by the statement "I need to cash a check" or "I want to feed the ducks." Often, the context for a statement is implicit.

semantic heterogeneity

synonymy

homonymy

The distinction between syntactic and semantic heterogeneity, summarized in Figure 7.4, mirrors the distinction between data and information in Chapter 1. Syntactic heterogeneity is concerned with the properties of data, namely, the format and syntax of data. Semantic heterogeneity is concerned with the properties of information, primarily the context and meaning that we associate with data that results in information.

Unlike syntactic heterogeneity, which is largely a technical issue and can usually be solved by technical means, semantic heterogeneity is often difficult to reconcile. The meaning we attach to data is very important, but very difficult to encode. This problem is particularly acute for GIS because geospatial data is used in such a rich variety of different disciplines, application domains, and communities, each of which has its own idiosyncratic conventions and terminology. Despite this challenge, the problem of encoding and computing with semantically heterogeneous data is a topic we will return to in Chapter 9, in the context of the artificial intelligence and technologies for the Semantic Web.

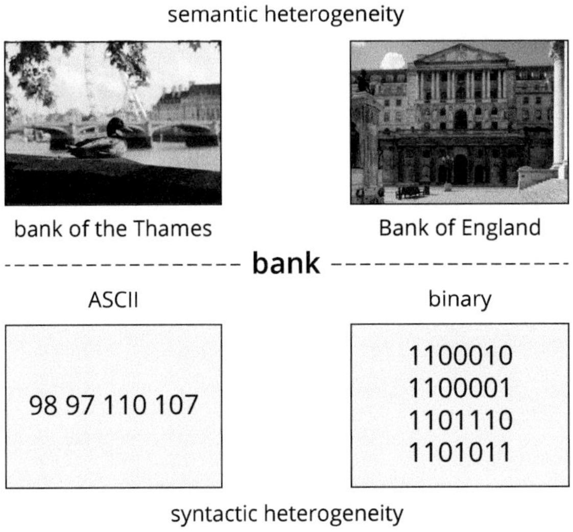

Figure 7.4: Syntactic versus semantic heterogeneity

7.2 *Distributed computing*

File transfer using exchange formats represent a flexible but highly decoupled and *asynchronous* approach to interoperability. File transfer is asynchronous in the sense that once a data set has been converted into a particular exchange format it may be seconds, hours, days, or years before that data is subsequently imported and used in a target application. Using digital communication networks, introduced in Chapter 1 (see Section 1.4.4) information systems are able to achieve more tightly coupled, coordinated, and synchronous forms of interoperability than are possible with file transfer.

The single most important concept in embarking on a discussion of computing using digital communication networks is the definition of a *distributed system*.

distributed system

> A distributed system is a collection of multiple autonomous modules connected via a digital communication network that can operate as if a single, coordinated system in order to complete some computing task.

A distributed system is therefore a special type of interoperable system because the different elements in a distributed system must be able interoperate with one another in order to coordinate on the completion of some task. Four features of a distributed system implied by the definition above are worth emphasizing at this point. To a greater or lesser extent, any distributed system will exhibit:

- *modularity*, being composed of multiple distinct computing elements or units, variously referred to as *nodes*, *modules*, or *components*;
- *coordination*, with different modules interoperating in a coherent and predictable manner;

- *transparency*, with computing tasks driven by users who may not be aware of the specific details of the configurations and interconnections between collaborating modules; and
- *autonomy*, with each coordinating module able also to operate independently or as part of other distributed systems.

The remainder of this chapter is concerned primarily with the degree to which different GIS architectures tackle these four features: modularity, coordination, transparency, and autonomy.

7.2.1 Two tier (client-server) distributed systems

The *client-server* architecture, sometimes also called the *two-tier* architecture, is perhaps the simplest of all distributed system architectures. Two-tier client-server systems are founded on a clear delineation between the responsibilities of two different types of collaborating module, called clients and servers. A *server* is a module that can offer a particular service to other nodes on the network. A module that consumes these services is called a *client*.

client-server

two-tier

server

client

The services provided by a server are defined by a server's *interface*. An interface is akin to a contract or agreement, clearly stating what it is a server can do for a client and how a client can access those services. A *protocol* is a standard (data) format for communication that specifies and implements a server interface. Figure 7.5 illustrates the essential features of a client-server distributed system.

protocol

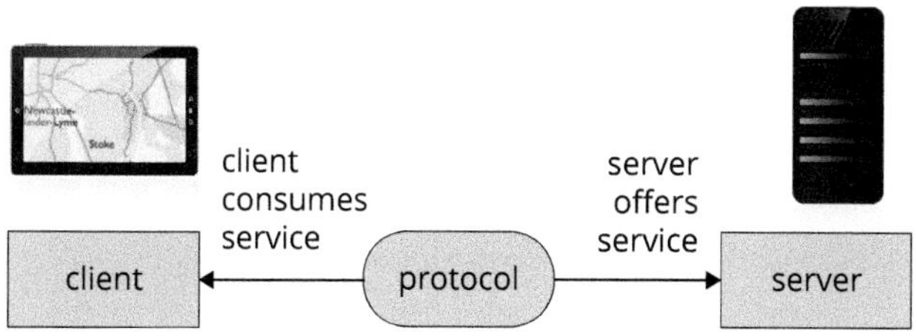

Figure 7.5: Client-server architecture

A familiar client-server distributed system is the WWW. Web browsers (clients) request web pages from web servers. The interface for web servers is defined by hypertext transfer protocol (HTTP). When you click on a hyperlink in a web page, your web browser sends an HTTP request for the required web page to the web server at the Internet address indicated by the hyperlink. Assuming the link refers to a valid address, the web server will respond with the appropriate web page also using HTTP. This kind of *request-response* protocol is common in client-server distributed systems.

request-response

Figure 7.6 illustrates the request-response protocol for an HTTP request using a standard visual language called a *sequence diagram*. In a sequence diagram, the different modules in a distributed system appear as labels above dotted lines representing time, advancing down the page. Arrows between modules indicate the passing of messages (data) from one module to another. For example, in Figure 7.6 the HTTP request for a web page is initiated by the

sequence diagram

client web browser. Receipt of that request message leads to the web server retrieving the requested page, returned in a response message to the client.

Figure 7.6: Sequence diagram for an HTTP client-server request-response

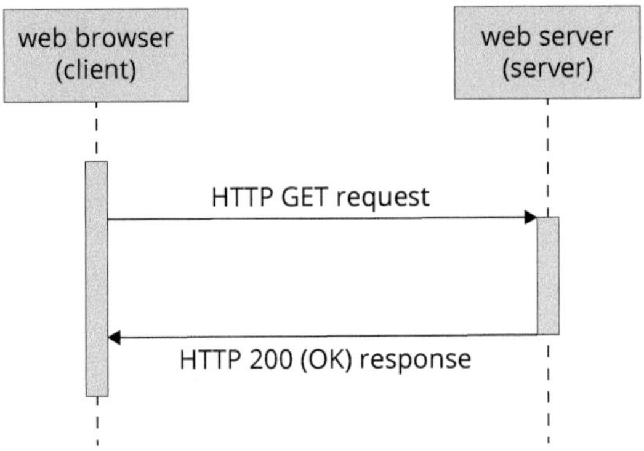

A server's interface, and the protocol used to access a server, specifies only what services a server offers, not how those services will be performed. The client-server architecture is not only interoperable but also modular, in the sense that clients and servers may be modified, upgraded, or even replaced as long as the interface remains constant. For example, when you upgrade or change your web browser software, it is still possible to access the same web pages on the WWW using the same addresses, because all web browsers honor the same protocol (HTTP) to access the web.

7.2.2 *Server- and client-side strategies*

server-side

client-side

A significant factor in shaping the characteristics of any client-server system is the allocation of processing responsibilities between the two tiers. In a *server-side* strategy, the server performs the bulk of the computation needed to complete a task. The terms *thin client* and *thick server* are often used to refer to the client and server in a server-side strategy. In a *client-side* strategy, the client performs the bulk of the computation needed to complete a task. Client-side strategies use *thick clients* and *thin servers*.

mainframe

Early GIS architectures were dominated by distributed *mainframe* architectures that relied on a strongly server-side strategy. The mainframe architecture (sometimes called the *terminal* network architecture) connects multiple "dumb" terminals (clients) to a powerful central mainframe computer (the server). Later as more powerful personal computers (PCs) became cheaper and more widely available, mainframe GIS declined in popularity and were replaced by desktop GIS software packages. These desktop GIS packages were able to perform all the required spatial data manipulation, analysis, and presentation operations, adopting a client-side strategy and accessing servers only as remote catalogs of spatial data sets.

Today, as the variety of computing devices and applications of GIS have proliferated, so have the mix of client-server strategies used in different scenarios. The design of any client-server distributed system must strike a balance between client-side and server-side responsibilities for completing

Box 7.2: XML and GML

The eXtensible Markup Language (XML) standard, developed by the World Wide Web Consortium (W3C), is not itself an exchange format; rather it is a standard *metalanguage* used for defining other languages and exchange formats, termed *vocabularies*. For example, Geography Markup Language (GML) is an XML vocabulary that has been developed as a exchange format for geospatial data. Like all markup languages, XML uses tags to annotate data with structure that can help in interpreting or analyzing the data. Tags provide a name for the thing the data is describing. XML tags may optionally have attributes that further refine the meaning of data. For example, data about the first edition of this book in XML might be represented as shown below.

```
<book>
 <author>Michael F. Worboys</author>
 <title edition="First">GIS: A Computing Perspective</title>
 <publisher location="London">Taylor and Francis</publisher>
 <year>1995</year>
</book>
```

An important feature of XML is that it is *human readable*. Without knowing the precise syntax of XML, most people would still be able to look at the XML data for the book (below) and understand what information is being represented. XML offers a number of further powerful features. For example, XML provides a mechanism to define what types of structures are allowed within a particular vocabulary (the Document Type Definition or DTD). Software that parses XML can use the DTD for a particular vocabulary to ensure the XML data conforms to the structure described in the DTD, termed *validation*. XML also provides a mechanism for defining *templates*, which can be used to transform XML from one vocabulary into another (eXtensible Stylesheet Language Transformation, XSLT).

the task. Because the server performs most of the computation, server-side strategies enable spatial data and processing to be accessed by clients with lesser computational capabilities (such as a tablet computer or smartphone) or using generic software (such as a web browser). Client-side strategies demand clients with higher computational capabilities, but conversely they require less powerful server systems and are able to offer increased flexibility and interactivity to the user.

Figure 7.7 illustrates the balance between server- and client-side strategies, highlighting the potential to allocate the bulk of presentation, application, and data layers to one or the other side of the interface, dependent on the capabilities and needs of clients and servers for particular applications.

In general, server-side GIS strategies also make more efficient use of network bandwidth than do client-side strategies, as the server only needs to respond with processed spatial data. Additional data that does not form part of the final response, but might have been used while processing the request, is retained by the server and does not need to be transmitted. Server-side strategies further enable higher levels of security for valuable or sensitive data because control over unprocessed data is retained by the service provider. Client-side strategies may leave service providers with less control over how their spatial data is used because of the potential for clients to store served data for subsequent purposes other than the specific request.

The balance between client- and server-side strategies for spatial data is exemplified between two complementary client-server protocols for web-based distributed GIS: WFS and WMS. *Web map service* (WMS) is a protocol for clients to request maps from a map server over the WWW. A client's

WMS

Figure 7.7: Server- and client-side strategies

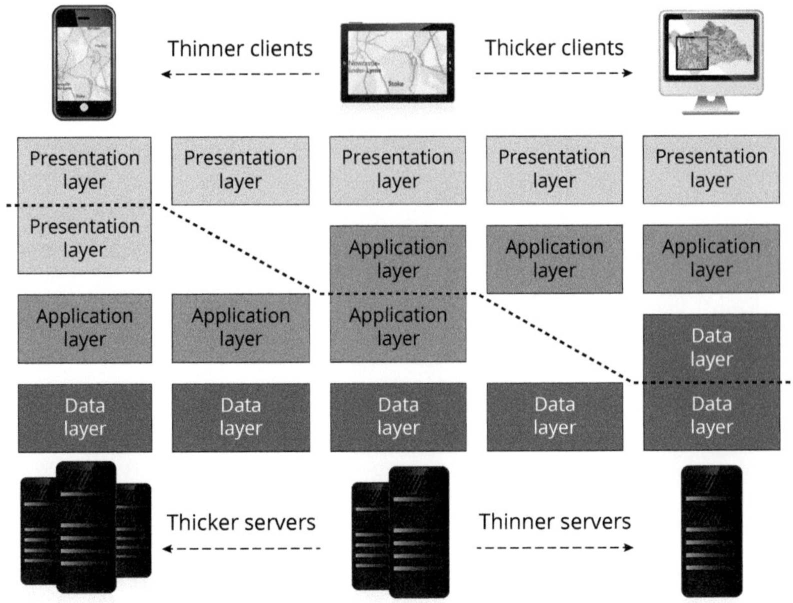

WMS request must contain the geographic bounding box for a desired map, and it must specify any required map styling options offered by the server. The map server then generates the requested map, responding with a rendered image that the client can display. Any changes to the map, such as different map styling or map extents, requires a new WMS request. *Web feature service* (WFS) is a related protocol that enables clients to request the underlying spatial data structured according to the simple feature model. A client's WFS request typically contains a geographic bounding box for the desired spatial data, and any filters specifying the types or attributes of requested features. The map server then responds with the requested data itself, in an XML format called GML (Geography Markup Language, see Box 7.2 on the previous page). With the supplied data, clients may construct and style their own maps to display, perhaps performing some spatial analysis, or indeed applying any desired spatial operations on the data.

WFS

Table 7.1 summarizes the key features of the server- and client-side strategies. Client-side strategies are ideal for specialized applications, when smaller numbers of advanced or expert users need flexible or personalized data manipulation capabilities accessed with more powerful computing devices. Server-side strategies are well suited to applications in which large numbers of non-specialist users only require access to simple processing and interaction capabilities, or if the underlying spatial data is sensitive or high value.

Table 7.1: Summary of server- and client-side strategies

	Server-side	Client-side
Client functionality requirements	Lower	Higher
User interface flexibility	Lower	Higher
Bandwidth usage	Lower	Higher
Data security	Higher	Lower

7.2.3 Multi-tier client-server architectures

A *multi-tier client-server* architecture, sometimes termed a *three-tier* or *n-tier* architecture, extends the basic two-tier client-server architecture with additional intermediate tiers. Multi-tier client-server systems are essentially composed of chains of two-tier systems. Figure 7.8 illustrates a multi-tier architecture with three tiers. As in Figure 7.5, client and server communicate with each other. However, in Figure 7.8 an intermediate "middle tier" acts as both a client (for the right-hand server using protocol B) and a server (for the left-hand client using protocol A).

multi-tier client-server

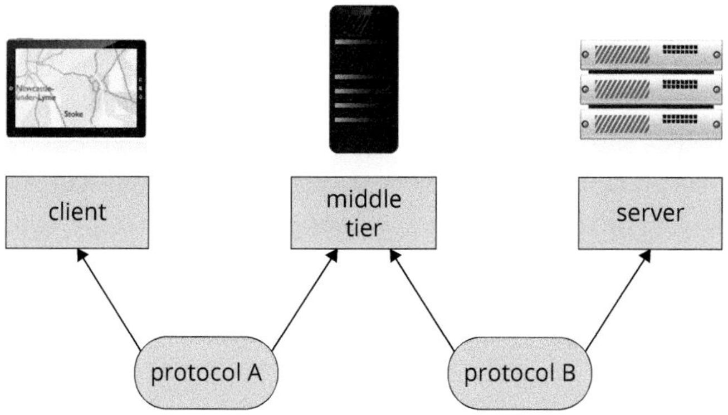

Figure 7.8: Multi-tier client-server architecture

Multi-tier architectures, such as that illustrated in Figure 7.8, are commonly used for making spatial data available over the WWW, sometimes termed *web mapping*. For example, a web browser (client) might request a web page containing a map from a geospatial web server (middle tier). The geospatial web server in turn requests the spatial data required from a spatial database (server), such as the DBMS discussed extensively in earlier chapters.

web mapping

The primary advantage of additional tiers within a client-server architecture is that they allow increased modularity, with specific functions allocated to particular tiers. In fact, today's sophisticated web mapping applications typically use client-server architectures involving at least four tiers. The sequence diagram in Figure 7.9 illustrates a typical multi-tier web mapping request-response cascade, involving a web browser, a web server, a spatial database server, and a specialized geospatial data server middle tier.

- An HTTP request for a web page containing a web map is initiated by a user via a web browser—for example, by opening a bookmarked page or clicking on a link embedded in another web page.

- The web server responds to the browser's first request with the indicated web page. The response will contain the web page itself—usually a mix of text-based HTML (hypertext markup language) and CSS (cascading style sheets)—in addition to client-side code that is designed to be executed in the browser to embed the map itself inside the displayed page. Not represented in Figure 7.9, the web server itself will often be a middle tier, requesting the web page markup, images, style sheets, and code from a file server or a regular, non-spatial database.

- Most web browsers are powerful enough to be able not only to display HTML, CSS, and images, but also to execute code embedded in web pages. In many cases, this enables sophisticated client-side map styling and rendering of spatial data in the browser itself. There exist today many powerful client-side code libraries specifically designed for web mapping.[4] Using code embedded in the web page returned from the first request in Figure 7.9, the web browser initiates a second request to a specialized web *geoserver*. The geoserver is designed to respond to web mapping requests, such as WMS, WFS, or WCS[5] requests.

[4] Two of the most popular libraries for displaying map data in a web browser are called *Leaflet* and *OpenLayers*.

geoserver

[5] Complementing WMS and WFS, WCS (web coverage service) is designed to serve raster data "coverages," such as remotely sensed satellite imagery.

- Receiving the WMS request directly from the web browser, the "geoserver" middle tier initiates a request to a spatial database server for the spatial data needed to satisfy the client's web mapping request. In Figure 7.9, the simple features returned by the spatial database are rendered as a map image in the geoserver, before being returned to the client as the response to the original WMS request.

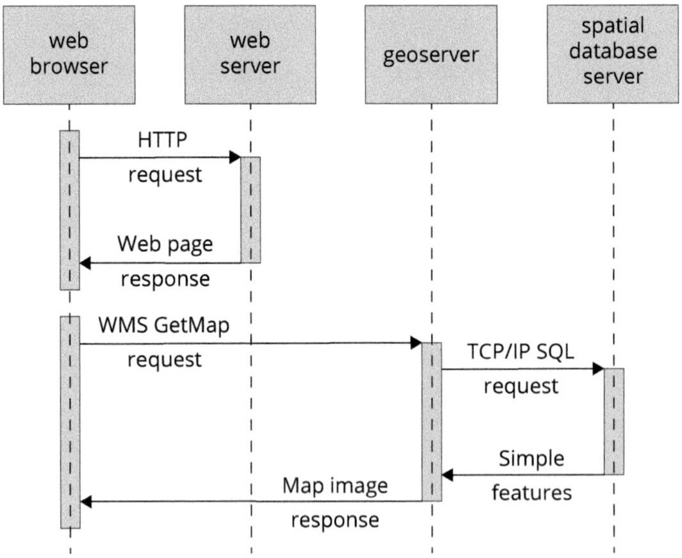

Figure 7.9: Sequence diagram for WMS request to a multi-tier web mapping architecture

In this manner, web mapping is enabled by a legion of servers, ready to respond to requests from an army of clients around the world. The approach is highly modular, since each component has a clearly defined role, and may be replaced by another component without affecting the functioning of the system, as long as the new component honors the same interface as the replaced component. The approach is also highly flexible, as different systems can be constructed from different configurations of the same modules. Indeed, the logical conclusion of this modularity is to decompose a distributed system into not simply three- or four-tiered components, but a great many components each providing a specific targeted service, often termed a *service-oriented architecture* (SOA, see Box 7.3 on the following page).

service-oriented architecture

7.3 Scaling and streaming

Any distributed system architecture, such as the client-server architectures introduced in the previous section, can be described with reference to three fundamental aspects:

1. a set of *modules* together with *interfaces* that describe what services each module offers and consumes;
2. one or more *protocols* for communication and exchange of data between modules;
3. a *configuration* defining how individual modules are organized and related to one another.

The web mapping service illustrated in Figure 7.9 and discussed above, for example, was enabled by four modules (a web browser, a web server, a geoserver, and a spatial database server); three communication protocols (HTTP, WMS, and TCP/IP); and a multi-tier configuration connecting a spatial database server, geoserver, a web server, and a web browser.

This section introduces two further aspects of distributed systems related to configuration as we build progressively larger and more complex services using distributed systems. First, Section 7.3.1 investigates the concepts needed to understand how GIS can *scale* to satisfy billions of service requests from around the world. Second, in Section 7.3.2 the question of coordination comes to the fore, addressing a key limitation of the client-server model—reliance on tightly coupled request-response protocols—using *stream computing*.

7.3.1 Scaling

Making GIS and web mapping available through distributed architectures opens up the possibility of providing that functionality not just to a few local users, but to millions or billions of users wherever they are in the world. Most estimates indicate that the leading web mapping services, for example, respond to more than a billion of mapping and routing requests every day. How is such mind-boggling performance achieved?

When scaling up any distributed system, there are two distinct options available (see Figure 7.10):

- *vertical scaling*, where the modules of the distributed system are running on more powerful processors; and
- *horizontal scaling*, where the modules of the distributed system are running on more numerous processors.

vertical scaling

horizontal scaling

Box 7.3: Service-oriented architectures

A service-oriented architecture (SOA) decomposes a distributed system into a potentially very large number of component modules combined into highly flexible configurations. SOAs share many concepts in common with the object-oriented model, where each object can be viewed as a module providing and consuming services as part of a larger system (see Section 4.1.2). Indeed, two of the earliest SOA technologies were called CORBA (the common *object* request broker architecture) and SOAP (the simple *object* access protocol). CORBA and SOAP provide the infrastructure for making standardized definitions of module interfaces and for distributed communication between software modules. They additionally provide a *registry* of objects (sometimes termed a *broker*), making it possible to discover what modules and services are available in order to aid construction of a new system configuration. The process of constructing a system from modules in this way is sometimes termed *orchestration* and the system itself is called a *composable* system. However, the requirement to describe the interface of each module in detail in CORBA, SOAP, and other similar technologies such as RPC (remote procedure call), carries with it the overheads of reconfiguring the system every time one of its component module's interface changes. A more popular approach today than CORBA or SOAP is to use a REST (representational state transfer) API (application programming interface) to construct an SOA distributed system without needing to provide as much detail about the available services. REST achieves this by treating all modules as repositories of data "resources" accessed via a simple interface with just four CRUD operations (recall, "CRUD" stands for creation, retrieval, update, and deletion—see Section 2.1.2 for further background on CRUD operations). Each CRUD operation is mapped to an HTTP method: POST for creation of resources; GET for retrieval of resources; PUT for updating resources; and DELETE for deleting resources.

Vertical scaling A basic mechanism for scaling vertically is to build bigger, more powerful computers that use multiple CPUs working in concert. The traditional von Neumann architecture of a computer, introduced in Section 1.4.1, is based on sequential processing of data, in which each machine instruction is executed in turn before control passes to the next instruction. Using multiple CPUs working in concert, termed *parallel computing*, it is possible to complete tasks more quickly than with a single CPU. How much faster a problem can be computed using parallel processing—sometimes termed the *speedup*—depends on the computing task at hand.

parallel computing

speedup

Some computing tasks can be easily decomposed into smaller tasks that can be processed in parallel and then recombined into a single answer (sometimes termed "embarrassingly parallel" problems). Other computing tasks, however, may need to be computed in serial or demand significant communication between parallel CPUs, reducing the benefit of parallelization. For example, zonal operations on raster data sets, encountered in Section 4.2.2, may parallelize efficiently because the computation can be easily decomposed by zone, such that the calculations related to each zone can be computed independently of that required for other zones. Many vector geometric algorithms, such as computing the intersection, union, or overlay between polygons (see Section 5.4.3), are much less amenable to parallelization, however. The complex structure of polygonal data and the interwoven nature of the computation does not readily lend itself to decomposition into independent calculations suitable for efficient parallelization.

One estimate of the expected speedup associated with a particular computing task is given by *Amdahl's law*:

Amdahl's law

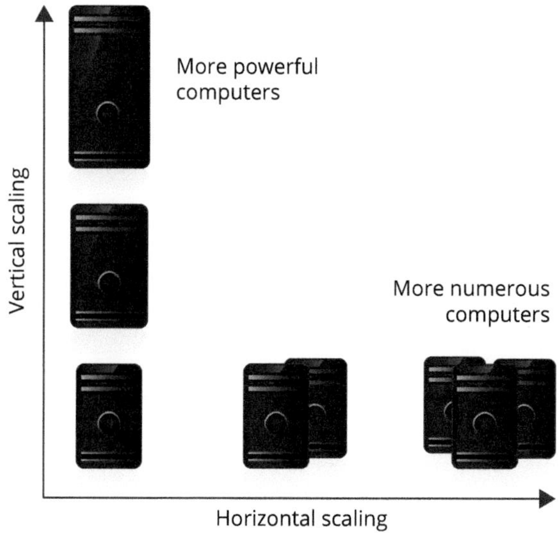

Figure 7.10: Horizontal and vertical scaling

$$s = \frac{1}{\frac{p}{n} + (1 - p)}$$

where n is the number of parallel processors used, p is the proportion of the task that is parallelizable, and s is the maximum theoretical speedup. In essence, Amdahl's law shows that the speedup achievable by parallelizing a computing task tends to be limited by the task's non-parallelizable steps. As the number of parallel processors increases, so the benefit of additional processors decreases and the speedup approaches $1/(1 - p)$, becoming dominated by the serial computations required by the task (see Figure 7.11).

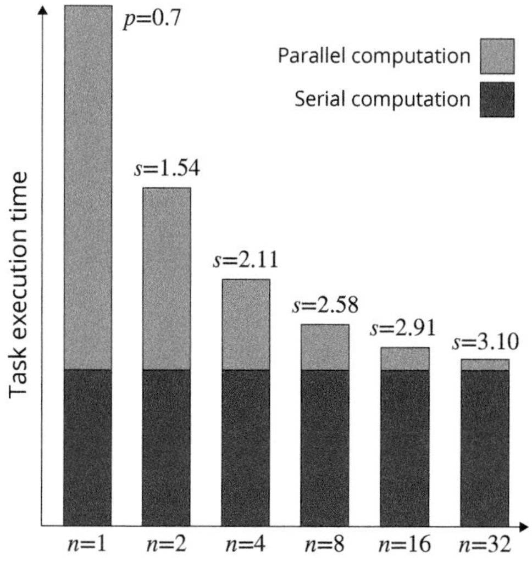

Figure 7.11: Illustration of Amdahl's law for a 70% parallel, 30% serial computing task ($p = 0.7$)

Most personal computers today rely on parallel processors, typically with 4, 8, 16, and sometimes more cores. *Cluster computing* is a form of parallel computing that connects multiple (multicore) computers together in a network to operate in concert as if a single computing unit. *Grid computing* allows

cluster computing

grid computing

looser connections between geographically dispersed computing facilities, themselves possibly clusters. Both cluster and grid computing have similar aims: to achieve parallel computing performance on tasks by treating the cluster or grid as if it were a single computing unit.

massively parallel processing

Some of the earliest applications of such *massively parallel processing* (MPP) were used to address spatial problems, such as processing Earth observation data. In the 1980s, NASA pioneered the application of MPP to processing the increasing volumes of high resolution satellite imagery that were being produced. The MP-1 mini-supercomputer at NASA Goddard Space Flight Center (pictured in Figure 7.12) combined over 16,000 processors in parallel. However, the difficulties in achieving significant speedup from massively parallel computing means that MPP today tends to be reserved for specialized supercomputing applications, rather than general-use GIS functionality.

Figure 7.12: MasPar MP-1 at NASA GSFC in 1993 (Source: NASA, Dave Pape)

cloud computing

Horizontal scaling The dominant mechanism for scaling up distributed GIS and web mapping applications today is *cloud computing*. Cloud computing is a form of horizontal scaling where additional computing resources can be added to a distributed system on demand and without requiring any technical knowledge of how the scaling is achieved. Behind the scenes, cloud computing services are powered by large computing clusters each comprising tens of thousands of CPUs. Users need not know anything about the underlying technology, networking, or even the location of the clusters in order to use them (termed "transparency" and discussed in more detail in Section 7.4). Instead, users can scale up applications from a few CPUs to a few thousand CPUs at the touch of a dashboard.

Three distinct models are often discussed in connection with cloud computing:

- SaaS — Software-as-a-service (SaaS) involves deploying entire software packages, such as GIS software, to the cloud in order to make them available to remote users.
- PaaS — Platform-as-a-service (PaaS) involves deploying software modules and tools, such as spatial databases, web servers, or geoservers, to the cloud, for example, to support multi-tier client-server web mapping applications, such as those described in the previous section.
- IaaS — Infrastructure-as-a-service (IaaS) involves providing access to virtual machines (a digital emulation of a physical computer) that can support any software or application, just as if it were running on a user's own physical machine.

The limits to parallelization inherent in vertical scaling still fundamentally apply to the clusters that enable cloud computing. The speedup to *individual* spatial computing tasks made possible with larger clusters is still limited by Amdahl's law, although the economies of scale of cloud computing has made immense computing power practicable even for everyday GIS applications. However, the primary benefits of horizontal scaling and cloud computing for GIS are more often connected with parallelizing multiple different spatial computing tasks, rather than individual tasks. A popular web mapping service may receive thousands of requests for maps or routes every second. Each request can be treated as a self-contained task that can then be relatively easily processed in parallel, independently of other requests. As long as access to the underlying data is managed carefully (for example, by replicating or caching frequently requested portions of data), horizontal scaling is able to satisfy many of the most popular navigation, routing, and web mapping applications today.

7.3.2 *Stream computing*

All the distributed systems encountered thus far can be characterized as being *tightly coupled*, in the sense that requests for services are invoked directly upon the modules providing those services. Tight coupling is essential to most "command-and-control" style applications. For example, when a human user requests a route from a web mapping service, they typically rely on receiving immediate response to that request.

However, tight coupling is not required by all applications. Tight coupling becomes a disadvantage for services that may be subject to strong peaks and troughs in demand or that may occasionally become temporarily unavailable. It also introduces dependencies between clients and servers through the interface definitions, as already discussed. Further, many applications are inherently more compatible with a "publish-and-subscribe" style of coordination between modules, rather than command-and-control coordination. For example, in real-time applications where data is being continually updated—such as traffic, weather, or emergency alerts—it is more efficient to "push" updates out to clients proactively, rather than issue periodic requests to "pull" the latest data from the server.

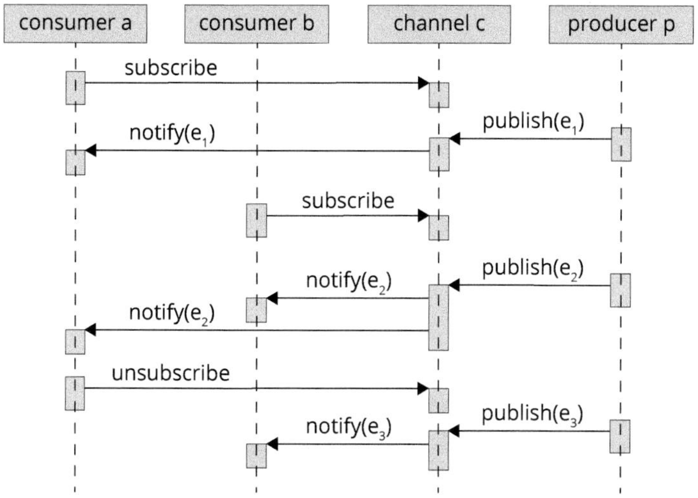

Figure 7.13: Sequence diagram for publish-and-subscribe interactions for three events (e_1, e_2, e_3) published to a channel with two subscribers

Figure 7.13 shows a sequence diagram for a simple publish-and-subscribe interaction involving one publisher (or *producer*) and two subscribers (or

producer

consumer
publish-and-subscribe model
publication channel
message broker

consumers). In the *publish-and-subscribe model* (sometimes shortened to the "pub/sub model"), consumers can choose to subscribe to a (publication) *channel* (also called a "topic" and sometimes referred to as a *message broker*). Once subscribed, consumers will be notified of any messages published to that channel. In Figure 7.13, consumer a first subscribes to channel c. Next, update e_1 is published to channel c by producer p, leading to consumer a being notified of e_1. Following the subscription of consumer b to channel c, the next update e_2 results in both consumers a and b being notified. After consumer a unsubscribes from channel c, only consumer b is notified of the final update e_3.

Perhaps surprisingly, publish-and-subscribe does not by default provide a *guarantee* of message delivery. The challenges surrounding message delivery guarantees relate to a classic problem in distributed systems, called the "Two Generals problem" (Box 7.4 on the next page). Instead, most publish-and-subscribe systems offer the following options:

at-most-once delivery

- *At-most-once delivery*: The default in publish-and-subscribe systems is that notifications are sent to each subscriber exactly once, but delivery is not guaranteed, so in some cases a notification might not be received.

at-least-once delivery

- *At-least-once delivery*: If delivery guarantees are required, it is possible to resend notifications until a confirmation of receipt is received from the subscriber, resulting in at least one but possibly more than one copy of a notification being received.

exactly-once delivery

- *Exactly-once delivery*: The most technically complex option, exactly-once delivery does guarantee unique notification receipt, but at the cost of significant communication and computational overheads.

Figure 7.14 illustrates the three different publish-and-subscribe delivery guarantees. In at-most-once delivery, message a is received by the consumer, but message b fails to arrive. In at-least-once delivery, message a again gets through, but message b is sent three times before a confirmation of receipt is received by the producer, and it is ultimately received twice by the consumer. In exactly-once delivery, message b is received and confirmed with one delivery. However, message a must be sent three times, each time checking for receipt confirmation, before the third copy is confirmed as received.

latency

Being loosely coupled, publish-and-subscribe systems do not normally attempt to provide any guarantees about notification delays, termed *latency*. Latency may naturally be highest for exactly-once delivery since the guarantee that a message will be received exactly once necessitates much more complex interactions between consumer, broker, and producer, including waiting for receipt confirmations. In many cases, publish-and-subscribe systems may provide limited guarantees about notification ordering, and whether messages will be received in the same order in which they were published.

event-driven architecture

The publish-and-subscribe model is an example of an *event-driven architecture* (EDA). There is a clear correspondence between clients and servers in

Box 7.4: The Two Generals Problem

The difficulties faced by distributed messaging systems are often illustrated using the "Two Generals Problem." As the story goes, long ago two allied armies had an enemy army surrounded. Only a coordinated attack by the allied armies could hope to defeat their shared enemy; the enemy's superior forces would surely triumph if the allies attack individually. Lacking modern radios and not wanting to reveal their position with fires or lights, the two allied armies' generals must rely on messengers to coordinate their attack. The first general dispatches a runner to the general of the second allied army with the message, "Let us attack at sunrise." Running the gauntlet of enemy scouts and sentries, the messenger is happy to arrive successfully. The second general then returns another runner with the message, "Message confirmed! We will attack at sunrise." The second message also gets through without mishap. At this point, we might hope everything is set for an allied victory. However, the second general cannot be *certain* that her confirmation was successfully received by the first general, and may decide to hold back her forces to avoid the chance of defeat. Anticipating this, the first general might send a third message: "Confirmation received! We will attack at sunrise." Again, the enemy scouts are sleeping at their posts, and the runner gets through a third time. But now the *first* general cannot know whether *their* confirmation message was received or not. So anticipating caution, the second general must now dispatch *another* confirmation message... As long as we cannot be certain that a message will be received, then no matter how many messages are sent, the two generals can never be certain they will be coordinated in their attack. The story illustrates how common knowledge can never be guaranteed in a distributed system in the presence of unreliable communication. Even if all messages are in fact received, the *possibility* of unreliable communication prevents the establishment of common knowledge.

client-server and service-oriented architectures (see Box 7.3 on page 278) and consumers and producers in the EDA. However, the loose coupling of EDA means that it is typically suited to quite different applications to client-server and service-oriented architectures, most especially applications where data is being continuously generated.

As such, EDAs are themselves an example of a broader class of distributed system, called *stream processing* systems. A *data stream* is a potentially unbounded, dynamic sequence of records that evolves over time. A channel in a publish-and-subscribe system is an example of a data stream. The data generated by sensors, such as satellite imaging systems or ground-based weather stations, typically also form data streams.

data stream

Online algorithms An important technique for computing with data streams is using *online* algorithms. An online algorithm (not to be confused with "online services" or "being online") is an algorithm that can process unbounded input dynamically, in series, as it arrives over time (i.e., as a data stream). In contrast, an *offline algorithm* requires all its data to be available at the beginning of its execution. Online algorithms are closely related to *incremental* algorithms, which accept input data in sequence and output an incremental solution after each new input. In fact, we have already informally encountered a number of incremental algorithms in this book, including the incremental Voronoi algorithm in Figure 5.35 and the B-tree insertion operation in Figure 6.8. The distinction between incremental and online algorithms is subtle: an online algorithm's input is always assumed unbounded and dynamic, while an incremental algorithm may be used to process a bounded, static data set incrementally. However, any incremental algorithm may be used for on-

online algorithm

offline algorithm

incremental algorithm

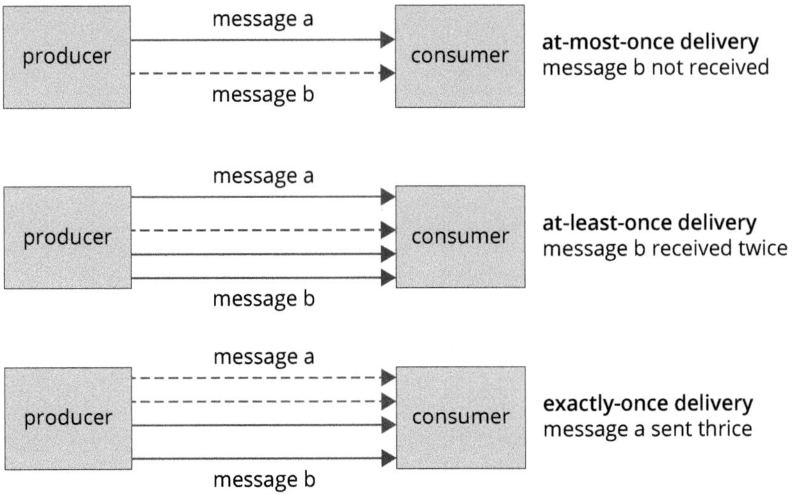

Figure 7.14: At-most-once, at-least-once, and exactly-once delivery in publish-and-subscribe systems

line processing, and any online algorithm may be used to process a data set incrementally.

To illustrate, Algorithm 7.1 presents an online Delaunay triangulation algorithm. By contrast, the (offline) divide-and-conquer Delaunay triangulation in Algorithm 5.5 begins by sorting the complete set of input points in ascending order of x-coordinate. The online Delaunay triangulation Algorithm 7.1 of course does not have that option—new points will arrive on the data stream over time, and so the complete set of points is unbounded and never known. Being an online algorithm, there are a few stylistic differences in Algorithm 7.1 to the offline algorithms encountered in previous chapters. First, to highlight that input data is being received incrementally, a new keyword **when** encloses the procedure that will be executed when new data is received on the stream. Second, rather than specifying an algorithm's **Input** and **Output**, as is conventional for an offline algorithm, an online algorithm must **Require** some preexisting (possibly empty) output from earlier increments and **Ensure** that output is updated when processing new inputs from the stream.

For brevity, Algorithm 7.1 makes the simplifying assumptions that there is at least one triangle already in the triangulation (see line **Require**) and that a new data point p lies inside one of the triangles already in the triangulation (Algorithm 7.1, line 1). Upon receiving a new point p on the data stream, Algorithm 7.1 finds the existing triangle abc in the triangulation that contains p, and adds three new edges ap, bp, cp to the triangulation (Algorithm 7.1, lines 2–3). The resulting structure will be a triangulation, but not necessarily a Delaunay triangulation satisfying the defining Delaunay condition that the circumcircles of every triangle contain no other points in the triangulation, discussed in Section 5.3.3 and illustrated in Figure 5.18 (for ease of reference, repeated in the margin in Figure 7.15). Consequently, the algorithm must check the Delaunay condition is satisfied for each of the new triangles abp,

Algorithm 7.1: Online (incremental) Delaunay triangulation

Require Delaunay triangulation T of points $p_1, ..., p_n$, where $n \geq 3$
1: **when** receiving(point p that lies inside the boundary of T)
2: find the unique triangle abc in T such that abc contains p
3: insert edges ap, bp, cp into T
4: swap(ab, p)
5: swap(bc, p)
6: swap(ac, p)
7: **end when**
8: **function** swap(edge qr, point p)
9: **if** qr is an exterior edge of T **then return**
10: find triangle dqr in T such that $d \neq p$
11: **if** d lies within circumcircle of pqr **then**
12: remove edge qr from T
13: insert edge pd into T
14: swap(dq, p)
15: swap(dr, p)
16: **end function**
Ensure Delaunay triangulation T of points $p_1, ..., p_{n+1}$

bcp, acp and their edge-adjacent triangles (lines 4–6 and the swap function line 11), swapping edges where the test fails (lines 12–13) and recursively checking new triangles adjacent to swapped edges (lines 14–15).

Figure 7.16 illustrates Algorithm 7.1 graphically, showing the insertion of a new point p into a Delaunay triangulation (Figure 7.16a). The first step adds edges fp, ip, and kp to the triangulation (Figure 7.16b). As a result of the insertion, point e lies inside the circumcircle of triangle fip, requiring edge fi to be "flipped" with ep (Figure 7.16c). Recursively checking the triangles created by that flip reveals a further violation, leading to flipping edge ef with edge bp (Figure 7.16d). Finally, checking edge ik reveals one last edge flip required with lp (Figure 7.16e), before the insertion update is complete (Figure 7.16f).

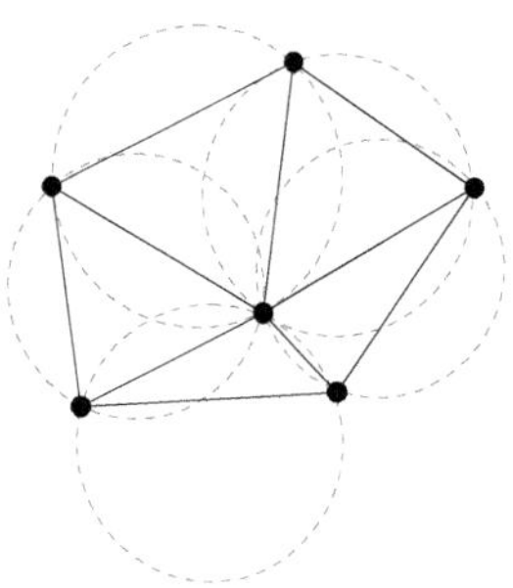

Figure 7.15: Circumcircles of a Delaunay triangulation, after Figure 5.18

7.4 *Location-aware computing*

The architectures encountered so far in this chapter are designed to operate as a single coherent whole, hiding from the user the specific characteristics of constituent modules and the system structure and organization. Such a distributed system is said to be *transparent*.[6] There are several different dimensions to transparency in distributed systems, including *migration transparency* (the ability to change, substitute, or upgrade individual system components without being visible to the user); *scaling transparency* (the ability to increase the computing resources available to the system transparently); and *concurrency transparency* (the ability for multiple users to share the same distributed computing resources without being aware of one another).

transparency

[6] Transparency in distributed systems is about *hiding* technical details from users, so perhaps might be better termed *opacity*.

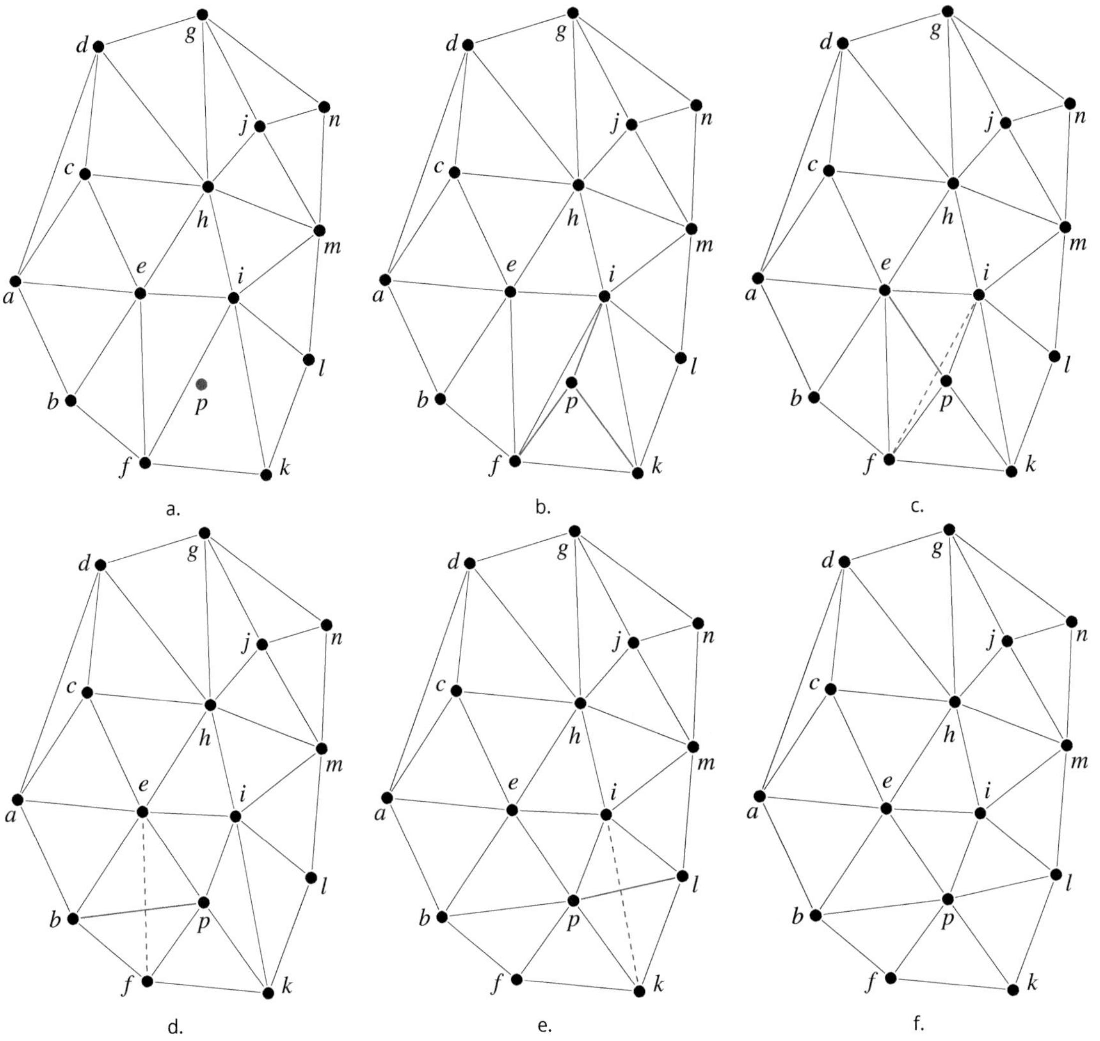

Figure 7.16: Insertion of a point p into a Delaunay diagram following the online algorithm in Algorithm 7.1

A further important form of distributed system transparency is *location transparency*: that the physical locations of individual computing nodes in the system are hidden from the user. Even though the architectures encountered so far in this chapter can be used to provide a range of spatial applications, such as web mapping and web GIS, the locations of the distributed system nodes supporting those applications remained transparent. A *location-aware* system, by contrast, actively exposes the relative or absolute location of some computing devices in order to provide more relevant information and services to a user. Location-aware systems are inherently distributed, connecting mobile computing devices with remote spatial data repositories through a digital communication network.

location-aware

In addition to being inherently distributed, location-aware computing draws on three other important computing traditions.

context-aware computing

- *Context-aware* computing concerns the use of sensors and other sources of information about a user's context to provide more relevant information and services. The term *context* here implies any information that can be used to characterize a user's physical, social, physiological, or emotional circumstances, for example, whether we are driving a car, in a meeting, tired, or angry. Location is a critical element of a user's physical context because where individuals are located strongly affects their access to their physical environment and so the information they may find useful. For example, a hungry London tourist may be interested in information about the best places to eat in London. The same information is unhelpful to someone looking for a good restaurant in New York. Location-aware computing is therefore a particular type of context-aware computing.

user context

- *Pervasive computing* (also termed *ubiquitous computing*) describes the idea that networked computers embedded throughout everyday objects can become unseen personal assistants, helping us with many of our daily tasks. The Internet of things (IoT), for example, is concerned with embedding digital sensing, computing, and network connectivity in everyday physical objects, from bicycles to light bulbs, and cameras to pet feeders. Location-awareness is frequently a critical component of applications of pervasive computing for mobile humans.

pervasive computing

- Finally, *mobile computing* is concerned with interaction and computing with devices that can move around with us as we go about our daily lives. As we have seen in earlier chapters, movement is a fundamental spatiotemporal phenomenon. Unsurprisingly, many applications of mobile computing rely on location awareness, using distributed systems to combine remotely stored data with information relevant to a user's changing location.

mobile computing

Figure 7.17 summarizes the interaction between these four topics: location-aware, context-aware, pervasive, and mobile computing. Several other areas of computing lie at the intersection of two or more of these topics. *Wearable computing*, for example, is concerned with digital devices that are worn about a user's body, accessible even when engaged in other activities. Wearable computers are inherently mobile, like their wearers. Wearable computers are often pervasive, such as glasses, watches, wristbands, or pins. Wearable computers may even be context- or location-aware, utilizing information about a user's immediate environment or physical or emotional state in order to provide more relevant information or services to that person.

wearable computing

Examples of different types of location-aware systems from Figure 7.17 include:

E911 system E911 (enhanced 911) is a program in the USA aimed at ensuring that mobile phone users who call the emergency services can be rapidly located. The E911 system is clearly *location-aware*. The system is also *mobile*, because cell phones may be carried at all times by users. Finally E911 is *pervasive*, because E911 technology is embedded within the cell phone system and aims to automatically locate users in an emergency without the need to explicitly call on the service.

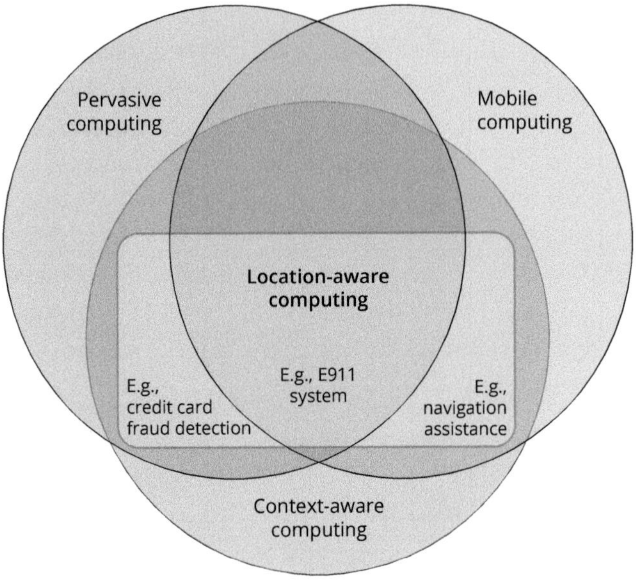

Figure 7.17: Relationship between location-aware computing and context-aware, pervasive, and mobile computing

Figure 7.18: Navigation assistance is a common use of location-awareness

Navigation assistance Navigation assistants, for example, on a personal smartphone, are among the most familiar applications of all location-aware systems (Figure 7.18). These systems are necessarily *mobile*, so that they can be carried around by a user, and *location-aware*, so that users can get navigation assistance from their current location. However, these systems are not necessarily *pervasive*, as they typically involve interaction with dedicated apps on a user's smartphone, rather than being embedded in other everyday objects.

Figure 7.19: Location-awareness is integral to many everyday objects and tasks

Credit-card antifraud system Credit card companies commonly detect fraud by analyzing the patterns of credit card usage. In particular, the location of each purchase is recorded by the credit card company. If a particular card is used in unexpected or widely different locations in rapid succession, this unusual pattern may trigger the credit card company to suspect fraud and block that card. Such a system is certainly *location aware*, because it tracks shoppers' locations via their credit card usage. It is also *pervasive*, because this protection is embedded within the credit card system (Figure 7.19). Fraud detection systems do not necessarily require mobile computing, however, since the computing devices involved in the system can be static (i.e., the network of retail "swipe" machines may be at fixed store locations).

Location-aware systems are able to alter fundamentally the way we interact with GIS. Instead of using a GIS at a desktop or workstation, location-awareness allows us to interact with spatial data and GIS at the same time as interacting with the geographic environments about which we are accessing information.

As with many important developments in computing, these new possibilities have been made possible by advances in technology:

- The availability and variety of portable and embedded computing devices commonly encountered has increased dramatically over recent decades. Computing hardware, discussed in Section 1.4, continues to become smaller, cheaper, more powerful, more energy efficient, and increasingly embedded within everyday objects in our homes, offices, vehicles, and public spaces.
- The availability, extents, and bandwidth of wireless digital communication networks has also increased sharply. Wireless networks, also discussed in Section 1.4, are vital to mobile computing because they allow computing devices to communicate with one another without the need to be physically connected.
- The use, variety, and reliability of positioning sensors capable of determining a mobile user's location has increased markedly. Today, positioning capabilities are commonly embedded in many everyday mobile objects and portable devices. The key principles and characteristics of positioning technologies are the subject of the following section.

7.4.1 Location sensors

There exists a wide range of sensors that can be integrated with computing devices to obtain information about a user's context automatically; these include sound, light, temperature, motion, as well as location sensors. We can distinguish two fundamentally different approaches to positioning: *active* and *passive* location sensing. Figure 7.20 shows a taxonomy of the different location-sensing techniques.

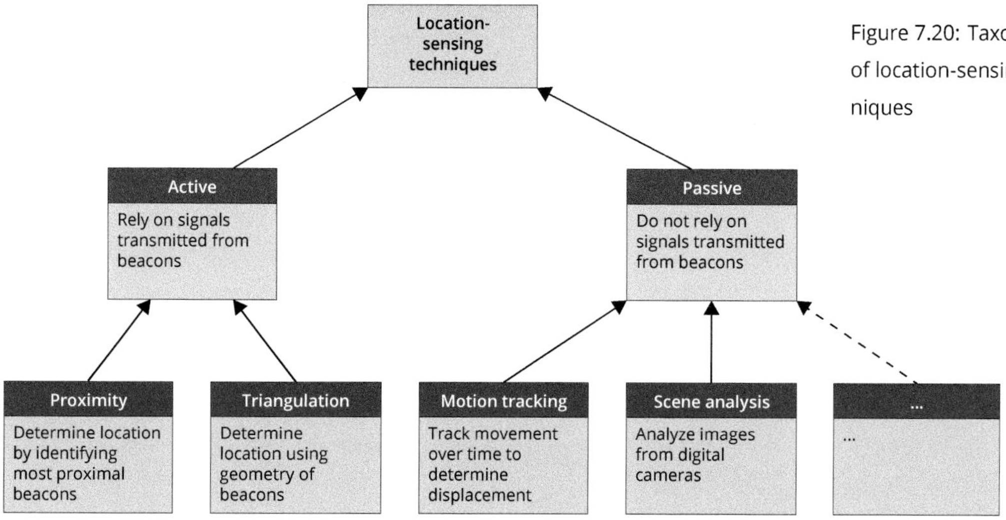

Figure 7.20: Taxonomy of location-sensing techniques

Active location sensing Active location sensing determines location using signals from transmitters or *beacons*. Two distinct approaches to determining

active location sensing

active location sensing: using the *proximity* of beacons or by *triangulation* of beacon signals. For example:

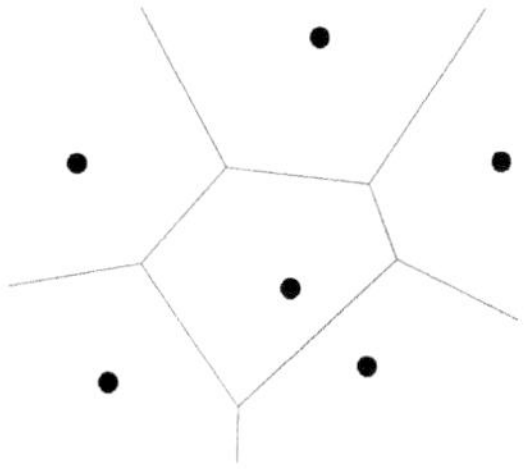

Figure 7.21: Voronoi diagram and Thiessen polygons, from Section 5.3.3

Proximity Proximity to beacon signals is used by all cell phone networks, for example, to locate cell phones. A cell phone network uses an array of wireless transmitters at dispersed locations to relay phone calls. At the most basic level, each transmitter serves a small geographical area where its signal is stronger than any other transmitter, termed a *cell*. Signal strength decays with distance from the transmitter, so in the simplest case the geometry of a cell forms a Thiessen polygon (see Chapter 5, Figure 5.17, and Figure 7.21 for ease of reference), each cell containing the proximal locations to the transmitter. A cell phone can determine in which cell it is located by identifying the transmitter with the strongest signal. The size of cells varies across a cell phone network, depending on terrain, transmission frequency, and expected volume of calls. Typically, terrestrial transmitters are spaced from about 200 m apart in urban areas to more than 5 km apart in rural areas.

lateration

angulation

Triangulation There are two forms of triangulation: *lateration* and *angulation*. Lateration is the process of computing the position of an object based on its *distance* from other known locations. Angulation, is the process of computing the position of an object based on its *angle* from other known locations. GNSS (global navigation satellite systems), such as GPS and Galileo, rely on lateration to determine location (discussed below). Angulation is used, for example, by some augmented cell phone networks to help determine location more precisely than with proximity alone.

Most active location-sensing techniques, such as GNSS, rely on a mobile sensor that receives signals transmitted by external beacons, usually at known locations. It is also possible to position using external sensors at known locations receiving signals from a mobile beacon. For example, *Active Badge* was an early example of an indoor positioning system that used an array of infrared proximity sensors, distributed at known locations throughout a building, to detect a coded infrared signal transmitted by a small badge worn by building occupants (Want, Hopper, Falcao, & Gibbons, 1992). However, mobile beacons place higher demands on the limited energy resources of mobile devices, and so most active positioning systems are designed around mobile sensors.

passive location sensing

Passive location sensing In contrast, passive location sensing techniques do not rely on signals transmitted from locator beacons. Instead, passive location-sensors determine location indirectly, by relating sensor-based information to other spatial information about the environment. The geographic environment is highly heterogeneous, so almost any type of sensor measurement can potentially be used in some form of passive location sensing (hence the dotted arrow in Figure 7.20). Imagine being transported from your home to an unknown location on the Earth's surface. As long as your watch accurately records the time at your home, and you are able to "sense" the time at your location (perhaps by observing the position of the sun) you can obtain a crude estimate of your location in terms of your longitude. With the development of reliable, portable clocks in the 18th century, sailors used a similar method to help them navigate.[7] Two common types of passive location sensing are:

[7] Described in the highly recommended book *Longitude* (Sobel, 1995)

Motion tracking Speed and direction sensors are often used in robotics to track the velocity of a mobile device or robot over time. This information can be used to calculate a robot's location at a particular point in time relative to its starting point. Motion tracking techniques are sometimes referred to as *dead reckoning*.

dead reckoning

Scene analysis A digital camera is a sensor array that can be used to determine location in certain circumstances. Scene analysis can be used either to determine the location of objects or people in a digital image, or to determine what the location of the camera must have been in order to have taken the picture.

Most active location sensors aim to measure *absolute* location with reference to some fixed framework, such as a global geodetic datum or a local coordinate system. Sensing absolute (coordinate) location is also commonly known as *positioning* or *localization*.[8] In contrast, some location sensors—in particular, some passive location sensing technologies—aim simply to detect *relative* location in terms of the spatial relationship to some other object or location (such as *inside* a building or room, *north* of a dock, or relative *distance* to another person).

absolute location

positioning

[8] The term "positioning" is preferred to "localization," since localization is also used in other contexts to refer to the adaptation of an information system to a particular region or language.

relative location

Global navigation satellite systems A GNSS receiver uses radio wave signals, transmitted from an array of GNSS satellites orbiting the Earth (Figure 7.22), to calculate location of the receiver on the surface of the Earth. GNSS rely on lateration (defined above as the process of computing the position of an object based on its distance from other known locations).

For example, assume for simplicity that the world is planar, and that we wish to discover the identity of an unknown city in the USA. If we are told that this city is 2900 km away from Los Angeles, then the locus of the city's possible locations is a circle with radius 2900 km centered on Los Angeles. If the city is also known to be 1000 km from Washington, only two possible locations fit this information: Chicago, IL and Montgomery, AL. Finally, if the city is known to be 1900 km from Miami, then only one location fits this information: Chicago (see Figure 7.23).

Figure 7.22: Illustration of a GNSS satellite in the EU Galileo constellation

GNSS uses lateration in three dimensions, whereby the locus of locations a fixed distance from a point forms a sphere rather than a circle. As a result, four spheres are needed to pinpoint a location in 3D space. Since the Earth itself is roughly spherical, most GNSS receivers can obtain a fix using signals from only three satellites by assuming that the receiver is located on the surface of the Earth.

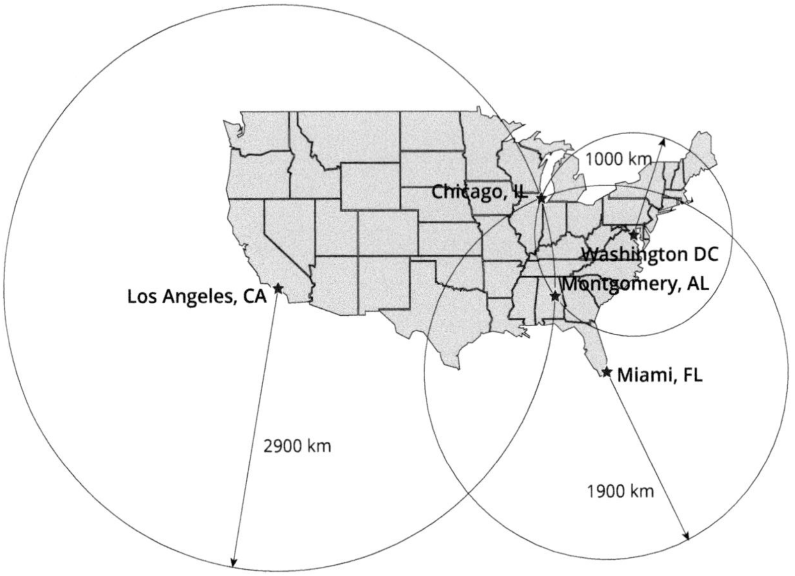

Figure 7.23: Lateration of location, based on distance from Los Angeles, Washington, and Miami

GNSS receivers calculate the distance to each satellite by measuring how long it takes a radio wave signal to travel from each satellite. Each satellite in the GNSS constellation continually transmits a radio signal with the exact time and that satellite's position. The lag between the time stamp on the satellite signal and the current time at the GNSS receiver is used to calculate the distance to each satellite, assuming the signal travels at or near the speed of light. GNSS satellites measure time using highly accurate atomic clocks. While the clocks in GNSS receivers are not so accurate, any inaccuracies will affect all measurements equally and so can be accounted for in the GNSS lateration algorithm.

GNSS is a revolutionary but complex technology. Using GNSS for location-aware computing carries some disadvantages. The precise frequency of transmission varies slightly as a result of the high speed of orbiting satellites. Each satellite uses a different encoding for its signal, which can last up to 30 seconds before starting again. As a result, the process of searching for signals and achieving an initial fix is complex, increasing the cost, size, and power consumption of GNSS receivers. For example, the *time to first fix* (TTFF) of a GNSS receiver usually lasts at least 20 seconds, and in some cases may require several minutes. In some applications, such as emergency services, this delay can be critical.

time to first fix

Sensor accuracy and precision Two important features of any location-sensing technique are its *accuracy* and *precision*. Sensor accuracy is the closeness of data from a sensor to the correct value. Sensor precision is the level of detail of the data generated by a sensor (see Chapter 10 for further information about accuracy, precision, and other aspects of uncertainty in location). For example, active proximity location-sensing techniques are often accurate, but their precision depends upon the density of beacons. In a basic cell phone network, where beacons may be located several kilometers apart, location precision may be low because it may not be possible to determine exactly where a phone is located within a cell. In areas with a greater density of cell phone towers, each cell is smaller, and so active proximity location sensing can be more precise. Figure 7.24 illustrates the relationship over a region with varying tower density.

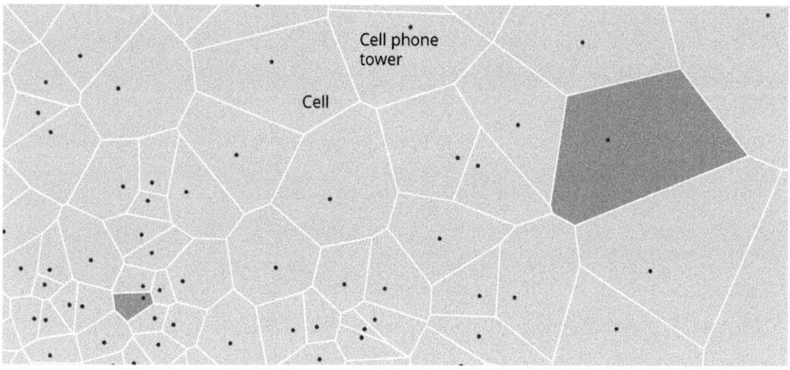

Figure 7.24: Precision of cell location, highlighting smaller, more precise location (southwest cell) and larger, less precise location (northeast cell)

Motion tracking techniques, on the other hand, tend to be precise but less accurate location sensors. Relatively small measurement motion tracking errors can be compounded over time, a process known as *error propagation*. For example, if you travel at 3.0 kmh^{-1} on a bearing of 068°, but your motion sensor inaccurately measures your velocity as 3.3 kmh^{-1} on a bearing of 074°, then the location calculated from the sensor data diverges from your actual location over time (Figure 7.25). After 5 seconds, the inaccuracy in location will be about 60 cm. After 5 hours, the inaccuracy in location will be more than 2 km.

error propagation

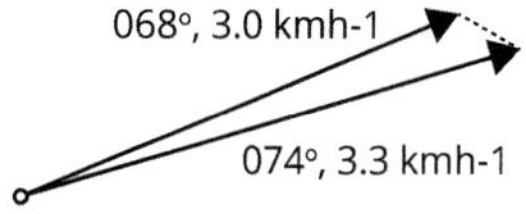

Figure 7.25: Error propagation in dead reckoning

GNSS is remarkable amongst location sensors in that it can achieve high levels of both accuracy and precision. However, in addition to the delay in obtaining an initial fix (TTFF), the high-frequency low-power GNSS microwave signals are easily attenuated by physical objects, such as buildings, trees, and water. As a result, without additional on-ground augmentation, GNSS signals cannot be received indoors, underwater, or in the shadow of obstacles such as buildings (so-called "urban caynons").

urban canyon

Improvements in accuracy and precision, as well as other benefits, can often be achieved by using combinations of different location-sensing techniques. For example, most in-car navigation systems use a combination of GNSS and motion tracking to accurately and precisely determine location. When a GNSS fix is available, the vehicle's location can be determined with high accuracy and precision. When local conditions, such as tunnels or urban canyons, block GNSS signals for short periods, tracking the speed and orientation of the vehicle can fill in the "gaps" left by GNSS positioning. Fusing this combined position with map-based information about the location of the road itself can further increase accuracy and precision, through a technique known as *map matching*.

map matching

A more in-depth discussion of location sensing lies beyond the scope of this book. For interested readers, there exists a wide variety of further reading available for those interested in more detail on positioning and location sensing, both from a computer science perspective (Hightower & Boriello, 2001) and a GIS perspective (Basiri et al., 2014).

7.4.2 *Location-based services*

location-based services

Figure 7.26: Many familiar apps today rely on location awareness

Figure 7.27: Contact tracing apps used during the 2019 pandemic are examples of location-based services

Location-based services (LBS) are specific applications that require location-aware computing to operate. Some examples of LBS have already been encountered above (in Figures 7.17 and 7.18, for example). Many of us use LBS every day—a quick glance at your smartphone's location services settings may reveal you are using a location-based service right now!

One way of classifying LBS applications is according to whether they rely on *current* location, *past* location, planned *future* location, or some combination. A simple location-aware street map or weather app may only be concerned with providing access to information and resources based on where a user is *currently* located (e.g., Figure 7.26). Apps such as automated road tolls and contact tracing (e.g., Figure 7.27) will additionally require *historical* information about a user's location, in order to retrospectively determine charges or potential exposure. Navigation, delivery, and logistics apps, for example, typically also support planning a user's future location based on their current as well as usually that person's historic location (e.g., Figure 7.18).

The 2019 coronavirus pandemic raised the profile of location-based services, even for people who were previously unaware. Most countries around the world encouraged their citizens to use smartphone "track-and-trace" apps to assist with contact tracing and combat the spread of the disease. In general, these apps relied on *relative* location sensing through *proximity* to short-range Bluetooth signals from other citizens' devices. In that case, the location information generated was devoid of absolute coordinate positions and could only be used to track location relative to other people.

A few countries (such as Israel) developed track-and-trace apps that used *absolute* positioning, based on each smartphone's in-built positioning capabilities (typically using GNSS). However, absolute positioning approaches were unpopular in most countries due to concerns about personal privacy. Location data is regarded by most people as sensitive and personal information. As a result, LBS always raise important questions about the control and security of any personal location information generated. Personal location data must be protected in a similar way to other private information, such as health data. However, location data has some particular characteristics that require special handling when using LBS—a topic we return to in Chapter 10 under the heading of *location privacy*.

7.5 Decentralized computing

A successful distributed system relies on achieving a satisfactory balance of *modularity*—the composition of multiple distinct units each with standardized, clearly defined functions; and *autonomy*—with each coordinating module able to operate more-or-less independently of the others. Recall the definition of a distributed system in Section 7.2, which explicitly captures this synergy as: "a collection of multiple *autonomous modules* [emphasis added] ... that can operate as if a single, coordinated system ..."

However, these two characteristics of modularity and autonomy are often in subtle competition with each other. Increasing autonomy of distributed system modules can lead to increasing difficulty in effective coordination of the whole distributed system. Conversely, increasing modularity by decomposing a system into its constituent functions can lead to increasing the interdependencies between the modules that implement these functions. For example, for surfing the WWW, a web server relies on web browsers that can request pages from the server and render them in a way users can read. Similarly, web browsers are useful because of the existence of web servers waiting to respond to requests for web pages. Web servers and web browsers depend on each other for their usefulness, decreasing their individual modular autonomy.

Achieving module autonomy at the same time as system modularity can become especially important for distributed systems made up of very large numbers of highly heterogeneous nodes. The Internet of things, for example, is the expanding global network of billions of physical objects with embedded computing, sensing, and digital communications capabilities.[9] In 2022, the Internet of things was estimated to comprise over 10 billion devices of myriad different kinds, from light bulbs to asthma inhalers and from security cameras to cars. A distributed system of this scale and diversity requires that each individual device can function autonomously as well as part of the collective.

Internet of things

[9] In my home today, for example, I can control all my lights, my heating and air conditioning, my TV and music systems, my alarm and security door, and many other home devices using a networked digital controller, such as my smartphone.

An important concept in building distributed systems with high levels of module autonomy is the idea of *decentralization*.

7.5.1 Decentralized systems

In a *decentralized* system, no single module or node has overall control or global authority over the system operation. However, while control is an important characteristic of decentralization, in practice most distributed systems satisfy that criterion. Only early mainframe architectures—introduced above, but rarely used in practice today—exert truly centralized control over a whole distributed system.

Instead, the defining characteristic of a decentralized system lies in the distribution of *information* throughout the system. In a decentralized system, information is always dispersed across the network, and never collected together in one location for centralized processing. In a decentralized system, defined below, individual nodes can hold pieces of the puzzle, but no single node can see the whole picture (Figure 7.28).

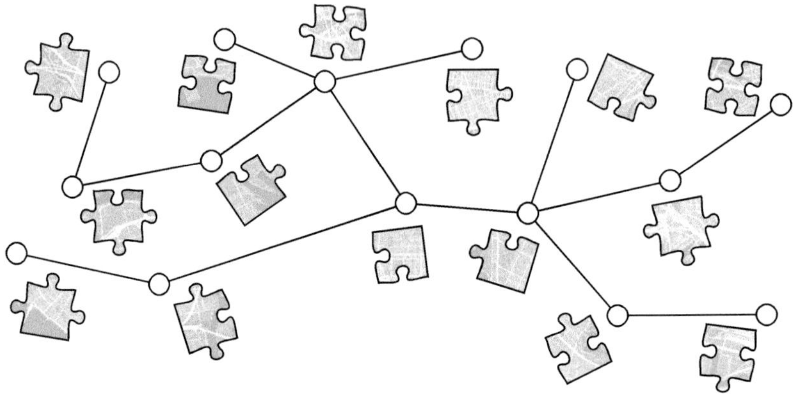

Figure 7.28: In a decentralized system no single node ever knows the entire system state, with information dispersed across the network

decentralized system

A decentralized system is a special case of a distributed system where no single component knows the entire system state.

By contrast, in all the distributed systems so far encountered, if required a single node may gather together all the information required for some computation from across the entire network (cf. Figure 7.28). Such a system is called a *centralized system* (Figure 7.29). For example, many countries' contact-tracing apps transferred users' location data to a central repository for analysis in order to detect and respond to disease outbreaks proactively. In response to privacy concerns, some national contact-tracing apps adopted a decentralized approach, where data remained on each user's device unless an early outbreak was (retroactively) detected.

centralized system

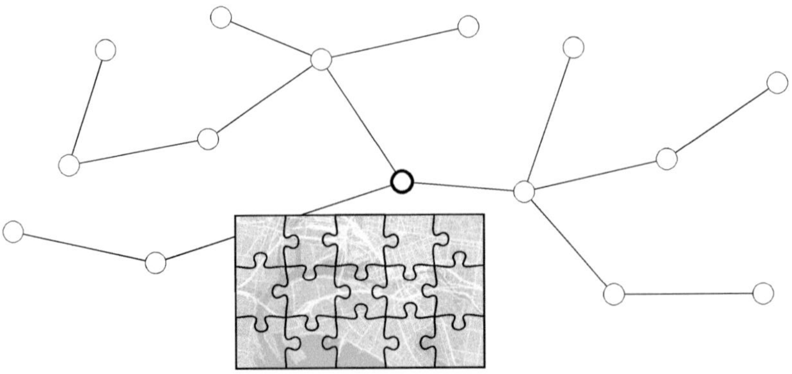

Figure 7.29: In a centralized system, a single node may have the authority if required to gather global information together from across the whole network

In addition to potential privacy implications, decentralized architectures bring inherent scalability advantages. Because a decentralized system carries no expectation that any single node has global oversight, new nodes can be added to the system at any time without any central registry, coordination, or control. This property is vital to the Internet of things, for example, where new IoT devices are continually being added without the need for any centralized tracking of which devices are where in the network. By contrast, in the event-driven architectures introduced in the previous section, new consumers must subscribe to published channels via a broker. The decentralized approach also ensures the network is less vulnerable to becoming overloaded with data as the system grows. Even though the number of connected devices in the Internet of things is expected to double over the next 5 years, reaching 30 billion devices by the end of this decade, we can still have confidence in continued IoT operation as it scales.

In addition to scalability, other advantages of decentralized architectures can include:

- *Increased robustness*: Just as nodes can be easily added to a decentralized system, so can they be easily removed. This property leads to distributed systems with increased robustness and fault tolerance. Decentralized systems do not have single points of failure, and nodes may fail or be removed at any time without interrupting the overall system operation.

- *Decreased communication*: Centralizing information and control typically requires more communication on the network. The increased autonomy of decentralized systems tends to lead to less communication, with nodes able to choose when and what to communicate. Communication—especially wireless communication—is one of the most energy-intensive operations for most computing devices. As a result, decentralized architectures are especially well suited to applications where node energy and communication bandwidth are limited, such as the Internet of things where many devices may be small and battery powered.

- *Increased security*: Decentralized architectures are often associated with increased levels of security. Attackers cannot exploit a single point of failure or a centralized controller. Further, individual nodes may opt not to communicate sensitive data. Consequently, decentralized architectures are well suited to applications where privacy is a priority, such as when a person's location needs protecting, as we have seen. We return to questions of privacy and in particular location privacy in Chapter 10.

- *Decreased latency*: Increased autonomy and decreased reliance on communication can also lead to decreased latency, in terms of the time taken by the system to respond to an event. Nodes in a decentralized system are empowered to act upon the available information, rather than having to await confirmation from a centralized controller or collation of complete information.

The Internet of things provides a ready example of a system that is well suited to decentralized architectures. The Internet of things depends on scalability and robustness but also benefits significantly from decreased communication, increased security, and decreased latency. Many IoT devices and applications also rely on spatial information to some extent (see, for example, Cao & Wachowicz, 2019). Indeed, there is another class of distributed system, related to the Internet of things, that explicitly relies on the generation, capture, and processing of spatial data called *geosensor networks*.

7.5.2 Geosensor networks

sensor mote

Geosensor networks are constructed from networks of sensors, introduced already in Chapter 1. A (wireless) sensor network is a wireless network of miniaturized sensor-enabled computers, called sensor nodes or sensor *motes* (Zhao & Guibas, 2004). Each mote can be armed with an array of on-board sensors. As we saw in Chapter 1, advances in MEMS (microelectromechanical systems) technology has led to a proliferation in low cost, miniaturized sensors, including temperature and moisture sensors, light sensors and cameras, microphones and vibration sensors, and chemical and gas sensors. A typical wireless sensor mote is pictured in Figure 7.30, showing the mote without its protective housing in order to expose the on-board sensors, the battery, and the wireless digital communications antenna.

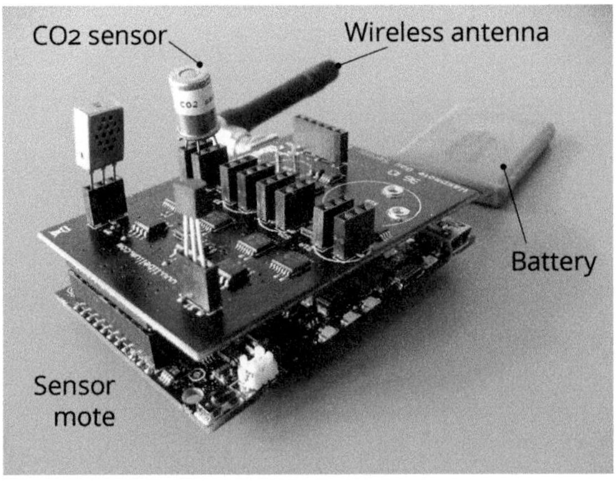

Figure 7.30: A wireless sensor mote, highlighting the battery, communications antenna, and on-board sensors

A wireless sensor network comprises hundreds or thousands of sensor motes, each able to use its sensors to detect changes in its immediate environment; process that sensed data using on-board memory and CPU; and share sensed data with other nearby sensors via wireless digital communications. Wireless motes are "untethered"—disconnected from wired networks or power supplies—and must rely on their own on-board power sources. Batteries are the primary power source for most motes, although in some cases energy harvesting from sources such as solar power can also be used. In all cases, the reliance on internal power sources limits motes to using short-range, low-power wireless communication to nearby motes, usually only a few meters or so. Figure 7.31 illustrates a wireless sensor network, highlighting the short-range communication links where only nearby motes are connected.

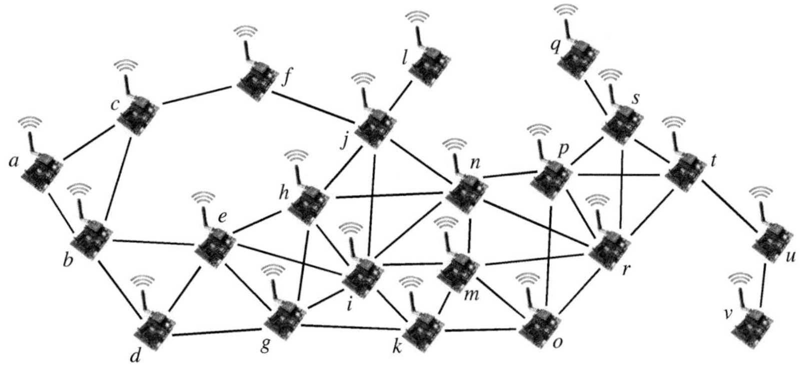

Figure 7.31: A wireless sensor network, highlighting local, short-range, peer-to-peer communication and multihop routing

"Flat" network structures such as that in Figure 7.31—where nodes connect directly to one another, rather than via a central hub or backbone—are called *peer-to-peer networks*. In a peer-to-peer network, not all (indeed few) pairs of motes are directly connected. When a mote needs to communicate over longer distances, it must use intermediate motes to relay its message, termed *multihop communication*. In Figure 7.31, messages between motes a and b can be sent directly (so-called "single-hop communication"). In contrast, sending a message from mote c to mote r will require multihop communication, relayed via motes f, j, and n, for example. In most cases there may be multiple different routes a multihop message can take (such as via b, e, i, m). The question of how to route messages efficiently and reliably within peer-to-peer sensor networks has been a major research topic in its own right (see Ganesan et al., 2004, for an overview of the key issues and concepts).

peer-to-peer network

multihop communication

Geosensor networks, then, are a special type of sensor network.

> A (wireless) geosensor network is a wireless sensor network that monitors phenomena in geographic space (Nittel et al., 2004).

geosensor network

In essence, a geosensor network is a wireless sensor network where motes can sense not only their environment, but also their (relative or absolute) spatial location. Geosensor networks use those locations in spatial algorithms and to perform spatial analyses. Motes may sense their location using any of the positioning techniques already explored in Section 7.4.1. Geosensor networks are fundamentally changing the way we use GIS technology to monitor natural and built environments, uncovering never-before-seen spatial and temporal detail (see Box 7.5 on the next page).

7.5.3 *Decentralized algorithms*

A critical question facing a geosensor network designer, and indeed the designer of any decentralized system, is how to construct useful applications when no single node has global knowledge of the entire system state. With

Box 7.5: Storm petrels on Great Duck Island

In a pioneering 2003 study, a network of more than 100 wireless sensor motes monitoring temperature, humidity, and light was deployed in seabird nesting grounds on Great Duck Island, an island off the east coast of Maine in the US (Szewczyk et al., 2004). The nests belonged to Leach's storm petrels, *Hydrobates leucorhous*—a seabird that prefers inaccessible and rocky offshore nesting sites. The difficulty in accessing remote nesting sites, combined with the sensitivity of nesting birds to disturbance by humans, made it especially difficult for scientists to monitor the breeding habits of this threatened species. The solution was to use the geosensor network, deployed in burrows well before the nesting season had begun, to monitor the birds once breeding had commenced. This information has helped provide new insights into breeding success and how the birds select a nest, informing conservation efforts to protect the birds and their habitat. Several further examples of environmental applications of geosensor networks can be found in Nittel (2009).

Leach's storm petrel, *Hydrobates leucorhous*

the exception of online algorithms, introduced in connection with stream computing in Section 7.3.2, almost all the algorithms encountered in this book assume global, centralized knowledge. Just as for online algorithms, algorithms in a geosensor network similarly cannot know in advance what data motes will encounter in the future. Importantly, individual motes in a geosensor network also cannot know what data motes at other locations are sensing. Only when information is explicitly routed through the network using multihop communication can motes gain information about sensed data at other distal locations.

As a result, decentralized algorithms share some features in common with online algorithms. Algorithm 7.2 provides one of the simplest algorithms for sharing information about a mote's detected events with the whole sensor network, called *gossiping*. As already encountered in connection with online algorithms (see Algorithm 7.1 on page 285), the **when** keyword is used in Algorithm 7.2 to enclose a procedure that is executed when some event occurs at a mote. Events may include a mote receiving a wireless communication from a neighboring mote or detecting a change in the sensed value of an on-board sensor.

gossiping

The first line of our decentralized algorithm defines the assumptions about the capabilities of motes in the sensor network, using the **Local** keyword. In Algorithm 7.2 we need to assume each mote has a:

- unique identity known to each mote m, written ident(m);
- sensor on-board each mote m, with sensed value sense(m);
- variable storing the state-of-operation for each mote m, written state(m); and
- set of zero or more immediate neighbors nghbr(m) of a mote m, each within direct (one-hop) communication range.

Algorithm 7.2: Gossiping in a sensor network

Local motes m_i in sensor network M; mote identity ident(m); mote sensor sense(m); mote state state(m); one-hop mote neighbors nghbr(m);
Global set M of sensor motes $m \in M$; all motes $m \in M$ as state(m) ←IDLE; gossip probability $b = 0.3$

```
when sense(m) < 0.0
    if state(m) = IDLE then
        broadcast (msge, ident(m), sense(m)) to all neighbors nghbr(m)
        set state(m) ←DONE
end when
when receiving (msge, i, s)
    if state(m) = IDLE then
        Randomly select a number x ∈ {0, 1} with probability P(x = 1) = b
        if x = 1 then
            broadcast (msge, i, s) to all neighbors nghbr(m)
end when
spontaneously
    if sense(m) ≥ 0.0 then
        set state(m) ←IDLE
end spontaneously
```

In understanding a decentralized algorithm, it is essential to be clear that each individual mote only has access to its *own* (i.e., local) information. A mote knows its own identity, its own sensor values, its own state, and which messages it receives from immediate one-hop neighbors. But no mote can know anything about any other motes in the network unless it is explicitly communicated to a mote in a message it has received. This constraint is at the core of geosensor network algorithms, and indeed any decentralized algorithm: each mote (or module more generally) is autonomous and operating in parallel with and independently of the other motes in the network.

Next, the **Global** keyword specifies any network-wide initialization or settings. Algorithm 7.2 lists:

- the set M of sensor motes;
- the initial state setting of each mote $m \in M$ as state IDLE; and
- a gossip probability b known to every mote $m \in M$.

The **Global** keyword is the only place in a decentralized algorithm where we are ever permitted to add such global constraints or settings. Global constraints can be thought of as the "factory settings" for the motes in the sensor network, initialized at the field station or in the workshop before deployment. Once the network is deployed and operating, there is usually no way to make global changes or assume shared knowledge across the network (except to manually visit each mote and reset it, equivalent to a new "factory reset"): this is what it means to be autonomous and decentralized.

Once deployed and operating, the body of Algorithm 7.2 is decomposed into three parallel procedures, as follows:

- Whenever a mote's sensor trips a threshold (line 1, a sensed value below 0.0—let's assume it is a temperature sensor tasked with detecting freezing temperatures), then the first procedure is triggered (lines 2–4). If the mote's state is IDLE (line 2), then it broadcasts to all its neighbors a message containing the mote's identity and its sensor reading (line 3), before transitioning into a DONE state (line 4). The change of state is to prevent a persistent low temperature reading triggering multiple messages and overloading the network.

- In turn, whenever a mote receives a message from a one-hop neighbor (line 6) there is a chance that it will rebroadcast that message (line 10), spreading the knowledge further through the sensor network. The probability of rebroadcasting ("gossip probability") was globally set to 30% for all motes—on average, 3 in every 10 received messages by a mote will be rebroadcast (lines 8–9). The reason for not rebroadcasting every received message is, again, to avoid overwhelming the network with communication. The gossip probability chosen is a heuristic: the aim is to set the probability high enough to ensure new information is relayed throughout the network, but not so high that rebroadcasting floods the network indefinitely.

- The third procedure happens **spontaneously**—i.e., whenever the mote is able or has free resources (lines 12–15). This procedure is designed simply to reset the mote state to IDLE (line 14), ready to broadcast new temperature alerts, if the sensor reading rises to 0.0 or above (line 13).

Figure 7.32 depicts a hypothetical execution of Algorithm 7.2 in a simple sensor network of 20 motes a–t (Figure 7.32a). The emergence of a "coldspot" (−0.8°C) is detected by motes f, h, and j, resulting in the broadcast of three alert messages (Figure 7.32b). The messages are received a total of eight times among six motes (motes i and m receive messages from both h and j, Figure 7.32a). At this point, Figure 7.32c shows the value of randomly selected number x (Algorithm 7.2, line 8) for each received message. The gossip probability $b = 0.3$ ensures most incoming messages will be ignored, but three motes (e, n, and l) by chance do "gossip" their message. Motes receiving the message multiple times must still respond to each message receipt event separately. Figure 7.32d–f show further snapshots in the execution, resulting in all motes except q having at least "heard" about the coldspot by Figure 7.32f.

While Algorithm 7.2 and Figure 7.32 illustrate the key features of a decentralized algorithm—autonomy and parallel execution across the network—the algorithm is deliberately simplified to aid explanation, and there are many ways it might be improved. We will look at just one such improvement in Algorithm 7.3 that explicitly uses spatial information, making this next example the basis of a true geosensor network.

georouting

Instead of gossiping in Algorithm 7.2, *georouting* uses knowledge of the coordinate locations of motes in a geosensor network to make routing dra-

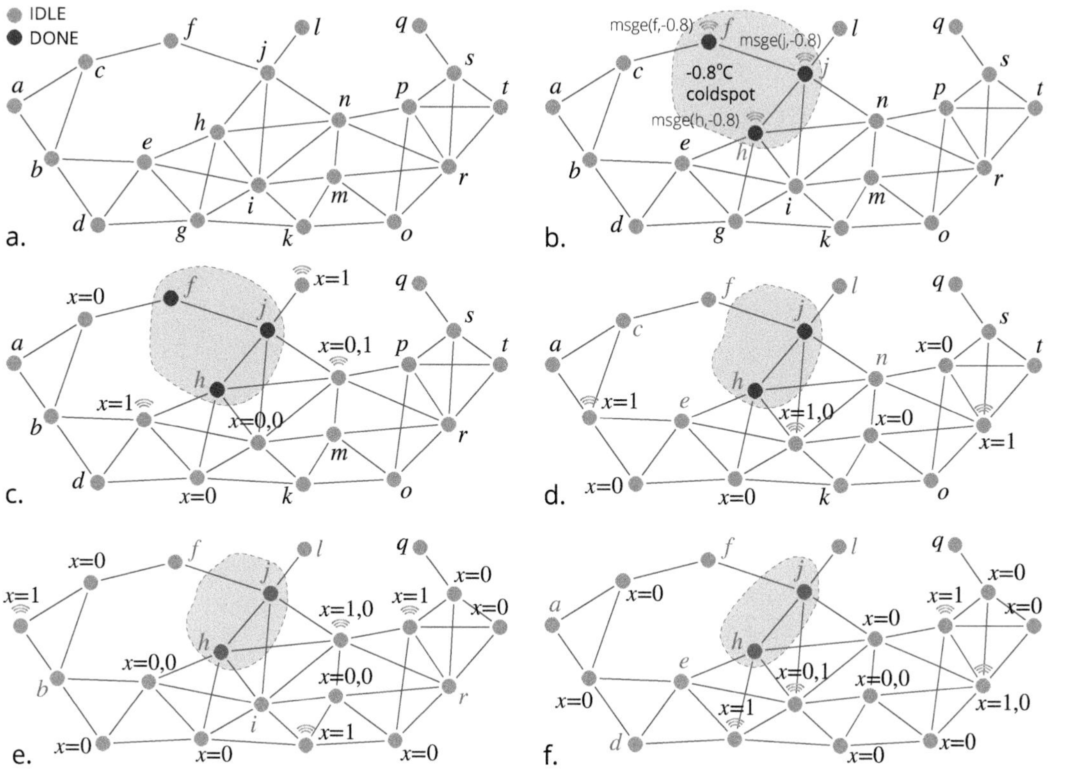

Figure 7.32: Gossiping in a sensor network equipped with Algorithm 7.2

matically more efficient. Algorithm 7.3 begins similarly to Algorithm 7.2, but with two important differences:

- The **Local** environment assumes the additional capability of motes to determine their own coordinate position, written locat(m), perhaps using an on-board GNSS sensor for example.
- The **Global** environment assumes that all motes are aware of the coordinate position g of a special "gateway" mote. Geosensor networks must often incorporate a high-capacity gateway mote, usually with wired power or additional batteries. The gateway is tasked with relaying important detected events to the outside world, via wireless WAN, for example.

Algorithm 7.3 also stores the identities and coordinate locations of any neighbors it discovers in a *hash table*—a data structure for storing lists of key-value pairs. The body of Algorithm 7.3 is again decomposed into a number of parallel procedures, briefly explained as follows:

hash table

- The procedure triggered whenever a mote's sensor trips the threshold (lines 1–6) is similar to the corresponding procedure in Algorithm 7.2. However, in Algorithm 7.3 the broadcast alert message also contains the identity of the mote's neighbor that is closest[10] of all its neighbors to the gateway mote (lines 3–4).
- Upon receiving an alert message from a one-hop neighbor (lines 7–10), a mote first checks if this message is "addressed" to it (line 8), ignoring any

[10] Recall from Section 3.1.1 that $|\vec{gp}|$ denotes the distance between the points (vectors) $\vec{g}$ and $\vec{p}$

Algorithm 7.3: Greedy georouting in a geosensor network

```
Local mote identity ident(m); mote sensor sense(m); mote coordinate position
    locat(m); mote state state(m); one-hop mote neighbors nghbr(m)
Global set M of geosensor motes m ∈ M; all motes m ∈ M as state(m) ←IDLE;
    empty hash table H; gateway mote coordinate position g
1: when sense(m) < 0.0
2:     if state(m) = IDLE then
3:         select j, p ∈ H such that for all k, q ∈ H then |gp⃗| ≤ |gq⃗|
4:         broadcast (msge, ident(m), sense(m), j) to all neighbors nghbr(m)
5:         set state(m) ← DONE
6: end when
7: when receiving (msge, i, s, t)
8:     if ident(m) = t then
9:         select j, p ∈ H such that for all k, q ∈ H then |gp⃗| ≤ |gq⃗|
10:        broadcast (msge, i, s, j) to all neighbors nghbr(m)
11: end when
12: when receiving (posn, i, p)
13:     store location p in hash table H with key i
14: end when
15: spontaneously
16:     broadcast (posn, ident(m), locat(m)) to all neighbors nghbr(m)
17:     if sense(m) ≥ 0.0 then
18:         set state(m) ←IDLE
19: end spontaneously
```

messages that are not. The mote then looks for the next neighbor closest to the gateway, before addressing a rebroadcast message to that specific target.

- The third procedure (lines 12–14) is half of the mechanism for ensuring that motes know who and where their neighbors are. Remember that the two procedures have no access to information about neighbors, unless it has been previously communicated. When a mote receives a **posn** message from a neighbor, it stores the identity and location of that neighbor, contained in the message payload, as a key-vale pair in its hash table *H*.

- The second half of the mechanism for informing motes about the locations of their neighbors is contained in the final procedure (lines 15–19). Motes **spontaneously** broadcast their own **posn** message to neighbors (line 16), before checking if a mote state reset to IDLE is indicated by the current sensor reading (lines 17–18) as for the previous algorithm.

Reflections

System architecture is a fast-moving area, with new technologies and trends emerging every month. Client-server systems remain at the core of most modern system architectures, but today's GIS have evolved significantly in their architectures over the past 15 years. The choice of topics in this chapter reflects an attempt to balance the stable core of system architectural knowledge against the background of rapid technological innovation, and the opportunities that will surely bring for GIS in the future. The final selection balances established GIS architectures (i.e., client-server), more recently adopted architectural innovations in GIS (i.e., cloud computing and stream computing), and newer architectures likely to grow in importance over the coming decade (i.e., decentralized and peer-to-peer architectures).

Nevertheless, there were other technologies and architectures that might also have been included. *Edge computing*, for example, aims to recruit the legions of consumer computing devices connected to the Internet into powerful, yet flexible loosely coupled applications. In common with the decentralized, IoT, and peer-to-peer architectures discussed in this chapter, edge computing aims to bring processing closer to the source of sensed data, in contrast with cloud computing.[11]

edge computing

[11] Bridging edge and cloud computing architectures, so-called "fog computing" aims to provide access to edge computing capabilities in the cloud, while managing the complexity of many small, autonomous, loosely coupled edge devices.

Although not an architecture so much as a technology, *blockchain* is a distributed ledger system that helps to create shared trust in transactions (such as financial or land transactions) without reliance on a central authority (such as a bank or national cadaster). Blockchain algorithms and systems are compatible with decentralized architectures, allowing high levels of autonomy in the nodes cooperating to create or validate a transaction. Blockchain is associated with a range of emerging applications in GIS (e.g., Cedeno Jimenez, Zhao, Mansourian, & Brovelli, 2022). One important example is *proof of location* which combines blockchain and decentralized location-aware architectures to support applications where proof of a person or object's presence at a particular location is required (such as, for example, many logistics or delivery applications, certain financial transactions, and a range of safety and insurance applications).

blockchain

proof of location

The selection of technologies in this chapter was also designed to illustrate the enduring and fundamental architectural principles of interoperability and modularity, together with the supporting issues of coordination, transparency, and autonomy. No matter the inevitable innovations in GIS architectures over the coming years, these principles will continue to provide a solid framework to classify and understand changes as they arrive (see Tanenbaum, 2013, for further exploration of these system principles).

Cartography and Geovisualization

8

Sections

Cartography and geovisualization is concerned with the design, evaluation, and implementation of effective user interfaces between people and GIS. Ultimately, GIS are only useful if they can support people in decision-making.This chapter is designed to help you develop the skills to:

- construct effective **data graphics** for presenting static, non-spatial data;
- understand, design, and critique **maps** used for presenting spatial data; and
- apply the principles of **interaction design** to GIS interfaces that go beyond presentation of spatial data to help in analysis, exploration, and decision-making with geographic information.

What is your favorite map? Even if you have never thought about this question before, you probably already have some ideas. Some people prefer the detail and precision of a traditional topographic map, perhaps showing an area with which they are familiar (e.g., Figure 8.1a). Others prefer the simplicity and practicality of a schematic map, such as is often used for public transport maps (e.g., Figure 8.1b). Perhaps you are attracted by the handmade beauty of a historical map, complete with its distortions and ornaments (e.g., Figure 8.1c). Or maybe you thought of a digital map: one that you can access from a computer or mobile device, centered on your location, enabling you to zoom and pan, select layers, or be guided along a route.

This chapter is about the visual display of geographic information, such as illustrated in the maps in Figure 8.1. In Section 8.2, we focus specifically on the art and science of map design and making, termed *cartography*. First, though, Section 8.1 reviews the basic principles of data graphics, which underpin the visual presentation of any information. Display is only part of the story, however. This chapter is also about *interaction* with geographic information. Here again, we shall start with the basic principles of interaction, covered in Section 8.3, before moving onto what makes interaction with geographic information special, in Section 8.4.

cartography

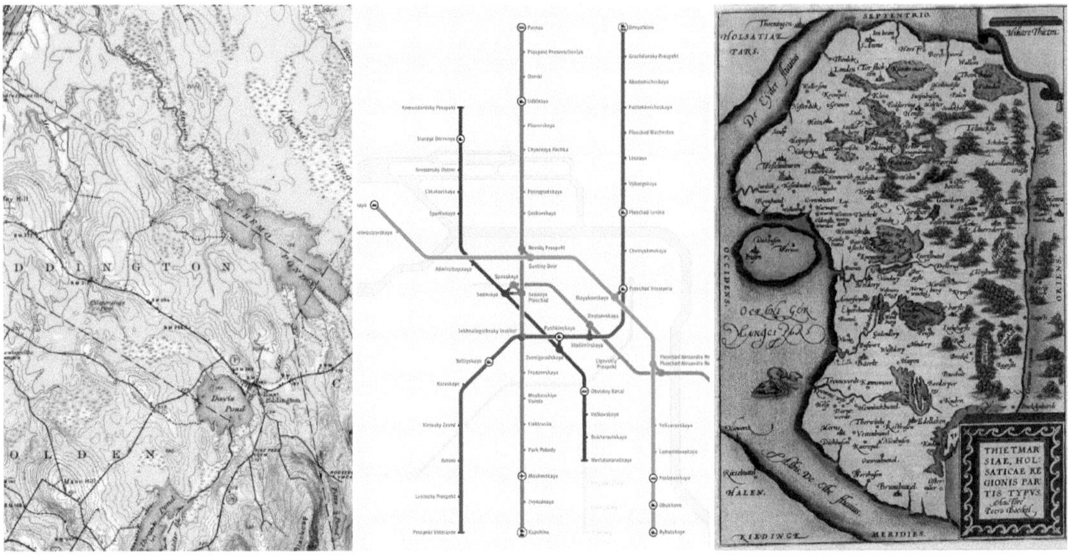

a. 1944 1:62500 topographic map (Source: U.S. Geological Survey)

b. St Petersburg subway map (Copyright Alex Florstein)

c. Ortelius' 1570 Thietmarsia map (Source: U.S. Library of Congress)

Figure 8.1: What is your favorite map? Portions of three different types of maps

8.1 Data graphics

Figure 8.2 shows two different graphical presentations of related data. Which do you prefer? Which do you think is better? For most people, the question of which graph is *better* is not so easy to answer as which is *preferable*. However, there are a number of principles one can apply to any graphical depiction of data for exactly this purpose. The term *data graphics* is used to refer to any graphical depiction of data, including graphs, charts, and maps.

data graphics

8.1.1 Principles of data graphics

Based on Tufte's famous book *The Visual Display of Quantitative Information* (Tufte, 2001), it is possible to identify four main principles of data graphics: density, integrity, correspondence, and aesthetics.

Data density We have all heard the adage "a picture is worth a thousand words." By the same logic, a data graphic that requires a description of *two* thousand words has a greater density of data than a similar graphic that can be described using only *one* thousand words. The principle of *data density* is concerned with presenting as much data as possible in the space available (see, for example, the "data-ink ratio" in Box 8.1 on page 310).

data density

Data density can be increased by increasing the amount of data in the graphic; it can also be increased by decreasing the amount of redundant lines, text, and marks. Comparing the data graphics in Figure 8.2, Figure 8.2b has much higher data density than Figure 8.2a. Figure 8.2b shows data for more

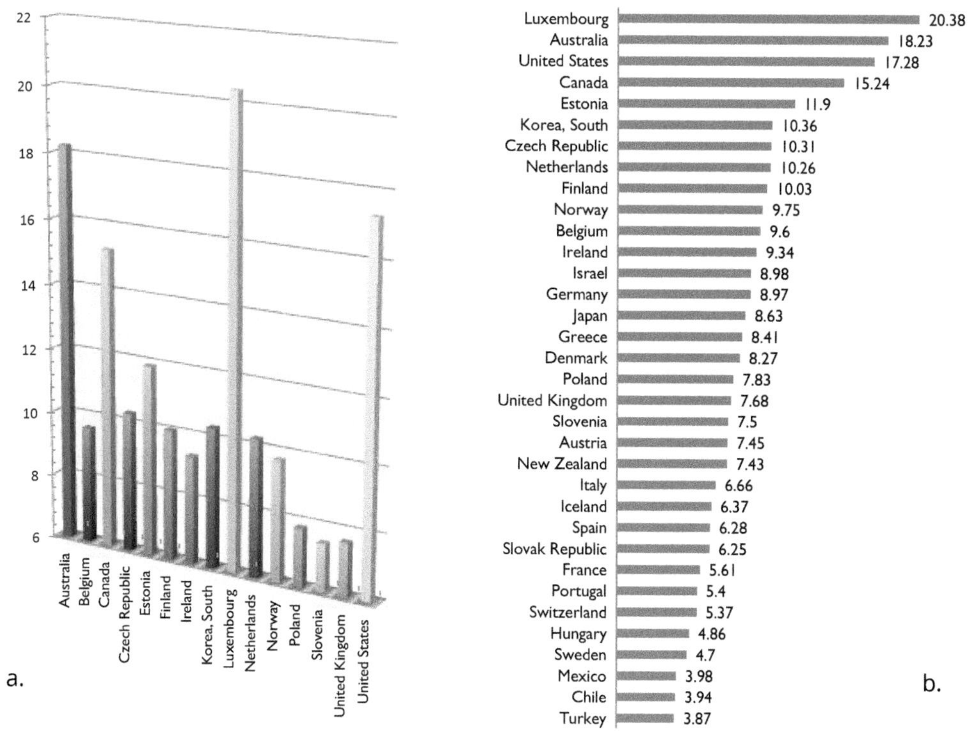

Figure 8.2: Two simple data graphics showing emissions of CO_2 (tonnes per capita) in some OECD countries

than twice as many countries as Figure 8.2a. The gridlines, ticks, box, shading, shadows, colors, and 3D effect in Figure 8.2a are all redundant: they can be removed without reducing the amount—and even increasing the legibility—of the data shown.

Data integrity Pictures can also lie. Any data graphic has the potential to distort data: to suppress or misrepresent facts and relationships in the data. The principle of *data integrity* is concerned with presenting the data itself, as far as possible, and avoiding embedding misleading messages in the data graphic.

data integrity

For example, closer examination of Figure 8.2 will reveal that the *y*-axis scale of Figure 8.2a begins at 6 tonnes of CO_2 per capita. This design choice tends to give the impression that countries like the United Kingdom have relatively low emissions, while in fact the UK is amongst the middle per capita CO_2 emitters in the OECD. The misleading impression is further advanced by the deliberate omission of the 15 countries with the lowest per-capita emissions.

Correspondence Data graphics are more than simple pictures that *show* something; data graphics should allow the viewer to compare, relate different items, and even infer causes and effects. The principle of *data correspondence* is connected with the aim of helping the viewer to understand patterns and form hypotheses about the relationships in the data.

data correspondence

Box 8.1: Data-ink ratio

A concrete example of the principle of data density is Tufte's *data-ink ratio*, in Tufte (2001). The data-ink ratio is the proportion of the ink used to print a graphic that is actually required to display the data, defined as:

$$\textit{Data-ink ratio} = \frac{\textit{Non-erasable data ink}}{\textit{Total ink required to print the graphic}}$$

Non-erasable data ink is ink that would lead directly to loss of data content in the image if removed (e.g., scales, shading, axis and grid lines, labels, and so forth). Although conceived in an era when most graphics were printed, one can easily imagine a translation of the ratio to the proportion of non-erasable pixels in a digital data graphic. Taking the data-ink ratio to extremes can lead to minimalist and even difficult to interpret data graphics. However, it remains an instructive exercise to try to maximize the data-ink ratio in your own data graphics.

In Figure 8.2b, for example, ordering the data by emissions, rather than alphabetically invites the viewer to identify high and low emissions, compare different countries, and contrast similar countries. The organization of data in Figure 8.2a makes identifying correspondences harder, such as identifying those countries with the highest per capita emissions.

Aesthetics Data graphics should be aesthetically pleasing. This principle of *aesthetics* is the hardest of our four principles of data graphics to assess objectively. It is possible you like the look of the data graphic in Figure 8.2a; I do not. However, aesthetics is not just about personal preference. Here again it is possible to identify and classify those features common to beautiful pictures and graphics. Evidence of care and attention to detail production; the use of alignment, centering, spacing; and proportion and balance in data graphics can combine to produce beautiful data graphics, in what Tufte (2001) referred to as "simplicity of design and complexity of data."

aesthetics

8.1.2 *Visual variables*

All data graphics, including maps, must be constructed from basic visual symbols, called *marks*. The different ways in which the characteristics of marks can be varied are termed *visual variables* or *graphic variables*. In this book we distinguish six visual variables: *position*, *size*, *pattern*, *color*, *orientation*, and *shape*, summarized in Table 8.1.

marks

visual variable

Position concerns where the marks appear, both relative to other marks as well as in absolute terms—where "on the page" or the screen. Pattern can be decomposed into two elements: *texture* (density) and *arrangement* (e.g., ordered or random) of marks. Color is usually decomposed into three elements: *hue*, *saturation*, and *value*, known as the HSV model. Hue refers to the shade of a color, for example, the amount of "redness" or "greenness." Value refers to the brightness of a color: its lightness or darkness. Saturation refers to the purity of a color: how intense a color appears. Saturation is often used in combination with value, but it may be used independently to control the prominence or impression of certainty of symbols. Shape concerns our ability to distinguish marks based on their geometric shapes. The shape of marks can be *abstract* or *mimetic*, depending on whether they resemble the thing they represent (mimetic) or not (abstract). Orientation concerns the

hue

value

saturation

Visual variable	Example
Position	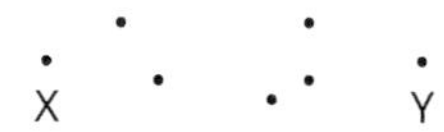
Size	
Pattern (texture, focus, and arrangement)	
Color (hue)	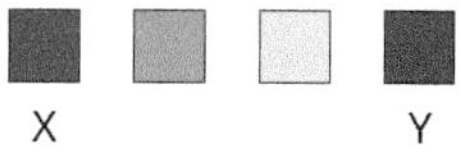
Color (saturation)	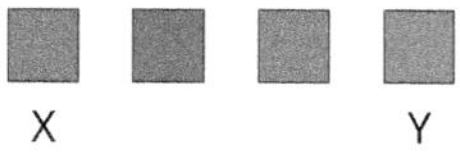
Color (value)	
Orientation	
Shape	

Table 8.1: Visual variables

the inherent directions in which a mark points for marks that possess strong rotational asymmetry (such as lines or arrows).

Interpreting the characteristics of marks, and the similarities and differences between marks, is the primary mechanism by which the viewer derives information from any data graphic. Our visual perception is capable of rapidly discerning fine distinctions between marks. Indeed, many distinctions can be detected by our visual system in a fraction of a second without even requiring our conscious attention. For example, you will immediately spot the "odd line out" in Figure 8.3a. Similarly, it is likely you are able to identify the odd line out in Figure 8.3b, even though the distinction between marks is fine (a difference in orientation of only 5 degrees). The characteristics of marks that "pop out" in our vision in this way are termed *preattentive features*, and include orientation, size, color hue and value, and many aspects of shape.

preattentive features

Other distinctions may require conscious attention to be discernible. For example, it will most likely take you much longer to identify the odd line out in Figure 8.3c. The combination of multiple distinguishing features of marks, such as color and orientation together, do not allow unique marks to "pop out" in the same way as single-featured marks.

Value message Certain visual variables can allow more than simple *difference* to be depicted; they may also allow *relative* values to be communicated. The

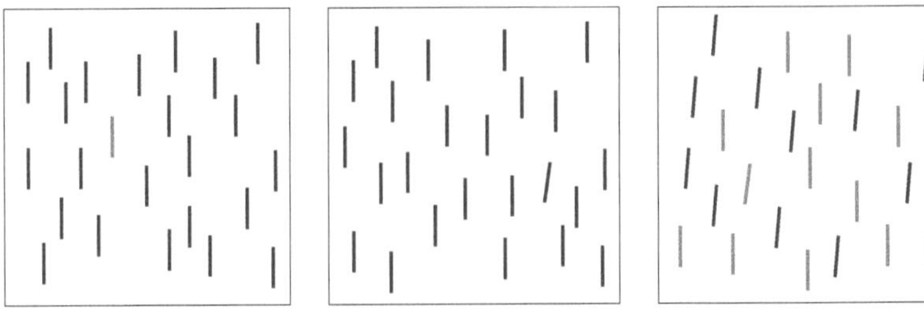

Figure 8.3: Preattentive features: Single and combined differences in visual variables

value message

association of a particular visual variable with greater or lesser relative value is referred to as the *value message* (sometimes called the *magnitude message*).

Some visual variables are best for communicating *ordinal* differences in value, allowing items to be ranked. For example, color saturation or value are especially useful for communicating relative ordering. Examining the color value example in Table 8.1, a natural interpretation is that there is more of something at Y than at X. The effect can be even more pronounced when combining changes in color value with reinforcing changes in saturation.

While color value and saturation can be used to communicate relative ordering, they are not especially good for communicating the *magnitude* of difference. A few visual variables can be used to communicate such *numerical* (i.e., interval and ratio, see Section 4.2.1) value messages—they can communicate not only whether there is *more* of something, but an impression of how *much* more there is.

Size is perhaps the best example of a visual variable associated with a numerical value message. Looking at the size example in Table 8.1, you can probably estimate how much more of something there would be at X than at Y. Most people's estimate of how *much* more is typically around 20–30 times more. In fact, the edge of the square at X is six times longer than that at Y, and so the area is 36 times larger. Humans tend to attach a value message in proportion to the length of lines, but in proportion to the area of areal symbols, such as squares (and in proportion to the volume of volumetric symbols, like cubes). However, the effects of such messages are somewhat muted because humans tend to consistently underestimate the relative size of areal and volumetric symbols, for example generally associating a value message of between 60–90% of the actual difference in area.

Table 8.2 provides a guide to the effectiveness of the six visual variables for communicating ordinal and numerical value messages, as well as their effectiveness for communicating simple difference without any value message (nominal level). Box 8.2 on the facing page also gives some background to another important principle in anticipating how people will interpret certain features of data graphics: Weber's law.

Semantics In addition to being used to distinguish different marks, variations in each visual variable are identified with common associations or

Box 8.2: Weber's law

Much of human perception is based on relative rather than absolute judgments. This principle is captured by the idea of a *just noticeable difference* (JND). The JND is the magnitude of the change in intensity of a stimulus (such as the length of a line) required such that the change can be perceived by a human. Perhaps surprisingly, numerous studies have shown that the JND depends not on the absolute size of the change, but rather on the size of the change relative to the original stimulus. Mathematically, if δI is the just noticeable change in the intensity of a stimulus I, then the proportion $\frac{\delta I}{I}$ is a constant! This means humans are able to perceive very small differences in very small stimuli, even though they may be unable to discern similar small differences in large stimuli. For example, most humans would be able to just discern apart a line of 10 mm length from a line that is 11 mm. However, discerning a millimeter difference in length in a line that is 1 m or even 10 cm is generally impossible for humans. Instead, one should expect it to require a line of 1.1 m or 11 cm, respectively, to ensure similar reliability in noticing the difference. Weber's law captures this general principle, that the perceived change in a stimulus is proportional to the intensity of the original stimulus.

	Numerical	Ordinal	Nominal
Position	Good	Good	Good
Size	Good	Good	Good
Pattern texture	Moderate	Moderate	Good
Orientation	Moderate	Moderate	Good
Color hue	Moderate	Moderate	Good
Color saturation	Moderate	Good	Poor
Color value	Moderate	Good	Poor
Shape	Poor	Poor	Good
Pattern arrangement	Poor	Poor	Moderate

Table 8.2: Effectiveness of visual variables for different measurement levels, after MacEachren (1995)

messages. For example, the color red is often associated with danger; green is for "go"; blue is associated with the sky or the sea. Shape is often used iconically: a square with a cross on top is conventionally used to indicate a church; a triangle with a dot in the middle is known by surveyors to symbolize a triangulation point (see Table 8.1).

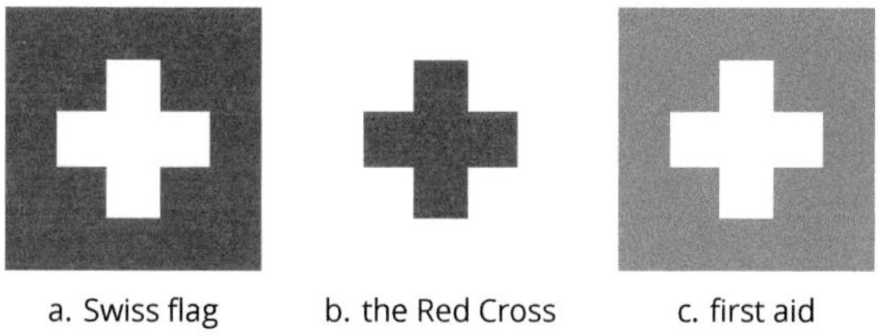

Figure 8.4: Semantics and visual variables

The semantic associations with symbols can be strong. Figure 8.4 shows a cross symbol in three different hue combinations (red and white, white and red, and green and white). These marks will be immediately familiar to most readers as the Swiss flag, the Red Cross symbol, and the first aid symbol. Awareness of the semantic associations of marks is important because ignoring those associations can confound the intended interpretation of a data graphic. If you relabeled all the first aid kits in your building with the Swiss flag, there would be serious potential for confusion despite the fact that both are striking and immediately recognizable in an emergency.

Box 8.3: The Peutinger table

The image below shows an excerpt from the Peutinger table, a 14th century copy of a fifth century map (source: Wikimedia, a photographic reproduction of a work of art in the public domain). The excerpt is recognizably centered on Sicily and southern Italy, enclosed by parts of the Dalmation and north African coastlines. The map is designed to provide a faithful representation of road connections and distances, but not the shapes and areas of the land masses. The results are akin to today's schematic maps, commonly used for transportation networks.

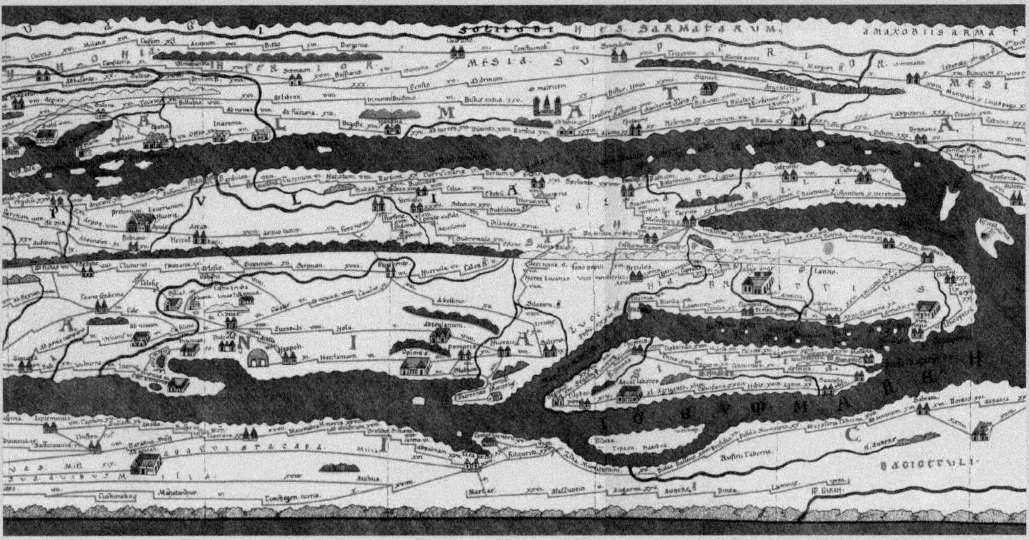

8.2 Cartography

Cartography is the art, science, technology, and history of maps and map-making. While study of the design of data graphics is a relatively recent endeavor, the design principles that govern maps have evolved over centuries of map use. Mappiform (map-like) depictions of the world date back at least 3000 years. Some surviving ancient maps, such as the Babylonian Imago Mundi (circa 600BCE) and the Bedolina Map (circa 800BCE), are symbolic; others, such as the Turin Papyrus Map (circa 1150BCE) are clearly intended as accurate depictions of aspects of the geographic world (see also Box 8.3).

The key feature of a map, as opposed to a data graphic more generally, is that the relative positions of marks on the map are used to indicate some aspect of relative geographic location. Maps are frequently also associated with being amongst the highest data densities of any type of data graphic.

In the context of a GIS, it is worth highlighting an important distinction, not always made clear, between the *spatial* and *graphical* representation of objects. Spatial objects exist to directly model the application domain, while graphical objects are the presentational form of spatial objects. Traditional hardcopy maps conflate the data storage and data presentation functions of a GIS. A fundamental difference between maps and GIS, then, is that in a map the spatial and graphical representations of objects may be the same; in a GIS spatial and graphical representations are always separate. In the remainder of this section we are concerned solely with the graphical representations of objects in a GIS or on a map, but not the underlying representation of the spatial data itself, in the machine (already discussed extensively in Chapters 3–6).

8.2.1 *Abstraction*

Maps are effective tools for presenting geospatial data because they provide abstract graphical representations of the geographic world, highlighting important information and suppressing unimportant information. Map users are therefore able to focus on the salient relationships represented in the map without becoming distracted by irrelevant details.

It is worth dwelling a moment on this point: maps derive their usefulness and power from what they leave out, as much as from what they put in. The geographic world around us is so complex, the data we can derive from it so dense, that maps become useful precisely because they do *not* show everything.

Figure 8.5 shows an example of a topographic map alongside an aerial photograph of the same region, Orono, Maine in the US. A *topographic map* depicts information about the land surface, including relief and features like rivers, roads, buildings, and so forth. Topographic maps are amongst the most complex and sophisticated types of maps. Despite this complexity, the topographic map omits much of the detail contained in the aerial photograph.

topographic map

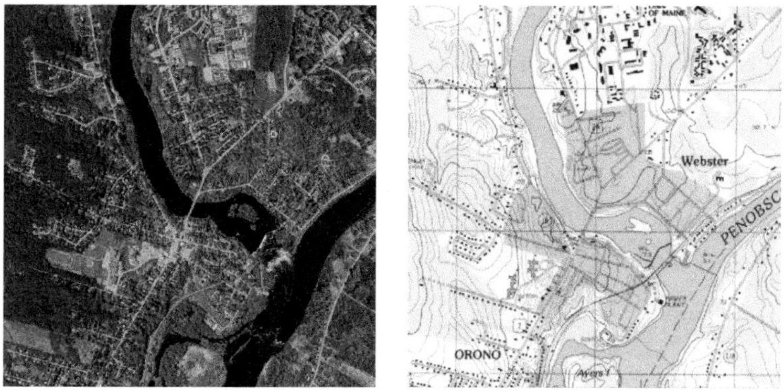

a. Aerial orthophoto b. Topographic map

Figure 8.5: Abstraction in topographic maps (Source: USGS)

The map achieves this abstract form by reducing the detail of both the semantic and geometric information it depicts. Reduction in semantic detail is achieved through *classification*: selecting a small number of aggregate feature types to portray on the map. In the topographic map in Figure 8.5b, the colors indicate just four main classes of land cover: water (blue), woodland (green), built-up urban (gray), and everything else (white). Other classes of land cover that might be evident in aerial photographs, such as running tracks, yards and gardens, and swimming pools, are omitted in favor of the *selected* classes of interest. Further, those chosen classes of interest are themselves the product of *aggregation* of more detailed classes. For example, it would be possible to split woodland into, say, scrub land, mature woodland, and park and amenity woodland, but at the cost of increased map complexity. Finally, reduction in geometric detail is achieved by omitting many small buildings, individual trees, and the detailed shapes of the regions one might identify from the aerial photograph. This process of geometric adaptation is called *generalization*, discussed in more detail in Section 8.2.4.

Highlighting the important spatial objects and relationships, using classification and generalization, ensures the topographic map is much more useful than the aerial photograph for a wide range of tasks, such as urban planning, hiking, or simply walking around town. Abstraction is one feature of maps that distinguishes them from other data graphics. Irrespective of how good data capture and graphics technologies become, there will always be a role for the map even in a world filled with images.

8.2.2 Types of maps

We have already encountered examples of one of the most recognizable types of maps: topographic maps (e.g., Figure 8.1a and Figure 8.5b). While there is no consensus on a single classification of the types of maps, in this book we distinguish two more major classes of maps. There are myriad subclasses, some of which we shall briefly touch on.

thematic map

Thematic maps Thematic maps aim to present spatially varying data on some specific topic or aspect of a geographic area. Consequently, thematic maps have more in common with the examples of data graphics we have already encountered than topographic maps. Figure 8.6 shows an example of one of the simplest thematic map types: a proportional symbol map. The map shows the number and distribution of reported UFO sightings in Ontario.

Hopefully, given what we have already learned about data graphics, your design skills are already on alert. Do you think it is a good map or not, according to our principles of data graphics? Figure 8.6 offers moderate data density. Despite relatively low levels of redundancy (color and coastline vignette are arguably unnecessary), relatively few data points are shown, at least by the standards of many maps. The map does allow for some spatial correspondences to be identified. It seems immediately clear that UFO sightings are more clustered around the major cities, Toronto and Ottawa. The size of symbols is used to communicate the value message: larger symbols mean more UFO sightings, and the relative symbol sizes approximately accord with the relative numbers of sightings. However, here lies a crucial problem with the map: data integrity. The relative sizes of the major cities and the UFO sighting symbols have been deliberately chosen to exaggerate the importance of the UFO sightings. Further, the lengthy time period over which the sightings were made (many years) has been suppressed. The misleading effect is enhanced with the hue semantics: dangerous (red) aliens marauding across Ontario's verdant (green) countryside.

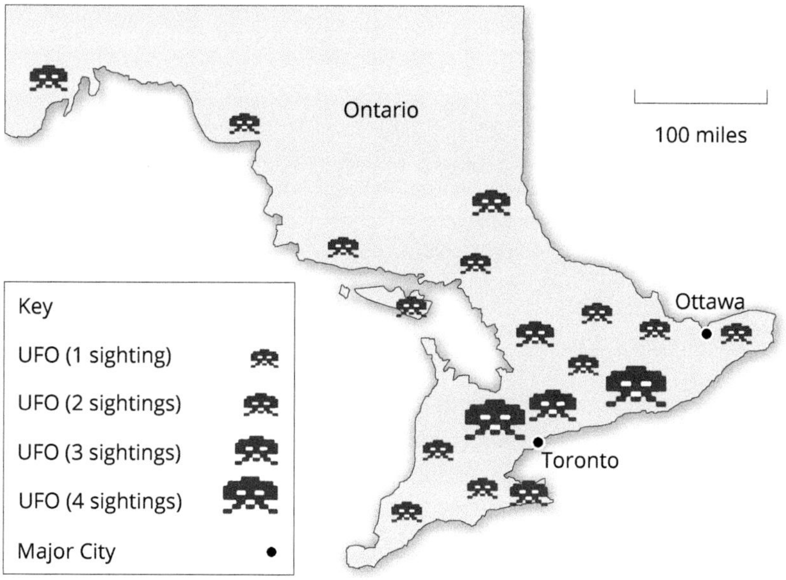

Figure 8.6: Proportional symbol map of UFO sightings in southern Ontario Province, Canada

An important lesson from Figure 8.6 is that, just like any data graphic, maps too can lie. Mark Monmonier's famous 1996 book *How to Lie with Maps* provides many examples of the manipulation and misrepresentation of information using maps.

One type of thematic map that is especially prone to mislead is the *choropleth map*. A choropleth map is a data graphic that presents data for defined geographic regions. Typically, choropleth maps use color or shading in depicting data values. For example, Figure 8.7 shows four related choropleth maps (a "small multiple" data graphic, see Box 8.4 on the following page) of the same data concerning carbon emissions in London boroughs. The maps visibly differ because of the different ways the underlying numerical data has been classified into five classes. The *equal interval* classification (Figure 8.7a) subdivides the entire range of the data into equally sized subranges. By contrast, a *quantile* classification (Figure 8.7b) generates classes that contain approximately equal numbers of members in the data. *Jenk's natural breaks* (Figure 8.7c) is a statistical procedure that identifies classes with high interclass but low intraclass variation. Finally, *mean-standard deviation* classification (Figure 8.7d) assumes the data is normally distributed, and it picks class breaks based on the mean and standard deviation of the data.

choropleth map

equal interval

quantile

Jenk's natural breaks

mean-standard deviation

Figure 8.7 is intended to highlight how the choices made by the map designer can communicate starkly different messages, even when showing the same underlying data. Some of the choropleth maps in Figure 8.7 give the impression of lower overall carbon emissions; some appear to highlight a concentration of high carbon emission regions in the north and west; some appear to single out particular regions as hotspots. Another important cause

Box 8.4: Small multiples

One technique for highlighting in correspondences in data graphics made popular by Tufte (1990) is the use of *small multiples*, also sometimes termed a trellis chart. A small multiple is constructed from an array of closely related data graphics, each usually using the same orientation, axes, scales, and symbology, but typically showing a different variable or alternative design. For example, looking ahead to Figure 8.7, the small multiple of four related maps of carbon emissions invites comparisons between the same regions in different maps, as well as between different regions in the same map. Aligning multiple related data graphics in small multiples is a powerful way to disaggregate complex data sets and help your audience identify relationships not easily discernible from a single graphic.

of misleading and mistaken maps is the so-called "modifiable areal unit problem" discussed further in Box 8.5 on page 321.

There are numerous other types of thematic maps. Figure 8.8 shows a famous example of a flow map: a type of thematic map that uses links to depicting flows between geographic locations. In the case of Figure 8.8, the flow map shows the exports of wine from France in 1864. Thicker lines provide the value message of greater exports. The map is carefully designed to avoid possible confusion from lines crossing each other or coastlines, and invite comparisons between countries and even continents.

cartogram

Finally, a *cartogram* is a type of thematic map that deliberately distorts geographic areas or distances in order to represent its thematic variable. Figure 8.9 shows a cartogram where the shape of each country is distorted in order that its area becomes in proportion to its population. The result is striking, highlighting in particular those densely populated "fat" countries (such as the UK, Japan) and sparsely populated "skinny" countries (such as Australia, Canada, and Russia).

Schematic maps Unlike the earlier maps in this section, the cartogram deliberately distorts geographic distances or areas in the pursuit of clearer communication of the map theme. A *schematic map* goes further still, suppressing metric geographic details in order to highlight *connectivity* between places. Because of the importance of connectivity, schematic maps are most familiar in the context of transportation maps.

schematic map

Figure 8.10 contains an example schematic map showing the Oslo metro network. The focus of such maps is the connections between stations. In common with most schematic transport maps, the lines are straight; the angles between are fixed to a small set of possibilities (in the case of Figure 8.10, restricted to only 90°); and the distances between stations are also generally fixed across the map, only lengthened for reasons of graphical clarity.

Schematic maps are excellent examples of the principle of abstraction. By deliberately omitting a wealth of metric details, schematic maps achieve much higher levels of usability for route planning than their planimetric counterparts. For contrast, Figure 8.11 shows the same Oslo metro network in planimetric view, with geographic distances and angles preserved. Even with

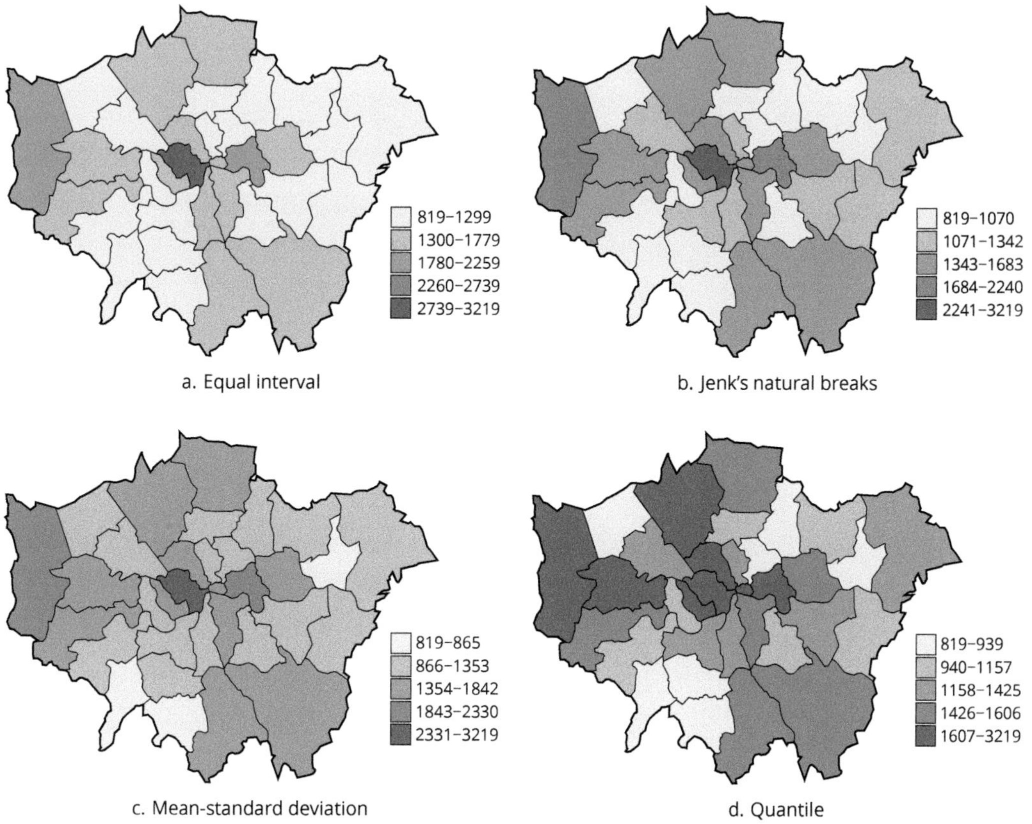

Figure 8.7: Choropleth maps of carbon emissions (kilotonnes of oil equivalent) in London, UK (Data source: london.gov.uk)

many stations and names omitted, the results would clearly in most situations be far less useful for getting around Oslo on public transport.

Schematic maps are sometimes also termed *topological maps*, because of their suppression of metric features in favor of the topological property of connectivity. However, it would not be true to say that schematic maps are purely topological. Comparing Figure 8.10 and Figure 8.11, it is clear that some metric properties are approximately preserved, such as the relative directions and broad shape of the network.

8.2.3 Map projections

Schematic maps and cartograms are examples of map types that deliberately distort metric features of geographic space in the service of clearer communication of data. But every map necessarily introduces some distortion of geographic space. This distortion arises because the Earth is not flat; maps are. It is impossible to project the surface of a sphere onto the plane without introducing some distortions. If you need to convince yourself of this, try to squash an orange peel onto a flat table surface: the results are always messy, requiring the peel to be stretched or ripped in some way.

Unlike the ad hoc and map-specific distortions of schematic maps and cartograms, study of the topic of *map projections* is concerned with the *systematic*

map projection

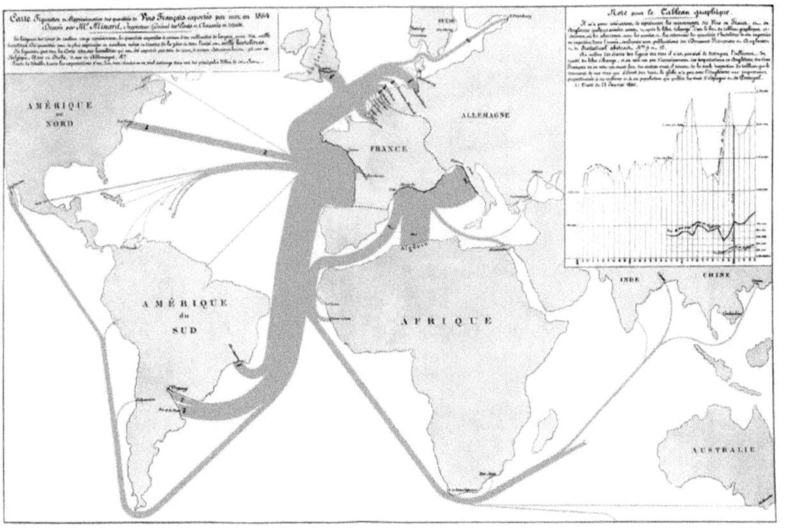

Figure 8.8: Minard's famous 1864 flow map of French wine exports (Source: Wikimedia, a photographic reproduction of a work of art in the public domain)

projective transformation of spherical coordinates (i.e., on the surface of the Earth) to planar coordinates (i.e., on a map sheet or a computer screen). Map projections were already encountered briefly in Section 3.2.3.

Figure 8.12 illustrates the three main types of map projection: projection onto a flat plane (azimuthal, Figure 8.12a); projection onto a cylinder (cylindrical, Figure 8.12b); and projection onto the curved surface of a cone (conical, Figure 8.12c).

Examples of each of the three types of map projection may be found in Figure 8.13. Figure 8.13a shows the globe projected using the Azimuthal Equidistant projection. A projection is termed *equidistant* if the projection preserves distances to some fixed point or line. In the case of the Azimuthal Equidistant, in Figure 8.13a, the distances from all mapped points to the map center (70°E, 20°N) are in proportion to one another. However, other geo-

equidistant projection

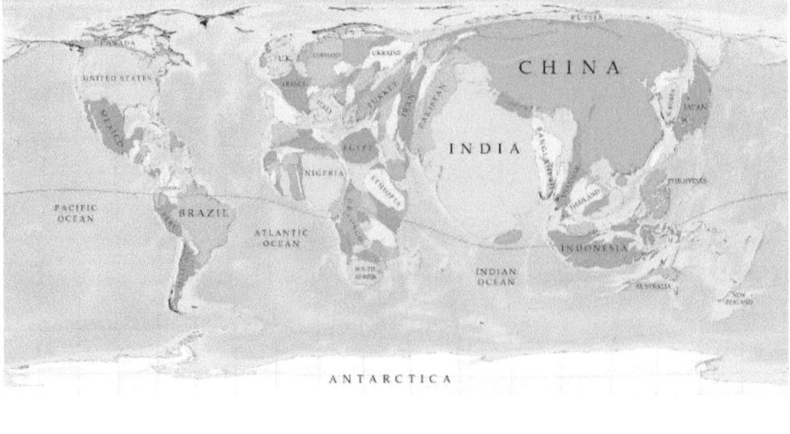

Figure 8.9: Cartogram of world population by country (Copyright Mark Newman)

Box 8.5: MAUP: Modifiable Areal Unit Problem

Maps frequently aggregate information from across a region of space, either for reasons of clarity or because data is only collected or made available on an aggregated, per-region basis. Census data, for example, is commonly aggregated over arbitrary administrative regions in order to protect an individual's privacy. However, many other data sources, such as the environmental data in Figure 8.7, is frequently captured, communicated, or presented over aggregate regions. But what would such maps look like if we had chosen different regions over which to aggregate data? As an example, the configuration below illustrates how the appearance of a map may change dramatically with a relatively minor modification to the areal units. In the figure, the locations of households and classification scheme remain unchanged. However, a mere 16° rotation of the gridded regions leads to a remarkably different pattern in the apparent spatial distribution of households across the aggregate regions.

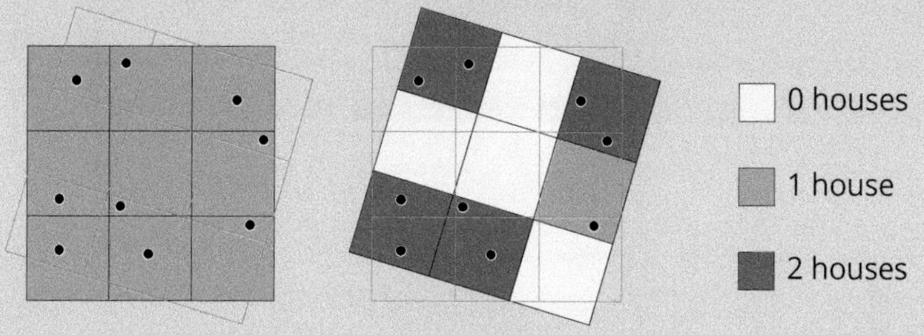

The *modifiable areal unit problem* (MAUP) summarizes this issue as follows: the results of any analysis upon spatial data that has been aggregated using arbitrary (i.e., "modifiable") regions (i.e., "areal units") is not guaranteed to be valid independent of those regions (i.e., "problem"). For example, in the illustration above, neither the apparently regular distribution of houses in the first grid nor the apparently irregular distribution in the second grid tells us anything about the actual underlying distribution of houses. A closely related problem to MAUP is the *ecological fallacy*: the mistaken idea that the properties of an aggregate region may be imputed to all the individual entities contained within that aggregate.

metric properties, including the shape and area of land masses, are distorted by the projection.

The Mercator map projection in Figure 8.13b exhibits different distortions. The Mercator projection is a cylindrical type of map projection. The Mercator projection does not preserve distances, and indeed it distorts the size and shape of larger objects. However, it does preserve angles and shapes locally, over small distances. Projections that preserve local angles are termed *conformal* projections. This property of conformality represented a breakthrough for maritime navigation in the 16th century, when the cartographer Mercator first introduced his eponymous projection.

conformal projection

The Albers Equal Area projection in Figure 8.13c is an example of a conical-type projection. As the name suggests, it is an *equal-area* projection. The Albers projection does not distort areas, with the sizes of countries and continents remaining in proportion to each other. Concomitantly, distortion will occur in shapes and distances using this projection.

equal-area projection

Map projections have a long history and have been topic of study and research in their own right. At the time of writing, there were more than 100 distinct projections documented and implemented in NASA's *G.Projector* software, a leading application for exploring different map projections. How-

Figure 8.10: Schematic map of Oslo metro network (Copyright Arsenikk)

ever, a more thorough investigation of map projections in this book would take us too far from our main themes. From the perspective of display and interaction, two messages are especially important to remember about map projections.

First, any 2D map must necessarily be subject to a map projection. Similarly, digital spatial data will either be supplied in spherical coordinates, in which case an appropriate map projection must be applied in order to display it on a 2D map; or spatial data will be supplied using projected coordinates, in which case the correct projection must be known and applied in order to relate that data to other data sets using coordinates in different projections.

Second, any map projection must necessarily introduce distortion into some aspects of the projected map, whether areas, distances, or shapes and angles. The principles of data integrity and correspondence demand that map projections are correctly applied to avoid unknowingly or misleadingly distorting maps.

8.2.4 *Map generalization*

map generalization

One further source of distortion in maps arises as a result of *map generalization.* Map generalization involves the distortion of geometric elements in a map in order to enhance its graphical clarity. The relationship between the size of a map and the size of the area that map depicts is captured using the

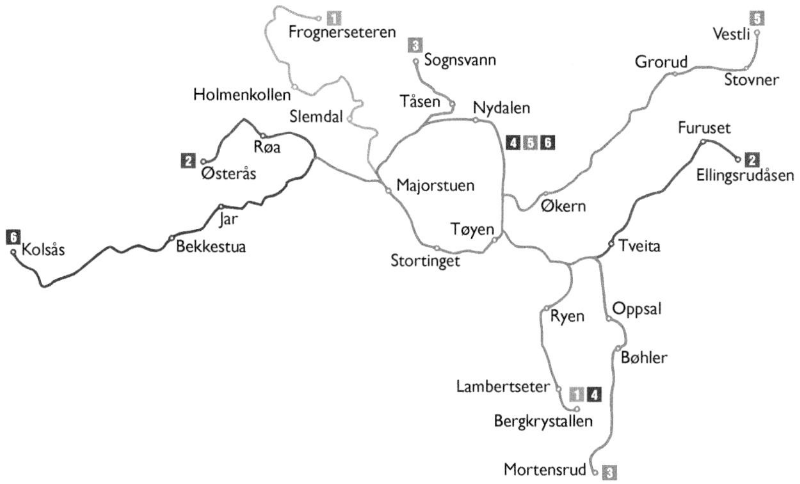

Figure 8.11: Planimetric map of Oslo metro network (Copyright Pneumaman)

representative fraction (see Box 8.6 on page 325). For example, a map where 25 m on the ground appears as 1 cm on a map has a representative fraction of 1:2500 (25 m = 2500 cm). In a map with a representative fraction of 1:63,360, each inch (2.54 cm) on the map represents a mile (approximately 1.6 km) on the ground (there are 63360 inches in a mile).

representative fraction

Maps must necessarily be much smaller than the real geographic spaces they portray. Consequently, distortions in the geometries of the features on the map are unavoidable if the map is to be legible. It is worth noting that because of the distortions introduced to at least some mapped distances by any map projection (even equidistant projections, see above), the representative fraction may vary across a map, especially a map with larger national or global extents.

There a many different map generalization procedures that have been used to enhance visual clarity. Here we categorize these into five distinct classes of map generalization operation:

- *Reduction*: Reducing geometric detail is perhaps the most fundamental class of map generalization operation. One example of reduction is *line simplification*. Recall that line simplification is the process of reducing

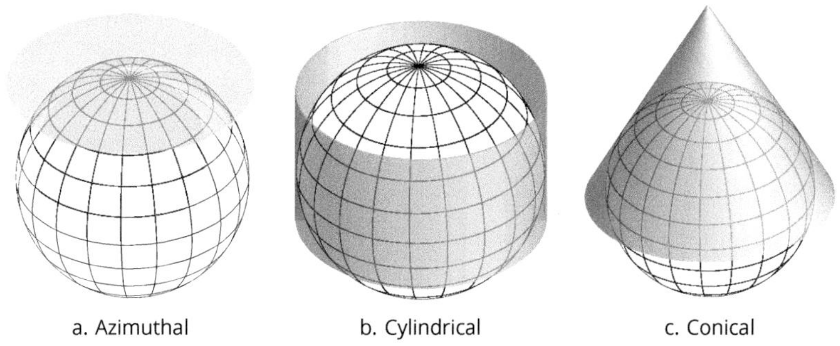

Figure 8.12: Types of map projection: azimuthal, cylindrical, and conical

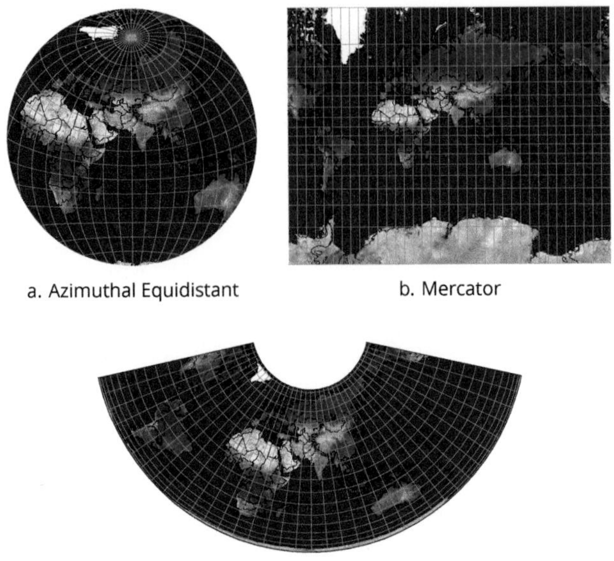

Figure 8.13: Example map projections

a. Azimuthal Equidistant

b. Mercator

c. Albers Equal Area

the level of detail in a polyline while retaining its essential character. The Douglas-Peucker algorithm is the most famous example of a line simplification algorithm (see Figure 5.7). Figure 8.14a shows another example of line simplification, termed *weeding*, resulting from removing all but each *n*th vertex in the polyline.

weeding

- *Merging*: Merging involves combining multiple distinct geometries into a single, clearer geometry. Figure 8.14b shows a group of buildings and the result of aggregation upon those buildings. Other types of merging operations found in the literature include *aggregation*, *agglomeration*, *amalgamation*, and *combination*.
- *Characterization*: Characterization involves changing the shape, size, or location of geometries in order to enhance clarity. Examples of characterization operations include *displacement* (moving geometries to ensure adequate visual separation between features, Figure 8.14c); *enhancement* (making geometries larger to ensure visibility, Figure 8.14d); and *typification* (changing or exaggerating the shape of a feature to enhance its symbolic or iconic characteristics, Figure 8.14e),
- *Collapse*: Collapse concerns the change in dimension of a mapped geometry, from 2D area to 1D line to 0D point. Figure 8.14f shows the runways at an airport collapsed from a polygonal area to two lines. Further collapse would result in the airport being marked by a point symbol.
- *Elimination*: Elimination involves the removal of whole geometric features in the pursuit of clarity (Figure 8.14g). Elimination is sometimes also referred to as *filtering*, *refinement*, or *selection*.

displacement

enhancement

typification

elimination

These five basic generalization operations are named for moving from more-detailed to less-detailed map features. For example, "simplification"

Box 8.6: Scale

The term "scale" is used in at least four distinct senses in geographic information science.

1. The term "scale" is frequently used as a synonym for "representative fraction": the ratio of the size of features on a map to the size of the same features in the world.
2. A related use of "scale" occurs in the phrases "large scale," "small scale," and so forth. In these cases, "small" and "large" refer to "small" and "large" *fractions*. Hence, a 1:1,000,000 map is usually termed "small scale" (small fraction), whereas a 1:1,000 map would be termed "large scale" (large fraction). Somewhat confusingly, this entails *large* denominators (the number of the right of the ":") equate to *small* scales; *small* denominators equate to *large* scales.

Together, these first two senses might more properly be termed "map scale."

3. Another sense of the term "scale" is as a synonym for relative (spatiotemporal) "extent." For example, the phrase "El Niño is a large-scale geographic phenomena" has no relation to maps; it is purely a statement about the spatial and temporal extents of El Niño (in implicit relation to other weather systems, say, "small-scale" rain storms). In natural language "scale" is often used in this sense, to refer to the relative size or extent of something.
4. Finally, and perhaps most importantly, "scale" is also used to refer to the dependence of phenomena upon level of detail. Many geographic phenomena and processes look different depending on whether they are being observed in more or less detail. For example, a topographic feature that looks like a "peak" (higher than its neighbors) at a fine level of detail may instead look like a "pit" (lower than its neighbors) as we "zoom out" to a coarser, more granular level of detail as neatly illustrated in Jo Wood's 1996 PhD thesis.

The key message is that the term "scale" should be approached with caution. Our advice is to avoid confusion by restricting your use of the term "scale" solely to its fourth and final sense above: dependence on level of detail. The first and third senses have ready synonyms to hand ("representative fraction" or "map scale", and "extent"), and the second sense is best avoided altogether, as it anyway frequently leads to confusion, even in discussions amongst experts.

suggests moving from more complex lines to ones with simpler geometries. Similarly, "collapse" suggests moving from higher dimension features to lower dimension features. However, it should be noted that there are a few occasions when specific generalization operations proceed in the opposite direction, generating more-detailed features from less-detailed features. Enhancement, for example, may in some cases be used to add more detail to an important feature to maintain clarity, even as the overall level of detail in a map may be being reduced (see, for example, the discussion of fractals in Section 3.4.3).

Naturally, it is also possible to combine multiple operations in a single map, or even in a single cartographic generalization process. Figure 8.15 shows two USGS maps of the same location in the docklands of Boston, Massachusetts. Looking at the map, one can see many examples of generalization, to assist with maintaining the clarity of the map when moving from the more-detailed 1:25000 map to the less-detailed 1:100000 map.

8.2.5 *Map labels*

One other essential feature of topographic maps in particular, but important on many other types of map, is the labels. Labels are used to identify particular instances of types of features (e.g., "Hospital," "Cemetery," or "Radio tower") or *toponyms*—names for places. Thus, the location of the label is con- toponym

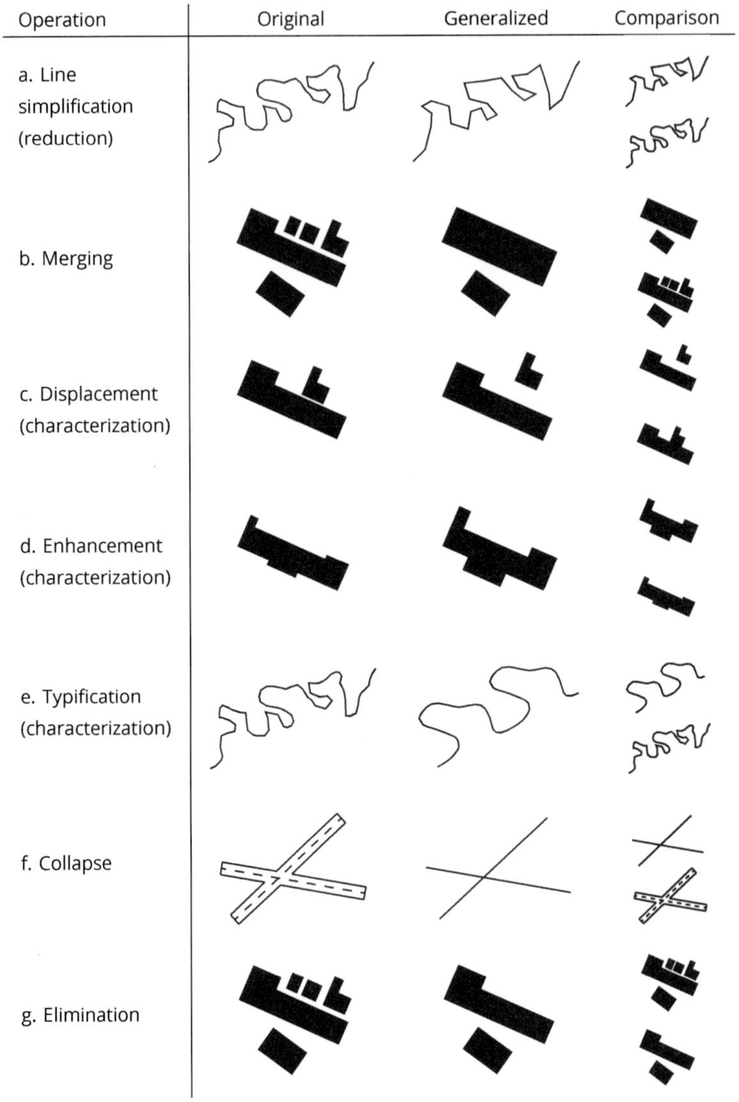

Figure 8.14: Map generalization operations

strained by the location of the feature or place it identifies. These constraints operate in three main ways:

- Point labels, where the label identifies a small point-like feature on the map, such as "Frog pond" in the northwest of Figure 8.15a or the "Light" on the harbor wall in the southeast. By convention, point labels are preferentially positioned slightly above and to the left of the point itself. However, other positions are also frequently used where other map symbols, labels, or geometries would interfere with preferred label position.

- Linear labels, where the name identifies a linear feature, such as a river or road. In Figure 8.15a, "Carson Beach" and "Street Beach" in the south of the map identify linear beach features. Linear labels typically run alongside the feature they label, preferentially conforming to the shape of the feature and avoiding backward sloping or upside-down text.

- Areal labels, where the name labels areas that are larger than can be handled with a simple point label. Areal labels are typically placed inside the areal feature itself, with the type spaced so as to provide the visual impression of "filling" the area. For example, "Pleasure Bay" and "Old Harbor" in the southeast of Figure 8.15a are toponyms that designate areas.

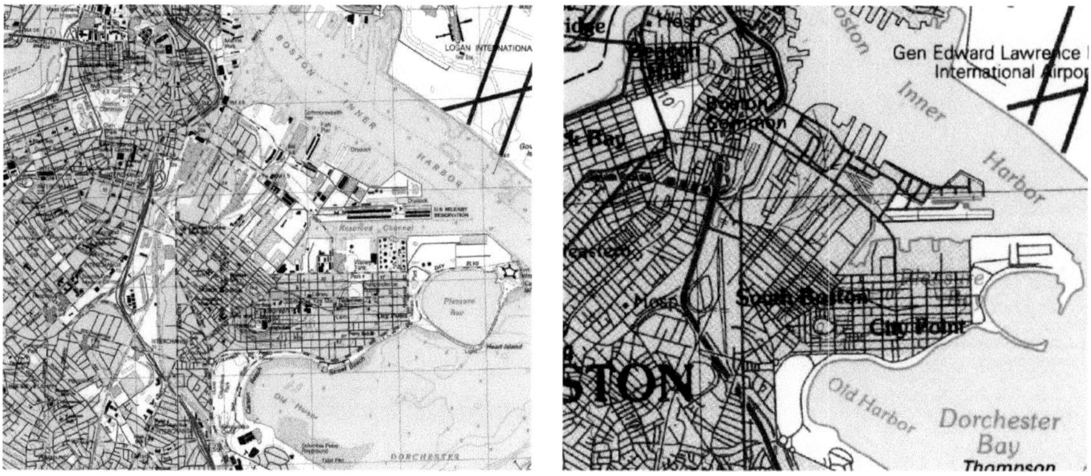

a. 1:25,000 map   b. 1:100,000 map

Figure 8.15: Example maps of Boston, Massachusetts, at different representative fractions (map scales) (Source: U.S. Geological Survey)

Some of the problems facing map labeling are similar to those facing the presentation of geometry or symbols. The labels must be legible and distinguishable, and positioned and sized such that each label is neither obscured by nor obscures other map features. When positioning a large number of labels, such as on a topographic map, no single, perfect solution will exist. Instead, clear and readable results require many compromises be struck between the positions and sizes of labels, symbols, and geometries across the entire map. Just as maps in general rely on selection, suppressing less-important features, so labels must be selected to identify only those objects and toponyms of greatest importance (cf. Figure 8.15a and b).

Other problems of map labeling are specific to the characteristics of text. The choice of type, for example, will have important impacts on the legibility and semantics associated with labels. In topographic maps, for example, the weight, size, width, slant, and typeface (e.g., Times versus Arial typefaces) are all carefully chosen to reflect categorical and hierarchical distinctions in the features labeled. Larger or more weighty type will carry a value message, such that features identified with bold type are more important than those identified with regular or lightweight type, for example. The use of capital letters similarly conveys a value message (e.g., "RIVER" versus "Stream"). Other aspects of shape provide both visual and semantic distinctions in type. For example, serifs (small lines attached to the ends of strokes in a letter, such as this text) tend to have connotations of heritage or formality. Sans serif fonts (used in figures and captions in this book) have connotations of modernity, and they tend to be more legible on maps (compare "Castle" and "Avenue"). The overall map style and cartographic identity will be strongly influenced by the choice of type. Most national mapping agencies have their own distinct and immediately recognizable type styles and conventions.

8.3 Interaction

Traditional maps and data graphics are mechanisms primarily for *presenting* information, for example, via digital display devices or hard copy maps. However, today's technology provides near-ubiquitous opportunities for *interacting* with data graphics and maps in ways not possible at the advent of GIS.

human-computer interaction

The term "human-computer interaction" (HCI) was first used during the 1980s to describe the interaction between computer systems and people. The term is also used to refer to the field of HCI: the study of the design, evaluation, and implementation of the interfaces between computing devices and people. Figure 8.16 summarizes the key components of HCI: the human, the computer, and the interaction between them.

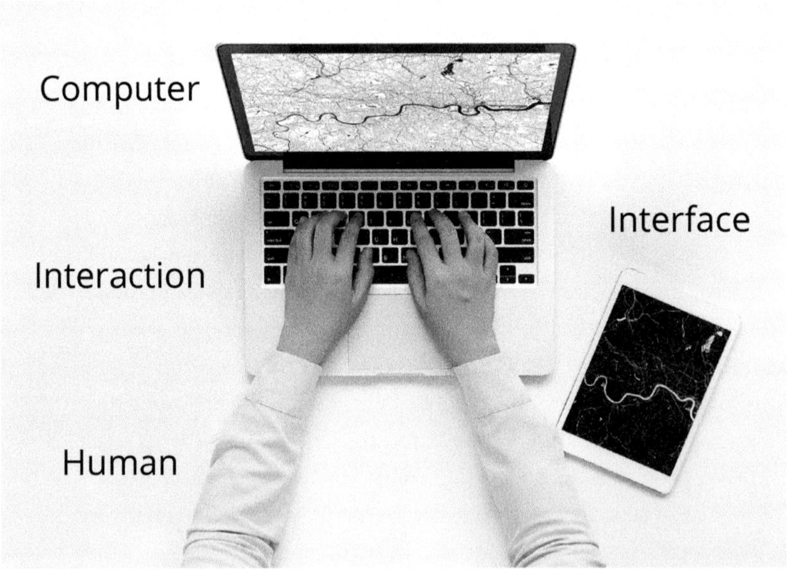

Figure 8.16: Components of human-computer interaction

dialog

A *dialog* is a process of interaction between two or more agents, whereby agents cooperate to resolve conflicts and complete some task. For example, humans commonly use dialog to communicate with each other. In a human conversation each interlocutor will listen to the other, responding to what has previously been said. The process of interaction between a computer and its human user also normally takes the form of a dialog, although HCI dialogs usually have a much more rigid structure than human conversations.

user interface

The dialog between humans and computers is mediated by a *user interface*. A user interface in HCI is the mechanism by which a human user accesses the functions offered by a computer.

8.3.1 *Input and output*

A plethora of devices exists to enable humans to interface with computers: keyboards, mice, trackpads, microphones, monitors, and speakers are amongst the most familiar. In order to make sense of all these devices, we must first note that humans and computers are able to send and receive information in several different modes, termed *input-output* (IO) channels. Computers output on IO channels that are detectable by human senses, such as sight, hearing, and touch.

IO channel

Humans have five senses that receive information from outside the body, termed exteroceptors: *sight* (visual sense), *hearing* (auditory sense), *touch* (termed "haptic", meaning related to touch), *smell* (olfactory sense), and *taste* (gustatory sense). A fire alarm, for example, provides us with auditory information about a dangerous fire, ideally long before we receive similar information with our visual or olfactory senses. Humans have two other senses relating to the movements of the body, termed *proprioceptors*: balance (*vestibular* sense) and *kinesthesia*, the ability to sense our own bodily movements and tensions.

exteroceptor

haptic

proprioceptors

Humans can input information to the computer on these same IO channels. For example, your computer receives information from you primarily on the haptic IO channel, usually via the keyboard, mouse, or touchscreen. In return, information is sent from your computer to you primarily on the visual IO channel, via its display screen, or perhaps on the auditory IO channel via its speakers. Visual information is the most commonly used sense in user interfaces devices, such as through computer monitors and screens. Auditory and haptic information often plays a secondary role in user interfaces, reinforcing visual information, for example, through speakers, microphones, keyboards, touchscreens, and vibrating tactile devices (but see Box 8.8 on page 332). Olfactory and gustatory senses play the least significant role in user interfaces. Proprioception is also used in human user interface devices, although normally in support of information on other IO channels, especially visual IO. For example, flight simulators can be mounted on a mobile base that mimic the movements of a real aircraft, stimulating the vestibular sense. Figure 8.17 summarizes all these concepts and terms.

The same IO channel can often be used for sending information in both directions—from computer to human and from human to computer—at the same time. For example, auditory information can be sent via a computer speaker and received by human hearing, or sent by the human voice and received by a computer microphone. To avoid confusion, the term *display* is often used to refer to output from a computer to a human for any IO channel, while *input* is normally reserved for input to a computer from a human (Figure 8.17).

display

input

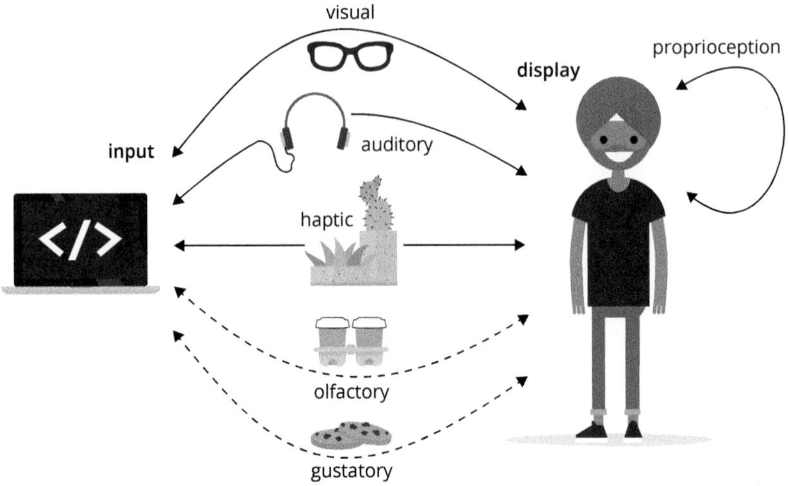

Figure 8.17: IO channels and human senses

A distinction is often made between *implicit* and *explicit* input. By default, user input is often *explicit*, with input entered consciously by the user via devices such as a keyboard or mouse. By contrast, *implicit* input occurs where the system automatically interprets information about a user's context or actions as input. For example, context-aware systems, such as LBS (introduced in the previous chapter), frequently make use of implicit input. A traveler wishing to find out about bus arrival and departure times might explicitly input their query to a transport authority website via their keyboard before leaving home. Alternatively, a location-aware mobile device could automatically interpret the user's location near a bus stop as an implicit query for information about the time of the next bus at that stop.

explicit input

implicit input

Some devices are capable of simultaneous input and display, frequently on different IO channels. Touchscreens, for example, receive input on the haptic IO channel at the same time as displaying information on the visual IO channel. Systems that enable display or input of information on more than one IO channel (mode) at a time are termed *multimodal* (Box 8.7 on the facing page).

multimodal

Irrespective of which IO channel a device uses, it must convert information between digital and analog formats. Humans are organisms that send and receive information in continuously varying analog formats; computers are digital machines that require discrete digital information to operate. For example, a computer mouse or keyboard converts the continuous analog movements of our fingers, hands, and arms into a discrete digital format that can be used by the computer. A speaker converts discrete digital audio information stored or generated by a PC into audible analog sound waves.

8.3.2 *Intuitive and expressive interfaces*

Two important characteristics of any user interface are how *intuitive* and how *expressive* is the interaction it supports. An *intuitive* interface is easy to use,

intuitive interface

Box 8.7: Multimodal display

Most GIS and mapping interfaces rely heavily on visual IO as the primary display mode. There are many situations in which visual display—or visual display alone—does not perform well in an interface. Haptic display, for example, is familiar in our vibrating mobile phones and computer game controllers, as well as used in some more specialized domains, such as Braille displays and embossers for the visually impaired. Using multiple IO channels, such as vision, haptics, and sound together for multimodal display carries three main advantages. First, multimodal interfaces may be accessible to a wider variety of users than unimodal interfaces, including visually or hearing-impaired users. Golledge, Rice, & Jacobson (2006) provide an overview of multimodality, and accessibility more broadly, in the context of GIS. Second, multimodal interfaces are able to operate in a wider variety of conditions than is possible with a unimodal interface (see Oviatt, 1999). For example, in vehicle navigation a visual map interface may be essential for certain situations, but when driving it would be dangerous to distract the driver's visual attention from the surroundings. Doyle, Bertolotto, & Wilson (2008) evaluated a multimodal GIS interface in connection with interaction errors and robustness. Finally, multimodal interfaces enable users to access complementary information on different IO channels concurrently. Experiments have demonstrated that such multimodal interfaces are preferred by users for more complex tasks, and they can lead to increased efficiency as well as decreased errors in user interaction, for example, as in Oviatt, Coulston, & Lunsford (2004).

requiring minimal effort to learn to operate. An *expressive* interface enables users to achieve specific tasks efficiently. Expressive interfaces allow users to specify commands more precisely, using a range of options that modify a command's behavior. The ideal interface style is both expressive and intuitive. Unfortunately, it is not usually possible to achieve both these goals. Different interface styles must strike a balance between being expressive and intuitive.

expressive interface

For example, a *command entry* interface allows a human user to issue commands to the computer directly. Common command entry interfaces include the R statistical language, the terminal in OSX and Linux operating systems, or the command prompt in Windows. Command entry interfaces are expressive because they typically offer many different options to modify the precise behavior of commands. However, such interfaces are rarely intuitive, because the human user needs to learn, master, and remember large numbers of commands and command options. At the Windows command prompt, for example, the command "`dir`" will result in the contents of the current directory being listed, while the command "`dir /a:d /p`" will list only the subdirectories of the current directory one page at a time.

command entry

By contrast, natural language interfaces enable computers to interact using written or spoken language. Natural language interfaces (such as Siri, Alexa, and Google Assistant) can be highly intuitive once some basic interaction principles have been understood (such as the need to employ the correct "wake word" to initiate interaction). However, natural language interfaces offer low expressivity in their limited ability to correctly interpret highly specific or complex commands. In addition to the many technical challenges in parsing human speech, human natural language is highly ambiguous. For example, the sentences "Time flies like an arrow" and "Fruit flies like a banana" are at first glance structurally similar. Only once the meaning of each sentence is grasped does it become clear that the two sentences are ambiguous and can assume quite different structures depending on the interpretation.

Box 8.8: Sound display

While most GIS and map interfaces rely heavily on visual display, sound can still play an important role in GIS and its applications. The process of using sound to represent data is termed *sonification*. *Sound symbols* are abstract sounds used to represent and distinguish information. For example, most people will be familiar with the Geiger counter often seen in movies. The Geiger counter sonifies the levels of ionizing radiation, with more frequent, louder, and higher pitch clicks indicating increasing levels of radiation. In fact, there are a variety of *sound variables* that may be used in sound symbolization, including loudness and pitch. Krygier (1994) provides an overview of sound symbolization for geographic information and enumerates the key sound variables. Sound is inherently dynamic, so many of the dynamic visual variables have counterpart sound variables, including duration, order, frequency (in this context meaning the periodic recurrence of sounds), and rate of change. An example of the use of symbolic sounds in a GIS interface to represent the reliability of geospatial information can be found in Bearman & Fisher (2012). Sound symbols can also be used metaphorically, just like visual symbols. Sound icons, often termed *earcons* (Gaver, 1986), are sounds that provide a metaphor for a command or function. For example, in desktop computing, the action of deleting a file is often accompanied by the sound of paper being scrunched up as if it were being discarded in a waste paper bin. Spoken language too is an increasingly common feature of many intuitive interfaces (see Section 8.3.2). Our sense of hearing also allows us to locate the source of sounds in 3D space. Slight discrepancies between the sounds heard in each ear enable us to determine the direction of a sound source, analogous to retinal disparity in vision. Other audio cues, like loudness, help to determine distance. Many digital sound systems take advantage of this facility and produce sounds that appear to be located in space. However, the precision with which humans are able to determine position in space, termed *spatial acuity*, is lower for hearing than for vision.

In the most commonsense interpretation, "flies" acts as a verb in the first sentence and a noun in the second sentence, while "like" acts as an adverb in the first sentence and a verb in the second sentence.

Figure 8.18 summarizes the balance between intuitive and expressive interfaces with some familiar examples. One of the most familiar examples is the *WIMP*—windows, icons, menus, and pointers—interface found on most personal computers today (see Box 8.9 on the next page).

WIMP

8.3.3 *Designing intuitive interfaces*

We have seen that interfaces must strike a balance between intuitiveness and expressiveness. More expressive interfaces can typically only be achieved by decreasing the level of interface intuitiveness. However, the converse is not the case: decreasing the expressiveness of an interface will not, on its own, necessarily lead to a more intuitive interface.

An important question, then, is how do we design interfaces that are more intuitive? This question is not as easy to answer as one might at first think. Even for the most familiar and everyday physical objects, such as doors and light switches, poor interface design is all too common. The bathroom tap in Figure 8.19, found in Canberra Airport, is sufficiently unintuitive that it is accompanied by four lines of operating instructions. If everyday objects such as bathroom faucets—invented thousands of years ago in Roman times—can still confound interface designers today, how much more challenging may we expect GIS and digital mapping interfaces to be?

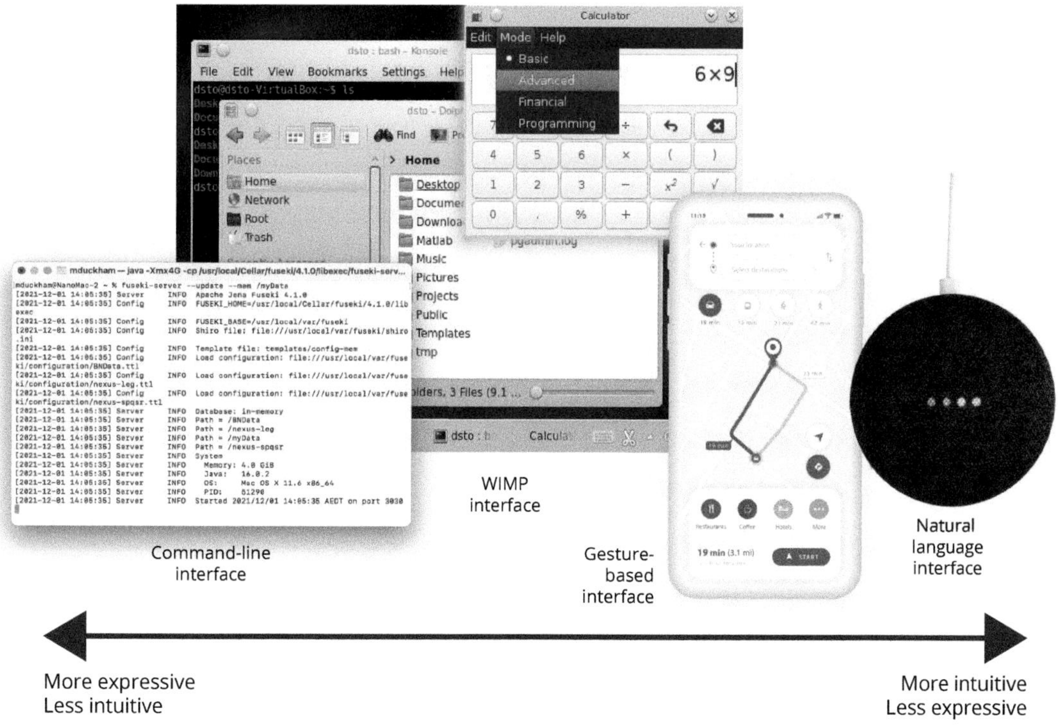

Figure 8.18: Intuitive versus expressive user interfaces

Hence, the question of what features of an interface lead it to be more or less intuitive lies at the core of HCI. In his classic book, *The Design of Everyday Things*, Don Norman sets out his influential approach to analyzing those features of any user interface—not just computer interfaces—that make them intuitive (Norman, 1988). This approach can be summarized into five fundamental principles of intuitive interfaces, summarized in Figure 8.20: visibility and affordances, mappings and metaphors, simplicity, feedback, and error handling.

Visibility and affordances Intuitive interfaces are "self-describing": simply by their form one should immediately see the main components and gain an

Box 8.9: WIMP interfaces

Originally developed in the 1970s, WIMP remains the most common interface style for todays desktop and laptop computers. WIMP stands for *windows*, *icons*, *menus*, and *pointers*. Windows are used as independent containers for particular processes or applications. Icons are small pictures that symbolize a command or function, such as a pair of binoculars or magnifying glass icon to symbolize a "search" command. Menus organize commands into logical groups and hierarchies (submenus) that are easier for users to access and remember. Finally, pointers enable users to directly manipulate objects, such as opening a file by clicking on it, or deleting a file by dragging it to a trash icon. WIMP interfaces have been successful because they are moderately intuitive at the same time as being moderately expressive. While WIMP is still common today, many devices offer other, more intuitive direct manipulation interaction styles, such as gesture-based interaction familiar in mobile devices such as phones and tablets.

Figure 8.19: A bathroom faucet (tap) as an interface. The instruction reads: "Sensor tap: This tap is 'hands free,' operated by sensor. To activate, please wave hand in front of the sensor, located to the right of the spout below"

visibility

affordance

intuition about what the interface does and how the components operate. *Visibility* is connected with ensuring that an interface makes the most important components most visible. *Affordances* on the other hand provide the user with clues about how to operate the components. In the physical world, chairs are for ("afford") sitting on, ladders are for climbing, handles are for grasping. However, even in the world of computer interfaces similar affordances can be found: buttons are for pushing, boxes are for checking, and sliders are for sliding.

The red button in Figure 8.20 is both highly visible and has the right affordances for pressing. By contrast, the mechanism for operating the faucet in Figure 8.19 is completely hidden from view and affords little aside from the incorrect action of grabbing the long spout.

Mappings and metaphors Users of an interface are not empty vessels. Users bring to any interaction a wide range of expectations drawn from experiences of other interfaces and of the physical world around them. Thus, intuitive interfaces should be compatible with such common experiences and harmonize with the mental models users bring to the interaction.

mappings

Mappings concern the relationship between interface controls and the physically or conventionally expected effects of those controls. For example, when viewing a map, pressing the down cursor or swiping up on the touchscreen might be expected to display an area to the south of the current view. Using a less obvious mapping (such as the keyboard "s" key) or even a counterintuitive mapping (such as the left cursor key or a pinch gesture) is guaranteed to make the interface less intuitive.

The control knob in Figure 8.20, for example, accords with a common mapping: turning clockwise to increase, counterclockwise to decrease. Again, the faucet in Figure 8.19 provides much less intuitive mappings, relying instead on a hand wave to start the flow of water.

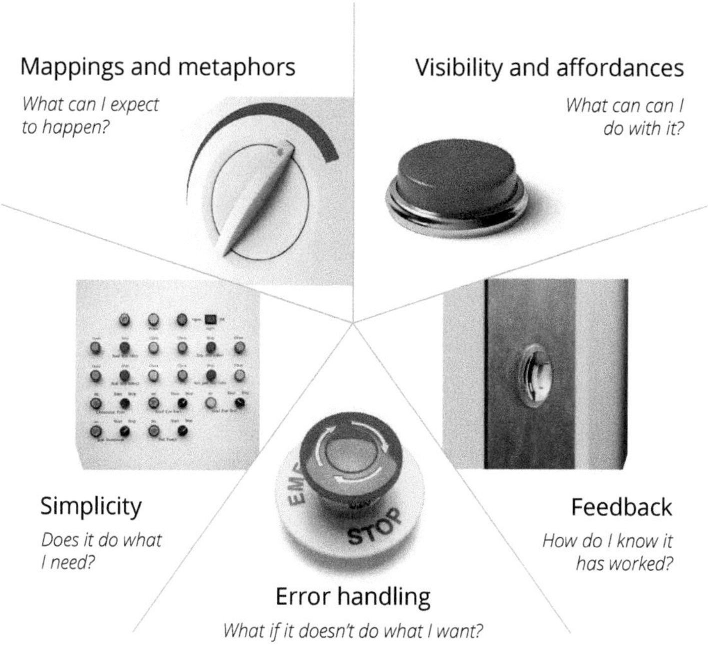

Figure 8.20: Five fundamental principles of intuitive interfaces

An important mechanism for supporting mappings is the use of *metaphors*. In language, a *metaphor* is the use of a word or phrase denoting one kind of idea or object in place of another, in order to make a figurative comparison between the two (see, for example, Box 4.6 on page 158). Metaphors are everywhere and are widely used to help humans understand an unfamiliar target domain with reference to some familiar source domain. The ingrained nature of metaphoric understanding in humans makes metaphors a powerful vehicle for human-computer interaction. For example, your personal computer relies extensively on the *desktop metaphor*. The desktop metaphor suggests a likeness between the computer interface and an office desktop, with windows taking the place of sheets of paper, calculators, clocks, and so forth on a physical office desktop. Using the desktop metaphor with a WIMP interface (Box 8.9 on page 333) helps make coordinating and switching between different software applications more intuitive.

metaphor

Indeed, many GIS interfaces are built around the use of the *map metaphor*: they are designed to exhibit map-like characteristics, utilize map symbolization, and echo the physical (static) characteristics of hardcopy paper maps. Maps and map symbolization are immediately familiar to many people. Consequently, adopting the map metaphor usually leads to intuitive GIS interfaces, at least for those people who are already comfortable with conventional maps (see Figure 8.21).

map metaphor

Simplicity Intuitive interfaces are simple. One characteristic of simplicity is providing only those components in an interface that are strictly necessary to the functions of a system, and providing only those functions that are strictly necessary to anticipated user needs. The term *feature cascade* is sometimes used

feature cascade

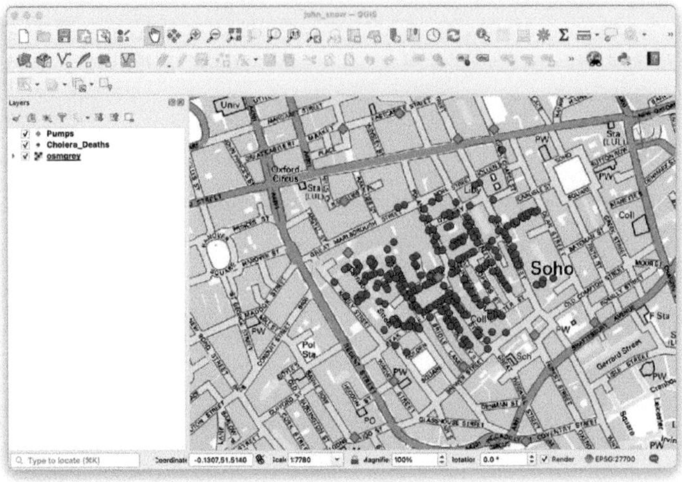

Figure 8.21: The map metaphor is common in GIS interfaces

to describe the undesirable tendency of more and more features to be added to interfaces, making them more and more complex and unwieldy.

The instructions accompanying the faucet in Figure 8.19 reveal an underlying complexity beyond what should be necessary simply to start and stop the flow of water ("... *wave* hand *in front* of the sensor, located *to the right* of the spout *below*"). While the control panel in Figure 8.20 is much less simple than the individual buttons, it also gives access to a much wider range of (labeled) actions.

feedback

Feedback Feedback is the process of accepting and responding to a user's actions with information about what a user has done, and what has been achieved. Feedback can be output on any IO channel, not only the IO channel used for the action. A visual confirmation popup, a satisfying sound, or a reassuring haptic buzz can all be used individually or even in combination to provide immediate feedback in response to a user's action, for example. The elevator button in Figure 8.22 (and Figure 8.20) immediately lights up (visual IO channel) in response to being pressed (haptic IO channel). A lack of clear feedback can be most frustrating, as will be familiar to any one with experience of ineffectually pressing an elevator button with a faulty light or fruitlessly attempting to activate a faucet sensor, such as in Figure 8.19.

Figure 8.22: Lift button, from Figure 8.20

To ensure the user's reasoning processes can continue uninterrupted, feedback needs to be rapid. As a rule of thumb, a response time of about 0.1 seconds or less is perceived by users as instantaneous. A response time of 1–2 seconds is still fast enough for a user's flow of thought to be uninterrupted but is a noticeable delay lessening the impression of interactivity. A response time of 10 seconds is about the limit for feedback, as a user's attention is likely to be transferred to other tasks if delays exceed this time.

Error handling Computers are machines and do only and exactly what they are instructed. Humans, however, are fallible and will make errors. Intuitive interfaces must allow, account for, and even anticipate mistakes by users. The emergency alarm button in Figure 8.23 (and Figure 8.20), for example, has an inbuilt reset function (twisting the activated button) to reset a mistaken activation.

Figure 8.23: Emergency alarm, from Figure 8.20

Forgiveness is the principle that things that are easy to do should be easy to undo. Conversely, things that are hard to undo should necessarily be harder to do in the first place. For example, an interface that allowed the irreversible deletion of all stored data from a computer's memory with a single button press would fail the forgiveness principle. In a similar vein, mistyping a word into a document is easy to do, and so it should be easy to undo (for example, using the delete key or an easily accessible "undo" menu item, gesture, or keystroke).

forgiveness

Mistakes can also be an opportunity for the user to learn about a system. Intuitive interfaces do not penalize or sanction users when they make mistakes, but instead they gently help users understand what to do differently next time.

8.3.4 Usability engineering

The process of developing interfaces that maximize usability is termed *usability engineering*. In general, all usability engineering techniques emphasize the importance of considering usability at every stage of system development. Developing usable interfaces is not an exact science: experience and judgment play a vital role in successful interface development. Most importantly, interface and interaction design is an iterative process, requiring continual cycles of prototyping, testing and critical evaluation, and redesign (see Box 8.10 on the next page).

usability engineering

The usability of an existing interface can be assessed empirically by evaluating its performance either in the laboratory or in actual use in the field. Interface usability may sometimes be quantified, especially where objective evidence is required for upgrading or replacing an existing interface. Five basic criteria are often used to quantify the degree of intuitiveness and expressiveness of an interface.

Time to learn How much time does it take for users to learn how to use a system?
Speed of performance How quickly can users carry out benchmark tasks?
Rate of errors How many and what type of errors do users make in carrying out benchmark tasks?
Retention over time How well do users retain their skills and knowledge over time?
Subjective satisfaction How much did users enjoy or dislike performing benchmark tasks?

In addition to empirical user evaluation, interface designers often use more qualitative analytical techniques to ensure usable interfaces. Box 8.10 on the following page briefly discusses two of the most common approaches to

Box 8.10: Design methodologies

A effective and easy-to-learn method to help novice interface designers develop innovative data graphics and information visualizations is the five design-sheet (FdS) methodology of Roberts, Headleand, & Ritsos (2015). The FdS method begins with a "brainstorm" sheet, where the designer focuses simply on quantity of designs, aiming to develop as many possible options as possible. Sheets 2–4 develop three promising but completely different design options in greater detail. Finally, the fifth sheet combines, refines, and realizes a single preferred design that aims to take advantage of the best balance of features of all the options explored. Another common interface design methodology is GOMS. GOMS recursively decomposes high-level interface functionality into smaller component functionality. In a GOMS model, the user *goals* are specified in terms of the basic *operations* a user can perform. GOMS also models the different *methods* for achieving a particular goal and attempts to predict which particular method will be *selected* and in which cases. Taking a design approach to problem-solving more broadly—not only interface design problems—is often referred to as "design thinking." Design thinking is usually explained with reference to five stages: first, *empathize* with the humans whose problem you are trying to solve; second, *define* a precise problem statement based on the information gathered in the first stage; third, *ideate*—generate ideas and look for alternative ways of approaching the problem (as for FdS "brainstorming"); fourth, *prototype* and experiment with scaled down versions of the ideas; and fifth, look for ways to *test* these solutions out. The design thinking process has direct analogy to the stages of the modeling process introduced in Section 1.3 (see Figure 1.15) and the five main stages of the system development lifecycle (summarized in Box 1.3 on page 19: requirements, analysis, design, implementation, and validation). Like modeling and system development, the design thinking process is never linear, with many cycles of iteration essential to success.

interface design—the five design-sheet method and goals, operations, methods, and selections (GOMS)—as well as to problem solving with design more broadly, often called "design thinking." In addition to the iterative nature of any design process, *prototyping* is one of the simplest but most important usability engineering techniques. Prototyping involves developing a "mock-up" (prototype) of the target interface in order to reveal problems that may be hard to detect with pen-and-paper designs. Prototypes may be used simply to test design ideas, termed a *throw-away* prototype, or developed as a preliminary version of the actual interface, termed an *evolutionary* prototype. Prototyping is the most informal and low-cost usability engineering technique and can be valuable for even the smallest projects.

prototyping

8.4 Geovisualization

geovisualization

Geovisualization (also termed *geographic visualization*) is the process of using computer systems to gain insight into and understanding of geospatial information. Experimental evidence indicates that humans use two fundamentally different types of information in thought processes: verbal and visual. *Verbal thinking* is important for reading and writing, engaging in conversation, and logical thought. *Visual thinking* is important for reasoning about groupings, parts, and spatial configurations of objects.

verbal thinking

visual thinking

scientific visualization

Scientific visualization is the process of using information systems to represent and interact with information in a way that enhances visual thinking. It is important to emphasize that the "visual" in visualization refers to "visual

Box 8.11: The cartography cube

The *cartography cube* has been used as a way of conceptualizing the relationship between conventional map use and geovisualization (MacEachren, 1994b). The cartography cube has three axes: *interactivity*, *goals*, and *audience*. Interactivity is the degree to which users can manipulate and redefine a map. Goals refer to the degree to which a map is designed to help users discover new information. Audience refers to the degree to which a map is targeted at a specialized audience. In general, conventional maps are located in the lower left-hand corner of the cube, presenting known information to a public audience using low levels of interactivity. Conversely, geovisualization techniques are located in the upper right-hand corner of the cube, using high levels of interactivity to help private audiences discover new information. For example, a conventional topographic map might be useful in *presenting* known information, such as the route to the top of a mountain, but on its own it is less likely to help users *discover* new information, such as understanding the land-forming processes within that region. A conventional topographic map is also *public* in the sense that anyone, from hill walkers to town planners, might use it. Many of the views of information we encounter using a GIS are *private* in the sense that they might only exist on our computer screen for a few seconds (or less) before being altered or replaced.

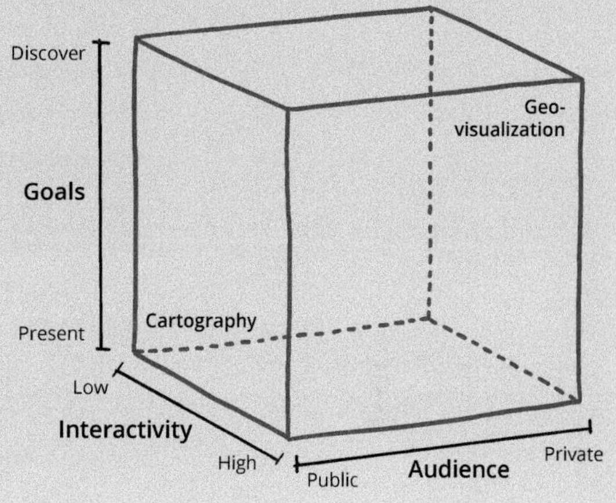

thinking," and not necessarily to the visual IO channel (visualization may be based on any IO channel, e.g., visual, auditory, haptic).

Geovisualization is a branch of scientific visualization that deals specifically with geospatial information (see Box 8.11). Geovisualization design principles emphasize the importance of using the dynamic, interactive, and multimedia capabilities of GIS to help users gain insight into geographic problems, sometimes referred to as *geographic thinking*. Geovisualization techniques can extend the map metaphor—which treats GIS and geospatial data as digital analogs of hardcopy paper maps—in several ways. Breaking away from the familiarity, but also the constraints of the map metaphor, can result in GIS interfaces that are more expressive and arguably more intuitive too. The following sections examine in more detail three important visualization techniques with particular relevance to geospatial information: animation, 3D displays, and geovisual analytics.

geographic thinking

8.4.1 Animation

Conventional paper maps are static and, aside from wear and tear, their appearance is fixed at the time of manufacture. In contrast, using the computer to display a series of static images can convey the impression of motion or change over time, a technique known as *animation*. Each static image within the sequence that makes up an animation is called a *scene*.

animation

scene

Animation may be used simply to highlight or emphasize features on a static map, for example, by using a flashing arrow to indicate "You are here!" on an animated tourist information map. However, the most important function of animation is for visualizing *change* in geographic phenomena. Static maps can sometimes be used to depict change, for example, using static symbols such as arrows to represent flow or movement. Animation is a more intuitive mechanism for visualizing complex dynamic phenomena than static maps. Weather reports, for example, often use a chronological series of maps of atmospheric conditions to show the movement of a storm front.

fly-through

attribute re-expression

Animations are not restricted to representing change over time. Animations can also be constructed from any sequence of spatial or attribute changes. One example of an animated spatial change is *fly-through* animations, such as those often used in architecture to show building designs, in which the user's perspective on a static environment or data set is gradually moved. Attribute change, also called *attribute re-expression*, involves an animated logical sequence of scenes constructed from ordered attributes. Figure 8.24 illustrates the difference between animations of chronological change and attribute change. In Figure 8.24a, the animation can be interpreted as showing a gradual northerly migration of the population over a 30-year period. In Figure 8.24b, the animation can be interpreted as highlighting a spatial variation in population age, with younger people comprising a greater proportion of the population in the southern regions.

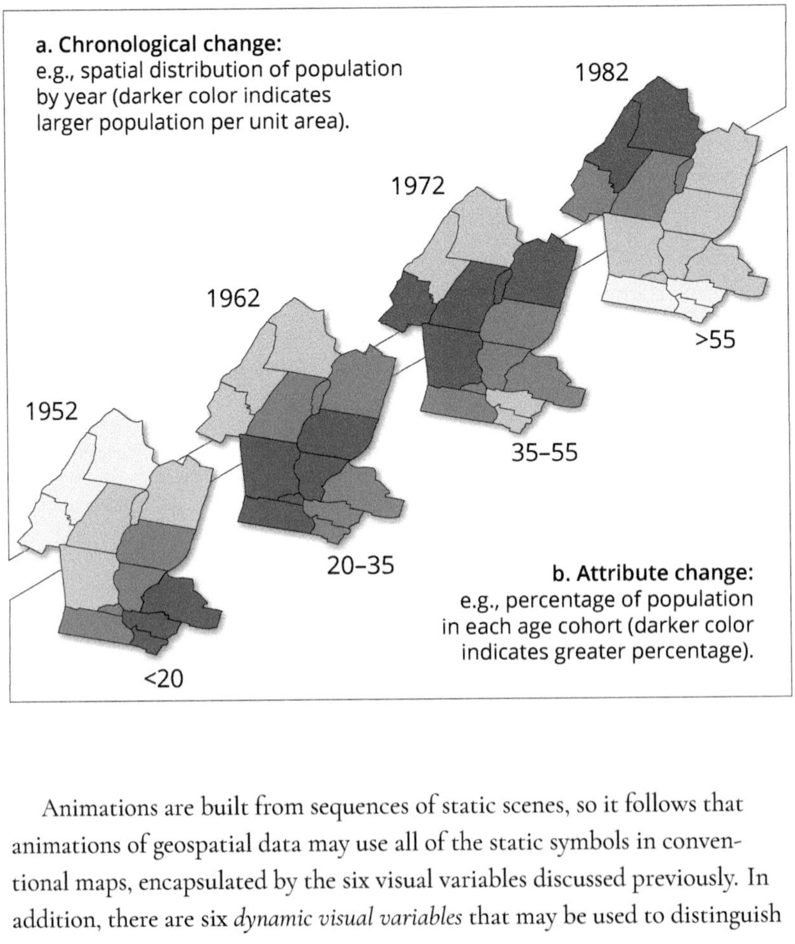

Figure 8.24: Chronological and attribute changes in animation

dynamic visual variable

Animations are built from sequences of static scenes, so it follows that animations of geospatial data may use all of the static symbols in conventional maps, encapsulated by the six visual variables discussed previously. In addition, there are six *dynamic visual variables* that may be used to distinguish different features in an animated map: moment, frequency, duration, magnitude of change, order, and synchronization (Di Biase, MacEachren, Krygier, & Reeves, 1992).

The point in time at which a change occurs in an animation is referred to as *moment*. Moment may be thought of as the temporal counterpart of the static visual variable, position. *Frequency* is the rate at which a change occurs in an animation, and it is akin to the static visual variable *pattern*. *Duration* refers to the length of time that each static scene is visible. Duration gives an animation its pace. Scenes that are visible for longer move at a slower pace and appear to have greater importance or relative value when compared with those faster-paced scenes that are visible for a shorter length of time.

moment

frequency

duration

The *magnitude of change* is the amount of change that occurs in moving from one scene to the next. Sequences of scenes in which the magnitude of change from one scene to the next is small yield smooth animations. Sequences in which the magnitude of change is large yield abrupt or jumpy animations. The ratio of magnitude of change to duration in an animation is termed the *rate of change*. The fifth dynamic visual variable is the *order* in which scenes appear. The order of scenes in an animation is normally dictated by the chronology of the data (for example, in animating the spread of an epidemic over time) or the natural order of the attribute used to construct the animation (for example, the age classes in Figure 8.24b). Finally, *synchronization* is closely related to order and refers to the relative timing of changes in two or more phenomena represented in an animation.

magnitude of change

order

synchronization

Order and synchronization are particularly significant in animation, as they can be used to suggest a causal relationship between phenomena. As long as the magnitude of change is sufficiently small, users may interpret features in one scene as being the cause of features in a subsequent scene. In an animated map of the spread of an infectious disease epidemic, for example, a region of disease that enlarges through a sequence of scenes would normally be interpreted as contagion, in which individuals with the disease in one scene are the cause of further cases of infection in subsequent scenes.

Table 8.3 summarizes the dynamic visual variables by showing changes as shapes located on an animation timeline, with time moving forward from left to right.

Animation in visualization should be used sparingly and with caution, however. Human visual perception is especially attuned to movement, and so incautious animation can be irritating and distracting to users. Further, humans are prone to pay attention only to certain parts, not all of an animation—an instance of a general feature of human perception termed *selective inattention* by psychologists. Selective inattention can cause users to miss important transitions in animations altogether, termed *change blindness* (explored in the context of animated choropleth maps by Fish, Goldsberry, & Battersby, 2011, for example). The sequential and timed nature of scenes can also make it difficult for users to gain an overview and identify insightful moments from longer animations. A 2002 review by Tversky, Morrison,

selective inattention

change blindness

Table 8.3: Dynamic visual variables, after MacEachren (1994a)

Dynamic visual variable	Example	Explanation of example
Moment		Moment of change is indicated with a circle
Frequency		Changes increase in frequency, from low to high
Duration		Result of square change lasts longer than circle changes
Magnitude	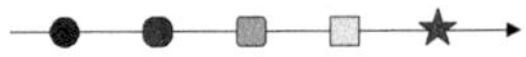	Change is gradual (circle to square) then abrupt
Order		Order of changes is apparently random
Synchronization	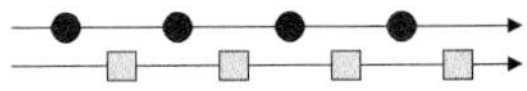	Changes in circle and square phenomena are closely related

& Betrancourt indicated that in practice animation fails to outperform a well-designed static data graphic in most cases.

8.4.2 3D displays

contour

hypsometric map

Conventional maps and most visual computer displays today are limited to two spatial dimensions; yet, the geographic world has three spatial dimensions. *Contours*, which connect points of equal height, are the most common mechanism for representing a topographic surface on a conventional 2D map. An example of contour lines is shown in Figure 8.25a, along with an example of a *hypsometric map* in Figure 8.25b. A hypsometric map uses a logical sequence of colors to indicate elevation. Such maps are only useful for displaying surfaces, not true 3D shapes like buildings, cliffs, or caves, because they can only show one (elevation) value at each location.

depth cue

Contour and hypsometric maps provide an abstract view of a 3D surface, with which a skilled mapreader is able to gain an excellent understanding of the shape of a surface. These types of maps are expressive but not particularly intuitive. Contour maps can be difficult for the novice mapreader to interpret. A more intuitive representation of 3D surfaces involves using psychological cues to give the impression of depth. There are several such *depth cues* that may be exploited in a 2D map or computer display. *Shading* may be used to give the impression that an object with height is casting a shadow. Figure 8.25c gives an example of hill shading, common in both conventional maps and computer displays. For topographic surfaces, the correct effect is most likely to be achieved when the source of illumination appears to be positioned in the northwest. Other sources of illumination are more likely to result in misinterpretation of the shape of surface, with valleys being mistaken for ridges, and vice versa.

Box 8.12: Stereoscopic displays

A *stereoscopic* display takes advantage of retinal disparity, by showing precisely controlled different images to each eye using special hardware, such as a headset, stereo glasses, or stereo display screen. The need for specialized hardware means that stereoscopic displays are used less frequently than conventional visual displays. Further, a variety of eye-related conditions that can interfere with stereo vision (including losing sight in one eye) mean that a significant minority of people lack the ability to see binocular depth cues. Happily, in most interfaces monocular depth cues are amply able to provide the required 3D effect. Arguably, the sporadic popularity—and more tellingly the quotidian unpopularity—of stereo 3D movies is a reflection of the adequacy of monocular depth cues for understanding 3D scenes in most situations, without the need for binocular depth cues.

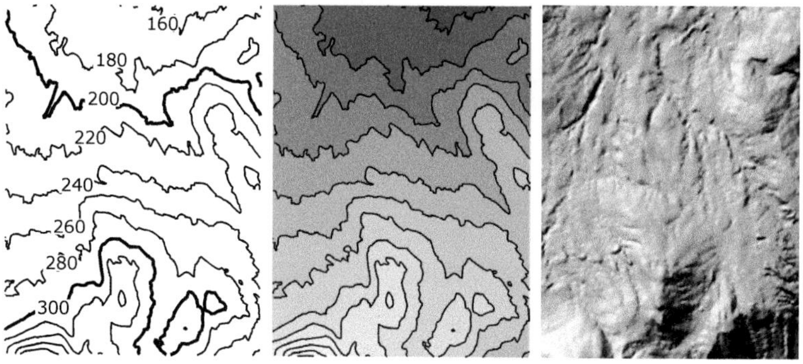

a. Contour b. Hypsometric c. Hill shading

Figure 8.25: Three representations of a terrain surface in two dimensions

Other depth cues include the *relative size*, *linear perspective*, and *interposition* of the objects depicted, illustrated in Figure 8.26. Familiar objects with a smaller relative size than expected appear to be farther away. For example, we would normally interpret Figure 8.26a as a moose that is farther away than the mouse, rather than as a mouse that is much larger than the moose. *Linear perspective* concerns the effect that parallel lines appear to converge into the distance, such as the edges of the road in Figure 8.26b. *Interposition* concerns the effect whereby distant objects are occluded by nearer ones, such as the cafe partially obscured by the hotel in Figure 8.26c.

linear perspective

interposition

a. Relative size b. Linear perspective c. Interposition

Figure 8.26: Relative size, linear perspective, and interposition depth cues

The impression of depth can also be achieved using animation, in which closer objects appear to move faster than more distant ones. This effect, which you can observe out of the window of any moving vehicle, is called *motion parallax*.

motion parallax

The depth cues described above are termed *monocular*, because they are effective with only one eye. Two other depth cues rely on *both* our eyes func-

monocular

binocular
convergence

tioning correctly (*binocular* depth cues). First, *convergence* occurs as our eyes move to fixate on near objects, such as your finger in front of your face. Convergence only influences depth perception for relatively close objects (less than a few meters away); eye convergence is negligible for objects in the middle to far distance. Second, as a result of the slightly separated positioning of our eyes, each eye receives a slightly different view of the world around us, termed *binocular* or *retinal disparity*. These slight discrepancies are used by the visual system to help determine how far away an object is, termed *stereopsis*. About 1 in 50 people, however, lack stereopsis abilities (termed *stereoblindness*) and many more (in some studies more than 1 in 5) have poor stereo vision. As a result, binocular depth cues should not be relied upon in geovisualization interfaces (Box 8.12 on the preceding page).

retinal disparity
stereopsis
stereoblindness

Both monocular and binocular 3D effects are commonly found in virtual and augmented reality interfaces. Virtual reality (VR) aims to immerse the user in a simulated or digitally generated interactive environment. VR environments are typically based on realistic 3D graphics and often delivered using stereo goggles or headsets. To complete the immersive effect, virtual reality environments usually also support visual display with other multimodal interfaces, in particular sound and haptics. In contrast to virtual reality, augmented reality aims to mix digital objects, accurately registered in 3D and in real time, into our view of the world around us. Augmented reality is more frequently unimodal visual display, because the sounds, smells, and feel of the world around us help make the experience more immersive.

virtual reality
augmented reality

Displaying physically 3D phenomena, such as topographic surfaces, is one obvious application of 3D interfaces in GIS. However, not all (or even any) of the dimensions need to be spatial to take advantage of 3D interfaces. In relation to geospatial information, the third dimension is often used to represent some other non-spatial data, a process known as *spatialization* (Skupin & Fabrikant, 2003).

spatialization

8.4.3 Geovisual analytics

visual analytics
geovisual analytics

Visual analytics is concerned with using interactive visual displays to support humans to identify and understand patterns and relationships in data. *Geovisual analytics* is the special case of visual analytics that deals with *spatial* data.

From a user's perspective, the purpose of all the different techniques encountered in this chapter can be seen as supporting a continuum of uses, from presentation of data at one extreme to supporting exploration and gaining insights into data at the other (cf. Box 8.11 on page 339). If conventional data graphics and traditional cartography have the primary purpose of "presenting knowns" in data, then geovisual analytics sits firmly at the opposite end of that continuum: "revealing unknowns" in data. When describing the goals of geovisual analytics, authors often use the terms *exploration* and *insight*, as well as terms like *discovery*, *hypothesis generation*, and *sensemaking*.

A range of different tools and techniques have been proposed to enable geovisual analytics. There are, however, a core of established interactive functions that underpin most geovisual analytics interfaces. The five most common basic interactions are discussed below, illustrated with a European Commission (EC) visualization tool for exploring European demographic statistics, the Eurosat RCI (regions and cities illustrated) geovisual analytics interface in Figure 8.27. The interface combines an interactive map view (left-hand side) with reconfigurable data graphics (right-hand side), such as scatter plots and bar charts. The key interactions the user can engage in are:

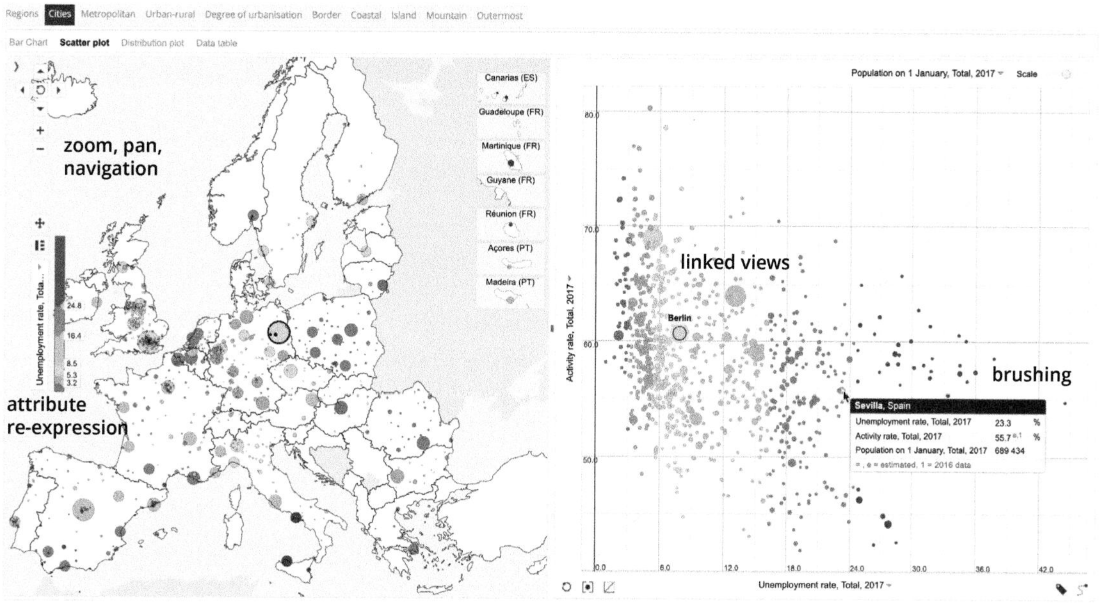

Figure 8.27: Eurostat RCI interface showing unemployment rates (color hue) and population (circle size) in European cities. The annotations highlight the geovisual analytics functions: panning, zooming, and navigation; attribute re-expression; brushing; and linked views (Source: European Union)

Panning, zooming, navigation Panning (change of viewpoint) and zooming (change of level of detail) are basic interactions for both spatial and non-spatial data. However, the ability to explore different views and pan and zoom around data can also be disorientating for users. Consequently, panning and zooming need to be paired with a high-level overview or a "home" function to ensure users can always re-orient themselves before they get lost in data. In Figure 8.27, pan and zoom can be achieved using the interface buttons in the top left corner or using familiar mouse gestures. Pan and zoom are complemented with a "reset" button that returns the user to the default view at any time. The scatter plot on the left-hand side also has its own independent rescaling functions, along with its own "reset" button.

Attribute re-expression Attribute re-expression is the ability to interactively change the way selected data is presented in the interface. For example, we have already seen in Figure 8.7 how changing the way data is classified can profoundly change the patterns that are evident. In Figure 8.27, the classification of unemployment rate (color hue) in the legend on the left-hand side can be altered interactively, with the user sliding class boundaries up and down. Doing this can be highly exploratory, highlighting previously unseen patterns, breaks, and relationships in data.

brushing

Brushing *Brushing* is a form of implicit input where a user's mouse hover is interpreted as a query for further information about any features located under the mouse pointer. In Figure 8.27, the mouse pointer's location above the scatterplot point related to Sevilla automatically brings up a tooltip box that contains more detail about the data for that city. The user need not explicitly type in a query or perform a mouse click. Simply lingering with the mouse automatically brings up the more detailed numerical data.

linked view

Linked views Linking multiple views involves changes to highlighted information in one view being mirrored by corresponding changes to other views. Linked views are especially well suited to exploring spatial data, with map views linked to non-spatial data graphics, such as bar charts or scatter plots. In linked views, a mouse click to highlight a feature in the map will instantly highlight the corresponding bar in the bar chart, for example. Similarly in Figure 8.27, highlighting Berlin's data point in the scatter plot immediately causes the corresponding data point in the map view to be highlighted with a thicker border.

Animation Animation has already been encountered in detail in Section 8.4.1. Animation can help users identify subtle patterns and interpret important changes not always visible in static views. As discussed, animations are often chronologically ordered but can also be constructed from other orderings, such as changes to attribute values (see Figure 8.24). Although no animation options appear in the interface in Figure 8.27, an animation toolbar does appear in the interface when displaying any time-varying data sets, such as population over time.

Reflections

The principles behind data graphics and HCI also lie at the foundations of cartography and geovisualization. Well-designed maps must also be well designed data graphics; effective GIS interfaces must additionally encompass good interaction design. In addition to the classic texts by Norman (1988) and Tufte (1990, 2001) already cited, there are some excellent textbooks on HCI and visualization topics from a computing perspective, with Munzner (2014), Preece, Rogers, & Sharp (2015), and Ward, Grinstein, & Keim (2015) especially recommended.

At the same time, geographic information scientists have unique expertise and perspectives to contribute to design conversations. Cartography draws on the world's oldest design traditions; maps were the first data graphics. Under the heading of "interfaces," cartography and geovisualization were identified back in Chapter 1 as one of the five key themes at the core of GI science. Unsurprisingly, then, there are a range of authoritative books and introductory texts on both cartography (e.g., Griffin, 2017; Peterson, 2020) and geovisualization (e.g., Andrienko & Andrienko, 2006; Çöltekin, Bleisch, Andrienko, & Dykes, 2017).

Figure 8.28: Don't do this! A spectral color ramp incorrectly used to depict height in a DEM

The essence of this design expertise is an understanding of the rules governing good information and interaction design combined with a spark of creativity and innovation. While there are few *inviolable* rules in design, there *are* correct and incorrect design decisions, good and bad designs, as we have seen. For example, Figure 8.28 shows one all-too-common mapping mistake—using color hue to represent a value message, in this case a spectral color ramp to depict elevation in a DEM. Hue is a poor choice for communicating numerical values such as height, as we have seen in Section 8.1.2. Not only does hue fail to communicate a value message (does red indicate a ridge or a valley, show more or less height?), but also the transitions between hues creates the impression of false boundaries in the surface (such as between blue and green and between green and yellow). Penn State University Professor Cynthia Brewer is the preeminent authority on color usage in maps, and her 2016 book on map design is essential reading for anyone planning to use color in maps.

Some rules are, however, made to be broken. Generations of geography students have been drilled on the need for "map furniture," for example. However, rigid adherence to the map furniture mantra—"scale bar, legend, north arrow, title"—is certainly not essential to effective communication with maps, and indeed it is more likely to leach the beauty, creativity, and joy from a design. A combination of proficiency and panache, finesse and flair is a hallmark of great design, and the most technically adept designers in GI science are frequently also the most creative (enjoy, for example, Andrienko et al., 2010; Bleisch & Hollenstein, 2018; Goodwin, Dykes, Slingsby, & Turkay, 2015; Harrower & Brewer, 2003).

9 Artificial Intelligence and GIS

Sections

Artificial intelligence (AI) encompasses a wide range of different algorithms and techniques commonly used in GIS for creating new knowledge from spatial data. More specifically, through this chapter you will discover how to:

- interpret, use, and design machine-readable **ontologies** to help increase the interoperability of spatial information;
- represent **qualitative spatial relationships**, such as left and right, cardinal directions, and region relations, within automated reasoning systems;
- apply the basic **machine learning** techniques that underpin many common **spatial analyses**; and
- understand the principles and limitations of advanced machine learning techniques, such as **artificial neural networks** and **deep learning**.

Artificial intelligence has been a central to the development of GIS since the earliest days of GIS research.[1] In a nutshell, *artificial intelligence* (AI) concerns the design and development of machines imbued with behaviors and abilities that might otherwise be regarded as requiring human intelligence (see Box 9.1 on the following page).

While the breadth and diversity of AI techniques today is vast, it is possible to identify two major threads that run throughout the history of AI: *knowledge representation and reasoning* and *machine learning*. Knowledge representation and reasoning (KR^2) is concerned with using automated logical reasoning together with symbolic representations of knowledge to solve AI problems. In contrast, machine learning (ML) is concerned with automated learning from data.

In this chapter, we begin by introducing the most successful and widely used KR^2 technique today: ontology engineering (Section 9.1). Whereas the emphasis in ontology engineering is more strongly on the representation of knowledge than reasoning about that knowledge, the emphasis is reversed in qualitative spatial reasoning (Section 9.2). Turning to ML in the second half of this chapter, Section 9.3 shows how many of the most familiar and frequently used spatial analysis tools and techniques are in fact founded on machine learning. Finally, Section 9.4 introduces one of the most important machine learning techniques in the recent decade: artificial neural networks and deep learning.

[1] In 1987, the second ever issue of the *International Journal of Geographical Information Science* (IJGIS), the flagship journal for GIS research, contained a research paper on a knowledge-based GIS called "KBGIS-II" with a facility to apply logical rules to spatial data, and even to learn new rules from data automatically (Smith, Peuquet, Menon, & Agarwal, 1987).

artificial intelligence

KR^2

machine learning

Box 9.1: The Turing test

The definition of AI as, effectively, a machine that can imitate human intelligent behavior is tied most closely to the work of British mathematician and computer scientist Alan Turing. In a landmark 1950 paper, Turing argued that a machine could be called "intelligent" if a human interrogator could not reliably distinguish the machine from a human interlocutor (Turing, 1950). Turing's thought experiment—which Turing called "the imitation game," also the title of the 2014 film about Turing's life and work—is commonly known in computer science as the *Turing test*. Turing made many other enduring contributions to the emerging fields of computing and AI, including an abstract mathematical description of computing machines and the types of problems that can be solved by computers, called a *Turing machine*. Sadly, Alan Turing died at age 41, just two years after he was convicted for "gross indecency"—a euphemism for homosexuality, then a crime in the United Kingdom—and forced to undergo a year of invasive hormone injections resulting in impotence and physical changes to his body. In 2009, the British government officially apologized for Turing's treatment, acknowledging at the same time the appalling treatment of innumerable gay men in Britain during the era.

9.1 Ontology engineering

ontology

The idea of *ontology* as the philosophical study and classification of what things exist was already encountered in Chapter 4 in the context of how we model geographical spaces. This section introduces the related idea of *an* ontology, which can be defined as an explicit, unambiguous, and machine-readable specification of the meaningful and relevant entities and relationships in an application.[2] The activity of designing and using computerized ontologies is termed *ontology engineering*.

[2] The topics of conceptual data modeling and entity-relationship models (Section 2.2.2) have clear commonalities with ontologies. However, ontologies are focused more generally on the stable meanings of words used to refer to things that exist in the world. Conceptual data models, by contrast, are slanted towards describing the data related to an application. Nevertheless, a good conceptual data model may also be the basis of a good ontology.

ontology engineering

The core idea behind ontologies[3] is that the meaning of words in natural language is relatively stable, while the meaning of items in digital data structures and data models can be highly transient and changeable. An ontology helps to make explicit the link between relatively stable human language words and relatively volatile data item definitions. As we shall see, using ontologies can have significant impacts on making GIS and spatial data more interoperable.

[3] We consistently use the count nouns—"an ontology" and "ontologies"—to refer to the "explicit specification of human concepts" sense of the word. The abstract, uncountable mass noun—"ontology"—is used to refer to the philosophical study sense of the word.

vocabulary

For example, Figure 9.1 depicts a simple and widely used ontology—or "vocabulary"—for describing locations in terms of spherical coordinates referenced to the WGS84 datum. A *vocabulary* is an informal term for a simple ontology, although there is no strict definition of what counts as "simple" in this context. We will use the term "vocabulary" in this chapter to refer to minimalist ontologies such as depicted in Figure 9.1.

Just as object-oriented classes capture *types* or *kinds* of (object) instances, so too do ontology classes. The vocabulary in Figure 9.1 has just two classes defined: SpatialThing, which has objects with a spatial extent as its instances; and Point which is a (subclass of) SpatialThing. Here the ontology engineering terminology deviates slightly from object-oriented terminology defined in Chapter 4. Classes in ontologies do not possess *behaviors* in the same way as object-oriented classes. However, the "is a" subclass relationship is used exactly as already encountered in the context of inheritance in OO models (Section 4.1.2): a Point "is a" (i.e., specializes a) SpatialThing.

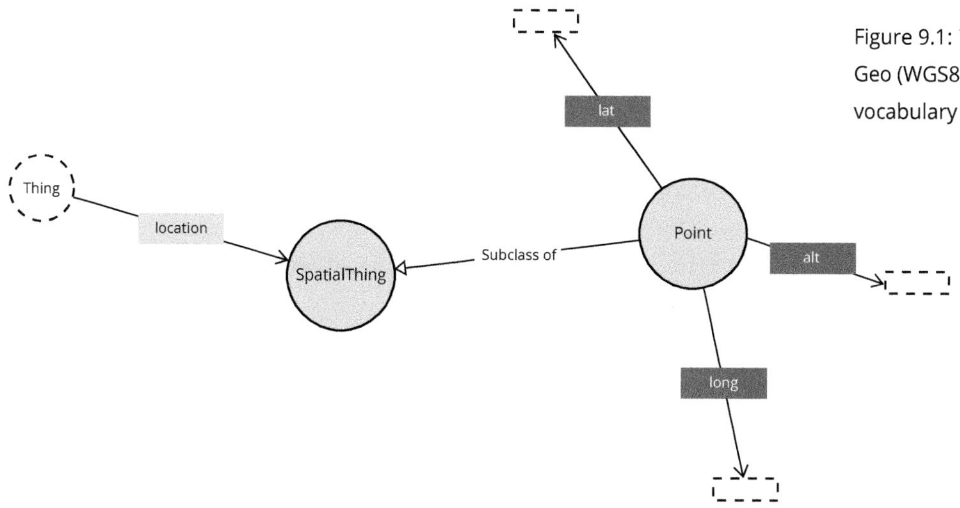

Figure 9.1: The Basic Geo (WGS84 lat/long) vocabulary

Classes may in turn possess *properties* in a similar sense to edge properties in the property graph model and akin to attribute (type) in ER modeling in Chapter 2. In an ontology, a property has the structure of a function in set-based representations (see Section 3.2.3). As for functions, ontology properties have a domain and a range.[4] In the Basic Geo vocabulary in Figure 9.1, the location property has any Thing as its domain (i.e., a location may potentially be a property of anything in an ontology) and SpatialThing as its range. In turn, each point (which is a SpatialThing) may have properties called lat, long, and alt representing the point latitude, longitude, and altitude. These properties have Point as their range (i.e., only points can have coordinates) but, interestingly, no defined range (empty dashed boxes in Figure 9.1).

[4] Recall from Section 3.2.3: the *domain* of a function is the set of things the function maps *from*; the *range* is the set of things the function maps *to*.

The reason for leaving the range undefined is to allow different users of the ontology to define for themselves the precise encoding and data type used for latitude, longitude, and altitude data (including text strings, floating point numbers, or double precision). Properties with no defined range in an ontology are called *annotation properties*, in contrast to *data properties* (which have a data type such as integer or text string as their range) and *object properties* (which have a class as their range, such as location in Figure 9.1).

annotation properties

Ontologies such as the Basic Geo vocabulary in Figure 9.1 can help increase interoperability because they provide a more stable, linguistic basis for describing the *meanings* behind spatial data. Whether I develop a relational database scheme with attributes called "latitude" and "longitude"; a graph database with nodes labeled "coordinate"; or object-oriented software that defines a class of "SpatialPoint" objects, as long as I associate those system entities with the appropriate classes or properties in the Basic Geo vocabulary then everyone should be able to understand what my data means, and link it into other spatial data sets in the right way. Happily, a suite of technologies has been developed to help make it easier to do exactly this—to make connections between data and the ontologies that describe stable, linguistic meanings for that data.

9.1.1 *Representing data using ontologies*

The discussion above introduces some of the key entities in an ontology including classes, subclasses, instances, and properties. In this section we explore briefly how one can use an ontology to represent the meanings behind data. The *Semantic Web* consists of a collection of technologies to link data to rich, machine-readable ontologies representing the meaning of that data.

Semantic Web

One of the most basic building blocks in the Semantic Web is URIs: *uniform resource identifiers*. A URI is a unique identifier for any digital resource. Web URLs (*uniform resource locators*) are a familiar example of URIs used to identify web pages. Indeed, most URIs today use the machinery of Web addresses to locate the resource. The URI of the SpatialThing class in Figure 9.1, for example, is http://www.w3.org/2003/01/geo/wgs84_pos#SpatialThing.

URI

URL

The connections between resources in the Semantic Web are captured using *triples*. Each triple can be thought of as an atomic fact of the form:

triple

subject predicate object

For example, the fact that "The Eiffel Tower is located in Paris" can be represented as a triple: "Eiffel Tower" (subject, "Tour Eiffel" in French), "located in" (predicate), "Paris" (object[5]). Facts can be combined together, linking information from diverse sources. For example, the triple below states a simple fact (triple) from GeoNames[6], an open source database containing myriad facts about millions of toponyms (placenames) from around the world.

[5] Note than the term "object" is used here in the linguistic sense of "the thing being acted upon in a sentence," rather than the object-oriented sense of "state + behavior."

[6] http://geonames.org

```
<https://sws.geonames.org/6254976/> # subject
    <http://www.geonames.org/ontology#name> # predicate
        "Tour Eiffel" . # object
```

The triple above is written in a language called *Turtle*. Simple triples in Turtle use the "subject predicate object" format for facts followed by a terminating full stop. Hence, the triple above asserts that the GeoNames instance 6254976 is named "Tour Eiffel".

Turtle

Rather than rewrite subjects multiple times, Turtle allows multiple predicates to be listed for the same subject by using a semicolon to separate predicates. Thus, the listing below:

```
<https://sws.geonames.org/6254976/> # subject
    <http://www.geonames.org/ontology#name> "Tour Eiffel" ; # predicate object
    <http://www.w3.org/2003/01/geo/wgs84_pos#lat> "48.85832" ; # predicate object
    <http://www.w3.org/2003/01/geo/wgs84_pos#long> "2.29452" . # predicate object
```

asserts the fact that the Eiffel Tower (or more precisely GeoNames instance 6254976 with name "Tour Eiffel") is located at latitude "48.85832" and longitude "2.29452". Note that the Basic Geo (WGS84 lat/long) vocabulary (introduced in Figure 9.1) is used to ensure that the semantics of the location coordinates are clear.

Further, commas can be used to separate multiple objects for the same predicate. For example, the fact that the Eiffel Tower has different official names in English, German, and French can be asserted using the statement:

```
<https://sws.geonames.org/6254976/> # subject
    <http://www.geonames.org/ontology#name> "Tour Eiffel" ; # predicate object
    <http://www.w3.org/2003/01/geo/wgs84_pos#lat> "48.85832" ; # predicate object
    <http://www.w3.org/2003/01/geo/wgs84_pos#long> "2.29452" ; # predicate object
    <http://www.geonames.org/ontology#officialName> # predicate
        "Eiffel Tower"@en, "Eiffelturm"@de, "Tour Eiffel"@fr . # object object object
```

The Turtle listing in Figure 9.2 below shows 11 of more than 70 facts about the Eiffel Tower in the GeoNames database. The listing contains additional information about a location map and further semantic information about the Eiffel Tower stored on the public data catalog DBpedia[7]. The first listed predicate (simply a) is shorthand for asserting the type of the subject. In this case, the GeoNames instance 6254976 is an instance of type Feature. To shorten the Turtle listings still further, the first three lines of the listing in Figure 9.2 define three *prefixes* (including gn for the GeoNames ontology and geo for the Basic Geo ontology).

[7] DBpedia (https://www.dbpedia.org/) is a Semantic Web database of information from Wikipedia, the crowdsourced encyclopedia.

```
@prefix gn: <http://www.geonames.org/ontology#> . # GeoNames prefix
@prefix geo: <http://www.w3.org/2003/01/geo/wgs84_pos#> . # WGS84 Basic Geo prefix
@prefix rdfs: <http://www.w3.org/2000/01/rdf-schema#> . # RDFS prefix

<https://sws.geonames.org/6254976/>
    a <http://www.geonames.org/ontology#Feature> ;
    gn:name "Tour Eiffel" ;
    gn:officialName "Eiffel Tower"@en, "Eiffelturm"@de, "Tour Eiffel"@fr ;
    gn:countryCode "FR" ;
    geo:lat "48.85832" ;
    geo:long "2.29452" ;
    gn:locationMap <https://www.geonames.org/6254976/tour-eiffel.html> ;
    rdfs:seeAlso <https://dbpedia.org/resource/Eiffel_Tower> .
```

Figure 9.2: Representation of the Eiffel Tower on GeoNames in Turtle

Eagle-eyed readers may notice that the Turtle in listing Figure 9.2 includes a new third prefix: rdfs. RDFS is a standard ontology engineering vocabulary called "RDF schema," itself based on RDF: the "resource description framework." RDF describes the fundamental data model for triples that underpins the Semantic Web. Indeed, the language Turtle is also based on RDF (Turtle stands for "terse RDF triple language"). Figure 9.3 encodes exactly the same information about the Eiffel Tower as the listing in Figure 9.2 but using RDF/XML format. The RDF/XML format is not as human-readable as Turtle, but it is often used in the Semantic Web because XML is a general purpose document format understood by many different computer applications.

RDF

Figure 9.3: RDF/XML representation of the Eiffel Tower on GeoNames, identical in content to the Turtle representation in Figure 9.2

```
<rdf:RDF xmlns:gn="http://www.geonames.org/ontology#" # GeoNames prefix
      xmlns:geo="http://www.w3.org/2003/01/geo/wgs84_pos#" # WGS84 prefix
      xmlns:rdf="http://www.w3.org/1999/02/22-rdf-syntax-ns#" # RDF prefix
      xmlns:rdfs="http://www.w3.org/2000/01/rdf-schema#"> # RDFS prefix

  <rdf:Description rdf:about="https://sws.geonames.org/6254976/">
      <rdf:type rdf:resource="http://www.geonames.org/ontology#Feature"/>
      <gn:name>Tour Eiffel</gn:name>
      <gn:officialName xml:lang="en">Eiffel Tower</gn:officialName>
      <gn:officialName xml:lang="de">Eiffelturm</gn:officialName>
      <gn:officialName xml:lang="fr">Tour Eiffel</gn:officialName>
      <gn:countryCode>FR</gn:countryCode>
      <geo:lat>48.85832</geo:lat>
      <geo:long>2.29452</geo:long>
      <gn:locationMap
          rdf:resource="https://www.geonames.org/6254976/tour-eiffel.html"/>
      <rdfs:seeAlso rdf:resource="https://dbpedia.org/resource/Eiffel_Tower"/>
  </rdf:Description>
</rdf:RDF>
```

The Semantic Web is designed to be agnostic to specific encodings, so several such different languages are commonly used in connection with ontologies. Ultimately, however, all are founded on subject-predicate-object triples, and so they can be represented using graphs, often termed *knowledge graphs*, such as that pictured in Figure 9.4.

knowledge graph

9.1.2 *Querying ontologies*

Section 2.4 introduced the graph database and explored in more detail the property graph model. Although property graph databases can also be used to store RDF data, the addition of node and edge properties in the property graph model is not required for storing triples. Instead, RDF data in the Semantic Web is usually stored in the second major class of graph database: the triple store. A *triple store* is a database specifically designed to store and retrieve only RDF subject-predicate-object triples.

triple store

As we saw in Section 2.4, there is currently no single standard for querying property graph databases, with languages such as Cypher among the leading influences on an emerging future standard. Happily, though, triple stores do enjoy a standard query language: SPARQL.[8] SPARQL shares many superficial similarities with the standard relational database query language SQL. However, in detail SPARQL is more akin to other graph query languages such as Cypher, as might be expected.

SPARQL

[8] SPARQL is an acronym for "SPARQL protocol and RDF query language"—confusingly, "SPARQL" is an abbreviated word in its own acronym.

Echoing SQL, SPARQL queries are constructed around SELECT ... WHERE statements. Unlike SQL queries, the SELECT clause is followed by one or more variables that will appear in the query results. The WHERE clause then lists patterns to match against the stored graph, much as is found in other graph

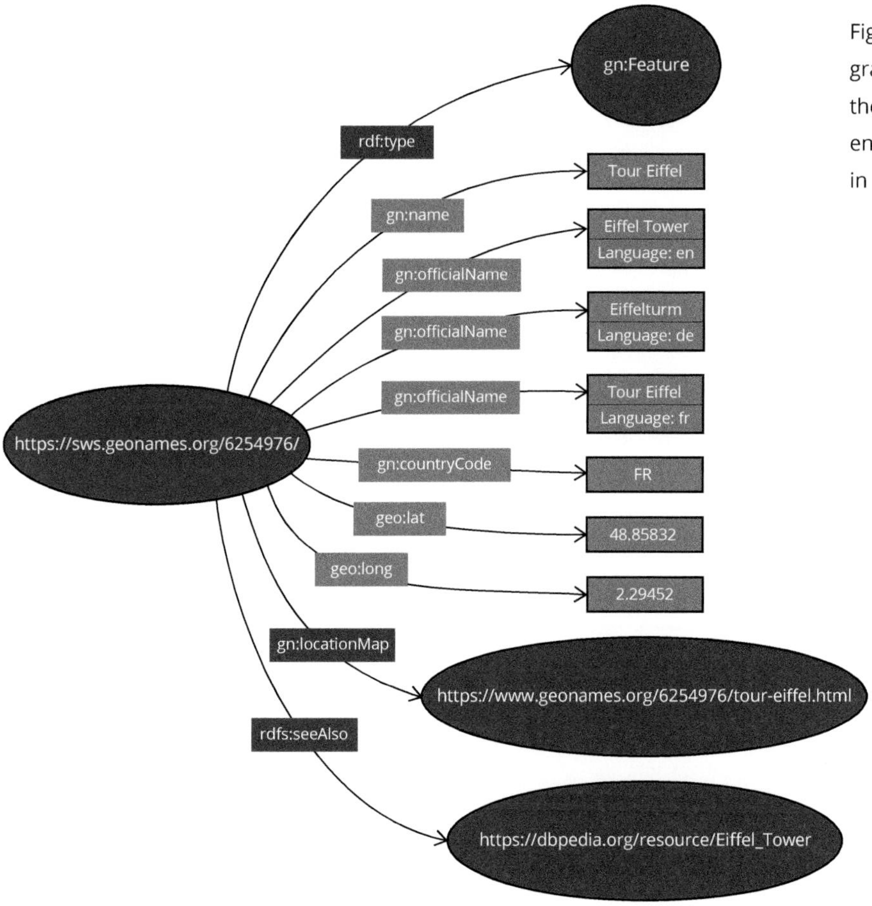

Figure 9.4: The knowledge graph representation of the Eiffel Tower triple data encoded in Turtle and RDF in Figures 9.2 and 9.3

query languages such as Cypher. For example, the SPARQL query in Figure 9.5 retrieves the latitude and longitude of the Eiffel Tower from the graph represented in Figure 9.4 and encoded in Turtle and RDF/XML in Figures 9.2 and 9.3, respectively.

In the query in Figure 9.5, following the definition of the gn and geo prefixes, the SELECT clause specifies two variables whose values will appear in the query output: ?latitude and ?longitude. Variables in SPARQL must always be denoted using the leading ? symbol. In the WHERE clause, three matches are defined. First, the query searches for matches with some variable ?x with gn:name as "Eiffel Tower". The terminating full stop indicates the end of a triple pattern, similarly to the Turtle encoding. Next, two further query matches look for the same subject ?x, but storing the object linked by predicate geo:lat in the variable ?latitude and the object linked by predicate geo:long in the variable ?longitude. The final output of the SPARQL query in Figure 9.5 is given in Table 9.1.

?latitude	?longitude
"48.85832"	"2.29452"

Table 9.1: Output of SPARQL query in Figure 9.5

Figure 9.5: SPARQL query to retrieve the latitude and longitude of the Eiffel Tower

```
PREFIX gn: <http://www.geonames.org/ontology#> # GeoNames prefix
PREFIX geo: <http://www.w3.org/2003/01/geo/wgs84_pos#> # WGS84 Basic Geo prefix

SELECT ?latitude ?longitude WHERE {
    ?x gn:name "Tour Eiffel" . # GeoNames instance named "Tour Eiffel"
    ?x geo:lat ?latitude . # Retrieve latitude
    ?x geo:long ?longitude . # Retrieve longitude
}
```

With the breadth and depth of the Semantic Web growing daily, SPARQL brings the power to query vast networks of spatial knowledge. For example, using the ontology to bridge directly from the GeoNames database to DBpedia, Figure 9.6 gives a SPARQL query to search for other buildings in Paris and sort them by completion date (listed in Table 9.2).

Figure 9.6: SPARQL query to retrieve GeoNames buildings in Paris by construction date (see Table 9.2 on the next page)

```
PREFIX gn: <http://www.geonames.org/ontology#> # GeoNames prefix
PREFIX geo: <http://www.w3.org/2003/01/geo/wgs84_pos#> # WGS84 Basic Geo prefix
PREFIX rdf: <http://www.w3.org/1999/02/22-rdf-syntax-ns#> # RDF prefix
PREFIX dbp: <http://dbpedia.org/property/> # DBpedia property prefix
PREFIX dbo: <http://dbpedia.org/ontology/> # DBpedia ontology prefix
PREFIX dbr: <http://dbpedia.org/resource/> # DBpedia resource prefix

SELECT DISTINCT ?name ?completed ?latitude ?longitude WHERE {
    ?x gn:name ?name . # Retrieve GeoNames name
    ?x geo:lat ?latitude . # Retrieve WGS84 latitude
    ?x geo:long ?longitude . # Retrieve WGS84 longitude
    ?x dbo:location dbr:Paris . # Retrieve location in Paris
    ?x rdf:type dbo:Building . # Retrieve instances of Buildings
    ?x dbp:completionDate ?completed # Retrieve completion dates of Buildings
}
ORDER BY DESC(?completed) # In order of completion date, most recent first
```

Note, however, that the SPARQL query Figure 9.6 does not perform a spatial operation when satisfying the search criterion "in Paris." Instead, the query uses the fact that the DBpedia database includes a predicate dbo:location and the object dbr:Paris. Just like our basic SQL queries, SPARQL does not include spatial operators, although attempts have been made to extend SPARQL with such spatial operations (see Box 9.2 on the facing page).

9.1.3 *Reasoning with ontologies*

To this point, our discussion of ontologies has only addressed the *representation* of human knowledge, with subject-predicate-object triples as the foundation of that representation. An essential ingredient in any KR^2 AI system is the ability also to *reason* with these representations.

Box 9.2: GeoSPARQL

GeoSPARQL is an emerging standard for representing geospatial data on the Semantic Web as well as an extension to SPARQL to enable spatial queries to be constructed. The representation of geospatial data in GeoSPARQL is based on the *simple feature model* already introduced in Section 4.1.3 and summarized in Figure 4.14. The spatial operations available in GeoSPARQL queries include basic spatial filters such as: within-distance queries for retrieving objects within a certain distance of a point; spatial point and range queries; and querying topological relations based on an extension to the 4-intersection model (4IM) already discussed in Chapter 3. The leading triple store databases also offer basic spatial indexes, such as geohashing introduced in Box 6.3 on page 238. However, today's spatial query capabilities for the Semantic Web are overall relatively basic.

?name	?completed	?latitude	?longitude
Tour Eiffel	1889	48.85832	2.29452
Bastille Opera	1990	48.85272	2.36999
Hyatt Regency Paris Etoile	1974	48.88070	2.2845
Tour Montparnasse	1973	48.84193	2.32207
Centre Georges Pompidou	1971	48.86060	2.35237
Hôtel de Ville de Paris	1892	48.85644	2.35244
Arc de Triomphe du Carrousel	1808	48.86173	2.33291
Panthéon	1790	48.84624	2.34613
Élysée Palace	1722	48.87060	2.31709
Palais du Luxembourg	1645	48.84844	2.33726
Hôtel de Ville de Paris	1357	48.85644	2.35244
...	...	...	...

Table 9.2: Output of SPARQL query in Figure 9.6 on the facing page

In fact, behind the scenes reasoning was woven in to the Semantic Web technologies we have already encountered. The SPARQL query in Figure 9.5 can satisfy its query simply by retrieving the required stored data from GeoNames, such as that listed in Figure 9.2 and Figure 9.3. The query in Figure 9.6, however, went much further linking seamlessly to a completely different data source, DBpedia. How did it do that?

The answer lies in the DBpedia knowledge graph. The worldwide network of structured, open access, Semantic Web data (including databases such as DBpedia and GeoNames) is termed *linked open data* (LOD). DBpedia contains over 3 billion such facts, with almost 400 facts about the Eiffel Tower alone, including in particular the (Turtle) fact:

linked open data

```
@prefix owl: <http://www.w3.org/2002/07/owl#> # OWL prefix
@prefix dbr: <http://dbpedia.org/resource/> # DBpedia resource prefix

dbr:Eiffel_Tower owl:sameAs <https://sws.geonames.org/6254976/>
```

OWL (Web ontology language) is an extension of RDF that additionally enables the description of logical relationships and constraints between concepts as triples. In the listing above, the OWL predicate `own:sameAs` asserts that the DBpedia Eiffel Tower instance `dbr:Eiffel_Tower` is one and the same as the GeoNames Tour Eiffel instance `<https://sws.geonames.org/6254976/>`. This simple but critical connection enables a SPARQL query engine to begin to draw inferences that bridge across and span the breadth of knowledge graphs.

OWL

To understand better how we reason with ontologies, and indeed across a swathe of KR^2 systems, we need first to delve into a little classical logic and deduction.

syntax

semantics

denotational semantics

Logic and deduction An important logical distinction is made between *syntax* and *semantics*. Consider the sentence, "The Eiffel Tower is in Paris." A *syntactic* analysis would focus on the terms of the sentence: the sentence contains two proper nouns (starting with uppercase letters), a verb, and a spatial preposition. The *semantics*, or meaning, of the sentence can only be determined by its context. If "Eiffel Tower" is the name of the iconic French landmark, and "Paris" refers to France's capital city, then the sentence has a clear meaning and is in fact true. However, if "Paris" refers to the small town of just over 1000 inhabitants in Monroe County, Missouri in the US, again the semantics are clear although this time the sentence is false. In this *denotational* view of semantics, the meaning of a term is determined by reference to the domain entity or relationship it denotes. The meaning and possibly the truth value of a complex proposition are determined by the meaning and structuring of its terms.

Consider the three sentences:

The Eiffel Tower is a building in Paris.
All buildings in Paris are French.
———
The Eiffel Tower is French.

conclusion

premise

The proposition "The Eiffel Tower is French" below the line (the *conclusion*) follows from the sentences above the line (the *premises*) by *deduction*. We can note that the process is quite general, and independent of the meanings of the individual terms. The general form of the deduction above is:

x is a *y*.
All *y*s are *z*s.
———
x is a *z*.

syllogism

deductive validity

This particular deduction is an example of a classical *syllogism*, whose principles were codified by Aristotle in around 350 BC. The deduction is said to be *valid* in the sense that it preserves truth: if the premises of a valid deduction are true, then the conclusion is also expected to be true. Note that not all forms of reasoning are deductively valid, however (see Box 9.3 on page 361).

In general our logical premises may be constructed from at least five different building blocks:

Constants which denote specific individuals, such as "Paris" or "France";
Variables which stand as placeholders for individuals, such as x, y, and z;
Connectives such as $\neg$ (not), $\wedge$ (and), $\vee$ (or), $\Longrightarrow$ (implies), and $\Longleftrightarrow$ (equivalence, if and only if);
Quantifiers namely $\forall$ ("for all ...") and $\exists$ ("there exists ..."); and

Predicates such as "Building(x)" (x is a building) or "In(y,z)" (y is located in z), which describe the properties of, or relationships between entities (as already encountered in the context of triples).

Together these building blocks can be combined into *sentences* as the basis of *first order predicate logic*. We can see them in action in our deductive inference rewritten formally as:

sentence

first order predicate logic

$$\begin{array}{l}\text{Building(Eiffel Tower)} \wedge \text{In(Eiffel Tower, Paris)}\\ \forall x\ \text{Building}(x) \wedge \text{In}(x, \text{Paris}) \implies \text{French}(x)\\ \hline \text{French (Eiffel Tower)}\end{array}$$

The collection of RDF triples that populate our knowledge graphs constitute a *knowledge base* of simple logical sentences. Using OWL, that knowledge base can be augmented with richer logical rules, such as the OWL owl:sameAs predicate discussed above. Other OWL and RDF predicates, such as owl:disjointWith, owl:ReflexiveProperty, and rdfs:subClassOf likewise lead directly to additional logical sentences in the knowledge base. Deductive reasoning can then be used to derive answers to questions that are not explicitly stored in our knowledge base. Deduction is a purely syntactic process and is therefore amenable to computing and automated reasoning. In fact, automated deductive reasoning capability is integrated into most SPARQL query engines. As a result, reasoning about queries usually happens "on the fly," as a query is executed.

knowledge base

For example, the following query asks the knowledge base "Is the Eiffel Tower a person?" In SPARQL, ASK queries are similar to SELECT queries, but only return "true" or "false" answers.

```
PREFIX gn: <http://www.geonames.org/ontology#> # GeoNames prefix
PREFIX rdf: <http://www.w3.org/1999/02/22-rdf-syntax-ns#> # RDF prefix
PREFIX dbo: <http://dbpedia.org/ontology/> # DBpedia ontology prefix

ASK {?x gn:name "Tour Eiffel"; # Retrieve the GeoNames "Tour Eiffel" instance
        rdf:type dbo:Person . # Is "Tour Eiffel" a dbo:Person?
}
```

The query above evaluates to "false," even though this information is nowhere explicitly stored in our knowledge base. Instead, the query follows a chain of deductive reasoning to arrive at the answer, intuitively as follows:

```
# GeoNames instance 6254976 is named Tour Eiffel
<https://sws.geonames.org/6254976/> gn:name "Tour Eiffel"

# DBpedia Eiffel Tower is the same instance as GeoNames 6254976
dbr:Eiffel_Tower owl:sameAs <https://sws.geonames.org/6254976/>

# DBpedia Eiffel Tower is also an instance of the class of buildings
dbr:Eiffel_Tower rdf:type dbo:Building

# Buildings are a subclass of architectural structures
```

```
dbo:Building rdfs:subClassOf dbo:ArchitecturalStructure

# Architectural structures are never people
dbo:ArchitecturalStructure owl:disjointWith dbo:Person
```

In contrast, if we ask SPARQL the question, "Is the architect of the Eiffel Tower a person?" then we can deduce the affirmative with the query below:

```
PREFIX gn: <http://www.geonames.org/ontology#> # GeoNames prefix
PREFIX rdf: <http://www.w3.org/1999/02/22-rdf-syntax-ns#> # RDF prefix
PREFIX dbo: <http://dbpedia.org/ontology/> # DBpedia ontology prefix

ASK {?x gn:name "Tour Eiffel" ; # Retrieve the GeoNames "Tour Eiffel" instance
        dbo:architect ?y . # Retrieve the "Tour Eiffel" architect
    ?y rdf:type dbo:Person . # Is the "Tour Eiffel" architect a dbo:Person?
}
```

The mechanics of how a SPARQL reasoning engine is able to perform this "on-the-fly" reasoning—required to satisfy queries like those above—is beyond the scope of this book. One may imagine, though, that the number of possible deductions from even a modest set of premises may be very large. The ability to automatically reason with the billions of facts in LOD (linked open data) today represents a major achievement in AI, and indeed in human history. However, first-order logic is sufficiently expressive as to make it provably impossible to define a general automated procedure to answer all questions we might conceivably ask of our first order predicate knowledge base.[9] SPARQL reasoning engines deal with this issue by restricting the scope of automated logical reasoning to carefully constructed "fragments" of first order predicate logic that can be efficiently answered by machines. Thus, it is always possible to construct queries about our knowledge bases that cannot be answered. However, in practice today's Semantic Web technology can be very effective in reasoning about LOD.

[9] Indeed, Alan Turing's work, also discussed in Box 9.1 on page 350, has played an important role in proving this limitation and more generally in understanding the limits of automated logical reasoning.

9.1.4 *Principles of ontology engineering*

Before moving on to explore more logical reasoning with spatial data, we dwell briefly on the question of what makes a *good* ontology, and how to design one. There are many situations where it is useful to be able to develop an ontology for a specific domain. The Semantic Web is a vast and rapidly expanding knowledge base. In many cases, adding new facts to this global knowledge base is as simple as describing your data using the terms and concepts already defined in existing ontologies. However, in other cases, the required terms and concepts may not yet exist and may necessitate new terms and concepts to be defined.

In general we can identify at least five desirable properties of "good" ontologies:

Box 9.3: Induction and abduction

We have seen the deductive reasoning process "preserves truth" and is said to be (deductively) valid as a consequence. Deductive validity is the main reason why deduction is so important and underpins most automated reasoning today. However, there are two other forms of inference that are not guaranteed to preserve truth but are nevertheless important in automated and in particular human reasoning. *Induction* is the reasoning process of inferring general rules from evidence. Suppose we have as before the premise that "The Eiffel Tower is a building in Paris" and now add as a premise that "The Eiffel Tower is French." An example of induction would be to use those premises as evidence to support the inference that "All buildings in Paris are French." Of course, more evidence beyond a single building—perhaps adding the Arc de Triomphe and the Pompidou Center—may in turn give us more confidence that our inductive inference is correct. However, in most cases no matter how much evidence we add, we can never be *certain* that the induced conclusion is valid. There is always the possibility that a counterexample lurks just around the corner (such as the American Embassy in Paris, which like most embassies has extraterritorial status and consequently is not French). *Abduction* is the process of selecting the most likely explanation from the available premises. For example, if we know that "The Eiffel Tower is French" and "All buildings in Paris are French" we might adduce that "The Eiffel Tower is a building in Paris." Once again, such inference does not preserve truth—consider, for example, the same abductive reasoning using the premises "Marie Antoinette is French" and "All buildings in Paris are French". However, inductive and abductive reasoning are used to great effect almost continuously by humans. So advancing the ability of computers to achieve similar feats is one of the long-term goals of AI.

Congruity The most important principle for any ontology is to *reuse* existing vocabularies and concepts: good ontologies aid in linking data together by building upon established terms, definitions, and concepts.

Modularity Complementing congruity and reuse, good ontologies are *modular* in the sense that they describe carefully scoped and bounded domains in a way that is accessible and understandable by others, not only by the individuals or expert communities who developed them.

Consistency Good ontologies are *consistent*, avoiding circularity in definition and internal contradictions, and helping users to classify instances unambiguously.

Clarity Ontologies should avoid basic *errors* in definition, such as failing to distinguish between entities versus their identifiers (cf. Section 2.2.2); objects in the world versus the digital artifacts that refer to them; and things in the world versus our knowledge of those things.

Openness Ontology engineering is fundamentally about describing the semantics of data in a way that promotes reasoning about linked data. Hence, making this data *openly* available is fundamental to the vision of linked open data (LOD) and the Semantic Web.

These principles can help designers in deciding between different options when constructing ontologies and LOD. Figure 9.7 contains a simple example of using linked data to represent a famous meeting in history between nursing pioneers and polymaths Mary Seacole and Florence Nightingale. These contemporaries, whose lives chart many interesting parallels as well as divergent experiences, are known to have met only once, briefly in Üsküdar in Turkey during the Crimea War on 8 March 1855. In constructing an ontology of the meeting, the most important step is to reuse established ontologies. The sim-

[10] http://semanticweb.cs.vu.nl/2009/11/sem/

upper level ontology

ple event model[10] (SEM, prefix sem) is a leading ontology for describing the general structure and vocabulary of events (Hage & Ceolin, 2013). Ontologies such as SEM that are designed to be used across many different domains are called *upper level ontologies*. The SEM is built around four key classes: Place, Time, Actor that describe the where, what, and who of an Event. The knowledge graph in Figure 9.7 contains a new class, Meeting, as a subclass of Event (a meeting "is an" event). As in previous examples, the knowledge graph also links to the GeoNames (prefix gn) and Basic Geo (prefix geo) ontologies to describe the meeting location (SEM Place), Üsküdar, and DBpedia (prefix dbr) to identify the historic figures: Mary Seacole and Florence Nightingale.

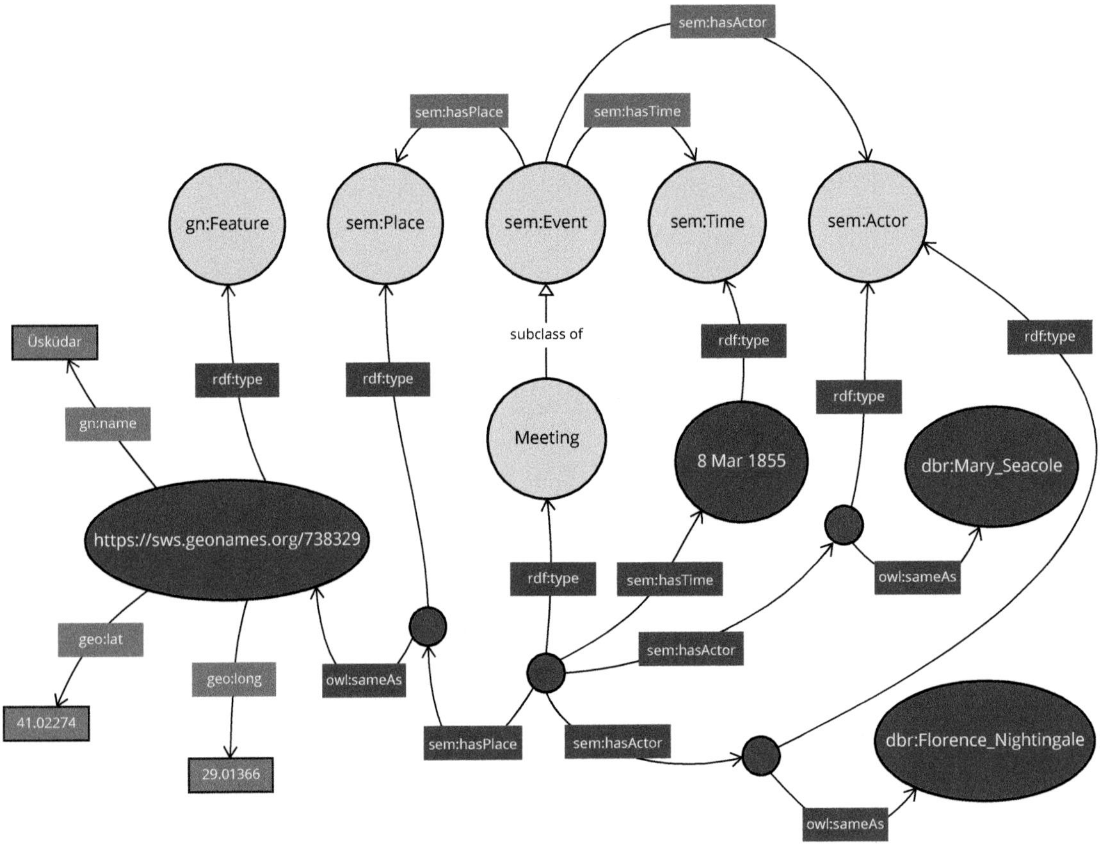

Figure 9.7: A knowledge graph representation of the meeting between Mary Seacole and Florence Nightingale on 8 March 1855 at Üsküdar (Scutari) in Turkey

When building ontologies for specific domains, such as historic meetings, it is generally advised to follow a structured process, such as below, after (Noy & McGuinness, 2001):

1. *Reuse*: Every ontology engineering project should begin by identifying the existing upper-level and domain ontologies relevant to the domain.
2. *Enumerate*: Next, enumerate the terms relevant to the domain, taking special care to be consistent in usage, for example, alert to synonyms and homonyms.

3. *Structure*: In step three, structure the terms into a hierarchy, taking care to ensure each instance of a subclass *A* "is an" instance of superclass *B*.
4. *Relate*: With the structure taking shape, define the properties of each class and the relationships to literals and other classes.
5. *Iterate*: Ontology engineering is an iterative process of revision and refinement, which benefits from a diverse team of developers that includes subject matter experts, ontology engineers, and generalists.

The textbook by Kendall & McGuinness (2019) is highly recommended to those readers interested to delve deeper than would be possible in this book's whistle-stop tour of ontology engineering.

9.2 Qualitative spatial reasoning

A distinction is often made in reasoning between *quantitative* and *qualitative* approaches. An approach is generally referred to as *quantitative* if it is based on analysis of numerical (interval or ratio) data. By contrast, an approach is usually referred to as *qualitative* if it is based on analysis of classifications and ordering (nominal and ordinal data). Hence, qualitative reasoning is concerned with the discrete, imprecise, and non-numerical properties of space and time.

quantitative

qualitative

There are in general three main reasons to be interested in qualitative representation and reasoning with spatial and temporal data:

1. Qualitative properties are *simple*, usually involving small, discrete representations that are inherently tolerant to imprecise human knowledge. For example, navigating instructions are much more understandable using "left" or "right" directions rather than precise bearings or angles (e.g., 276°).
2. Qualitative properties *supervene* on quantitative properties, in the sense that qualities can always be derived from quantities. For example, a bearing of 071° can be represented qualitatively as "East" or to the "right."
3. The boundaries between qualities usually correspond to salient discontinuities for humans. For example, "left" and "right" are intrinsically connected with our embodied physical experience of the world.

supervenience

Qualitative spatiotemporal reasoning (QSR) is an area of study within KR^2 that has developed, over the past 40 years or more, a variety of logical systems for representing and reasoning about qualitative spatial and temporal objects and relations.

qualitative spatial reasoning

9.2.1 Point algebra

The *point algebra* of Vilain & Kautz (1986) is certainly the simplest and arguably the most elegant of all QSR logics. The point algebra describes three qualitative relations between two points in time (Figure 9.8). We write

point algebra

before(A, B) if A comes before B in time; *after*(A, B) if A comes after B; and *same*(A, B) if A and B are the same point in time. In common with most qualitative representations, the different relations are chosen so as to be *jointly exhaustive pairwise disjoint* (JEPD): one and only one of these relations will necessarily hold between any two points in time.

jointly exhaustive pairwise disjoint

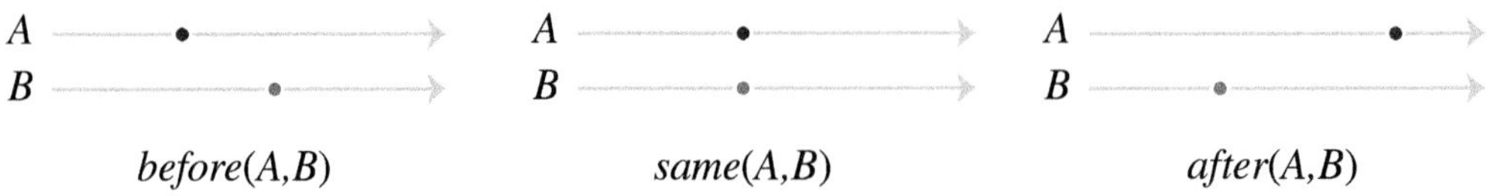

Figure 9.8: The three relations in the point algebra of Vilain & Kautz (1986)

As we will by now be expecting in KR^2, representation is only half the story. The aim of the representation is to enable logical reasoning over qualitative relations. In the point algebra, we may immediately notice that for any three time points A, B, and C if *after*(A, B) and *after*(B, C) then we can deduce that *after*(A, C). We might also notice some other inferences, such as if *after*(A, B) then *before*(B, A). Unfortunately, not all combinations of relations help drive inference. In particular, *after*(A, B) and *before*(B, C) then any of *after*(A, C), *before*(A, C), or *same*(A, C) are possible.

We can capture all such logical inferences in a single table, shown in Table 9.3. For example, if *before*(A, B) (first row of data in the table) then we can deduce the *converse*, that *after*(B, A) from the entry in row 1 column 5 in Table 9.3. If additionally we know that *before*(B, C), then we can infer that *before*(A, C) from row 1 column 2. In general, for three objects X, Y, and Z, the inference $R(X, Y) \wedge P(Y, Z) \implies Q(X, Z)$ is called the *composition* of relations R, P, and Q.

converse

composition

Composition	*before*	*after*	*same*	Converse
before	*before*	*before, after, same*	*before*	*after*
after	*before, after, same*	*after*	*after*	*before*
same	*before*	*after*	*same*	*same*

Table 9.3: Composition and converse for the point algebra

From these foundations, it is possible to construct and reason about more complex configurations of points. For example, Figure 9.9 shows a network of point algebra relations between five points, $A \ldots E$. Based on the supplied relations in Figure 9.9 (solid arrows) it is possible to deduce the other, unknown relations between points, including:

- *after*(B, A), the converse of the supplied *before*(A, B);
- *before*(A, E), the composition of *before*(A, B) and *before*(B, E); and
- *after*(D, A), the composition of *after*(D, C) and inferred converses *after*(C, B) and *after*(B, A).

Some relations though may still be undetermined, such as the relation between E and C in Figure 9.9, which may be any of *before*, *after*, or *same*.

The network of qualitative relations for a specific configuration of objects is called the *constraint network*. The task of automatically reasoning with such

constraint network

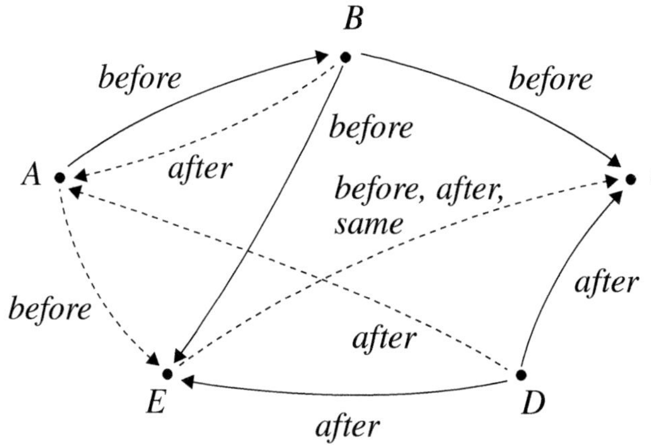

Figure 9.9: Partial constraint network for a point algebra configuration

qualitative relations can be reduced to the question of whether a set of given relations (constraints) is consistent with one another, termed the *constraint satisfaction problem* (CSP), and is computable in $O(n^3)$ time complexity for the point algebra.

constraint satisfaction problem

9.2.2 *Interval algebras*

A natural step following the point algebra is to consider the relations between time *intervals*, rather than points. Allen's *interval algebra* (1983, 1984) defines 13 JEPD relations between two time intervals (Figure 9.10).

interval algebra

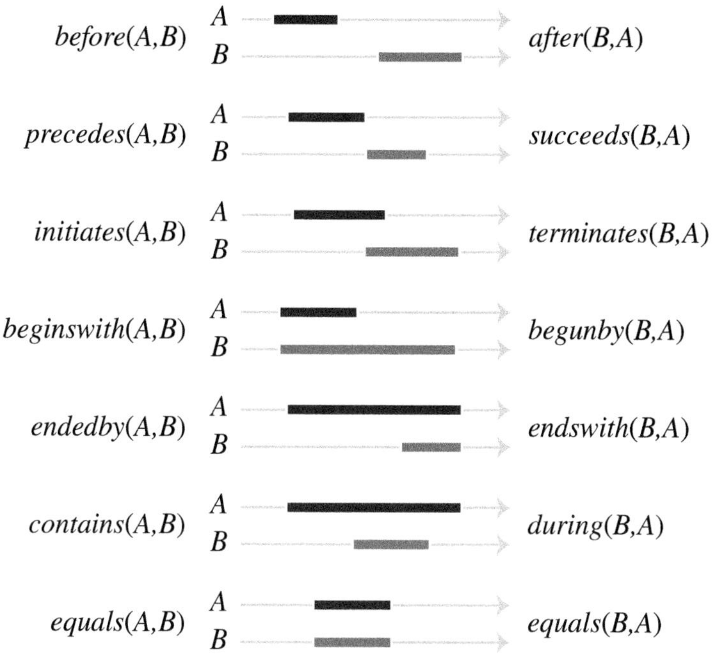

Figure 9.10: The 13 relations in the interval algebra of Allen (1983)

Although Allen's interval algebra defines many more relations than the point algebra, reasoning through constraint networks using composition and converse can proceed in much the same way as for the point algebra.

Figure 9.11 illustrates an example configuration with its associated constraint network. Note that the relationship between B and D is under-determined: it must be one of *before*(D,B), *precedes*(D,B), or *initiates*(D,B), but it is not possible to say which given the available information.

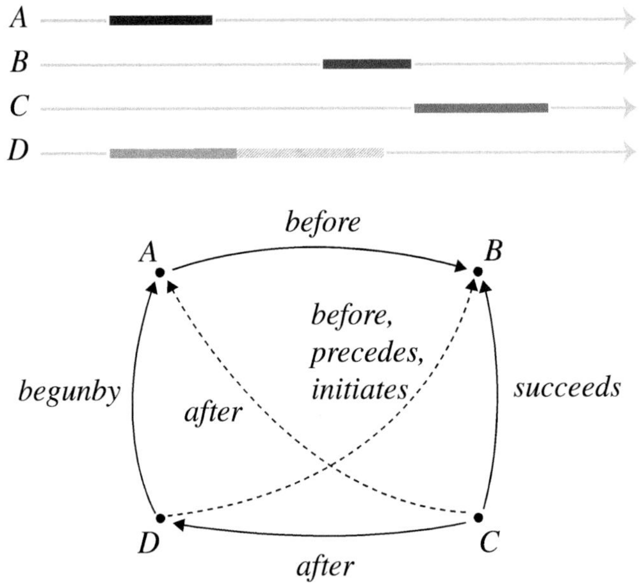

Figure 9.11: Constraint network for an interval configuration

The interval algebra can also be applied directly in two dimensions to give a set of JEPD region relations between two axis-parallel rectangles. The result, called the *rectangle algebra*, defines $13 \times 13 = 169$ relations as combined pairs of interval algebra relations. For example, the configuration in Figure 9.12 may be described as *initiates–contains*(A,B) (or conversely *terminates–during*(B,A)).

rectangle algebra

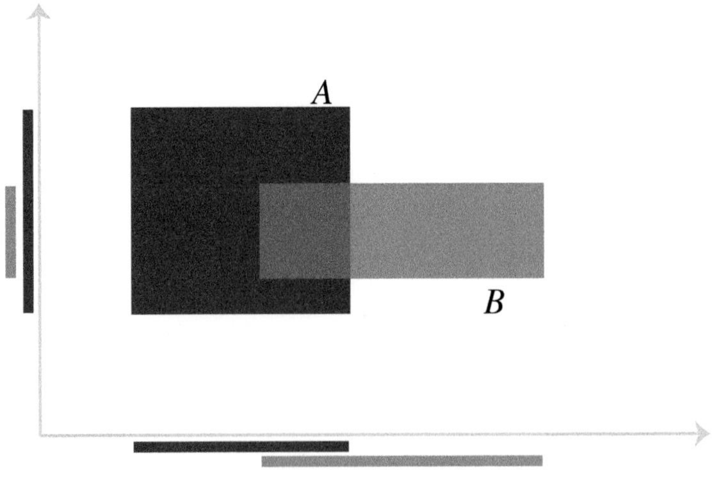

Figure 9.12: Example configuration of two regions in the rectangle algebra

Intuitively, such reasoning proceeds as for the point algebra, above. However, in fact the constraint satisfaction problem is not tractable for the full interval and rectangle algebras (nor indeed for the other full qualitative spatiotemporal logics below). A mathematical solution to this problem is to identify restricted subsets of relation sets that lead to logics that are again tractable. However, each qualitative spatiotemporal logic demands its own unique subsets, and the process of identification can present significant challenges to mathematicians. In practice, reasoning with the full logics usually proceeds by computing a "local" constraint satisfaction for all possible sets of three objects instead of seeking the "global" CSP solution. This heuristic approach, termed *path consistency*, is tractable and can be computed in $O(n^3)$ time. Unfortunately, the heuristic is not guaranteed to be correct in all cases.

path consistency

9.2.3 Region connection calculus

Perhaps the most famous of all qualitative spatial logics is RCC, the *region connection calculus*. In the context of RCC, the word "region" may be thought to include any definition of a spatial region, including both continuous areas and discretized representations, such as a polygon or set of pixels in a raster.[11]

region connection calculus

[11] In fact, regions in RCC can have much wider, and possibly non-spatial interpretations, including categories in conceptual spaces (Gärdenfors & Rott, 1995).

RCC is founded on *Clarke's calculus of individuals* which begins with a binary connection relation between regions. We write $C(X, Y)$ as shorthand for "region X is connected to region Y." The connection relation is reflexive and symmetric, satisfying the following axioms:

Clarke's calculus of individuals

1. For each region X, $C(X,X)$
2. For each pair of regions X,Y, if $C(X, Y)$ then $C(Y,X)$

The surprising part of Clarke's calculus is the next step. It turns out that many of the set-oriented and topological relations between spatial objects we encountered in Chapter 3 may be derived using only the minimal machinery above. Table 9.4 shows the construction of various RCC region relations based solely on the connection relation. The final column also highlights the correspondence to the eight 4-intersection model (4IM) relations already summarized in Table 3.3.

The eight RCC relations with corresponding 4IM relations (DC, EQ, PO, EC, TPP, $TPPi$, $NTTP$, $NTTPi$) form a JEPD set of relations, called RCC-8. Interestingly, there is another JEPD subset of five RCC relations, called RCC-5: EQ, PO, PP, PPi, and DR. Both RCC-8 and RCC-5 are widely used in spatial reasoning. The RCC-5 relations provide less fine-grained distinctions at the region boundaries (DR encompasses both DC and EC; PP encompasses TPP and $NTTP$; PPi encompasses $TPPi$ and $NTTPi$).

Reasoning in RCC is based on composition and converse, shown in Table 9.5, exactly as for all the spatial logics we encounter. Hence, from $PPi(X, Y)$ and $PO(Y, Z)$ we can deduce either $PO(X, Z)$ or $PPi(X, Z)$ holds, for example. Neither RCC-5 nor RCC-8 are tractable in general, and so the path consistency heuristic applied to the constraint network is commonly used in practice.

rowcolorLightestPrimaryRelation	Definition	Description	4IM relation
$DC(X,Y)$	$\neg C(X,Y)$	X and Y are disconnected	X disjoint Y
$P(X,Y)$	$\forall Z.C(Z,X) \implies C(Z,Y)$	X is a part of Y	
$O(X,Y)$	$\exists Z.P(Z,X) \wedge P(Z,Y)$	X overlaps Y	
$EQ(X,Y)$	$P(X,Y) \wedge P(Y,X)$	X equals Y	X equals Y
$PO(X,Y)$	$O(X,Y) \wedge \neg P(X,Y) \wedge \neg P(Y,X)$	X partially overlaps Y	X overlaps Y
$PP(X,Y)$	$P(X,Y) \wedge \neg P(Y,X)$	X is a proper part of Y	
$PPi(X,Y)$	$PP(Y,X)$	Y is a proper part of X	
$DR(X,Y)$	$\neg O(X,Y)$	X is discrete from Y	
$EC(X,Y)$	$C(X,Y) \wedge \neg O(X,Y)$	X is externally connected to Y	X touches Y
$TPP(X,Y)$	$PP(X,Y) \wedge \exists Z.EC(Z,X) \wedge EC(Z,Y)$	X is a tangential proper part of Y	X covers Y
$TPPi(X,Y)$	$TPP(Y,X)$	Y is a tangential proper part of X	X covered by Y
$NTPP(X,Y)$	$PP(X,Y) \wedge \neg\exists Z.EC(Z,X) \wedge EC(Z,Y)$	X is a non-tangential proper part of Y	X inside Y
$NTPPi(X,Y)$	$NTPP(Y,X)$	Y is a non-tangential proper part of X	X contains Y

Table 9.4: Defining region relations in RCC based on the connection relation

9.2.4 *Further qualitative spatiotemporal reasoning*

The approach to qualitative spatiotemporal reasoning described above—the definition of a qualitative representation combined with composition and converse operations for reasoning—has led to the development of a rich variety of qualitative spatial and spatiotemporal logics. For example, Figure 9.13 shows three related sets of JEPD qualitative relations for describing cardinal directions. Figure 9.13a is the projection-based *cardinal direction calculus* (CDC) for reasoning about cardinal directions between points. The representation partitions the plane into four semi-open regions (NE, NW, SE, SW) semi-bounded by four halflines (N, S, E, W), and centered on the reference point as the origin. Thus, in Figure 9.13a $E(B,A)$ (B is east of A) and $NW(C,A)$ (C is northwest of A).

cardinal direction calculus

Composition	EQ	DR	PO	PP	PPi	Converse
EQ	EQ	DR	PO	PP	PPi	EQ
DR	DR	EQ, DR, PO, PP, PPi	DR, PO, PP	DR, PO, PP	DR	DR
PO	PO	DR, PO, PPi	EQ, DR, PO, PP, PPi	PO, PP	DR, PO, PPi	PO
PP	PP	DR	DR, PO, PP	PP	EQ, DR, PO, PP, PPi	PPi
PPi	PPi	DR, PO, PPi	PO, PPi	EQ, DR, PO, PP, PPi	PPi	PP

Table 9.5: Composition and converse for RCC-5

Table 9.6 gives the complete table of composition and converse operations for the 9 projection-based CDC relations (eight directions plus the identity relation O). Hence, one can reason from Figure 9.13a and Table 9.6 that, for

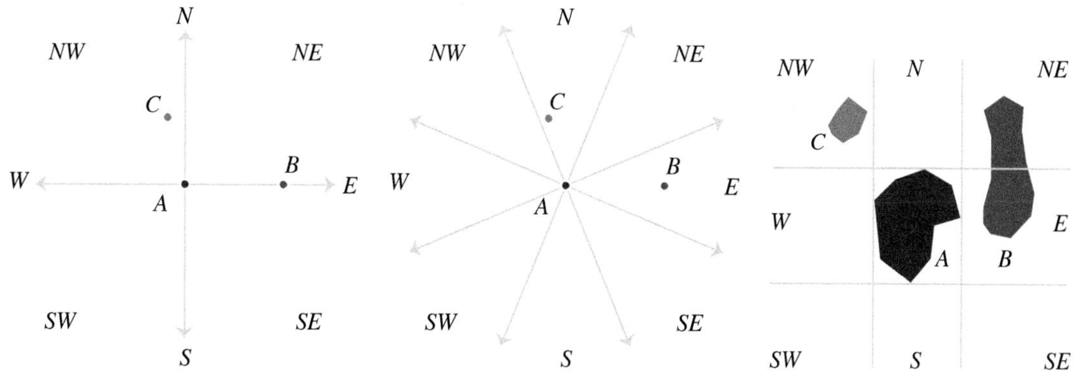

a. Projection-based CDC b. Cone-shaped CDC c. CDC for regions

Figure 9.13: Three different qualitative representations of cardinal directions, after Goyal & Egenhofer (2001) and Frank (1992)

example, $NW(C,B)$ since $NW(C,A)$ (given), $W(A,B)$ (converse of supplied $E(B,A)$, row 3 of Table 9.6) and applying the relevant composition rule (NW composed with W, row 8 column 7 of Table 9.6).

Figure 9.13b summarizes an alternative cone-shaped CDC representation which partitions the plane into 8 semi-open cone-shaped regions. A different composition table from Table 9.6 is of course required to reason with the cone-shaped CDC representation.[12] Figure 9.13c includes a third version of the CDC designed for reasoning about extended spatial regions. The representation uses four axis-parallel lines to partition the space based on the minimum and maximum x any y coordinates of the reference object. Because located objects may span multiple sectors (such as region B in Figure 9.13c which occupies both the NE and E rectangular areas relative to A) a more complex set of 218 direction relations must be defined that list which of the 9 rectangular regions for a reference object a related object occupies.

[12] A useful self-test exercise is to attempt to complete the composition table for the cone-based CDC yourself, as it raises some additional challenges.

Cardinal directions are defined based on an external frame of reference, independent of any intrinsic orientation of the objects themselves. Such extrinsic reference systems are referred to as *allocentric*. Figure 9.14 illustrates three *egocentric* qualitative direction representations that rely on the intrinsic direction of the reference object.

allocentric

egocentric

	N	*NE*	*E*	*SE*	*S*	*SW*	*W*	*NW*	*O*	Converse
N	*N*	*NE*	*NE*	$\mathcal{E}$	$\mathcal{U}$	$\mathcal{W}$	*NW*	*NW*	*N*	*S*
NE	*NE*	*NE*	*NE*	$\mathcal{E}$	$\mathcal{E}$	$\mathcal{U}$	$\mathcal{N}$	$\mathcal{N}$	*NE*	*SW*
E	*NE*	*NE*	*E*	*SE*	*SE*	$\mathcal{S}$	$\mathcal{U}$	$\mathcal{N}$	*E*	*W*
SE	$\mathcal{E}$	$\mathcal{E}$	*SE*	*SE*	*SE*	$\mathcal{S}$	$\mathcal{S}$	$\mathcal{U}$	*SE*	*NW*
S	$\mathcal{U}$	$\mathcal{E}$	*SE*	*SE*	*S*	*SW*	*SW*	$\mathcal{W}$	*S*	*N*
SW	$\mathcal{W}$	$\mathcal{U}$	$\mathcal{S}$	$\mathcal{S}$	*SW*	*SW*	*SW*	$\mathcal{W}$	*SW*	*NE*
W	*NW*	$\mathcal{N}$	$\mathcal{U}$	$\mathcal{S}$	*SW*	*SW*	*W*	*NW*	*W*	*E*
NW	*NW*	$\mathcal{N}$	$\mathcal{N}$	$\mathcal{U}$	$\mathcal{W}$	$\mathcal{W}$	*NW*	*NW*	*NW*	*SE*
O	*N*	*NE*	*E*	*SE*	*S*	*SW*	*W*	*NW*	*O*	*O*

Note: $\mathcal{N} = NE, N, NW$; $\mathcal{E} = NE, E, SE$; $\mathcal{S} = SE, S, SW$; $\mathcal{W} = SW, W, NW$;
$\mathcal{U} = N, NE, E, SE, S, SW, W, NW, O$

Table 9.6: Composition and converse for the projection-based cardinal direction calculus (CDC) in Figure 9.13a dipole relation algebra

- The *dipole relation algebra* considers the direction relations between pairs of directed line segments, called *dipoles* (Schlieder, 1995). The relation between a pair of dipoles A and B is determined by whether B begins and ends to the left or right of A, and whether A begins or ends to the left or right of B. For example, with reference to Figure 9.14a, the relative direction of B from A can be described as $\boldsymbol{rlrr}(A, B)$ (B begins to the right and ends to left of dipole A, and A begins and ends to the right of dipole B). The symbols s and e for "start" and "end" complete the possible relation options. For example, relative direction of A from C Figure 9.14a can be described as $\boldsymbol{errs}(C, A)$ (A begins at the end of C and ends to right of C, and C begins to the right of A and ends at the start of A).
- Figure 9.14b shows the single-cross calculus of Freksa (1992b). The single-cross calculus involves a *ternary* (three-place) relation, rather than the binary (two-place) qualitative relations we have so far encountered. The calculus models the qualitative direction of a point (C or D in Figure 9.14b) relative to an observer (at point A in Figure 9.14b) facing some landmark (at point B in Figure 9.14b). For example, the two relations described by Figure 9.14b are $\boldsymbol{rb}(A, B, C)$ (standing at A, C is to the right of and beyond B) and $lf(A, B, D)$ (standing at A, D is to the left and forward of B).
- Figure 9.14c shows the double-cross calculus of Freksa (1992b). The double-cross calculus further refines the single-cross calculus by enabling points that are forward of B but in front of A to be distinguished from points that are forward of B but behind A.

Figure 9.14: Three different egocentric direction relations, after Freksa (1992b), Zimmermann & Freksa (1996), and Wolter & Lee (2010)

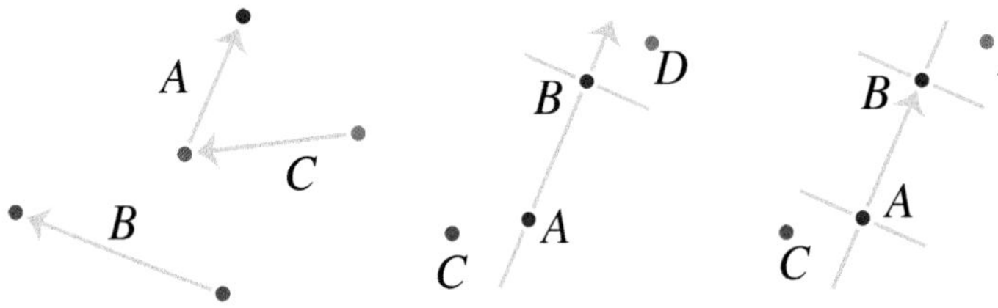

a. Dipole relation algebra b. Single-cross calculus c. Double-cross calculus

As before, reasoning using the dipole, single-cross, and double-cross representations is based on the definition of composition and converse operations. Indeed, there are many more logics for qualitative spatiotemporal reasoning, including reasoning about movement with the *qualitative trajectory calculus*, QTC, of Van de Weghe, Cohn, Tré, & De Maeyer (2006).

qualitative trajectory calculus

9.2.5 *Conceptual neighborhoods*

One final aspect of qualitative spatiotemporal reasoning is worth dwelling on briefly, before we leave KR^2 for machine learning: *conceptual neighborhoods*. Pairs of relations in a qualitative spatial logic can be distinguished based on whether or not it is possible for an instance of one relation to transition "smoothly" into another. For example, a point moving continuously on the

conceptual neighborhood

plane at a distance from the origin can transition directly from cardinal direction relation *N* to *NE*, but not directly from *N* to *E* (without first passing through relation *NE* or the origin *O*, see Figure 9.13). Similarly, two regions may smoothly transition from RCC-8 *NTPP* to *TPP*, but not from *NTTP* to *NTTPi* or *PO* without first passing through some other relations.

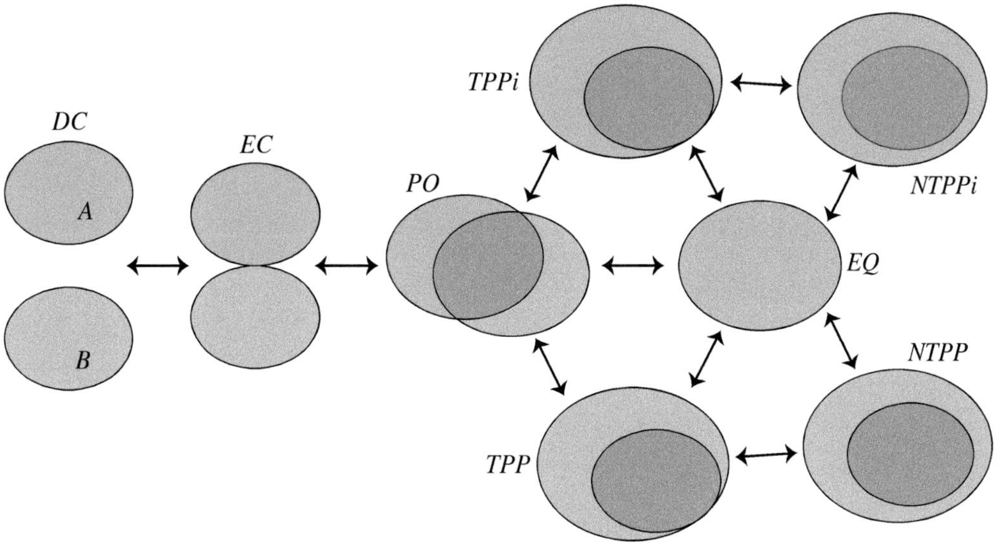

Figure 9.15: A conceptual neighborhood for the RCC-8

The idea of such conceptual neighborhoods was first described by Freksa (1992a): two relations are conceptual neighbors if they can be directly transformed into one another by continuously deforming them. Conceptual neighborhood graphs, such as that for RCC-8 in Figure 9.15, serve two important functions. First, they move qualitative spatial calculi towards describing change. Second, they provide further cognitive structure to underpin qualitative spatiotemporal logics, offering greater ability to capture human expectations about how similar certain spatial situations are.

Together, ontology engineering and qualitative spatial reasoning were selected as the two most important areas of KR^2 for GIS. However, it is important to be aware that there are many other KR^2 topics and techniques with historical (such as *expert systems*) or contemporary relevance to GIS (in particular, probabilistic reasoning techniques beyond the scope of this text, such as *Bayesian networks*, *Markov logic networks*, and *causal reasoning*), as well as closely allied topics such as agent-based modeling (see Box 9.4 on the next page). However, here we leave KR^2 to embark upon an exploration of the other major branch of AI: machine learning.

9.3 *Machine learning and spatial analysis*

The previous sections show how KR^2 uses logical reasoning and symbolic representations of data to mimic human problem-solving and decision-

Box 9.4: Agent-based models

Although not usually included in discussions of AI, agent-based models and agent-based modeling (ABM) have much in common with the more formal and logical techniques of KR^2. Agent-based models comprise three key elements. First, *agents* in an ABM are autonomous units or elements within a computer simulation. Agents in the simulation usually mirror (i.e., model) autonomous units or elements in the real world, such as vehicles, people, animals, trees, or wireless sensor motes. Second, agents in an ABM are placed into an *environment*, which similarly typically mirrors a real-world environment, such as a city or a region of land. Third, an ABM models the *interactions* that arise between agents and between agents and the environment. The ultimate objective of any ABM is a better understanding of interactions in the real-world through simulation. In particular, ABMs are used to model complex systems with emergent behaviors that go beyond our ability to understand using more analytical and numerical techniques. The emergent behavior of ant colonies is a classic example of a complex system that has been studied using ABMs. Understanding the emergent properties of ant colony behavior based on individual ant behavior—seeking food, returning it to the nest, and communicating and coordinating with other ants in the colony to achieve this—is extremely difficult to approach using purely analytical tools, such as mathematical equations. Instead, an ABM simulation of ants (i.e., agents), their environment, and the different ways they can interact with other ants (e.g., communicate) and their environment (e.g., move, leave pheromone trails) has provided significant insights into how relatively simple rules followed by individual ants can give rise to the awe-inspiring complexity of ant colony behavior (Wilensky & Rand, 2015). One of the most popular approaches to developing ABMs is the belief-desire-intention (BDI) approach to agent programming. Using BDI, each agent is equipped with *beliefs* about their environment, other agents, and the agent itself; goals the agent wishes to achieve, termed *desires*; and plans of actions that the agents can select from to achieve their goals, termed *intentions*. There are clear analogies between the knowledge represented by an agent's beliefs and the "reasoning" with desires intentions and intentions of agents in the simulation, and the more formal structures of KR^2. ABMs are widely used in GI science to help researchers understand many and diverse complex geographical systems (Crooks, Malleson, Manley, & Heppenstall, 2021; Heppenstall, Crooks, See, & Batty, 2012) as well as for understanding emergent computational system behavior, such as in geosensor networks (Duckham, 2013).

making capabilities. Machine learning (ML) applies statistical models to big data sets in order to recognize patterns in data and derive rules that describe data. Machine learning involves learning directly from data, without first imbuing the machine with explicit knowledge about the problem domain, in contrast to KR^2.

Recent years have seen a huge growth in the combination of machine learning with GIS (in particular, a growth in applications of *deep learning* discussed further in Section 9.4). One reason for this growth in machine learning is the growth in the variety and volume of spatial data. Many ML techniques are "data hungry," in the sense that the more data that is available to the machine to learn from, the better (in principle) the performance of that learning process. More spatial data means more opportunities for machines to learn from that data.

data hungry

Another reason for the growth in machine learning with GIS is the increasing availability and decreasing cost of high performance computing platforms, such as scalable cloud computing. Many ML techniques require significant computing resources to train and apply to data. As a result, the ability to access clusters of computers beyond those directly available on your desktop or in your office has transformed the range of ML techniques available to the broad spectrum of GIS professionals.

Advances in the field of machine learning itself have, of course, also contributed to this rapid growth in recent years. However, the central principles and techniques of machine learning have been well established for many years. In fact, it may be a surprise to learn that many basic spatial analyses found in every GIS are themselves examples of machine learning. Hence, many of the principles and techniques behind machine learning may already be familiar to you if you have already encountered some spatial analyses. Of course, the special nature of spatial data means that care needs to be taken when applying machine learning to spatial data. An understanding of the underlying principles of machine learning helps us to construct robust workflows and meaningful outputs when learning from spatial data.

This section unpacks the link between machine learning and traditional spatial analysis, touching on a range of fundamental spatial analysis and machine learning algorithms for spatial data. Along the way, we will revisit and consolidate some of the fundamental spatial concepts already introduced in previous chapters.

9.3.1 *Machine learning tasks*

Machine learning aims to equip computers to do something akin to a skill that comes naturally to humans: learning from experience. Specifically, a machine "learns" by identifying patterns in data. When the machine finds a pattern, it updates its model of the solution to reflect this new "experience." In this way, ML techniques aim to adaptively improve their performance as the number of samples available to learn from increases. At some point, when the machine has found enough patterns, it has built a model that is ready to be applied to new data it has not previously seen and to make predictions about that data.

The wide range of problems that machine learning can help solve can be categorized into three main tasks:

- *Clustering* is the task of learning "natural" groupings from within data called *clusters*, such that the similarity of data items within each cluster is greater than the similarity of items between clusters. clustering cluster
- *Classification* is the task of learning to predict qualitative class labels that can be associated with input data. classification
- *Regression* is the task of learning to predict quantitative values associated with input data. regression

Figure 9.16 summarizes the three techniques graphically, illustrating example inputs and outputs to typical algorithms. Clustering identifies groups of similar items within undifferentiated input data. Classification and regression both construct generalized models from the input data that enable either the qualitative category (classification) or quantitative value (regression) of previously unseen data to be predicted.

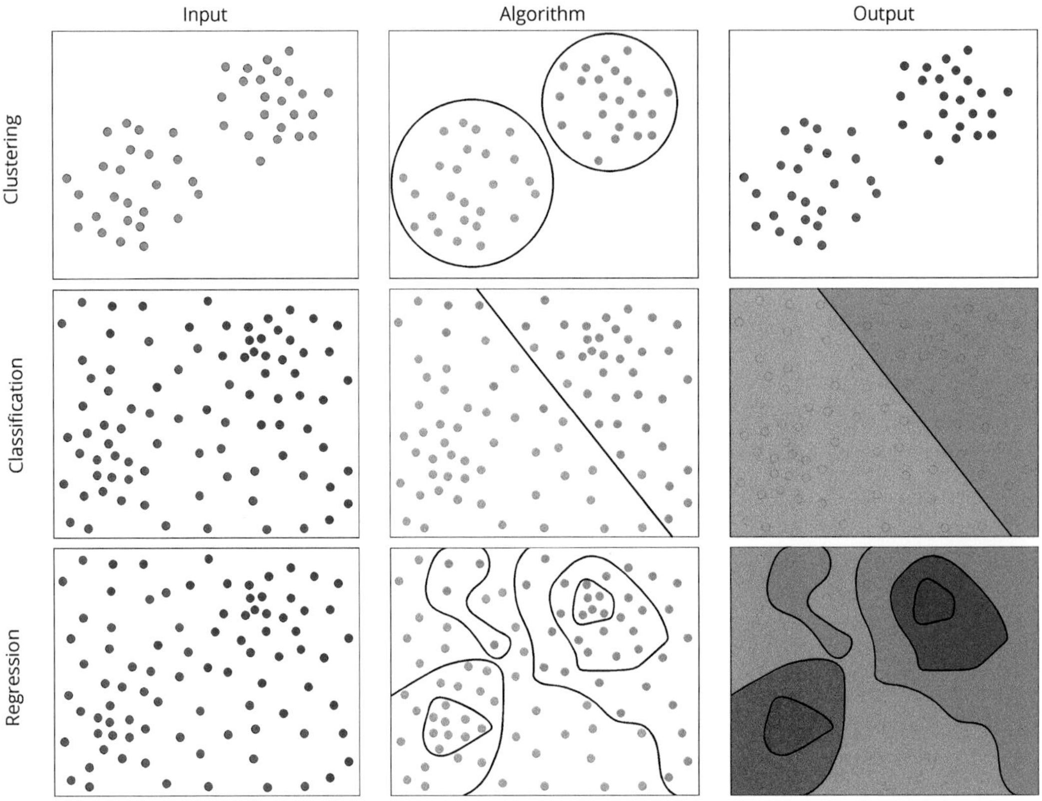

Figure 9.16: Three types of machine learning problems: clustering, regression, and classification

9.3.2 *Machine learning workflows*

As we shall see, there are many different machine learning techniques that each mimic human problem-solving abilities across one or more of the problem areas introduced above. Irrespective of the specific ML technique, we can summarize the machine learning workflow into four basic steps, shown in Figure 9.17.

data cleaning

data wrangling

Step 1: Preparation Data is the key to machine learning. However, raw data alone often needs processing, restructuring, and *cleaning*—the process of detecting and fixing or removing incorrect, duplicate, or incomplete records from a data set—before it is ready to serve as input to an ML algorithm. This first preparatory step of transforming the source data into a clean and compatible format and structure is termed *data wrangling* in Figure 9.17.

The preparation step also involves selecting the right ML technique or algorithm for solving the required problem. The latter parts of this section provide a tour of several of the most common ML techniques, starting simple and becoming increasingly complex. However, it is worth emphasizing that a more *complex* technique does not necessarily make a *better* technique. Rather, the challenge is to understand the strengths and limitations of the different choices, and how appropriate each ML technique is for use with different types of data and in solving different problems.

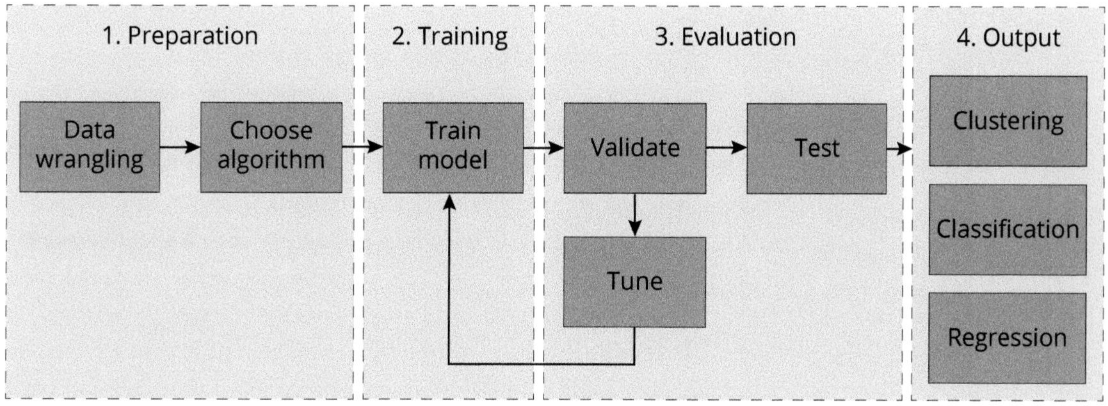

Figure 9.17: The machine learning workflow

Step 2: Training Step 2 in our workflow is at the core of machine learning: *training*. Irrespective of which ML technique is chosen, training involves conditioning an ML *model* to the data provided. Different ML techniques vary widely in how they approach this task. Two major classes of ML technique are often distinguished when discussing training: supervised and unsupervised learning.

training

Some ML techniques begin with a set of *labeled* data items with which to train the ML model. This *training data set* provides the machine with examples of the "right answer" used as the target for the training process. Traditionally, training data sets often needed to be manually labeled by a human operator. As a result, ML techniques that require a labeled data set for training are still termed *supervised learning*. Today, many data sets already include labels that can be used for training. Land cover data sets, for example, include land cover class labels that are often used in spatial applications of supervised machine learning.

training data set

supervised learning

In contrast, other ML techniques are conditioned directly on data without the need for any labels, termed *unsupervised learning*. However, even if human involvement was required at some point to generate the labels needed for supervised learning, in practice human operator intervention is relatively rarely required either for supervised or unsupervised learning techniques. Requiring no labels, unsupervised learning can be applied to a wider range of different data sets than supervised learning. However, without labels guiding the learning process, it is also more challenging to ensure accurate results and achieve meaningful outputs using unsupervised learning.

unsupervised learning

Step 3: Evaluation Whether supervised or unsupervised learning, step 3 *evaluates* how well the training process has performed. Evaluation can be separated into two components: *validation* and *testing*. Validation occurs *while* the model is being trained and is necessary in order to ensure that the learning process is correctly attuned to the specific characteristics of the input data and the problem being addressed. Validation is sometimes run on a reserved portion of the data—typically between 10–20% of the complete data set—called the *validation data set*. In other cases, validation simply involves checking output diagnostic parameters that accompany the training output, and provide a guide to the quality of the training.

validation data set

As a result of validation, some changes to the ML settings may be indicated to ensure the training process is performing as accurately as possible. Any top-level settings used to control the training process—such as the number of iterations used in training process or the target number of learned clusters or categories—are termed *hyperparameters*. In contrast, the internal parameters of the ML model adapted by the training itself to ensure a good fit to the input data—such as model coefficients or weights—are referred to simply as *model parameters*. The process of adjusting hyperparameters in response to validation is termed *tuning*.

hyperparameters

model parameters

tuning

Finally, testing occurs *after* the model has been built. Testing is used to check that the final output has captured something generalizable about the data, and it is not simply biased towards the specific idiosyncrasies of the input data set. Testing usually aims to compare the output with some external "ground truth" data to evaluate how well the trained model has learned about the salient characteristics of the input data. In most cases, testing is run on another reserved portion of the data, typically 10–30% of the complete data set, called the *test data set*. In order to be able to rely on the results of testing, it is essential that the test data set is kept completely separate and independent from the training data set (and from the validation data set too, if used).

test data set

Step 4: Output A successful machine learning workflow will construct a trained and evaluated model that represents a solution to one of the three general problem types already enumerated: clustering, classification, and regression. In connection with clustering problems, the primary output of the machine learning workflow is the identified clusters, learned from the input data by the tuned ML model. ML techniques for solving classification problems, on the other hand, output not a data set, but the learned model itself. Assuming careful validation and testing, this output model can then be applied to the problem of classifying related data not previously seen by the model. In general, all machine learning techniques generate one of these two types of output: either a data set that encapsulates the solution to the problem, or a trained model that can subsequently be applied to related data to generate a solution automatically.

9.3.3 *Evaluation and overfitting*

The evaluation step of the machine learning workflow is critical to ensure that machine learning outputs are accurate and meaningful. Machines do not possess the commonsense of humans when they learn. Evaluation provides the safeguards to ensure the machine is not simply learning nonsense, and identifying patterns where none exist. Without careful evaluation, the machine can learn patterns from noise, inaccuracies, or randomness in data, where in truth no meaningful patterns exist.

Validation metrics provide an indication of the internal quality of the learned model, in terms of the fit between the input and the output. Test metrics provide an indication of the generalizability of the model, in terms of the fit between the output and a reserved test data set. However, similar measures are often used for evaluation, whether validation or testing. Statistical measures of spread such as standard deviation, mean square error (MSE), and R^2 help to evaluate the quality of learning with quantitative data, such as regression and some types of clustering.

When evaluating learning with qualitative output, such as some other types of clustering and classification, commonly used metrics are based on errors of *commission* (items placed in a class X that should not have been) and errors of *omission* (items not placed in a class X that should have been).

errors of commission

errors of omission

Crosstabulating all of these errors by class results in a *confusion matrix*, such as that shown in Table 9.7. In Table 9.7, the confusion matrix shows that 44 red dots were correctly classified, with 3 errors of omission (red dots classified as blue) and 5 errors of commission (blue dots classified as red) in the classification example in Figure 9.16. With just two classes, all the red errors of commission are also blue errors of omission (and mutatis mutandis, all red errors of omission are blue errors of commission). More complex error interrelationships arise with more than two classes.

confusion matrix

	red	blue	total
red	44	3	47
blue	5	35	38
total	49	40	87

Table 9.7: Example confusion matrix for the classification in Figure 9.16

Overall classification accuracy is often captured as the number of correct classifications as a proportion of the total number of classified data items. For instance, taking Figure 9.16 and Table 9.7, the classification accuracy is $(44 + 35)/87 = 90.8\%$. For individual categories, three further metrics are widely used:

- True positive rate (TPR), also called *sensitivity* or *recall*, measures of the data correctly classified as a proportion of all the data classified. For example, the TPR for red in Table 9.7 is $44/47 = 93.6\%$.
- Positive predictive value (PPV), confusingly also called *precision*, is a measure of the data correctly classified as a proportion of the data that should have been so classified. For example, the PPV for red in Table 9.7 is $44/49 = 89.8\%$.
- The F1 score is a mean of TPR and PPV, calculated as $2 * \frac{\text{TPR}*\text{PPV}}{\text{TPR}+\text{PPV}}$. For example, the F1 score for red in Table 9.7 is $2 * \frac{0.936*0.898}{0.936+0.898} = 91.7\%$.

true positive rate

positive predictive value

F1 score

TPR and PPV are always reported together because they complement each other. For example, a model that classifies everything as red will always achieve a PPV of 100% (since everything that should be classified as red will assuredly *be* classified as red) but a low TPR (since many things that should not be classified as red assuredly will be). Conversely, a model that classifies almost everything as blue with result in a low PPV but a high TPR. The F1 score provides a convenient summary of the combination.

overfitting

The term *overfitting* is used to describe the common error in machine learning where the model corresponds too closely to the input data set, at the expense of learning something generalizable about the data. Overfitting occurs when the model learns about inconsequential noise in the data set, rather than about consequential patterns. Validation and testing at the foundations of detecting overfitting: a model with high validation metrics but low testing metrics is the hallmark of an overfitted ML model.

9.3.4 Clustering algorithms

k-means clustering

One of the most frequently used clustering algorithms is *k-means clustering*, which partitions an input data set into a specified number of k clusters. The clusters are constructed such that the arithmetic mean (centroid) of each cluster is closer to every input data point in the cluster than to any data point in another cluster.

The k-means clustering Algorithm 9.1 begins by randomly selecting a set of k "cluster head" points from within the domain of the input point set P. Algorithm 9.1, line 1, uses the MBB (minimum bounding box) to constrain the set of initial cluster heads. The algorithm then computes the subset of points nearest to each cluster head, termed the *assignment* or *allocation step* (Algorithm 9.1, line 3). The set of k subsets of points (blocks) forms a partition of P. Next, the algorithm calculates the arithmetic mean (centroid) of each block, termed the *update step* (Algorithm 9.1, line 4). The alternating assignment and update steps iterate until the blocks stabilize to the optimal solution, found when the cluster heads no longer move after each iteration.

Algorithm 9.1: Computing the k-means of an input point set P

```
Input Set of points P, desired number of clusters k
1: randomly select k points c_1, ..., c_k (cluster heads) such that c_i ∈ MBB(P)
2: repeat
3:     partition P into k blocks, where block i contains all points nearest to c_i
4:     recompute cluster heads c_1, ..., c_k as the centroid of points in block i
5: until cluster heads c_1, ..., c_k are optimal (i.e., no longer change)
Output Blocks P_1, ...P_k that partition P
```

Figure 9.18 summarizes the algorithm diagrammatically for an input data set of 37 points and a cluster number $k = 3$ (Figure 9.18a). Three randomly chosen cluster heads are introduced in the assignment step (Figure 9.18b). The update step (Figure 9.18c) then updates cluster heads as the centroid of the identified clusters. Once the cluster heads reach their optimal location, the clusters themselves will stabilize (Figure 9.18d), and further iterations will not result in any further movement of cluster heads. Algorithm 9.1 is known as *naïve* k-means clustering because it is intractable even for 2D sets of input points. Happily, there are many more efficient heuristics used in practice to solve k-means.

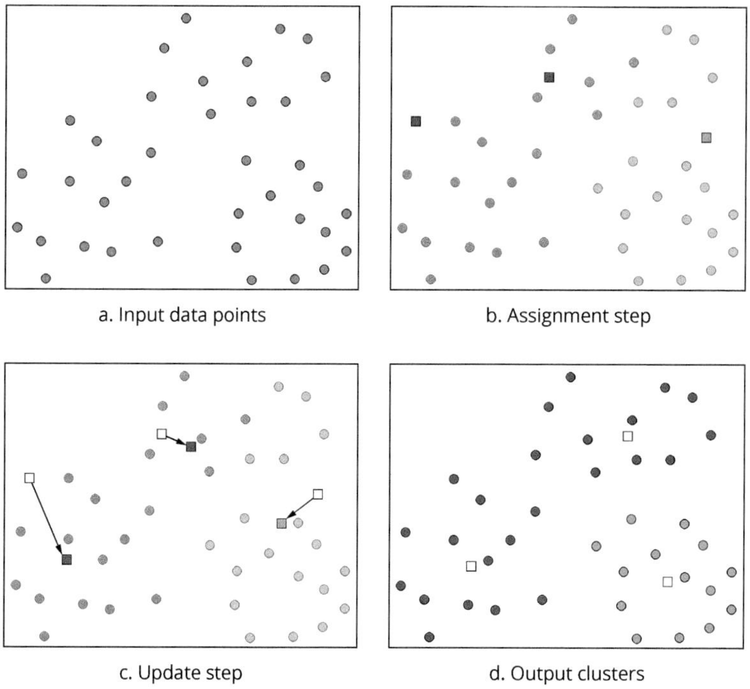

Figure 9.18: Four stages of the naïve k-means clustering algorithm, in Algorithm 9.1. Cluster heads are indicated with square points.

The k-means clustering algorithm "learns" the optimal location clusters that minimize within-cluster and maximize between-cluster distances. Tuning the algorithm usually involves tweaking the number of clusters k, in response to diagnostic parameters such as the statistical spread of points around cluster heads. For example, plotting the sum of squared distances between points and their cluster heads against different k can help identify the smallest k that captures most of the variation in point locations (termed the *elbow method* of validation). Testing can be performed on a reserved, randomly selected subset of the data, with the output clusters compared in terms of their closeness, giving some indication of the level upon which the identified clusters are dependent on the specific input data points.

elbow method

As might be expected, the results of k-means clustering are strongly dependent on the required number of clusters, k. A further weakness of k-means clustering is a sensitivity to outliers, which will always be allocated to a cluster and so tend to "drag" cluster heads away from a "natural" centroid. Many other clustering algorithms have been proposed that seek to identify clusters with different characteristics. One of the clustering algorithms most frequently used with spatial data is DBSCAN (*density-based spatial clustering of applications with noise*). DBSCAN identifies groups of more densely packed points, and it is able to exclude outlier points that do not fit well in any cluster (Figure 9.19). In addition to the set of input points, DBSCAN requires two input parameters: a minimum number of points in any cluster (to ex-

DBSCAN

clude very small clusters of, say, 2 or 3 points), and a neighborhood size ϵ that controls the maximum threshold distance from a point to some (at least one) other point in the same cluster.

A simplified version of DBSCAN is shown in Algorithm 9.2. A first incomplete cluster is initialized with a single, randomly chosen seed point (Algorithm 9.2, lines 2–3). DBSCAN then finds the yet-unvisited neighbors within distance ϵ of any points in the incomplete cluster (lines 5 and 15–17). Identified neighbors are added to the cluster (line 12), as well as being removed from the set of unvisited nodes (line 13). The cluster is complete when no further neighbors of nodes in the cluster can be found (line 6). Clusters bigger than the minimum cluster size m are stored in the result set before iterating; clusters smaller than m are discarded as outliers (lines 7–8). Finally, the algorithm randomly selects another unvisited point to seed a new cluster (lines 9–10) iterating until all nodes have been visited (line 14).

Algorithm 9.2: DBSCAN algorithm

Input input nodes V, minimum cluster size m, neighborhood distance ϵ
1: set unvisited points $U \leftarrow V$
2: randomly select points n in U
3: update $U \leftarrow U - \{n\}$ and set new cluster $N = \{n\}$
4: **repeat**
5:     $M \leftarrow$GetNeighbors(U,N,ϵ)
6:     **if** $M = \emptyset$ **then**
7:         **if** $|N| \geq m$ **then**
8:             store copy of N in result set R
9:         randomly select point n in U
10:        update $U \leftarrow U - \{n\}$ and set new cluster $N = \{n\}$
11:    **else**
12:        update $N \leftarrow N \cup R$
13:        update $U \leftarrow U - R$
14: **until** $|U| > 0$
15: **function** GetNeighbors(U, N, ϵ)
16:    **return** $\{u \in U | \exists n \in N . \delta(u,n) \leq \epsilon\}$
17: **end function**
Output Result set of clusters R

9.3.5 *Classification algorithms*

Decision tree learning is a foundational ML technique for classification. Despite being a machine learning technique, decision tree learning has much in common with the knowledge representation approaches to AI encountered earlier in this chapter. *Decision tree learning* involves the construction of a *decision tree* that models the key decisions behind the classification process. A decision tree is a directed acyclic graph (see Section 2.4.1) where:

decision tree learning

- each non-leaf node in the decision tree represents a logical test that can be performed on the data;

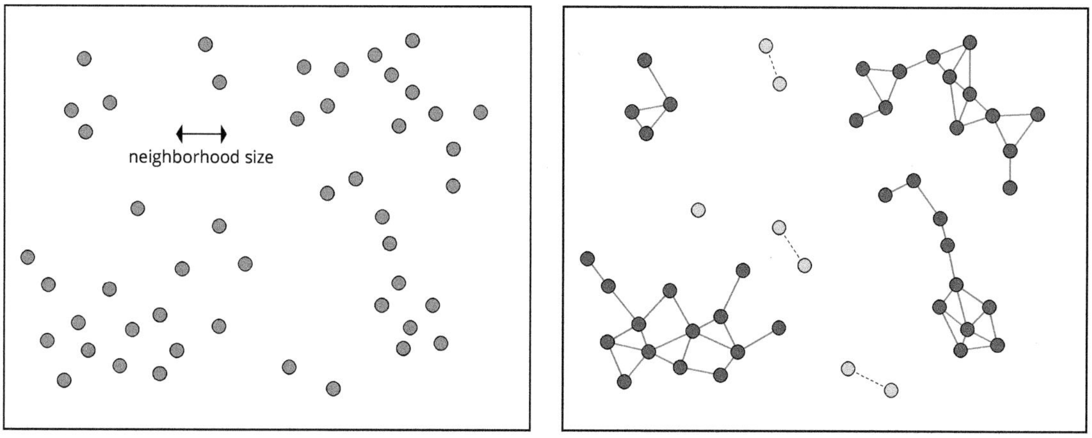

a. Input data points

b. DBSCAN clusters and outliers

Figure 9.19: DBSCAN clustering on a set of input points with minimum number of points equal to 3 and neighborhood size indicated

- a directed edge represents one potential outcome of the test indicated by the source node of that edge; and
- each leaf node represents an outcome of the classification process.

For example, Figure 9.20 shows a decision tree related to the thermal comfort of walking tours in a city. In recent years, thermal comfort has become an important issue in urban planning and design, as more walkable cities with greater thermal comfort help promote healthy active lifestyles and lower reliance on car use. The decision tree in Figure 9.20 indicates, for instance, when the weather is cool, then thermal comfort is assured. However, on a hot day, even a shady route cannot provide thermal comfort on a long walk.

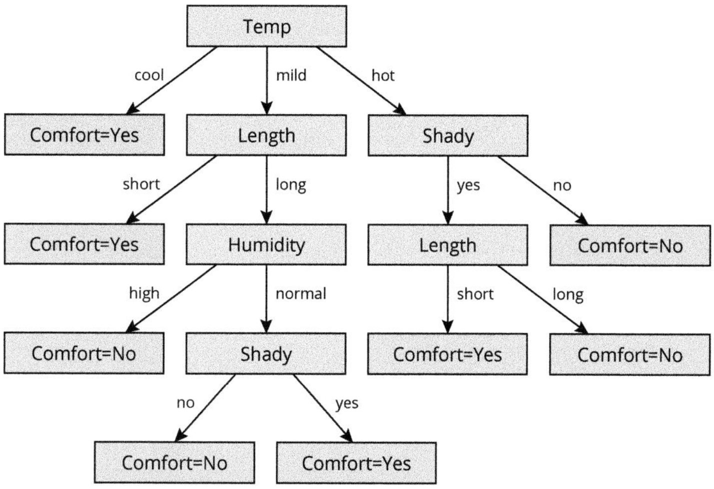

Figure 9.20: A decision tree for thermal comfort in city walking tours

Although the decision tree is a symbolic representation of knowledge that sits comfortably in the KR^2 branch of AI, decision trees can be learned from data. Table 9.8 provides the data that was used to automatically construct the decision tree in Figure 9.20. Each record in Table 9.8 provides information about the thermal comfort of individual instances of different walks under different weather conditions.

Table 9.8: Example data about thermal comfort of walking tours through a city

Temp	Humidity	Shady	Length	Comfort
Cool	Normal	Yes	Long	Yes
Mild	Normal	Yes	Short	Yes
Cool	Normal	No	Long	Yes
Mild	High	No	Long	No
Mild	High	Yes	Long	No
Hot	High	Yes	Short	Yes
Mild	Normal	Yes	Long	Yes
Hot	High	Yes	Long	No
Mild	High	No	Short	Yes
Mild	Normal	No	Long	No
Cool	Normal	Yes	Short	Yes
Hot	High	No	Short	No
Mild	Normal	No	Short	Yes

ID3

The decision tree learning algorithm used to create the decision tree in Figure 9.20 from the data in Table 9.8 is among the most famous of all machine learning algorithms, called *ID3* and developed by Quinlan (1986). The ID3 algorithm listed in Algorithm 9.3 takes as input a table of data (such as that in Table 9.8) and a target attribute, which indicates the classification outcomes (in this case the Comfort attribute). Algorithm 9.3 initializes by creating a new, empty root node (line 1) before calling the recursive BuildSubtree function to build the subtree underneath that root. The BuildSubtree function first checks if the table contains multiple values for the target attribute (line 4). If the table contains only a single target attribute value, then the algorithm has identified decision tree leaf node that classifies with that value. If instead the table contains multiple target attribute values, then the algorithm needs some way of selecting the best attribute in the table to use next to classify the data. The ingenious and elegant answer to this problem is at the heart of the ID3 algorithm and uses the important concept of *information entropy*.

It is not difficult to appreciate that data and information are distinct concepts (refer to Section 1.3.2). The string 01101101 contains 8 bits (binary digits) of data, but no information unless we have some means of interpreting it. A red light at a railroad crossing can convey a lot of information to a motorist with only a small amount of data (assuming that the light is either on or off, then only a single bit of data). Data is easy to measure: it may be measured in bits and bytes. A more difficult question is, "How do we measure information?" Indeed, "What is information?"

One way to describe information is as a *signal* communicated on a *channel* from a *transmitter* to a *receiver*. This simple model, termed Shannon-Weaver information theory, can help in constructing a metric for the *quantity* of information communicated, not just the quantity of data. *Information entropy*, also called *Shannon entropy*, is a measure of the quantity of information in a signal in terms of its capacity to "surprise" the recipient. "Surprise" in this

information entropy

context is used to mean the smallness of the chance that a particular signal will be received. The smaller the chance, the greater the surprise, and the higher the information is valued.

For example, in Maine in the winter, there is almost always snow on the ground. A weather forecast of winter snow in Maine, therefore, is not unexpected and so does not contain much information. However, the same forecast of winter snow in Melbourne would be unusual, to say the least, and therefore might be said to contain much more information (and see Box 9.5 on the following page).

Algorithm 9.3: ID3 decision tree learning algorithm

```
Input table tab, target attribute tar
create empty root node n
BuildSubtree(tab, tar, n)
function BuildSubtree(table tab, target tar, node n)
    if only one row in: SELECT DISTINCT tar FROM tab then
        set n as: leaf node with value from SELECT DISTINCT tar FROM tab
    else
        find attribute att in table tab with the largest information gain for tar
        set n as: non-leaf node with decision attribute att
        set lst as: the list of all attributes in table tab excluding tar
        for all attribute values val in: SELECT DISTINCT att FROM tab do
            set sub as: SELECT lst FROM tab WHERE att=val
            set n' as: a new empty child node
            create new edge from n to n' labeled with decision att=val
            BuildSubtree(sub, tar, n')
end function
Output Decision tree with root node n
```

Calculating information entropy, then, involves calculating how surprising a particular signal is expected to be, to a receiver. For some variable X with values $\{x_1, ..., x_i\}$ the information entropy $H(X)$ is calculated in Equation 9.1 as follows:

$$H(X) = -\sum_{x \in X} p(x) \log_2 p(x) \qquad (9.1)$$

In Equation 9.1, $p(x)$ is the probability of x, interpreted in ID3 as the proportion of all records in the table with attribute value x. For example, in Table 9.8 the attribute Comfort (8 Yes values, 5 No values, 13 records in total) has information entropy $8/13 \log_2(8/13) + 5/13 \log_2(5/13) = 0.946$. Deciding which attribute is most useful to use as a decision node in the decision tree can then be reformulated as the question of which attribute is associated with the largest *information gain* (i.e., the greatest reduction in surprisal value). Using information entropy, information gain $IG(X, a)$ associated with attribute a and variable X is calculated in Equation 9.2 as follows:

information gain

$$IG(X, a) = H(X) - \sum_{T \in \tau} p(T) H(T) \qquad (9.2)$$

where τ is interpreted as the set of tables resulting from a split using attribute a and $p(T)$ the number of records in table $T \in \tau$ as a proportion of those in X.

Box 9.5: Information quantity and value

The Shannon-Weaver concepts of transmitters and receivers of signals via channels are of course metaphors. The image of information leaving the source, being transmitted through a medium (channel) to a receiver, perhaps also being the subject of degradation through noise, is compelling. In fact, it is so compelling that sometimes it is easy to forget that it is only an image, and an image with limitations. In particular, in Shannon and Weaver's elaboration of the metaphor, "value" becomes synonymous with "quantity" of information, which is measured by information entropy. However, it is easy to find intuitive examples where value is not measured this way. Suppose a burglar has determined by observation that almost every night the owners of a house forget to set their house alarm system and only remember on those rare occasions when they take their dog for a walk. On the night planned for the burglary, the lack of a dog being walked provides unsurprising but highly valuable information to the burglar. Context is the key here, and an understanding of context needs to contribute to any metric of information value. A stronger position would be that context is a necessary facilitator of information flow, and without context there is no flow (see Sperber & Wilson, 1995).

For example, the information gained by selecting attribute Temp is:

$$
\begin{aligned}
&\left(\frac{8}{13}\log_2\frac{8}{13}+\frac{5}{13}\log_2\frac{5}{13}\right)-\\
&\left(\frac{3}{13}\left(\frac{3}{3}\log_2\frac{3}{3}+\frac{0}{3}\log_2\frac{0}{3}\right)+\right.\\
&\frac{7}{13}\left(\frac{3}{7}\log_2\frac{3}{7}+\frac{4}{7}\log_2\frac{4}{7}\right)+\\
&\left.\frac{3}{13}\left(\frac{1}{3}\log_2\frac{1}{3}+\frac{2}{3}\log_2\frac{2}{3}\right)\right)=0.204
\end{aligned}
$$

By contrast, none of the other attributes in Table 9.8 exhibit such high information gains (the information gain for Humidity is 0.046, Shady is 0.076, Length is 0.115). Algorithm 9.3 uses such calculations of information gain (line 7) as the basis for selecting the attribute to use in each decision node (lines 8–9). Then, by splitting the table into smaller tables using the selected attribute (Algorithm 9.3, line 10–11) the algorithm continues building the decision tree recursively with a new decision edge and child node (lines 12–14).

Decision trees, such as that in Figure 9.20, are powerful and intuitive ML classification tools. However, decision trees also tend to suffer from overfitting, especially if grown to their fullest extent, necessary to classify every data item from the input data. To address this limitation, a *random forest* is a collection of decision trees grown from randomized samples of the input data, illustrated in Figure 9.21. Each sample typically uses about two-thirds of input data. However, samples are often "topped up" to be the same size as the input data set with randomly sampled duplicate entries, a sampling process called *bootstrapping*. Validation of each tree can use the data omitted from its generating sample, termed *out-of-bag estimation* (OOB).

random forest

bootstrap sampling

out-of-bag estimation

A random forest may take many hundred such random samples in order to grow hundreds of trees. After training this way, classification of unseen data can proceed using majority voting from across all the trees in the forest. Each new record encountered is classified by all the trees in parallel, with the class most frequently represented across the forest being assigned to the output.

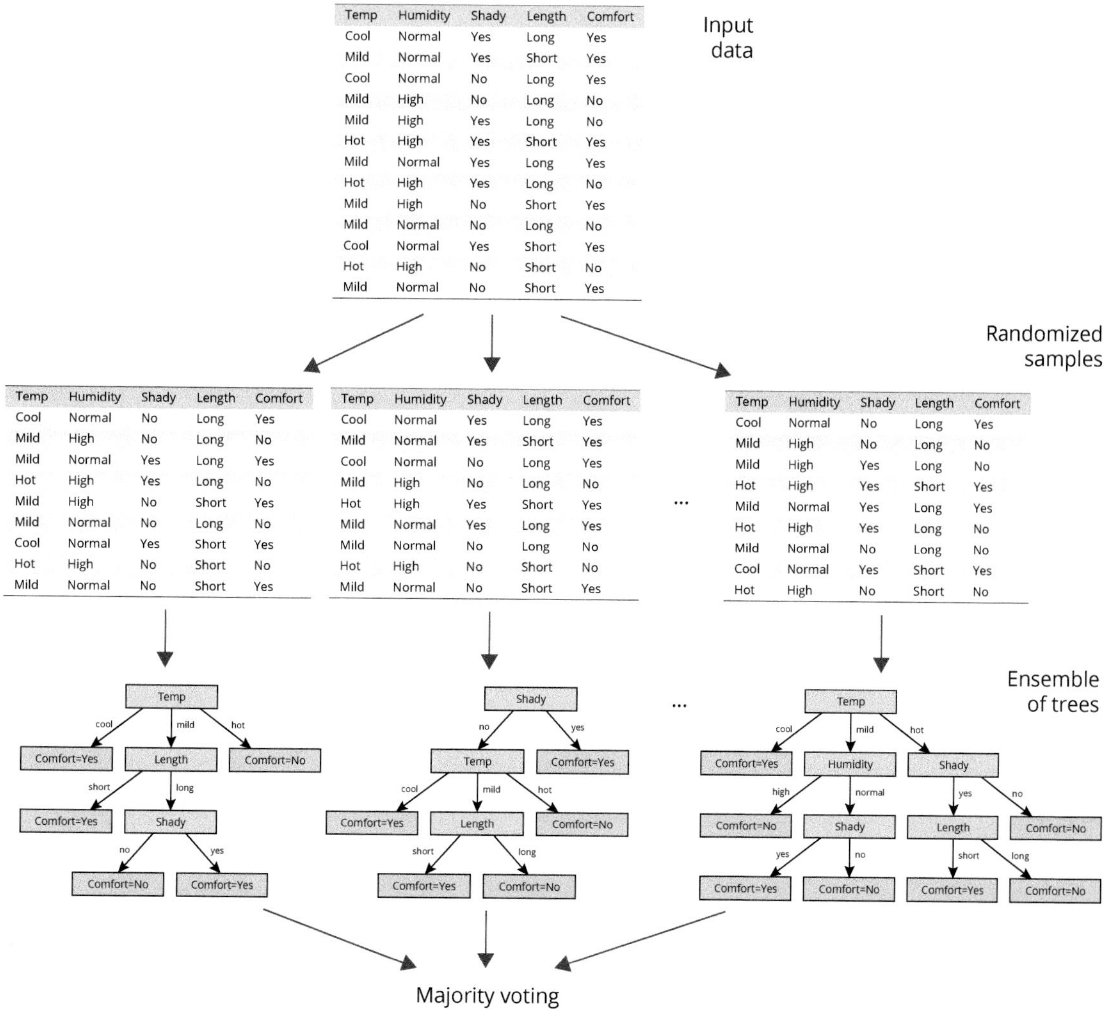

Figure 9.21: Random forest ensemble of decision trees without bootstrapping

The approach of learning multiple models from data and then recombining the set of model outputs into a final output is a common technique in machine learning, termed *ensemble learning*. In general, ensemble learning tends to increase overall stability and accuracy of learning outputs. Accordingly, random forests are an example of an ensemble learning technique, with the particular objective of reducing the tendency to overfitting common in single decision trees.

ensemble learning

Decision tree learning can also be adapted to handle continuous-valued data, by classifying quantities using data classification techniques such as quantiles, equal intervals, and mean-standard deviation encountered in the previous chapter. Moving beyond classifying quantitative data to true predictions with continuous values, however, requires regression algorithms.

9.3.6 Regression algorithms

Regression aims to fit a model to quantitative data, such that the model can subsequently be used to predict values that are not part of the input training data set. One of the most important techniques for spatial regression is *Kriging*, named after mining engineer and statistician Danie Krige.[13] At the root of Kriging is a construction called the *semivariogram*. The semivariogram provides a model of the spatial autocorrelation (degree of spatial interdependence) in a data set (Section 4.2.1). The semivariogram captures the degree to which "near things are more related" (Tobler's first law), and it uses that as its basis for regression.

Kriging

[13] Both the name "Krige" and the technique "Kriging" are pronounced in English with a hard "g."

semivariogram

Figure 9.22a summarizes the key features of the semivariogram as follows:

- On the x-axis, the semivariogram measures the distance between pairs of points in spatial data, termed *lag*.
- On the y-axis, the semivariogram measures the average of variation between points at that corresponding lag. Variation is measured as *semivariance*: half the average squared difference between pairs of points at a specified lag.
- The intercept of the curve with the y-axis is termed the *nugget*, and it represents the variability between points at zero lag. The curve will not usually intercept the origin. The non-zero nugget is interpreted as the level of error in measurements, or of variation below the smallest measurement granularity.
- Semivariance is expected to increase with increasing lag (because near things are more related), but only up to a certain level, termed the *sill*.
- The lag beyond which semivariance no longer appreciably increases is called the *range*. The range is the limit of autocorrelation, beyond which distance all point pairs are equally uncorrelated.

lag

semivariance

nugget

sill

semivariogram range

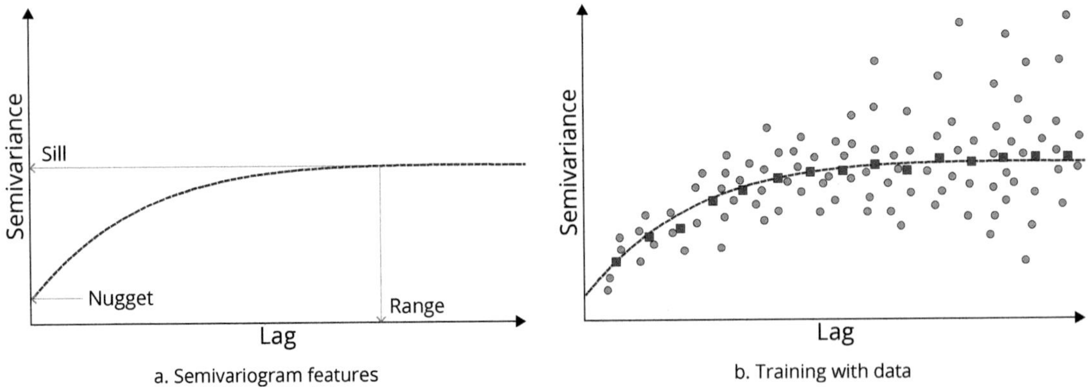

Figure 9.22: The semivariogram, the basis of Kriging

Fitting (i.e., training) a semivariogram to a data set N of points involves first computing the semivariance between the sets of points at different lag distances. For a set of pairs of points $N(h)$ at approximate lag distance h (i.e., within some small tolerance), the semivariance γ is computed as:

$$\gamma(h) = \frac{1}{2N(h)} \sum_{(i,j)\in N(h)} \left(Z(p_i) - Z(p_j)\right)^2$$

where $Z(p_x)$ is the observed value at point p_x. Figure 9.22b plots the semivariance of a set of points against lag (circular data points) with the curve (dashed line) fitted to the data using regression. Often the curve may be fitted to averaged values (square data points in Figure 9.22b) for clarity. There are many different types of curve that may be fitted to the data, from simple linear relationships to sophisticated Gaussian models and sigmoid functions. One of the most commonly used models is a modified quadratic, called a *spherical model* (as used in Figure 9.22b). The objective in all cases is to achieve the best fit between model and data. However, the choice of curve type typically has a significant impact on the training results.

The output of this training process is a model of the autocorrelation in the data, based on the structure of the semivariogram. Specifically, the predicted value $\hat{Z}$ at some new location p_0 is estimated as the sum of known values at all the other input data points, weighted by inverse distance and according to the semivariogram (i.e., so nearby data points have a much stronger influence than more distal data points). This estimate is captured by the equation:

$$\hat{Z}(p_0) = \sum_{k=1}^{|N|} \lambda_k Z(p_k)$$

where λ_k is the set of weights derived from the semivariogram based on the distance of each input data point p_k from p_0. The model can be evaluated in the usual way, for example, by reserving a test data set unused in the training and computing the MSE or R^2 deviations between test data and predictions.

Empirical Bayesian Kriging The approach described above is the most basic form of Kriging, called *simple Kriging*. An important limitation of simple Kriging is that it assumes *statistical stationarity*: that the statistical properties of the training data are constant across the study area. In contrast, we know to *expect* spatial data to exhibit *nonstationarity*—where variability itself varies spatially and there exists no such thing as an "average location" (see Box 1.7 on page 34).

statistical stationarity

Several more sophisticated extensions of simple Kriging have been developed to provide more sophisticated predictions that relax the strict requirement for stationarity, such as *ordinary Kriging*. One of the most widely used such extensions is called *empirical Bayesian Kriging* (EBK). EBK generates not one semivariogram, but many semivariograms from overlapping, randomly selected subsets of input data.

empirical Bayesian Kriging

Thus, training generates a *spectrum* of semivariograms, each one an estimate of the "true" semivariogram. The spectrum of semivariograms forms an ensemble that can be recombined into a single output model. As before, the ensemble approach tends to provide more reliable predictions for each new location. Figure 9.23 shows the results of EBK regression analysis of point-based observations of temperature across the state of Oregon, US.

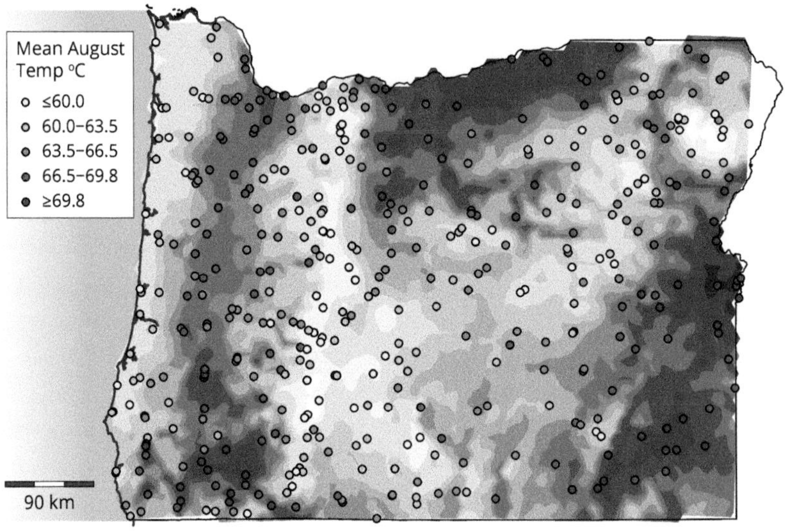

Figure 9.23: Empirical Bayesian Kriging of temperature points across the state of Oregon, US

Like all Kriging, EBK accommodates autocorrelation in spatial data. The temperature surface output by EBK in Figure 9.23 captures the spatially autocorrelated continuous variation in temperature points.[14] However, unlike simple Kriging, EBK also allows for moderate levels of nonstationarity, because the estimate of the value at a location can be based on recombining a subset of the spectrum of semivariograms from nearby locations. Hence, the level of autocorrelation used to generate the surface in Figure 9.23 from the input temperature points may vary somewhat across the area.

[14] The input points and the output surface have been classified in Figure 9.23 for cartographic clarity, but the underlying input and output data is continuous.

Geographic weighted regression One further, important regression technique capable of accommodating both autocorrelation and nonstationarity is called *geographically weighted regression* (GWR). Whereas Kriging trains a model to estimate a variable at unknown *locations*, GWR trains a model to estimate the spatially varying and unknown *relationship* between different variables.

geographically weighted regression

GWR is a spatial extension of multiple linear regression, where training data is used to construct the relationship between a dependent variable and one or more independent variables. Predicting a dependent variable y based on p independent variables $x_1 ... x_p$ with multiple linear regression has the general form:

$$y = \beta_0 + \sum_{k=1}^{p} \beta_k x_k + \epsilon$$

where ϵ is an error term and β_k are the parameters of the regression that need to be estimated. GWR additionally accounts for local, spatial variations in the relationship between variables, using the form:

$$y(u) = \beta_0(u) + \sum_{k=1}^{p} \beta_k(u)x_k + \epsilon$$

where $\beta(u)$ indicates the estimated parameter at a location u.

Multiple linear regression estimates a single, *global* relationship between the dependent variable and each independent variable based on the entire training data set. By contrast, GWR estimates the *local* relationship between dependent and independent variables for each location in the training data set. Figure 9.24 shows an example of the output of a GWR model for a regression on the Oregon temperature data already encountered in Figure 9.23. It is expected that temperature is not only autocorrelated but correlated with other environmental variables, such as elevation. Further, the degree to which temperature is correlated with elevation is expected to vary spatially, with some locations exhibiting stronger correlation than others. The GWR output in Figure 9.24 shows the spatially varying strength of this relationship. Higher coefficients (stronger relationships) are found at higher elevations; lower coefficients tend to cluster nearer to the coast, where proximity to the sea tends to dominate local weather conditions rather than elevation.

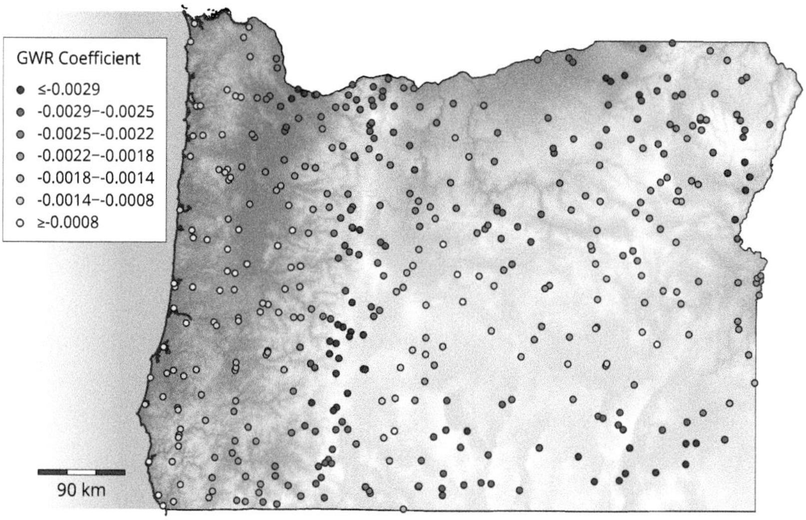

Figure 9.24: Output coefficient for geographically weighted regression of mean August temperature (dependent variable) against elevation (independent variable, lighter green indicates higher elevation) across the state of Oregon, US (cf. Figure 9.23)

The extent to which the prediction at a location is influenced by other nearby spatial locations is controlled by the *kernel bandwidth* or simply *bandwidth*. Larger bandwidths take into account more observations in the prediction, in the extreme approaching the global linear regression results. Smaller bandwidths are more strongly influenced by nearby locations, in the extreme tending towards extreme overfitting. The predictions at a particular bandwidth are usually calculated using a distance decay function, such as a Gaussian or bisquare distribution. In that way, closer locations within the bandwidth influence the estimate for a location more than more distal locations within the bandwidth.

kernel bandwidth

Bandwidths themselves can be expressed in (Euclidean) distance, termed a *fixed kernel*, or in terms of the number of neighbors, termed an *adaptive kernel* (since in this case the distance used will adapt to the density of input data points). Figure 9.25 shows an example of an adaptive bisquare kernel containing 19 points at two locations a and b. Figure 9.24 was constructed using an adaptive bandwidth of 30 neighbors, for example.

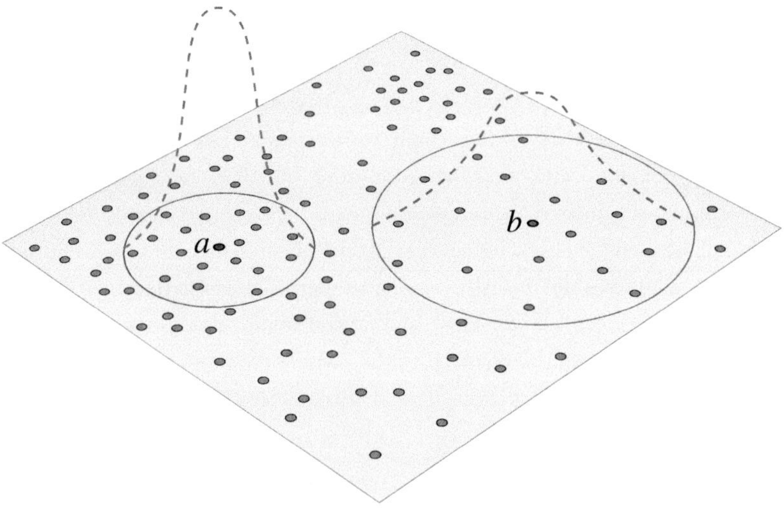

Figure 9.25: Bisquare adaptive kernel containing 19 neighbors at estimated locations a and b

The first step of any GWR modeling, then, is to select an appropriate bandwidth. Bandwidth may be estimated by an experienced human analyst or an optimized bandwidth can be computed. The most common optimization procedure is *cross-validation*, where the chosen bandwidth minimizes the sum of the square differences between the estimated and known dependent variable. More formally, the selected bandwidth h minimizes CV—the difference between observed dependent variable at each location y_i and that estimated from the neighbors within distance h, excluding the observation y_i itself, denoted $y_{\neq i}(h)$:

$$CV = \sum_{i-1}^{n} (y_i - \hat{y}_{\neq i}(h))^2$$

The R^2 value is commonly used for evaluating the quality of a spatial regression such as GWR. The overall R^2 for the regression in Figure 9.24 is 0.74—moderately high, but perhaps might be improved using an optimized bandwidth procedure. In addition, Figure 9.26 shows the spatial variation in R^2 values output by the geographically weighted regression in Figure 9.24. Figure 9.24 highlights the spatially varying quality of the regression, of lowest quality in the southern central region, but performing well in other areas, especially the northwest of the state.

It lies beyond the scope of this book to attempt a more complete treatment of spatial analysis. The leading textbook specifically on this fascinating

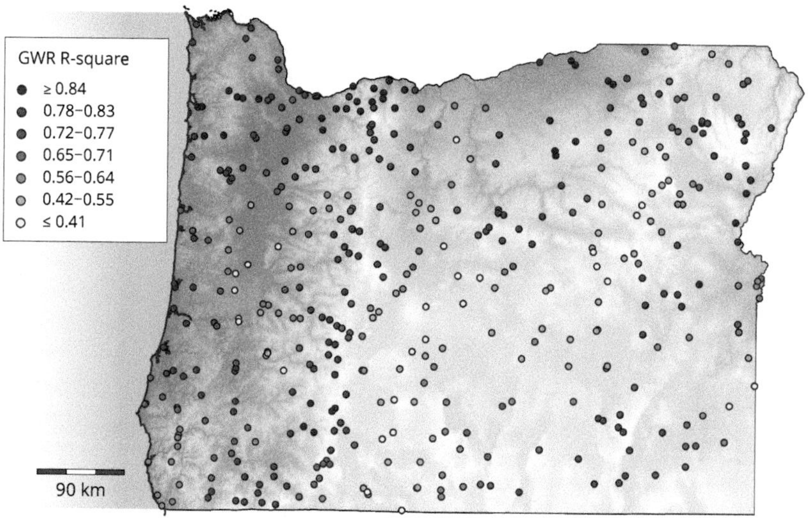

Figure 9.26: Output R^2 for geographically weighted regression of mean August temperature (dependent variable) against elevation (independent variable, lighter green indicates higher elevation) across the state of Oregon, US (cf. Figure 9.24)

topic is O'Sullivan & Unwin (2010). Other GIS textbooks, such as Burrough et al. (2015) and Longley et al. (2015), devote significant portions of the contents to the topic. A great number of other, application-oriented texts adopt a discipline-specific view, such as spatial analysis for social scientists (e.g., Rey & Franklin, 2022), or a technology specific perspective, such as spatial analysis with R (e.g., Comber & Brunsdon, 2020). The central message of our treatment of this topic, however, is that the most venerable and celebrated spatial analysis techniques are also machine learning techniques. Learning patterns from data is at the root of spatial analysis too.

9.4 *Deep learning*

The previous section provides an introduction to the huge variety of machine learning techniques that exist today, by focusing on that subset of clustering, regression, and classification techniques most frequently used with spatial data. There is one further family of machine learning techniques, however, that has had arguably more impact on GIS than any other: deep learning.

9.4.1 *Artificial neural networks*

Deep learning is founded on *artificial neural networks*. An artificial neural network (ANN), or simply a *neural network*, is a machine learning classification or regression algorithm and is commonly represented using a weighted, directed graph. The simplest possible ANN has multiple input nodes connected to a single output node by weighted directed edges as in Figure 9.27. The output node computes the weighted sum of inputs, and it then applies an *activation function* to determine what output is generated. A simple activation

neural network

activation function

function is to apply a threshold, such that a 1 is output if the weighted sum plus some bias b is above the threshold, -1 otherwise.

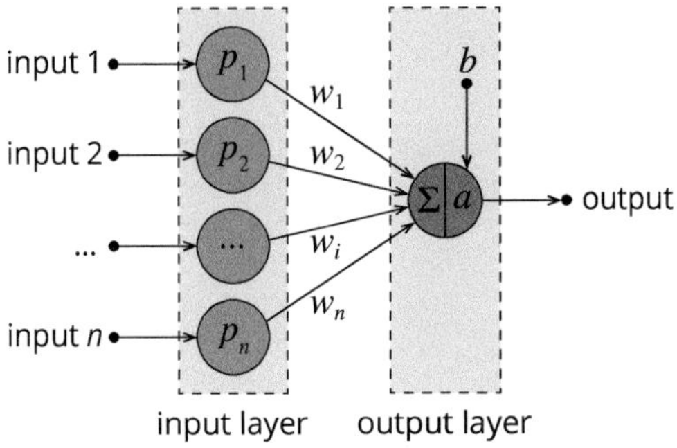

Figure 9.27: A single-layer feed-forward ANN or perceptron

The structure of the ANN in Figure 9.27 has just one layer of activating (output) nodes. The structure also only ever feeds data forward from inputs towards outputs, known as a *feed-forward network*. Accordingly, this structure is known as a *single-layer feed-forward neural network*, or more simply as a *perceptron*. The terms "perceptron" and "neural network" were coined to highlight the biological inspiration of animal neurons and brains behind these computational structures.

feed-forward network

perceptron

Despite their simplicity, even perceptrons alone can provide basic machine learning classification capabilities. Consider the classification problem in Figure 9.28a, where a training data set of points in Euclidean space has been classified into two categories shown in red (corresponding to -1) or blue (corresponding to 1) based on an implicit linear boundary between them.

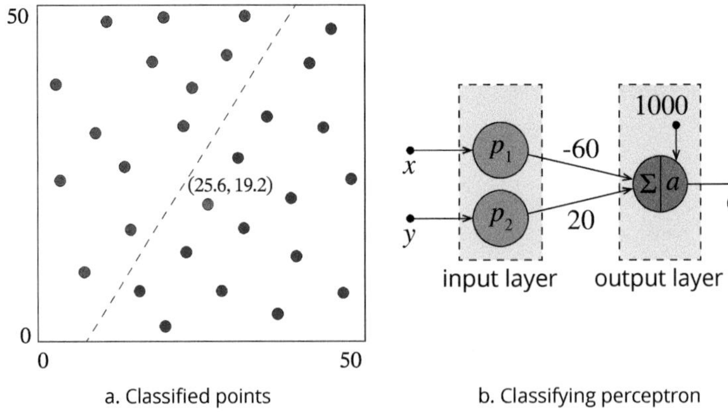

Figure 9.28: Simple classification using a trained perceptron

The perceptron in Figure 9.28b has two inputs, corresponding to the x and y coordinates of input points. The activation function, $a : \mathbb{R} \rightarrow \{-1, 1\}$, combines the weighted sum with the bias as follows:

$$a(v) \mapsto \begin{cases} -1 \text{ if } v + b > 0 \\ 1 \text{ otherwise} \end{cases}$$

For example, applying the weights $w_1 = -60$ and $w_2 = 20$ and a threshold of $b = 1000$, the perceptron in Figure 9.28b can correctly classify all the points in Figure 9.28a. A new point (shown in gray) with coordinate (25.6, 19.2) is classified by the perceptron as $a(-1152) = 1$ (i.e., classified as blue since $-60 * 25.6 + 20 * 19.2 + 1000 \leq 0$).

Implicitly, the perceptron is performing binary classification based on the dashed line indicated in Figure 9.28a. Training such a perceptron is effectively a matter of using the training data to learn the weights and bias needed to correctly classify the data. There are many sophisticated strategies used to train neural networks efficiently. However, a simple training strategy that illustrates the key features of neural network training is shown in Algorithm 9.4.

Algorithm 9.4: Simple perceptron training

Input n perceptron weights $w_1, ..., w_n$, activation threshold b, training data table T with $n + 1$ attributes (n features, 1 classification)

1: initialize weights $w_1, ..., w_n \leftarrow 0.0$
2: initialize activation threshold $b \leftarrow 0.0$
3: **repeat**
4: randomly select and remove one row $\langle p_1, ..., p_n, o \rangle \in T$
5: compute weighted sum $v \leftarrow \sum_{i=1}^{n} w_i . p_i$
6: classify output as $o' \leftarrow \begin{cases} 1 \text{ if } v \leq b \\ -1 \text{ otherwise} \end{cases}$
7: **if** $o' \neq o$ **then**
8: update all weights $w_i \leftarrow w_i + o'.p_i$
9: update threshold $b \leftarrow b - o'.v$
10: **until** all rows in T have been visited

Output Trained weights $w_1, ..., w_n$ and threshold b

The simple training strategy in Algorithm 9.4 begins by initializing the weights and bias with arbitrary values (lines 1–2). Next, we iterate through the training data set, selecting and removing data items to process in random order (lines 3–4). The selected training data item is classified using the (initially arbitrary) weighted sum and output in lines 5–6. If the computed classification o' does not match the training data ground truth classification o each weight w_i is updated to $w_i + o'.p_i$ (line 8) and the bias b is updated to $b - o'.v$ (line 9). Otherwise, the weights and bias are left unchanged.

Table 9.9 steps through the first 13 iterations of Algorithm 9.4 using the training data set and untrained perceptron from Figure 9.28. The example rapidly converges on the approximate weights and threshold close to those used above ($w_1 = -58.2$, $w_2 = 20.3$, $b = 926.1$).

9.4.2 *Multilayer perceptron networks*

The perceptron discussed above illustrates several features common to all ANNs, such as the graph-based structure, the classification itself using weighted sums, and the learning of weights and biases from training data. However, as might be expected, such simple ANNs have significant limitations. In particular, perceptrons are only able to learn a certain specific type

Table 9.9: Example training of perceptron in Figure 9.28b using data in Figure 9.28a and Algorithm 9.4

x	y	o	class	w_1	w_2	b	v	o'	predicted
13.1	26.4	-1	red	0.0	0.0	0.0	0.0	1	blue
19.1	48.3	-1	red	13.1	26.4	0.0	1525.3	-1	red
19.3	3.0	1	blue	13.1	26.4	0.0	332.0	-1	blue
39.0	13.3	1	blue	-6.2	23.4	332.0	69.4	-1	red
47.3	24.6	1	blue	-45.2	10.1	401.5	-1889.5	1	blue
15.4	6.5	1	blue	-45.2	10.1	401.5	-630.4	1	blue
31.1	17.4	1	blue	-45.2	10.1	401.5	-1230.0	1	blue
2.8	38.5	-1	red	-45.2	10.1	401.5	262.3	-1	red
31.3	46.8	-1	red	-45.2	10.1	401.5	-942.1	1	blue
44.3	36.6	1	blue	-13.9	56.9	-540.6	1466.8	-1	red
43.1	32.3	1	blue	-58.2	20.3	926.1	-1852.7	1	blue
23.3	38.0	-1	red	-58.2	20.3	926.1	-584.7	-1	red
3.4	22.7	-1	red	-58.2	20.3	926.1	263.0	-1	red
...	...	...	...	...	...	...	...	...	...

of classification (i.e., those where classification can be achieved with a single straight line or hyperplane, termed *linearly separable*) and are not able to classify reliably in the presence of noisy data.

More sophisticated ANNs, however, can overcome these limitations, and indeed provide the basis of "universal" classifiers, able to approximate any continuous function. There are four main features that more sophisticated ANNs add to the perceptrons already encountered.

multilayer perceptron network

hidden layer

First, many perceptrons can be connected together into a single, combined neural network with multiple layers, termed a *multilayer perceptron network* (MLP). Figure 9.29 shows an MLP with three activating layers. In between the (inactive) input layer and the output layer, two "hidden" layers provide many more connections leading to many more possible paths between input and output. Hidden layers stage the processing of input data, with successive layers usually having fewer nodes—termed the *width* of the layer—than its predecessor. The proliferation of paths made possible by the multilayer structure enables much more sophisticated classifications to be learned than achievable with a single perceptron.

Figure 9.29: A three-layer perceptron network (weights omitted)

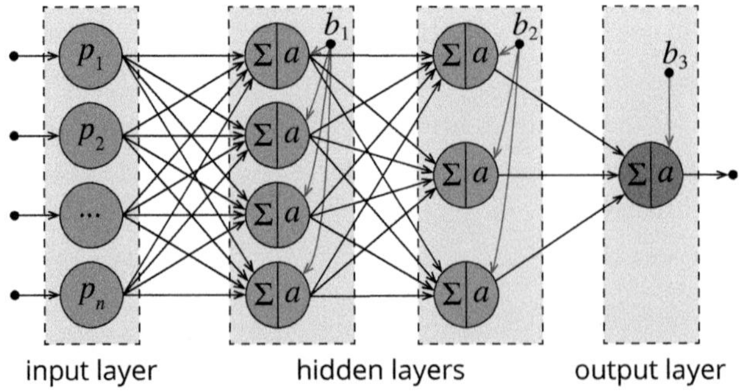

Second, MLPs can employ more sophisticated activation functions than the simple step function encountered in the previous section. Figure 9.30 shows the two most common activation functions used in preference to the simple step function. The *sigmoid function* in Figure 9.30b is effectively a continuous differentiable analog of the step function. Returning a continuous value in the range [0.0, 1.0], rather than binary values such as {0.0, 1.0}, ensures small changes to weight and bias parameters do not lead to sudden dramatic changes in outputs, as can happen with binary step functions. The *rectified linear unit function* (ReLU) function in Figure 9.30c is also continuous, but not differentiable. Both sigmoid and ReLU functions are nonlinear, enabling the neural network to approximate nonlinear classifications. Being piecewise linear, however, the ReLU activation function can be computationally more efficient in practice.

sigmoid function

rectified linear unit function

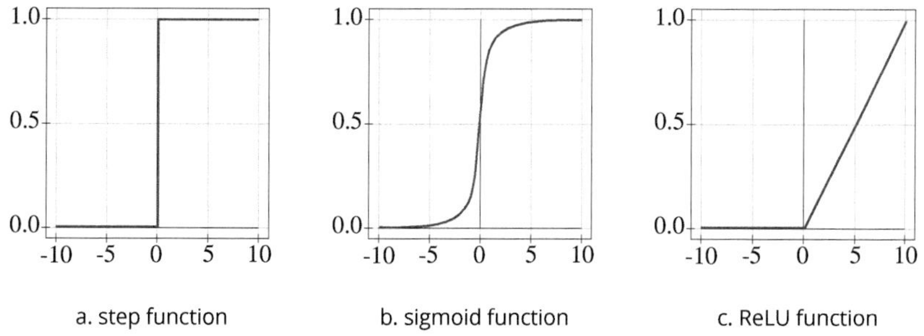

Figure 9.30: Three common activation functions

Third, the added hidden layers in MLPs demand more sophisticated mechanisms for updating weight and bias parameters during training. In a single-layer perceptron, all classification errors can be attributed to the parameters associated with the output layer. In MLPs, updates to parameters across the hidden layers are required in proportion to the contributions of hidden nodes to output errors. A process known as *back-propagation* is used to achieve this and effect parameter updates in the hidden layers too during training.

back-propagation

Fourth, the neural networks we have encountered thus far are feed-forward and fully connected, with every node in a layer with forward connections to every node in the next layer. As we shall see in the next section, these features too are extended by true deep learning techniques.

9.4.3 Deep learning models

We have seen above that the number of nodes in a layer provides a measure of the *width* of a neural network. The *depth* of an ANN is defined by the number of layers in a feed-forward network, and more generally by the length of the path from input to output in neural networks where feedback loops are possible. *Deep* learning is particularly concerned with the construction, training, and use of *deep* neural networks. While there is no universally agreed upon definition of what counts as "deep," generally *deep learning* is taken to mean any ANN with paths of three or more activating nodes from input to output. Hence, at a minimum a feed-forward MLP with two hidden layers would count as deep learning.

deep learning

convolutional neural network

Convolutional neural networks Perhaps the most important family of deep neural network structures for spatial data is the *convolutional neural network* (CNN), particularly used with image-based and raster spatial data. Large images may contain millions of pixels, which would translate into billions of weights in a fully connected MLP. Even a small 8 channel, 1024 pixel square satellite image would require more than 8 million weights ($8 \times 1024 \times 1024 =$ 8,366,608) for each node in an MLP.

Such wide, fully connected MLPs are not computationally efficient. More importantly, nor are they statistically efficient because we expect that most of the meaningful weights will relate nearby pixels, due to autocorrelation. CNNs achieve increase both computational and statistical efficiency by effectively restricting the connections between neural network nodes to those representing nearby localities. To achieve this, two new types of layers are introduced by CNNs:

convolutional layer
kernel
pooling layer

1. a *convolutional layer* which combines values from pixels within a specified neighborhood using an image processing *kernel*; and
2. a *pooling layer* which combines values from nearby pixels using maximum, minimum, median, or other aggregate function of all the pixel values within a specified neighborhood.

Figure 9.31 illustrates the operations embedded into the structure of these two types of layers. Starting with a simple 8×8 pixel image on the left of Figure 9.31, and placing the 3×3 kernel over each pixel in turn, a new 6×6 image is computed as the pixel-wise weighted sum of kernel and pixel values.[15] In Figure 9.31, the kernel configuration shown is known as a *high-pass filter* and is tuned to detect edges in the image. Other kernel configurations will tend to highlight other features, allowing each convolutional layer to be tuned to activate in response to certain types of image features.

[15] For example, placing the kernel in the top left of the input image in Figure 9.31, the top left pixel of the convolutional layer is computed as: $4 \times 4 + 4 \times -1 + 4 \times -1 + 3 \times -1 + 3 \times -1 = -2$.

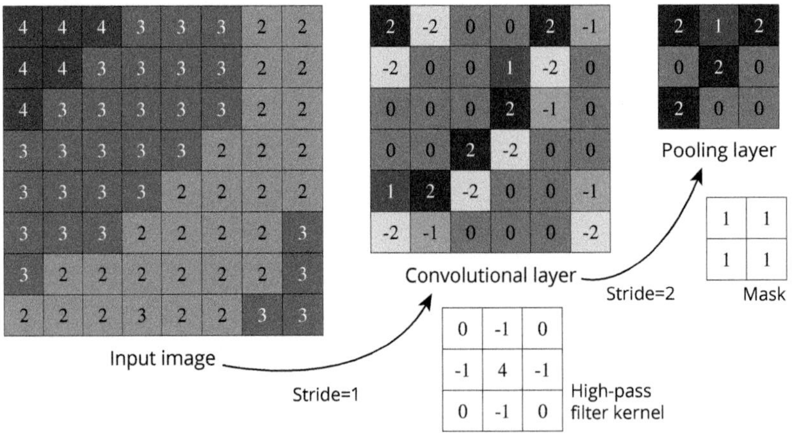

Figure 9.31: CNNs: convolution of an image using a 3×3 high-pass filter with stride 1 followed by max pooling using a 2×2 mask with stride 2

stride

The progress of the kernel across the image is controlled by a model parameter, termed *stride*. A stride of 1 means the kernel is applied to every available pixel; a stride of 2 applies the kernel to available pixels in every other row and column; a stride of 3, available pixels in every third column and row, and so on.

Similar to convolution, pooling is also based on aggregation of neighboring pixels. Pooling applies an aggregation operation—such as maximum, minimum, or modal pixel value—to the set of all pixels in a neighborhood, called a *mask*. In Figure 9.31, the result of the **max** pooling operation using a 2 × 2 mask with stride 2 is shown in the final pooling layer. Pooling assists in summarizing and downsampling an image, reducing the size of subsequent layers and thus the number of parameters to train.

mask

Figure 9.32 summarizes a generalized CNN architecture. The input image is used to construct a set of convolutional layers (just three in Figure 9.32), each derived using different kernels. Pooling layers downsample into a set of smaller layers. Most CNNs apply a series of convolution and pooling layer pairs, not shown in Figure 9.32. The final pooling layers are then recombined by feeding into one or more fully connected hidden layers and ultimately the output layer, as we have seen already used with MLPs. Training a CNN begins in the usual way with back-propagation through the hidden layers. Pooling has no trainable parameters, and so no learning takes place with pooling layers. Importantly, back-propagation through convolutional layers updates the *kernel* weights rather than individual ANN edge weights, which dramatically reduces the number of parameters to learn, when compared with a fully connected MLP for example.

Figure 9.32: Summary architecture of a CNN, with three (3 × 3) convolutional filters, 2 × 2 pooling filters, and two fully connected hidden layers

Graph neural networks CNNs operate on the implicit neighborhood structure of raster images. The kernels and masks that underpin the convolutional and pooling layers in CNNs are the mechanism by which that neighborhood is encoded within a CNN. *Graph neural networks* (GNNs) generalize the concept of the CNN to the arbitrary neighborhood structures found in abstract graphs. Rather than use pixel neighborhoods, GNNs are structured around learning with encoded graph neighborhood structures.

graph neural network

Graphs are fundamental to GIS, as we have seen in Chapter 2, Chapter 3, and indeed in this chapter too (in connection with knowledge graphs, Section 9.1). New applications of GNNs are emerging every day, and we do not delve further into this fast-moving area. Nevertheless, it is possible to identify three broad types of tasks tackled using GNNs:

- node-level tasks, such as node classification and clustering;
- edge-level tasks, such as edge classification and link prediction; and
- graph-level tasks, such as graph classification and graph matching.

recurrent neural network

Recurrent neural networks Recurrent neural networks are our final stop on this tour of deep learning. Unlike feed-forward networks, a *recurrent* neural network (RNN) contains feedback loops, where node outputs may connect back into the current or previous layers as well as forward towards the output (see Figure 9.33). Looped connections equip the neural network with an analog of working memory, where previously processed training data may influence subsequent training steps. As a result, RNNs are especially well adapted for learning from data where the sequence or timing of data items is significant.

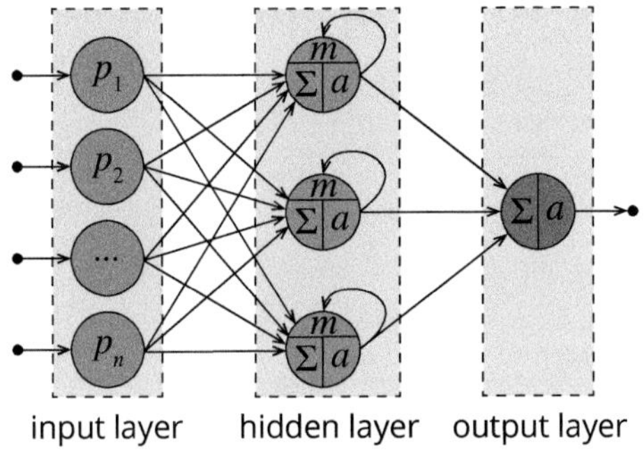

Figure 9.33: RNN: recurrent neural networks allow loops, where feed-forward and feedback links between nodes are permitted

In practice, to enable recurrence, nodes accepting feedback connections are equipped with a memory to store the node state from one iteration to the next. The capacity to store node state between iterations is indicated with an m in the nodes with feedback links in Figure 9.33. As for CNNs and GNNs, the back-propagation algorithm has to be modified to accommodate successive parameter updates, termed *back-propagation through time* (BPTT). In short, BPTT effectively "unrolls" the RNN into a sequence of MLPs (as shown in Figure 9.34), back-propagating parameter updates through the chain, before rolling back up the RNN ready for the next iteration.

back-propagation through time

Deep learning has become fertile ground for research and development of new models and techniques, of which GNNs and RNNs are just two examples. Other examples include GANs—*generative adversarial networks*—which pit two neural networks one against the other. The two networks compete against each other in training: one (the *generator*) learns to generate new data with the same statistical properties as the training data; the second (the *discriminator*) learns to spot the differences between real and generated data sets. Further innovations and new applications are to be expected in the coming decade, especially as bigger and bigger data sets containing billions of

generative adversarial network

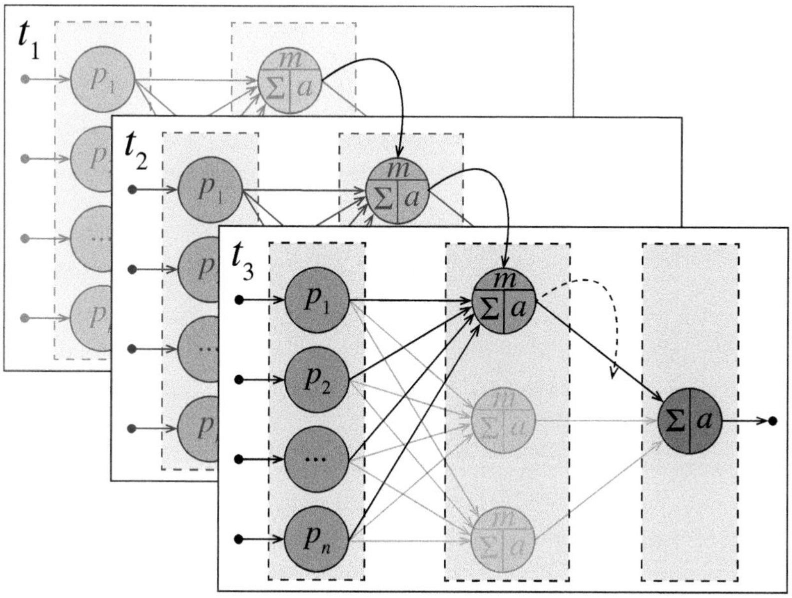

Figure 9.34: Back-propagation in RNNs can be achieved by unrolling the network into a chain of MLPs

items, such as the data sets used to train neural networks for natural language processing (NLP) like GPT (Brown et al., 2020), become more and more common.[16]

[16] ChatGPT—released after this sentence was written, but before this book was published—serves to underline the rapid pace of innovation in deep learning. The remarkable ability of ChatGPT to generate seemingly intelligent text guarantees that its release in late 2022 will forever remain a milestone in the history of machine learning. Trained on vast corpora of human written text and tasked with prediction (regression) of word frequencies, ChatGPT and similar tools are, nevertheless, statistical models constructed using deep learning techniques fundamentally no different to those explored in Section 9.4. The ability of such statistical models of word frequencies to mimic human writing led renowned AI researcher Emily Bender and coauthors to coin the term "stochastic parrots" (Bender, Gebru, McMillan-Major, & Shmitchell, 2021).

Reflections

As we have seen, AI is not *only* about machine learning; and neither is machine learning—nor AI more broadly—a *new* frontier for GIS. This chapter unpicks the deep and longstanding connection between AI and GIS. The strong link between machine learning and spatial analysis, highlighted in Section 9.3, is especially significant. In addition to the spatial analysis textbooks already cited (including Burrough et al., 2015; O'Sullivan & Unwin, 2010), Li et al. (2016), Li (2020), and Li & Arundel (2022) offer accessible, recent introductions and summaries of the variety of spatial analysis techniques in the context of AI and big data. The synergies between GIS and AI, and the unique problems and opportunities thrown up by their combination, have led to its emergence as a distinct area of inquiry, sometimes called *GeoAI* (Mao, Hu, Kar, Gao, & McKenzie, 2017).

GeoAI

Undoubtedly, recent years have seen unprecedented growth in the development and practical application of deep learning in particular. These advances are enabling similarly rapid advances in applications to GIS. CNNs (convolutional neural networks), for example, already have well-established spatial applications to feature extraction from satellite imagery (e.g., Romero, Gatta, & Camps-Valls, 2016). Such applications are helping to automate spatial analyses that previously required many months of manual labor (for example, to

Figure 9.35: Deep learning to segment street panoramas into landscape types, after Sun et al. (2021)

automatically segment street view panoramas into landscape types—sky, trees, buildings, roads, grass pixels, Figure 9.35). Further spatial application areas of deep learning—such as urban analytics (see Li, Zhao, & Zhong, 2022) and "GeoQA" (geographic question-answering systems, e.g., Mai, Janowicz, Zhu, Cai, & Lao, 2021)—are likewise certain to continue to expand.

Nevertheless, the inherent structure and rich semantics of spatial and temporal information also demands abstract and symbolic representations and reasoning, such as ontologies and linked open data (Mai, Janowicz, Yan, & Scheider, 2019). While the ML and KR^2 branches of AI adopt markedly different approaches to intelligence, both are well suited to GIS applications. Indeed, combinations of ML and KR^2 are possible and offer the potential to capitalize on the advantages of both (e.g., Du, Wang, Ye, Sinton, & Kemp, 2022; Duckham et al., 2022; Li, Ouyang, & Zhang, 2022). However, as we witness wider use of powerful GeoAI techniques in society, the challenges are not only technical, they are ethical. The next chapter tackles head-on not simply questions of what we *can* we do with GeoAI, addressed in this chapter, but also questions of what *should* we do.

Spatial Information in Context

10

Sections

For better and for worse, we live in an information age. Spatial information and the spatial technologies encountered in previous chapters have a pivotal role in society, today and into the future. In this chapter you will learn how to:

- describe the structure and types of uncertainty and how uncertainty impacts the fitness-for-use of spatial data;
- distinguish between the different approaches to logical reasoning with uncertain data, such as rough sets, fuzzy sets, and probability;
- discuss location privacy and anticipate what can go wrong when we lose it; and
- recognize the social context, implications, and biases in using GIS, and analyze the role of spatial data and GIS in advantage and marginalization in our society.

KNOWLEDGE is power, so the oft-quoted aphorism tells us. Spatial information underpins knowledge—and thus power—in society today, and has done for many centuries. The history of positioning and mapping, for example, is intimately intertwined with the history of power in society. The world's first national mapping agency, the UK's Ordnance Survey[1], was founded to create maps for the British Army. The first ever *GNSS* (global navigation satellite system), the US global positioning system (GPS), was developed by the US military originally to solve problems of long-range missile guidance. One of the earliest GIS software packages still in active development and use today—GRASS GIS[2]—began life as a US Army Corps of Engineers project in 1982.

[1] The literal meaning of "ordnance" in English is "military supplies," especially weapons and ammunition.

GNSS

[2] https://grass.osgeo.org

Understanding the role of spatial information in society is therefore not just a core element of GIS expertise; the inherent connection to power makes this understanding an ethical responsibility for geographical information scientists. Other texts have already explored the role of GIS applications in society (such as Chrisman, 2006; Nyerges et al., 2011). Instead, taking a "computing perspective," this chapter looks at the fundamental characteristics of spatial information that have "special" implications in the wider context, including the wider societal context.

We begin with one of the most longstanding and important of all topics in GIS: uncertainty (Section 10.1 and Section 10.2). Section 10.3 then looks at the role of location in understanding and managing personal privacy, and the close connections between privacy and uncertainty. Finally, Section 10.4

takes a step outside of the technical and procedural aspects of GIS to place GIS maps and mapping in the context of society today, reflecting on the implications for geographic information scientists.

10.1 Uncertainty

uncertainty

Uncertainty is an important and unavoidable property of the world, information about the world, and our cognition about the world. The term *uncertainty* is often used in two related but distinct senses. First, "uncertainty" can refer to our *state of mind*. We might say we are uncertain about an object's location, for example, meaning that we are unsure in our own minds about where the object is located. Second, though, "uncertainty" may be applied directly to the world, or to data or information about the world. If we say that an object's location is *itself* uncertain, it is usually taken to reflect on the limitations of our measurements of that location.

Figure 10.1 illustrates the distinction using the specific example of the cities of Aalborg in Denmark and Linköping in Sweden. The statement "I am uncertain whether Aalborg is north of Linköping" is a reflection of my uncertain state of mind (Figure 10.1a). Although my knowledge of geography is not sufficient to be certain whether Aalborg or Linköping is further north, the data is clear on this.[3] In contrast, the statement "It is uncertain whether the population of Aalborg is greater than Linköping" is a statement about the limitations of data itself. The best population data we have (Linköping population of 119,016 measured on 31 December 2020 census; Aalborg population of 119,862 measured on 1 January 2022 census) is not sufficiently up-to-date or contemporaneous to determine with certainty which of the two cities is larger.

[3] In fact, Linköping is further north, at a latitude of about 58.4 degrees north compared with about 57.0 degrees north for Aalborg.

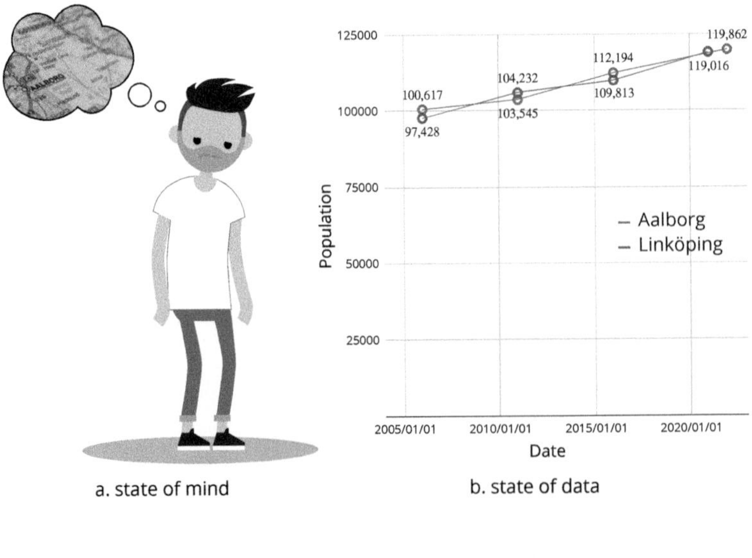

a. state of mind

b. state of data

Figure 10.1: Uncertainty as a state of mind (e.g., I am uncertain whether Aalborg is north of Linköping); and applied directly to data (e.g., it is uncertain whether the population of Aalborg is greater than Linköping)

Consider the relatively basic proposition of mapping the boundary of a forest using a GIS. A number of questions about the exercise become immediately apparent:

Uncertain specifications What exactly do we mean by "the forest boundary"? Is it related to the density of trees? Or perhaps the height? Maybe we refer to a local authority's fiat boundary? And how do we make sense of changes over time in each of these?

Uncertain measurements How good must our measurements of boundary location be? What if the measuring instrument is incorrectly calibrated, or we misuse it (for example, misread the dial or use it at the wrong temperature)? Are we aiming to measure to the nearest meter? Millimeter? Nanometer?

Uncertain representations How should we discretize the continuous forest boundary in a way amenable to digital representation (cf. Section 4.1.1)? Do we use a raster image? Or straight-line vectors between sampled boundary points? Or perhaps interpolate more sophisticated curves between boundary points?

10.1.1 The ontology of imperfection

Uncertainty arises because of *imperfection* in our knowledge about the world. Knowledge about the world is gained through observation. Observations are imperfect in the sense that they can never fully or correctly reflect all aspects of reality. The word "imperfect" here is not intended to be used with any value-laden connotations—imperfection is neither better nor worse than perfection. Rather, imperfection is integral to the very fabric of our data, representations, and analyses of the geographic world, and it needs to be faced head-on.

imperfection

Imperfection is therefore where we begin our journey into uncertainty. Figure 10.2 shows a classification of three different types of imperfection. Imperfection is at the root of our classification, with two distinct, orthogonal subclasses: *inaccuracy* and *imprecision*.

- Inaccuracy refers to a lack of *correctness* in data, in the sense that our observations or representations do not accord with reality. Inaccuracy is synonymous with *error* (see Box 10.1 on the following page).
- Imprecision concerns a lack of *specificity* or *detail* in data.

inaccuracy

error

imprecision

To see that inaccuracy and imprecision are independent and orthogonal, consider the following statements:

- "This chapter was written in Australia" is accurate, but imprecise as it does not provide much *detail* about where in Australia it was written.

- "This chapter was written in the village of Madeley, in Staffordshire, England" is much more precise than the previous statement (it provides much

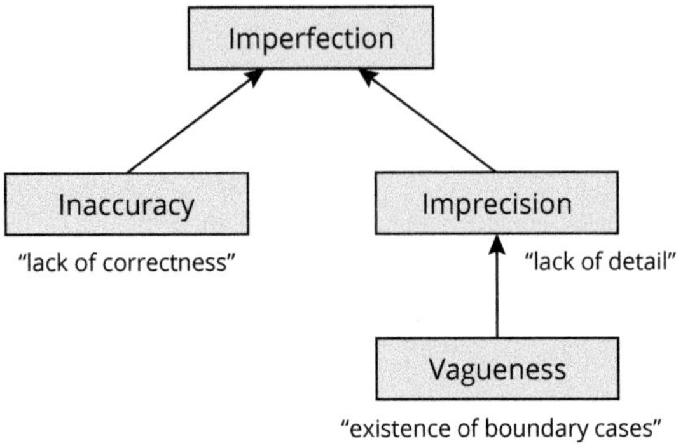

Figure 10.2: Ontology of imperfection

more detail about the location), but it is hopelessly inaccurate. (In fact, Mike wrote the *first* edition in Madeley, but this chapter was not part of the first edition).

vagueness

A third subclass of imperfection—*vagueness*—concerns the existence of *borderline cases* for a concept. For example, the location "Qomolangma" (also known as "Mount Everest") is a vague concept. There exist many locations that are definitely part of Qomolangma (such as the summit and the north face) and many locations that are definitely not part of Qomolangma (such as Madeley and Melbourne). Crucially, there are also many borderline locations in the Himalayas where it is not clear whether they form part of Qomolangma or not. The term *crisp* is often used as an antonym for vagueness, to refer to concepts or regions where there are no borderline cases. Similarly, a *crisp set* is simply a normal, classical set, in which an element is either a member of the set or not (no borderline cases).

crisp

crisp set

Not all imprecision is vagueness. For example, we could assert that a person is located in the USA—not a vague statement (assuming for the moment a crisp boundary to the USA) but nevertheless imprecise. However, every vague statement must also be imprecise, because of the lack of specificity at the boundary. Hence, vagueness is shown as a special type of imprecision in Figure 10.2.

Box 10.1: Errors of omission and commission

An incorrect data item is an *error*; error and inaccuracy are synonymous. There are two types of errors (inaccuracies) that we can distinguish, similar to the distinction between Type I and Type II errors in statistical significance testing. An *error of commission* is made when we conclude that a proposition is the case, when in fact it is either undetermined or not the case. An *error of omission* occurs when we do not declare a proposition to be the case, when in fact it is the case. For example, a remotely sensed image may be used to determine land cover. For a particular land cover class, such as "forest," those regions that are actually forest but have been misclassified as some other land use class, such as "agricultural crops," constitute the errors of omission for the forest class. Conversely, those regions that are not in actuality forest but have been misclassified as forest constitute the errors of commission for the forest class. In the case of land cover maps, errors of omission and commission are inversely related to each other, because *every* location must be a member of exactly *one* land cover class. Thus, those "forest" regions that are misclassified as "agricultural crops" constitute errors of omission for "forest," but the same errors are errors of commission for "agricultural crops."

Reasoning with vagueness Reasoning using vague assertions is surrounded by difficulties, exemplified by the *sorites paradox*. The sorites paradox concerns the logical problem that arises when attempting to reason about the vague concept of a "heap" (etymologically, "sorites" derives from *soros*, Greek for "heap").

sorites paradox

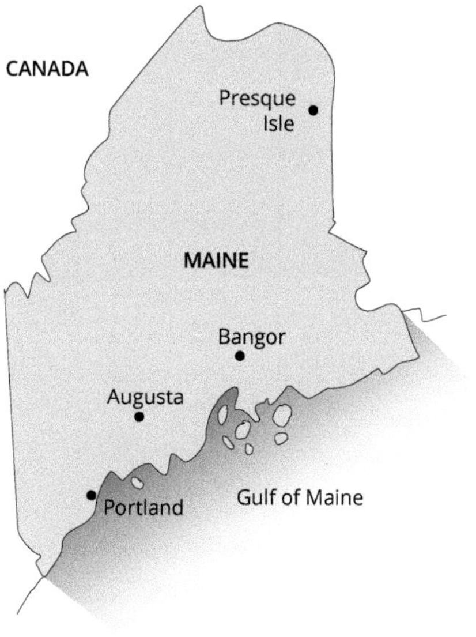

Figure 10.3: The sorites paradox: where is "southern Maine"?

This paradox can be illustrated in another way, using a geographically vague concept such as "southern Maine" (Figure 10.3). We may be certain that Portland, Maine, is in southern Maine, and that Presque Isle is not. Because the geographical region called "southern Maine" has no clear boundary, it seems reasonable to assert that a single step taken by a person walking north along the road from Portland to Presque Isle cannot make the difference between that person being in southern Maine or not. Given that no single step can make such a difference, it follows that a traveler starting at Portland and walking to Presque Isle could reasonably infer at each step that they were still in southern Maine. Eventually, the traveler would arrive both at Presque Isle and at a paradox: the original assertion that Presque Isle is definitely not in southern Maine is contradicted by the step-by-step reasoning process during the walk from Portland to Presque Isle. The sorites paradox shows that our normal modes of reasoning, in this case the principle of mathematical induction, cannot be applied in the face of vagueness.

Granularity and indiscernibility Granularity is closely related, but not identical, to imprecision. Granularity refers to the existence of clumps or grains in data, in the sense that individual elements in the grain cannot be distinguished or discerned apart from one another.

granularity

For example, imagine a remotely sensed image of a region of the Earth. In this image, any two locations on the ground that fall within the same pixel cannot be distinguished by the observation. Granularity in an observation makes differences *indiscernible* between things that are in principle distinguishable from one another (Figure 10.4).

indiscernibility

Indiscernibility is a key concept in modeling granularity and imprecision. Intuitively, a finite collection of elements is indiscernible with respect to a particular observation if any pair of elements in the collection cannot be distinguished from each other by the observation.

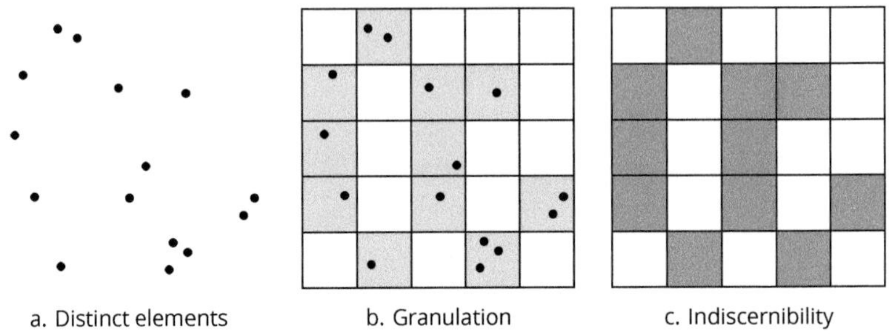

Figure 10.4: Granularity and indiscernibility

Figure 10.5: Close-up of the eight class land cover map already encountered in Figure 4.15 (Map source: US NLCD 2019 Land Cover)

Indiscernibility may occur not only by granulation of the *physical* space but also the *thematic* space. The granularity of a classification imposes a granulation on the different thematic types. For example, a land cover classification, such as that already encountered in Figure 4.15 (reproduced again in Figure 10.5), leads to thematic indiscernibility within the different classes of land cover.

frame of discernment

quotient set

In the case of pixelation of an image (as in Figure 10.4) and a flat thematic classification (as in Figure 4.15), indiscernibility is often assumed to be an equivalence relation (reflexive, symmetric, and transitive; see Section 3.2). More formally, let ρ be a binary equivalence relation on set X. Any equivalence relation on a set partitions that set into equivalence classes. In our case, the classes will consist of blocks (subsets) of members of X that are indiscernible from each other. The blocks are pairwise disjoint and completely cover X. The set of blocks is termed the *frame of discernment* of X with respect to ρ, and is written X/ρ termed the *quotient set* of X. With reference to Figure 10.4, Figure 10.4a represents the set X. The frame of discernment is the grid in Figure 10.4b, which results in the shaded blocks (equivalence classes) in Figure 10.4c.

An equivalence relation is not always a good model of indiscernibility, however. Many land cover classifications are hierarchical, where a location may belong to multiple related classes (e.g., agricultural land cover may also be pasture or cropland, which may in turn be cereal crops) and so do not form an equivalence relation. Spatially too, the sensor arrays used to generate remotely sensed satellite imagery often exhibit more complex forms of indiscernibility, where the transitivity property in particular may not be satisfied. A relation that is symmetric and reflexive but not transitive is termed a *similarity relation*.

similarity relation

10.1.2 Dimensions of data quality

Data quality refers to the characteristics of a data set that may influence the certainty associated with decisions based upon that data set. The accuracy and precision of a data set are clearly elements of data quality because each may influence the confidence a decision maker would have in decisions based on that data set. *Lineage* is another element of data quality, which describes the *provenance* of a data set (e.g., its source, age, intended use). The term *metadata* is often used to indicate data that provides information about a data set. Data quality elements such as accuracy, precision, and lineage are examples of metadata. Table 10.1 summarizes some important data quality elements.

data quality

lineage

metadata

Element	Concise definition
accuracy	Correctness of data, in accordance with reality
bias	Existence of systematic distortions within data
completeness	Exhaustiveness of data, in terms of the range of features represented
consistency	Level of logical contradictions within data
currency	How "up-to-date" data is
granularity	Existence of clumps or grains within data
lineage	Provenance of data, including source, age, and use
precision	Level of detail or specificity of data
reliability	Trustworthiness or degree of confidence a user may have in data
timeliness	How relevant data is to the current needs of a user

Table 10.1: Some data quality elements

Accuracy, granularity, lineage, and precision have been discussed above. *Bias* concerns the existence of systematic distortions within a data set. Such distortions may be introduced deliberately, or as an unforeseen consequence of observational, data collection, or analysis techniques. It is worth noting that precision, accuracy, and bias all have statistical counterparts (see "Statistical precision and accuracy" Box 10.2 on the next page). *Completeness* describes the exhaustiveness of a data set, in terms of what types of features are included and excluded from the data set.

bias

completeness

Consistency concerns the existence of logical contradictions within the data set. Consistency is violated when information is self-contradictory, either explicitly or by implication. Any sources of imperfection in data can lead to inconsistency. For example, assume the actual population of Bangor, Maine is 32,095. Two inaccurate estimates of the population are likely to be inconsistent with each other (for example, inaccurate population estimates of 34,371 and 31,473 inhabitants). Similarly, two imprecise observations of the population, if wrongly interpreted, may be inconsistent with each other (for example, a population of 32,000, accurate to the nearest 1000 inhabitants, and a population of 30,000, accurate to the nearest 10,000 inhabitants). Finally, if we assume for a moment that "Bangor" is a vague concept referring to the Bangor region (Bangor city itself has a precisely defined administrative boundary), then disagreements about where exactly the boundary lies will lead to different measurements of population. This results in inconsistency:

consistency

Box 10.2: Statistical precision and accuracy

Many of the data quality elements described in this chapter have closely related specific statistical counterparts. Precision is measured statistically using standard deviation:

$$\sigma = \sqrt{\frac{1}{n}\sum(x_i - \overline{x})^2}$$

where n is size and $\overline{x}$ is the mean value of some set of observations $\{x_i | i = 1...n\}$. Accuracy is measured statistically using root mean square error (RMSE):

$$rmse = \sqrt{\frac{1}{n}\sum(x_i - t)^2}$$

where t is the "true" value for the set of observations, or some value taken to be true. Standard deviation measures the spread of a set of observations of a value, while RMSE measures the discrepancy between a set of observations of a value and the true value. Although standard deviation and RMSE are conceptually distinct, numerically they are often equivalent, as the mean $\overline{x}$ of a set of observations is normally used as an estimate of the true value t of a set of an observable phenomenon. This numerical equivalence of statistics is one of the most common reasons precision and accuracy are confused. Conceptually, though, the two remain quite distinct: statistical precision gives the spread of a set of observations as an estimate of the *specificity* of a measurement (i.e., standard deviation); statistical accuracy measures the deviation of a set of observations from the *correct* measurement (i.e., RMSE).

some inhabitants excluded from the Bangor region in one observation may be included in another.

currency

The *currency* of a data set indicates a temporal relationship between data and its source, that is, how "up-to-date" it is. Depending on how time-critical the application is, currency is more or less of an important factor. *Timeliness* is more concerned with temporal relevance, and whether the data arrives at the right moment. Current data may not be timely, and vice versa.

timeliness

reliability

Reliability refers to the trustworthiness of data, or the degree of confidence a user may have in a data set. To an extent, reliability will be dependent on the other elements of data quality. For example, low accuracy data sets are likely to be unreliable. However, the relationship is not entirely straightforward. A biased data set, for instance, may still be "reliable" in the sense that the systematic distortions may be uniformly applied and predictable.

10.1.3 *Uncertainty in context*

Although uncertainty makes decisions harder and outcomes less clear, uncertainty has an important role in helping our society to understand the limitations of data and the caveats surrounding GIS analysis. For example, where uncertainty results in inconsistency, that may prompt us to review, revise, or refine our beliefs. In general, three types of action may be prompted by the discovery of new information that is inconsistent with existing beliefs (see also Figure 10.6):

least-squares adjustment

Resolve inconsistency Most conventional approaches to inconsistency aim to resolve inconsistency in some way when it occurs. For example, *least-squares adjustment* is a common statistical technique for resolving inconsistencies in survey measurements.

Retain inconsistency The disadvantage of resolving inconsistency is that information is usually lost during the resolution process. For example, once least-squares adjustment has been applied, the original observations are often discarded, making it difficult to resolve any subsequent inconsistencies that may arise. Retaining inconsistent information opens up new possibilities, such as performing "on the fly" least-squares adjustment on inconsistent measurements stored in the database, recomputed as new survey measurements are included.

Initiate dialog Finally, inconsistency may lead to a *dialog* between the agents responsible for the different information sources. An important example of dialog occurs when inconsistencies prompt the acquisition of new information, such as further data collection, in an attempt to resolve the original inconsistencies.

a. Resolve b. Retain c. Dialog

Figure 10.6: Responses to inconsistency

fitness-for-use

More broadly, one response to encountering uncertainty, data quality metadata, or inconsistency in data sets is to assess the *fitness-for-use* of the data set in our specific application. For example, a scientific study of habitat change in a national park might require a temporal series of spatial data with relatively high precision and accuracy. Detailed lineage and completeness metadata may also be needed to enable a comparison between (possibly changing) habitat classifications used through time. If instead the application is a tourist map to help inform and educate visitors to the national park about habitat types and history, data sets with much lower accuracy and precision may be entirely appropriate. Even with data quality metadata, assessing whether a particular data set is fit for some specified use may be complex, requiring experience and judgment; without such metadata, the task is impossible.

10.2 Logic and imperfection

explosive logic

The classical logic we encountered in Chapter 9 is *explosive*: it has the important but sometimes frustrating property that anything may be inferred from a contradiction. For example, consider the following three premises and conclusion:

Bangor, Maine has a population of 32,000 inhabitants.
Only cities with more than 100,000 inhabitants are large.
Bangor is a large city.

Bangor is a sausage.

The conclusion "Bangor is not a large city" is a valid deduction from the first two premises above. However, this contradicts the third premise "Bangor is a large city." The explosive nature of classical logic means anything can be deduced from this implicit contradiction, such as the nonsense conclusion "Bangor is a sausage."

As a consequence, inconsistency must be avoided or resolved at all costs in classical logic. This section begins to outline some alternative logical approaches to representing and reasoning under uncertainty. As we have seen, uncertainty can take many guises, and the variety of its logical representations mirrors this.

10.2.1 Possible worlds

possible worlds

As we have seen, in classical propositional and first-order logic, statements are either true or false. Under conditions of uncertainty, we may not be able to determine whether a statement is true or false. A simple way to extend propositional and first-order logic to model uncertainty is by considering all the *possible worlds* that an agent might know about. As a simple example, consider an area of land divided into two regions: A and B (Figure 10.7). Suppose that we are interested in knowing whether each region is forested or not. For the moment, we assume that the property of being forested is not vague, and therefore must be either true or false (but not both) for any given region. The states of possible knowledge can be expressed using the following statements:

p: "Region A is forested"
q: "Region B is forested"

For this domain there are four possible worlds, enumerated below:

World W_1: Statement p is true, statement q is true
World W_2: Statement p is true, statement q is false
World W_3: Statement p is false, statement q is true
World W_4: Statement p is false, statement q is false

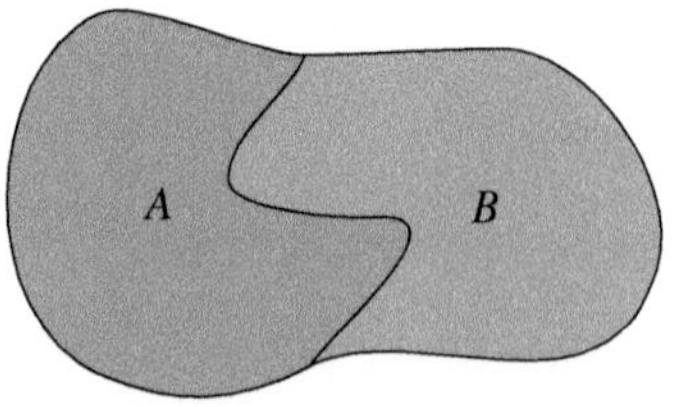

Figure 10.7: Two regions A and B

The statements p and q are either true or false (but not both), so exactly one of the possible worlds may hold at any particular time. Our state of knowledge is represented by a set of worlds: those worlds that are feasible, given what we know. If we are in a state of complete ignorance, then all worlds are possible, represented by the set $\{W_1, W_2, W_3, W_4\}$. In contrast, complete knowledge may be represented by a singleton set such as $\{W_1\}$ (i.e., both regions are forested). In general, the larger the set of possible worlds, the more imprecision exists in our state of knowledge. The empty set represents a state of *inconsistent* information. Indirect information might lead to increased precision in our state of knowledge. For example, if from a state of complete ignorance, we learn that both regions are of the same land type, our knowledge may be represented by set $\{W_1, W_4\}$.

In the case above, we assumed that the land types in the two regions were independent of each other, and therefore that the truth values of the two statements were *a priori* independent. This need not be the case as the following example shows. Figure 10.8 shows an additional region C, which contained in region A. To represent this, we add the statement:

r: "Region *C* is forested"

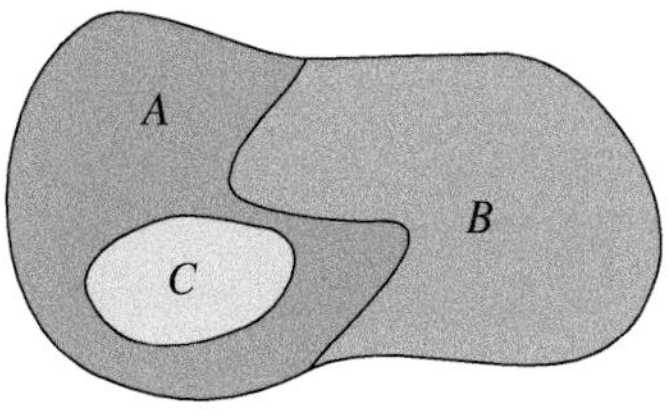

Figure 10.8: Three regions A, B, and C

Suppose we have the constraint that if region A is forested, then the region C, contained within A, must be also be forested (although the converse need not be true). Then, for the enlarged domain, there are six possible worlds, shown below. Note that as a result of the constraint above, two situations are not possible (those where p is true, r is false, and q is true or false) and so are omitted from the possible worlds below.

World W_1: p is true, q is true, r is true
World W_2: p is true, q is false, r is true
World W_3: p is false, q is true, r is true
World W_4: p is false, q is false, r is true
World W_5: p is false, q is true, r is false
World W_6: p is false, q is false, r is false

10.2.2 *Belief and knowledge*

The possible worlds semantics lead to a useful distinction between *knowledge* and *belief*. *Belief* is an agent's conviction in the truth of a statement. *Knowledge* is defined as *justified true belief*. For example, I might *believe* that pigs can fly, but I cannot *know* this, as it is in fact false. Further, while all knowledge is true belief, not all true beliefs are knowledge. For example, I might believe that there is life on distant stars, and this might indeed coincidentally be

belief
knowledge

the case; but that is not sufficient for knowledge—there must also be some justification in the form of evidence or inference.

modal operator

It is clear, then, that there is a difference between something being the case, and our belief or knowledge of it. One common device for constructing logics of belief is to precede a proposition by a *modal operator* to indicate our knowledge or belief of it. So, as before, if we have statement:

p: "Region A is forested"

then:

Kp is the statement	"I know that region A is forested"
Bp is the statement	"I believe that region A is forested"

Belief and knowledge might be related by formulas like the following:

$\neg K \neg p \rightarrow Bp$	If I don't know that p is not the case, then I can believe p.

Logics of knowledge and belief incorporate premises expressing properties of the knowledge and belief operators. For example, from the above discussion, we might have:

$Kp \rightarrow p$	If I know p, then p must be true.

and

$\neg Kp \rightarrow \neg p$	If I don't know p, then p cannot be true.

closed world assumption

The axiom $\neg Kp \rightarrow \neg p$ models the *closed world assumption* of database theory—if the database does not hold a fact, then the fact is not the case. In some cases, depending on the domain, there may be other more complex axioms that are obeyed, such as:

$Kp \rightarrow KKp$	If I know p, then I know that I know p.
$\neg Kp \rightarrow K\neg Kp$	If I don't know p, then I know that I don't know p.

introspection
positive introspection
negative introspection

These last two axioms encapsulate the concept of *introspection*. The former axiom, "I know what I know," is known as *positive introspection*; the latter axiom, "I know what I don't know," is *negative introspection*.

This modal formalism can be extended to model knowledge held by more than one agent. In this case, the modal operators are labeled by names of agents. For example, if a is the name of an agent, then $K_a p$ might express the fact that agent a knows proposition p.

10.2.3 Belief revision

belief revision
non-monotonic logic

If new information arises that contradicts our current beliefs, we may want to review, revise, or retract our old beliefs so as to make way for the new information. This process is called *belief revision*. Typically, belief revision systems are *non-monotonic* because the belief set may grow or contract as new information is added to the knowledge base.

A key question facing any non-monotonic belief-revision system is which beliefs to retract and which to retain. Unfortunately, because beliefs are often founded on other beliefs, the effects of removing one belief may cascade through the knowledge base in a way that is difficult to predict. A well-known example, due to Gärdenfors, concerns a knowledge base that contains the following facts and rules:

The bird caught in the trap is a swan.
The bird caught in the trap comes from Sweden.
Sweden is part of Europe.
All European swans are white.

Suppose we learn new information that, "The bird caught in the trap is black." This information is implicitly inconsistent with the knowledge base, as the knowledge base may be used to infer the statement, "The bird caught in the trap is white." The question facing any belief-revision system is which beliefs to retract in order to regain consistency. Discarding any one of the five pieces of information we now have (the four listed above, plus the new information) will result in a consistent system.

To decide what to keep and what to remove, we must have some way of choosing one piece of information over another. Techniques for deciding what to retract usually involve evaluating the strength of beliefs held, using a *preference relation*, often taken to be a partial order. For example, we might prefer a well-established piece of information (e.g., "Sweden is part of Europe") than a fact about the current situation (e.g., "The bird caught in the trap is a swan"). Sweeping statements such as, "All European swans are white" might also be called into question. A general principle often used in belief revision to revise beliefs so that the amount of change is minimized—the *principle of minimal change*. Another possible principle is to favor beliefs that arise from evidence obtained spatially or temporally close to the phenomenon, rather than farther away (the *nearness* principle). For example, when faced with two inconsistent pieces of information, we might prefer the more current information source, as it is by definition temporally closer to the present.

preference relation

principle of minimal change

Belief revision is related to the topic of *default reasoning*. In the example above, the first three facts relating to the bird caught in the trap being a swan from Sweden, which is a part of Europe, are potentially possible to verify. The fourth universal statement, the rule that "All European swan are white" would be difficult or impossible to verify beyond doubt (we would need to check every swan in Europe). Universal statements make powerful rules but may be subject to occasional counterexamples, such as our Swedish black swan. Instead of an unconditional universal, what we really want to say is:

default reasoning

All European swans are white (except if we have definite evidence to the contrary in the case of a particular swan).

While the default is that European swans are white, default reasoning allows the possibility that some counterexamples may exist. A default predicate is assumed to hold, unless any evidence turns up to the contrary, in which case it is retracted. Default reasoning is therefore a further example of non-monotonic reasoning.

10.2.4 *Three-valued and many-valued logics*

In the preceding discussion of possible worlds, knowledge, and belief, statements are either true or false, with uncertainty modeled as the state of an agent's knowledge or belief about the world. We might also need to allow for the possibility that a statement is neither true nor false. A statement could be neither true nor false for a variety of reasons, such as when there is insufficient evidence to decide, when it makes no sense to give a value, or when the statement is inherently vague. For example, the statement "Region *A* is forested" may not be decidable as true or false because no data exists to confirm or deny the truth of the statement (perhaps cloudy weather prevented satellite imagery of the location). Alternatively, the statement "Region *A* is forested" might be neither true nor false because the concept "forested" is vague: while there may be some trees in the region, it might be unclear what exactly constitutes the definition of "forested." In the former case, more evidence might allow a decision, true or false, to be made; in the latter case, no amount of evidence will suffice.

three-valued logic

A seemingly simple solution is to introduce a third truth value, resulting in *three-valued logic*. We could call the third truth value "unknown" or "undetermined," and indicate with a symbol such as *U* or '?'. In this case, we have three truth values: ⊤ (true), ⊥ (false), and ? (undetermined). Just as we can make truth tables for two-valued, classical logic (see Appendix B), so can we for three-valued logic. Depending on the interpretation of '?', we can arrive at different truth tables. An example of a three-valued logic is the Kleene system. Truth tables for the logical operators ¬ (not), ∧ (and), ∨ (or), ⟹ (implies), and ⟺ (equivalence) are shown in Figure 10.9.

∧	⊤	?	⊥
⊤	⊤	?	⊥
?	?	?	⊥
⊥	⊥	⊥	⊥

∨	⊤	?	⊥
⊤	⊤	⊤	⊤
?	⊤	?	?
⊥	⊤	?	⊥

¬	⊤
⊤	⊥
?	?
⊥	⊤

⟹	⊤	?	⊥
⊤	⊤	?	⊥
?	⊤	?	?
⊥	⊤	⊤	⊤

⟺	⊤	?	⊥
⊤	⊤	?	⊥
?	?	?	?
⊥	⊥	?	⊤

Figure 10.9: Truth tables for Kleene's three-valued logic

In Kleene's system, uncertainty is interpreted as a limitation on reasoning or computing resources. So, a statement takes truth value '?' whenever its truth or falsity cannot be determined by the resources at hand, even though we know it has to be either true or false (that is, inherent vagueness is not considered in Kleene logic).

For example, suppose we have statements *p* and *q* as before, and that "forested" is assumed to be a crisp property of a region. Suppose further that because of cloudy conditions we do not know whether or not region A is forested, written $F(A) = ?$. Now suppose we have clear evidence that region B is forested ($F(B) = \mathsf{T}$). The truth values of the following statements can be determined from the truth tables in Figure 10.9:

Regions A and B are both forested	$F(A) \wedge F(B) = ?$
Regions A or B, or both, are forested	$F(A) \vee F(B) = \mathsf{T}$
If A is forested, then B is forested	$F(A) \implies F(B) = \mathsf{T}$

Kleene logic is one example of a three-valued logic. Three-valued logic is itself a type of *many-valued logic*. Some many-valued logics allow an infinite number of intermediate values to be assigned to statements that lie between absolute truth and absolute falsity. Such logical systems are often termed *degree theories* because values may be interpreted as "degrees of truth." Fuzzy set theory, discussed in the following section, is an important example of a degree theory.

many-valued logic

degree theory

10.2.5 Fuzzy sets

A common approach to uncertainty in GIS is the application of *fuzzy set theory*. In the case of crisp (classical) sets, an element either belongs to the set or not. With fuzzy sets, a membership function grades the levels of belief in whether an element belongs to the set or not. Although fuzzy set theory appears quantitative, often using real numbers between 0 and 1 as measures of degree of membership, it mostly makes use only of the ordinal properties of these numbers. In this account we will concentrate on the case where membership values are in the real interval $[0, 1]$. In general, membership values may be taken from an algebraic structure called a lattice.

Formally, let X be a universe of discourse. A *fuzzy membership function* is a function μ from X to the real interval $[0, 1]$, $\mu : X \rightarrow [0, 1]$. A *fuzzy set* A in X is a set of ordered pairs $(u, \mu_A(u))$ for all $x \in X$, where μ_A is a fuzzy membership function.

fuzzy membership function

fuzzy set

An example of a fuzzy set that describes a region with an uncertain boundary is shown in Figure 10.10. Each pixel is labeled with a fuzzy set value, indicating the strength of belief that each pixel belongs to the region. The degree of belief might arise, for example, from evidence about the pixel gained from Earth observation.

Just like crisp sets, fuzzy sets have properties and operations, some of which are listed below.

- Fuzzy set A is *empty* if $\mu_A(x) = 0$ for all $x \in X$.
- Fuzzy set A is *contained* in B if $\mu_A(x) \leq \mu_B(x)$, for all $x \in X$.
- Fuzzy sets A and B are equal if $\mu_A(x) = \mu_B(x)$, for all $x \in X$.
- The *complement* of fuzzy set A is the set A' with membership function $\mu_{A'}$ such that $\mu_{A'}(x) = 1 - \mu_A(x)$, for all $x \in X$.
- The *union* of fuzzy sets A and B is the set $A \cup B$ with membership function $\mathbf{max}(\mu_A(u), \mu_B(u))$, for all $x \in X$.

Figure 10.10: Fuzzy region as pixel values

0.0	0.0	0.0	0.0	0.0	0.0	0.0	0.0
0.0	0.1	0.2	0.2	0.2	0.1	0.1	0.0
0.1	0.2	0.4	0.4	0.6	0.2	0.1	0.0
0.2	0.3	0.4	1.0	0.8	0.4	0.3	0.2
0.1	0.3	0.8	1.0	0.8	0.4	0.3	0.2
0.1	0.2	0.6	0.6	0.6	0.3	0.3	0.1
0.1	0.1	0.1	0.2	0.1	0.1	0.1	0.0
0.0	0.0	0.0	0.1	0.1	0.0	0.0	0.0

- The *intersection* of fuzzy sets A and B is the set $A \cap B$ with membership function $\mathbf{min}(\mu_A(u), \mu_B(u))$, for all $x \in X$.
- The *support* of fuzzy set A is the crisp set containing all elements with non-zero membership of A, support$(A) = \{x | \mu_A(x) > 0\}$.
- For $0 \leq \alpha \leq 1$, the *α-cut* of fuzzy set A is the crisp set given by $A_\alpha = \{x | \mu_A(x) > \alpha\}$.

α-cut

What we have given above for union, intersection, and complementation operations are special cases of more general categories. For example, the intersection between two fuzzy sets may be realized not only by the minimum operation, but by any operation that is a *triangular norm* ("T-norm"—in short, obeys the triangle inequality).

triangular norm

The idea of a fuzzy region is rather more general than our example indicates. Define a *fuzzy region* as a fuzzy set whose support is a region. Fuzzy regions have more structure than the more basic fuzzy sets. For example, assuming the regions are based on a square cell grid, as in the example in Figure 10.10, then the cells have many topological and geometrical properties and relations, such as adjacency, area, distance, and bearing. Fusing fuzzy set theory with geometrical or topological structure enables operations such as fuzzy connectivity, fuzzy convexity, fuzzy area, and fuzzy perimeter, to be defined, sometimes termed *fuzzy geometry* or *fuzzy topology*, respectively.

fuzzy region

fuzzy geometry

By way of example, let R be a fuzzy region based on a square cell grid with fuzzy membership function μ_R. Then the fuzzy area of R, $\mathbf{farea}(R)$, may be defined as the sum of the $\mu_R(x)$, for all $x \in X$. In the example of Figure 10.10, the fuzzy area of the region is 14.1. This notion of fuzzy areas satisfies only some of the basic properties that we would expect an area measure to satisfy. For example, while it is the case that:

$$\mathbf{farea}(A \cup B) = \mathbf{farea}(A) + \mathbf{farea}(B) - \mathbf{farea}(A \cap B)$$

it is *not* necessarily the case that:

$$\mathbf{farea}(A) = \mathbf{farea}(A \cap B') + \mathbf{farea}(A \cap B)$$

10.2.6 Rough sets

Rough sets are motivated primarily by the need to represent and reason about granularity and indiscernibility in information, as discussed in Section 10.1.1. Unlike those techniques presented above, which model our belief in whether a statement is true or not, rough set theory supports reasoning under conditions of granularity and indiscernibility.

An important question when discussing granularity is how to represent subsets of X at the level of granularity imposed by the indiscernibility relation ρ. Rough sets provide a framework in which this question may be answered. First, let set A be a subset of X. Then we define the following two constructs:

$$\overline{A} = \{b \in X/\rho \mid b \cap A \neq \emptyset\}$$
$$\underline{A} = \{b \in X/\rho \mid b \subseteq A\}$$

$\overline{A}$ is called the *upper approximation* to set A, while $\underline{A}$ is termed the *lower approximation* to set A (with respect to the indiscernibility relation ρ on set X). The pair $\langle \underline{A}, \overline{A} \rangle$ is called the *rough set* (with respect to the indiscernibility relation ρ on set X). Note that $\underline{A}$ is always a subset of $\overline{A}$ in X/ρ.

upper approximation
lower approximation
rough set

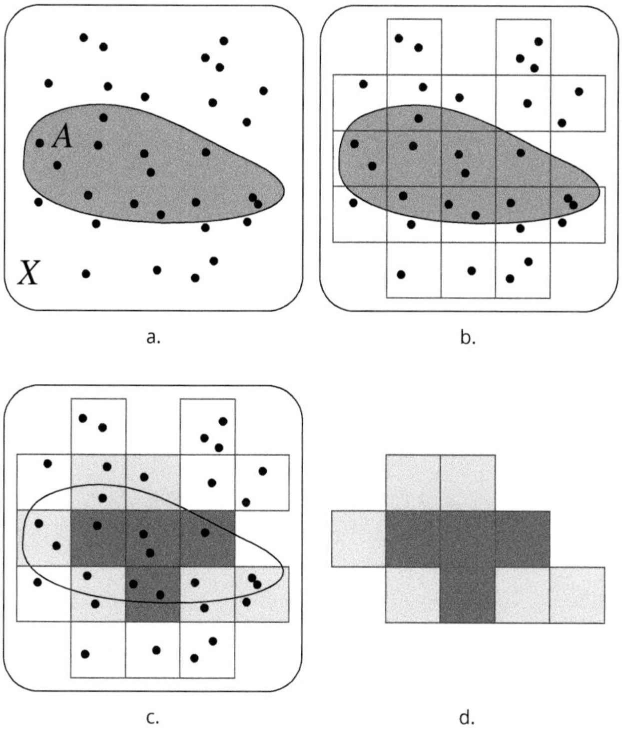

Figure 10.11: Rough set construction, showing: a. subset A of X; b. the blocks of the partition induced by the indiscernibility relation ρ; c. the construction of $\underline{A}$ and $\overline{A}$; d. and the rough set $\langle \underline{A}, \overline{A} \rangle$

Figure 10.11 shows an example of the construction of a rough set. Subset A of X is shown in Figure 10.11a, where the points indicate the elements of X. In Figure 10.11b the blocks of the partition induced by the indiscernibility relation are shown. Figure 10.11c shows the construction of $\underline{A}$ and $\overline{A}$, where the blocks of $\underline{A}$ are shown as darker and the blocks of $\overline{A} \backslash \underline{A}$ as lighter color. Finally, Figure 10.11d shows the rough set $\langle \underline{A}, \overline{A} \rangle$ ($\underline{A}$ is the set of darker blocks, and $\overline{A}$ is the set of all blocks).

Just as for crisp and fuzzy sets, rough sets form an algebra with rough intersection and union defined. Rough set analysis also gives a way of deducing functional dependencies in an information system—an example of a basic data mining technique.

10.2.7 Probability

Probabilistic methods form the most important and widely used quantitative approach to uncertainty. This section sketches a rough outline, concentrating on the Bayesian approach.

aleatory

Chance arises when observing the outcome of an *aleatory* (random) experiment, such as the throw of a die. If X denotes the set of possible outcomes, we can specify a chance function $\mathbf{ch} : X \to [0, 1]$. The valuation $\mathbf{ch}(x)$ gives the proportion of times that a particular outcome $x \in X$ might in principle occur, and is determined either from a frequency analysis (throwing the die many times and seeing what happens) or determination from the nature of the experiment (assuming that the die is fair and so 50-50 heads to tails). The function **ch** should satisfy the constraint that the sum of chances of all possible outcomes is 1. For a subset $S \subseteq X$, $\mathbf{ch}(S)$ is the chance of an outcome from set S.

So, the rules that a chance function must obey are:

$$\mathbf{ch}(\emptyset) = 0 \tag{10.1}$$
$$\mathbf{ch}(X) = 1 \tag{10.2}$$
$$\text{If } A \cap B = \emptyset, \text{ then } \mathbf{ch}(A \cup B) = \mathbf{ch}(A) + \mathbf{ch}(B) \tag{10.3}$$

Also, given n independent trials of a single aleatory experiment, the chance of the compound outcome $\mathbf{ch}^n(x_1, ..., x_n)$ is given by:

$$\mathbf{ch}^n(x_1, ..., x_n) = \mathbf{ch}(x_1) * ... * \mathbf{ch}(x_n) \tag{10.4}$$

Suppose next that an aleatory experiment has been only partly completed, and the outcome has been determined to be in the set $V \subseteq X$. If $U \subseteq X$ is the outcome set under consideration, we write the chance of U following the partial determination that the outcome is in V as $\mathbf{ch}(U|V)$ (read "the chance of U given V" or the *conditional probability* of U given V). Then:

conditional probability

$$\mathbf{ch}(U|V) = \frac{\mathbf{ch}(U \cap V)}{\mathbf{ch}(V)} \tag{10.5}$$

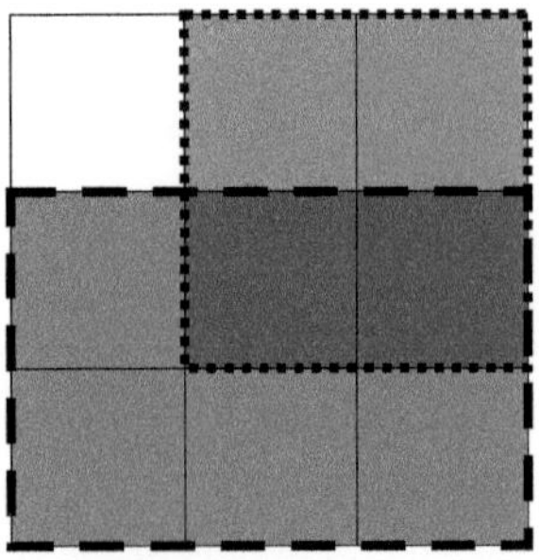

Figure 10.12: Conditional probability example with regions A (dotted boundary) and B (dashed boundary)

For example, suppose we are looking at the pixels in Figure 10.12 for evidence of a landslide. Assume that the extent of the landslide is small compared with the pixel size, and so the landslide has occurred in exactly one of the pixels. Suppose also, we have evidence from other sources that exactly one landslide has occurred somewhere within the study region. Then, our initial estimate of the landslide being in region A is 4/9 by Equation 10.3 above. Now, suppose some new geological evidence becomes available that tells us that the landslide can only be region B. Let U be the outcome set given by the statement "the landslide is in region A" and V be the outcome set given by the statement "the landslide is in region B." Then, by Equation 10.5, the chance that the landslide is in region A now becomes:

$$\mathrm{ch}(U|V) = \frac{\mathrm{ch}(U \cap V)}{\mathrm{ch}(V)} = \frac{(2/9)}{(6/9)} = 1/3$$

10.2.8 Bayesian probability

This example leads neatly into a brief word about a theory of belief, developed by the English clergyman Thomas Bayes (1702–1761). The notion of chance introduced above is objective, in that it depends not on our cognitive state but on external conditions. The chance of a fair die landing a six is 1/6 independent of our belief. Indeed, the rational way to set our belief level is based on the laws of chance in this case—that is why the laws are generally useful. However, if we make "degree of belief" our prime object of consideration, the laws of chance 10.1–10.3 above, may be rephrased as follows.

A *degree of belief* with respect to a set X of possibilities (outcomes) is given by a belief function $\mathbf{Bel} : X \to [0,1]$. For a subset $S \subseteq X$, $\mathbf{Bel}(S)$ is the degree of belief in an outcome from set S. Bayesian theory rests on the following rules that a Bayesian belief function is expected to obey.

degree of belief

$$\mathbf{Bel}(\emptyset) = 0 \quad (10.6)$$
$$\mathbf{Bel}(X) = 1 \quad (10.7)$$
$$\text{If } A \cap B = \emptyset, \text{ then } \mathbf{Bel}(A \cup B) = \mathbf{Bel}(A) + \mathbf{Bel}(B) \quad (10.8)$$

Suppose we begin with a Bayesian belief function $\mathbf{Bel} : X \to [0,1]$ and then learn that only a subset of possibilities $V \subseteq X$ is the case. Then, replace $\mathbf{Bel}$ with a new Bayesian belief function $\mathbf{Bel}_V : X \to [0,1]$ given by:

$$\mathbf{Bel}_V(U) = \frac{\mathbf{Bel}(U \cap V)}{\mathbf{Bel}(V)} \quad (10.9)$$

or in the language of conditioned belief, and parallel to Equation 10.5:

$$\mathbf{Bel}(U|V) = \frac{\mathbf{Bel}(U \cap V)}{\mathbf{Bel}(V)} \quad (10.10)$$

Applying Equation 10.10 symmetrically gives:

$$\mathbf{Bel}(V|U) = \frac{\mathbf{Bel}(U \cap V)}{\mathbf{Bel}(U)} \quad (10.11)$$

Eliminating $\mathbf{Bel}(U \cap V)$ from Equations 10.10 and 10.11 gives:

$$\mathbf{Bel}(U|V) = \mathbf{Bel}(U) * \frac{\mathbf{Bel}(V|U)}{\mathbf{Bel}(V)} \tag{10.12}$$

posterior belief

prior belief

Equation 10.12 is a widely used form of the Bayesian formula. Our *posterior belief* $\mathbf{Bel}(U|V)$ is calculated by multiplying our *prior belief* $\mathbf{Bel}(U)$ by the likelihood that V will occur if U is the case. $\mathbf{Bel}(V)$ acts as a normalizing constant that ensures that $\mathbf{Bel}(U|V)$ will lie in the interval $[0, 1]$.

10.2.9 Dempster-Shafer theory of evidence

Although at first sight Bayesian theory seems natural, there are some problems. Consider the following example:

> b_1: there is life in the Sirius system
> b_2: there is not life in the Sirius system

By Equation 10.8, $b_1 + b_2 = 1$, even though there is little evidence either way. So how do we assign belief levels? If we have no evidence either way, the natural thing is to assign equal belief levels, and so $b_1 = b_2 = 0.5$. In this way, high levels of belief are assigned, based on little evidence. Such a model does not appear to accord with our intuition.

The Dempster-Shafer theory of evidence provides a way of pooling the total evidence available. It takes account of evidence both for and against a belief. As with Bayesian theory, a real number between 0 and 1 indicates the degree of support a body of evidence provides for a belief. The Dempster-Shafer theory focuses on the combination of degrees of belief or support from distinct bodies of evidence. Dempster's *rule of combination* provides a method for changing beliefs in the light of new evidence. The example in the previous paragraph pointed to a possible problem when the degrees of independent and exhaustive beliefs are forced to sum to one. To develop this further, take our usual statement:

> p: "Region A is forested"

and as before, assume that the property "forested" is crisp (so it is definitely the case that either A is forested or not, but not both). Let us consider two cases:

Case 1 (information scarcity) There are roughly equal small amounts of evidence on both sides.

Case 2 (information glut) There are roughly equal large amounts of evidence on both sides.

credibility

plausibility

Bayesian analysis will give $\mathbf{Bel}(p)$ and $\mathbf{Bel}(\neg p)$ the value roughly 0.5 in both cases, using $\mathbf{Bel}(p) + \mathbf{Bel}(\neg p) = 1$, and so it fails to make a distinction between the cases. The way forward is to make a distinction between *credibility* and *plausibility*, or in negative terms, a distinction between *disbelief* and *lack of belief*. The credibility of a state of affairs is the amount of evidence we have in its favor. The plausibility of a state of affairs is the lack of evidence we have against it. In terms of the belief function:

$$\text{credibility}(p) = \text{Bel}(p)$$
$$\text{plausibility}(p) = 1 - \text{Bel}(\neg p)$$

Now, the analysis of case 1 is that the credibility of both p and $\neg p$ is small, but the plausibility of both p and $\neg p$ is large. For case 2, the credibility of both p and $\neg p$ is larger, but the plausibility of both p and $\neg p$ is smaller. This is the starting point for the Dempster-Shafer approach. Using Dempster's rule of combination, evidence for and against a state of affairs can be combined.

We leave our tour of uncertainty and the logics of imperfection at this point, and segue in the next section to a new topic—location privacy. Although the link between the relatively applied topic of privacy and the frequently abstract topic of the logics of imperfection may not be immediately apparent, in fact uncertainty is fundamental to GIS approaches to privacy, as we shall see.

10.3 Privacy

Privacy is internationally recognized as a basic human right, for example, in Article 12 of the United Nations Universal Declaration of Human Rights. The right to privacy is usually taken to encompass several different types of privacy (Figure 10.13), including communication privacy (the right to private correspondence), territorial privacy (the right to control intrusions into our homes and workplaces), and bodily privacy (the right to protect ourselves from intrusive procedures).

From the perspective of GIS, however, we are primarily concerned with the right to *information privacy*. Information privacy is usually defined as:

information privacy

> the claim of individuals ... to determine for themselves when, how, and to what extent information about them is communicated to others. (Westin, 1967)

In short, information privacy is concerned with questions of who controls the collection, communication, storage, and use of information about an individual.

Many, if not most spatial data sets will have no associated privacy concerns because the information they contain does not relate to individuals. Land cover or transport network data, for example, usually contains no information about individual people, and so it can be used without regard to any privacy implications.[4]

[4] While such data may not be subject to privacy restrictions, it may of course still be subject to other restrictions, such as intellectual property (IP).

Privacy becomes a central issue when a spatial data set contains *personal data*. Data is considered to be personal if a person's identity is apparent or can reasonably be ascertained from that data. Medical records, for example, are unequivocally personal data, as they relate to an identified person's medical history. Importantly, even if a person's name and identifiers (such as health care numbers) are removed from a medical record (a process called *anonymization*), such data would usually still be regarded as personal and thus private. A person's medical history is frequently unique and often sufficiently distinctive

personal data

anonymization

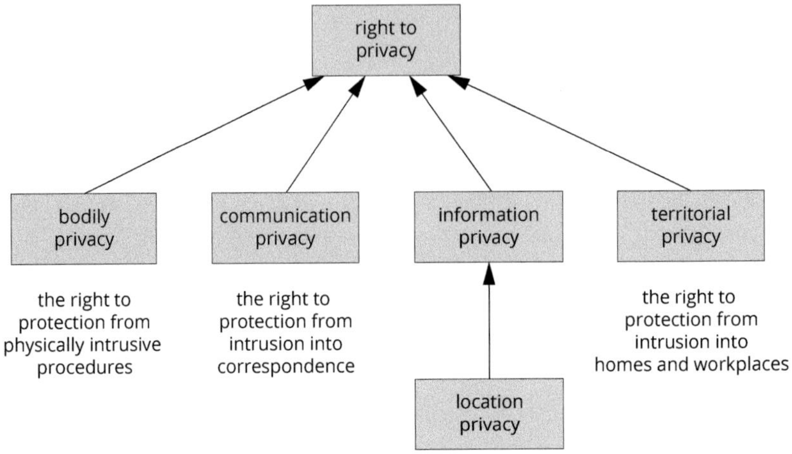

Figure 10.13: The different types of right to privacy

to be able re-identify an individual from their anonymized medical history alone.

As we shall see, several types of spatial and spatiotemporal data, including address data and movement traces, can similarly serve to uniquely identify individuals, and so they must be treated as personal data. Spatial or otherwise, the handling of all personal data is underpinned by five widely accepted principles, called *fair information practices.*

10.3.1 Fair information practice principles

data protection

Data protection aims to help individuals, organizations, and governments avoid harmful breaches in location privacy such as those in Box 10.3 on the facing page. Although approaches to data protection differ around the world, most approaches can be summarized with reference to five general principles that underpin data protection, called *fair information practices* (Figure 10.14).

fair information practices

1. *Notice and transparency*: Individuals must be made aware of who is collecting personal information about them and why.
2. *Consent and use limitation*: Individuals must consent to personal information being collected. The use of personal information must be limited only to those specific purposes for which consent has been given.
3. *Access and participation*: Individuals must be able to access and review any stored personal data which relates to them. Individuals may then require that any errors in their personal data be corrected.
4. *Integrity and security*: Collectors must ensure personal data is accurate and up-to-date and protect against unauthorized access, disclosure, or use.
5. *Enforcement and accountability*: Collectors must be accountable for any failures to comply with the other four principles.

In many countries around the world, fair information practices are enshrined in law. Privacy breaches may be punishable with fines or even imprisonment in serious cases (see Box 10.3 on the next page). Fair information

Box 10.3: AOL query logs

In 2006, the Internet search company AOL publicly released a data set of 20 million records detailing the web search queries of hundreds of thousands of AOL users over a three-month period. The queries ranged from the informative ("madonna tour dates"), through the banal ("steak and cheese") and illegal ("buy ecstasy"), to the disturbing and melancholic ("i hate my job"). The release was a misguided but well-intended attempt to create a shared resource for information retrieval (IR) researchers. As the public outcry grew, AOL acknowledged the mistake. The release was retracted three days later, but not before the data set had become widely copied, shared, and reposted across the Internet. The data set had been anonymized, without any user's personal details attached. However, against each query record, the data did include a unique AOL identifier for each user. In privacy research, a unique but anonymized user identifier is termed a *pseudonym*. For example, the search queries of user 4417749 included "tea for good health," "retirement communities for single women," "landscapers in lilburn, ga," and "jarrett arnold." Following a short investigation by The New York Times, a journalist identified and approached Thelma Arnold, a 62-year-old widow from Lilburn, Georgia in the United States, who consented to being publicly identified as the AOL user with pseudonym 4417749. AOL was sued for the incident, and Thelma and other users whose privacy had been breached were awarded damages. A number of AOL staff were dismissed or resigned as a consequence, including the AOL Chief Technology Officer (CTO). However, the case highlights the subtleties and hazards connected with personal data. Although AOL never intended to cause harm to its own users, once a privacy breach has occurred then no legal proceedings, employment terminations, financial penalties, or damages can ever undo the impacts on individuals.

practice laws typically include exceptions to privacy protections, where other considerations may take priority over personal privacy. For example, emergency services or law enforcement may be able to obtain access to personal data if necessary to avert serious harm to an individual or where serious criminal activity is suspected.

10.3.2 Location and identity

Many different types of information can be used to identify you. Your name, for example, is often used as an identifier. In general, names do not uniquely identify people, as many people may share the same name. Consequently other identifiers, such as your social security number, your tax office number, or your Twitter handle, are designed to be unique, such that no two people share the same identifier. With today's technology, a person's physical attributes, such as genetic profile, facial or gait recognition, are increasingly being used to identify them. Importantly in the context of GIS, a person's precise and accurate location is a physical attribute of a person that uniquely identifies them—potentially more so even than their genetic fingerprint. Consider this: there is no one else located exactly where you are located right now, even if you have an identical twin sibling with the same genetic profile!

A history of a person's location over time may similarly be used to identify the individual responsible for that movement trace. Hence, a person's current or past locations are usually regarded as personal data using the definition above (i.e., data where identity is apparent or can reasonably be ascertained). Consequently, anonymization of location data (i.e., removing personal identifiers, such as names or identification numbers from locations) may not be sufficient to prevent re-identification and safeguard an individual's privacy.

Figure 10.14: Fair information practice principles

Other significant locations may also readily be used to identify a person, such as their home address or even their work address. Publicly available data from a land registry, business registry, or white or yellow pages telephone directory, for example, enable a person's identity to be inferred from a home or work address in many cases.

location privacy

Location privacy, then, is a special case of information privacy that concerns the right of individuals to control and protect information about their location over time (see Figure 10.13).

10.3.3 Privacy and spatial aggregation

Location is especially sensitive personal data because it can be readily used to link an individual to other data about them, or to infer new knowledge about what a person is doing, likes, or what interests them. An individual's location at a crime scene—perhaps identified through their cell phone positioning or a facial recognition using a surveillance camera—might be legitimately used by the police to solve serious crimes. The same location information, however, might also be unfairly used by a hostile security agency to uncover an individual's personal affiliations or beliefs, such as through their attendance at a political demonstration or rally.

census

demographics

census blocks

One important source of personal spatial data is the national *census*, where countries periodically collect detailed socioeconomic data about citizens based on household addresses. A census surveys citizens on a wide range of sensitive *demographics* data, such as age, income, employment, ethnicity, education, religion, home location, and so on. Census data is typically aggregated across small spatial regions called *census blocks* to protect the privacy of the individuals living in those blocks. Census block sizes vary by country, but typically contain a minimum of at least 50–100 residents, if anyone at all lives there.[5] Figure 10.15 shows the census block in one of the world's most densely populated spaces, Manhattan Island in New York, containing approximately 900 residents per block on average.

[5] Census blocks with zero residents are also acceptable from a privacy perspective, since no people implies no privacy concerns.

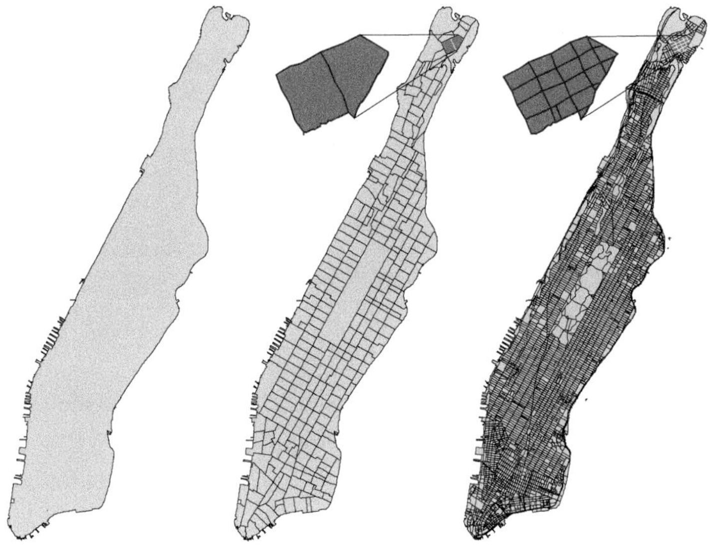

Figure 10.15: Spatial aggregation across census tracts and blocks in Manhattan Island, NY, with two tracts and 21 blocks highlighted in the neighborhood of Inwood

Census blocks ensure a certain level of privacy protection for those people living within the block. Rather than reporting, say, household income for a specific address, average household income for all households in the census block can be reported. As a consequence of spatial autocorrelation (near things are more related than far things), such aggregate statistics will usually still provide a meaningful estimate of the income of each household in that area sufficient for many spatial analyses.

Indeed, spatial aggregation is a general technique that can be used to protect individuals' privacy when using personal location data. In addition to census data, spatial aggregation is routinely used to protect privacy while analyzing data about everything from incidence of disease or health conditions to energy smart meter usage, and from crime statistics to counting votes in an election. An interesting question then becomes: what level of spatial aggregation is appropriate in order to protect an individual's privacy for a particular data set?

In general, the larger the area over which data is aggregated, the more people included in the aggregate statistic, and the greater privacy protection afforded to the populace of that area. Of course, this relationship is not straightforward and varies with population density and interacts with the specific block size and shape (see, for example, Box 8.5 on page 321 on the modifiable areal unit problem or MAUP). In densely populated locations, a small increase in spatial area may lead to large increases in the number of individuals included; similarly, sparse populations require large increases in area for even small increases in included population.

Box 10.4: Differential privacy

The growing range of big data sets available, combined with today's high-performance computing capabilities, have led to new opportunities for attackers to subvert privacy protections. For example, imagine an attacker wants to uncover the income of one particular person living in census block containing 100 people. The census agency might publish an aggregated data table containing the average of household incomes in that census block. If the attacker can somehow discover the average income of the *other* 99 people in that block, however, it becomes straightforward for them to deduce the *actual* income of that particular person of interest. This type of strategy, termed a *differentiated attack*, is especially hard to defend against because it is difficult to control to what *other* information an attacker may have access. In 2021, a threat analysis by the US Census Bureau found that it was possible to re-identify more than 50 million individuals using differentiated attacks that combined the 2010 US census with an array of other big data sets available today. In anticipation, the 2020 US Census saw the introduction of a new privacy protection technique called *differential privacy*. Differential privacy involves deliberately adding noise to aggregate query results (one form of a privacy protection technique called *obfuscation*) in a statistically reliable way that guarantees that no two data sets can ever be used to re-identify a single individual. Groundbreaking work by computer scientist Cynthia Dwork (2006) provided the mathematical foundations of differential privacy that enable us to know with certainty exactly how much noise will be needed to guarantee that differential attacks against an individual can never succeed.

The relationship to personal privacy may be further complicated by the distribution of observed attributes through that population, and the statistics used to aggregate the data. For example, if the reported statistic is that 99% of people in a block are, say, unemployed or have a criminal record, then that information might still reasonably be regarded as insufficiently private, no matter how large the block area or the included population size.

Counterbalancing the push towards larger aggregations that more securely protect individual privacy is the utility of the data sets that results from aggregation. At the most coarse-grained level, a census data set aggregated to the national level, for example, will certainly maximize the privacy of the citizens in that country. However, it will also not provide any useful data for analysis of the spatial patterns of demographics within the country.

census tracts

To address this, most national statistics agencies provide census data at a number of levels of aggregation, from census blocks of a few hundred people to *census tracts* of a few thousand people (see Figure 10.15), to administrative regions with tens or hundreds of thousands of people. National statistics agencies carefully structure their data spatially to provide a stable hierarchy of spatial units that protect individual privacy at the same time as enabling spatial analysis that can yield meaningful insights into demographics. Census data is applied to an enormous range of different applications, from public health interventions to private marketing campaigns. As an extra layer of protection, only trusted research statisticians or agency staff are authorized to access the finer-grained, more disaggregated levels of census data. Further sophisticated techniques are increasingly needed to prevent privacy attacks the combine census data with other data sets in an attempt to re-identify individuals (see Box 10.4).

10.3.4 *Privacy, imprecision, and inaccuracy*

The spatial aggregation of data discussed above is an example of using imprecision to protect privacy. Aggregating data across all the people in a region simultaneously decreases the spatial detail (i.e., to where, precisely, the data refers) and the detail about each individual's identity (i.e., to whom, exactly, the data refers). In general, a data set where any single individual's records cannot be distinguished from data about *k* other individuals in the data set is said to be *k-anonymous.*

k-anonymity

It is also possible to deliberately decrease the precision of an individual's spatial location in order to protect their privacy, while still retaining information about their identity. For example, most smartphones are able to track a person's location to a high level of precision using a combination of GNSS positioning and other positioning systems, such as WiFi fingerprinting. In most instances, smartphone users can protect their location privacy by choosing to limit the level of precision of location data available to third-party apps in their smartphone's privacy settings.

Just as imprecision can be used to protect a person's location privacy, so can inaccuracy. For example, location privacy can be protected by deliberately adding perturbations (noise) to a person's location data. In general, the process of deliberately adding imperfection—whether imprecision or inaccuracy—to spatial data in order to protect an individual's privacy is termed *obfuscation* (sometimes also called *location masking* or *geomasking*). Figure 10.16 contrasts these three approaches to location privacy protection: using inaccuracy in location, imprecision in location, and *k*-anonymity.

obfuscation

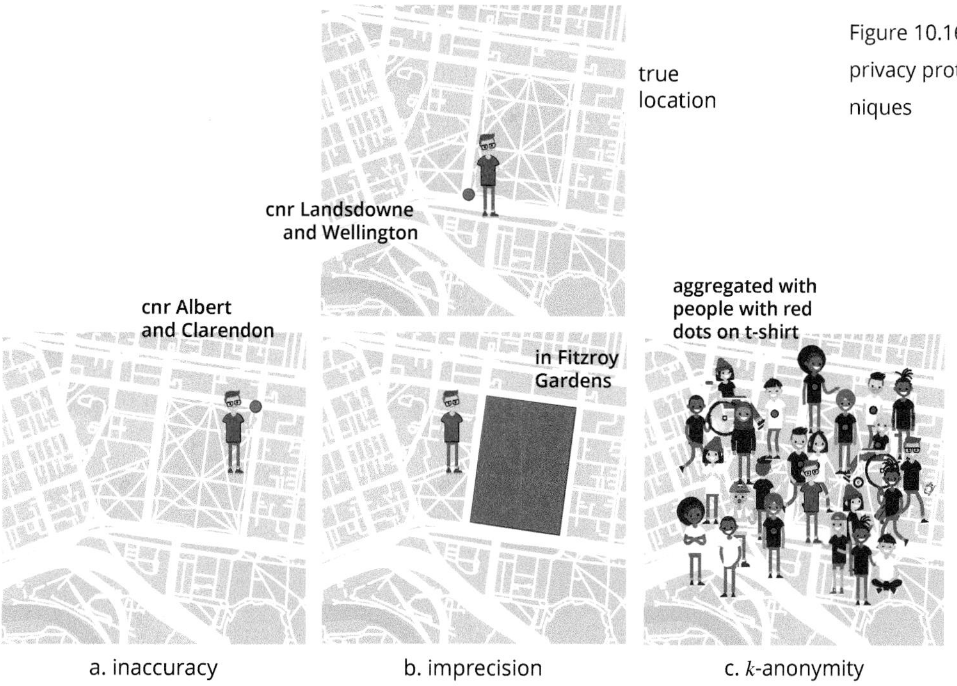

Figure 10.16: Location-privacy protection techniques

In Figure 10.16, the true location of the person with the red t-shirt is at the street corner (of Landsdowne and Wellington) indicated with the red dot. Using obfuscation, that person might protect their privacy by deliberately reporting their location incorrectly (as the nearby corner of Albert and Clarendon in Figure 10.16a). Although this reported location is inaccurate, it is perhaps close enough to still be useful (for example, as a basis for an LBS query for the nearest restaurant). Alternatively, reporting location as an area rather than a point (for example, as Fitzroy Gardens, Figure 10.16b) is an instance of location obfuscation using imprecision. Finally, location privacy protection using *k*-anonymity can be achieved by aggregating location records and location queries across groups of nearby people (e.g., Figure 10.16c).

In all cases, the degree of privacy can be controlled by adapting the level of inaccuracy (e.g., increasing distance from the true location) or imprecision (e.g., increasing size of reported region or number *k* of indistiguishable people). However, increasing levels of inaccuracy and imprecision will also make it more difficult to use the supplied location, for example, in an LBS. There is always a balance required between the desired level of privacy (requiring increasing levels of inaccuracy and imprecision) and the desired level of utility of the resulting service (requiring increasing levels of accuracy and precision). My location to within a city suburb may be sufficient to respond to the location-based query, "What Nepalese restaurants could I dine at this evening?" However, getting useful directions to my chosen restaurant might require me to reveal my location to within a city block or even to within a few meters.

A further challenge of using inaccuracy for obfuscation is the potential for a hostile observer to identify patterns in the perturbations and so infer a more accurate and less private location than desired. For example, obfuscation based on randomly perturbing a person's location provides limited protection from privacy attacks. Even with such noise added, a hostile observer can average successive reported locations of a stationary person to deduce increasingly accurate estimates of that person's location. More sophisticated privacy attacks can be also used to subvert location obfuscation based on random perturbations even for moving objects.

There exists a range of further techniques in the literature to achieve location privacy under different assumptions and constraints. For readers interested to explore the topic further, see, for instance, Duckham & Kulik (2005), Ardagna, Cremonini, Damiani, Capitani di Vimercati, & Samarati (2007), and Damiani (2014).

10.4 Doing GIS critically

The focus of this book's "computing perspective" has been firmly on data and information, technology and models, algorithms and architectures, interfaces and users. One other vital component to GIS is *you*: the geographic information scientist. This final section looks briefly at that human context, and in particular at the role and responsibilities of the GIS analyst in critically reflecting on what we do.[6]

Critical reflection on GIS, its uses, and its implications for society has been an important strand of research in GI science for many years (see, for example, Elwood, 2010; Harvey, Kwan, & Pavlovskaya, 2005; Schuurman & Pratt, 2002). Nadine Schuurman coined the term *critical GIS* to describe this study of the impacts of GIS on people and society (Schuurman, 2000).

[6] As already highlighted in Chapter 1, the small amount of space explicitly devoted to critical perspectives in this book should in no way be taken as a reflection of the importance of this topic; rather, such critical reflection should be taken to implicitly pervade all the topics, concepts, and techniques covered here.

10.4.1 *Critical GIS*

critical GIS

A central aim of critical GIS is to make the relationship between "truth" and GIS explicit. As we have seen throughout this book, GIS comprise a suite of complex concepts and technologies with vast data sets captured by an array of sophisticated sensors and techniques. However, that complexity and sophistication—the myriad simplifications, assumptions, and decisions that are encapsulated in a GIS—are often hidden from view when we are making a map or analyzing spatial data. Critical GIS is first and foremost a reminder not to forget that fact: not to conflate GIS with reality, spatial data with "the truth."

Instead, GIS can be seen as a system of knowledge. GIS necessarily support and prioritize particular *ways of knowing* the world and suppress and marginalize others. For example, GIS embed ways of knowing that can be at odds with Indigenous peoples' understanding of the world (Reid & Sieber, 2020). GIS may enforce crisp boundaries or static snapshots of the world, where Indigenous communities negotiate fuzzy boundaries in fluid relationships to the land and one another. Working with the Cree Nation of Wemindji in Northern Canada, Reid, Sieber, & Blackned (2020) highlight the integral nature of time to the Cree, inseparable from space rather than as distinct dimensions in a GIS (cf. Section 4.3); and the importance of cyclic time to the Cree, rather than the linear time more commonly underlying GIS (cf. Section 9.2.2).

ways of knowing

Maps too are an expression of power, with a history deeply entwined with conquest, colonialism, and control. As we have already seen in this chapter, many geospatial technologies and national mapping agencies have their origins in the military. More subtly, GIS and the maps and analyses output by GIS can create their own "truth." Since ancient times, maps have been used to create the world, not simply document it—for example, in defining property rights, land cover, and regulating society (Harley, 1988). There are many examples of mapping exercises, sometimes ostensibly conducted to empower Indigenous inhabitants, that lead instead to dispossession, development, or exclusion of Indigenous people (e.g., Fox, Suryanata, Hershock, & Pramono, 2006). The mere introduction of GIS can be the cause disunity and division within communities and between citizens (Sieber, Robinson, Johnson, & Corbett, 2016).

As a consequence, geographic information scientists have a responsibility to attempt to anticipate the potential impacts of GIS use and mapping decisions, and to act upon that knowledge ethically. Expertise in GIS and spatial data carries with it a duty to exercise good judgment and act for the betterment of society. Most professional societies for GIS, surveying, and mapping, for example, have attempted to specify such "good behavior" through established codes of conduct (see below).

The idea of critical GIS as the anchor of a chain of responsible action is summarized in Figure 10.17. The remainder of this section looks a little deeper at some of the key features of such responsible action.

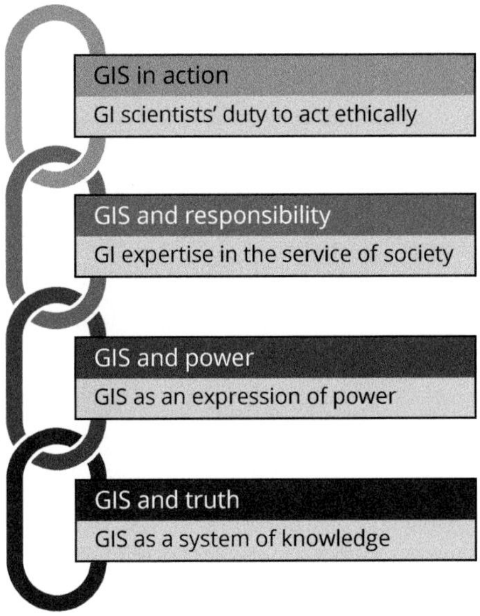

Figure 10.17: Critical GIS as the basis of a chain of responsible action for geographic information scientists

10.4.2 FAIR and CARE

An important principle in support of "doing GIS critically" is openness. The concept of open data has already been encountered in connection with ontology engineering (Section 9.1.4). Openness promotes sharing, which in turn raises quality. Openness is supported by good data management practices, codified in 2016 as the *FAIR* (findable, accessible, interoperable, reusable) data principles (Wilkinson et al., 2016).

FAIR

- *Findable* data should be accompanied by metadata, introduced earlier in this chapter, to ensure data can be described and searched for.
- *Accessible* data is available to download through standard protocols and accompanied by clear procedures and requirements for access.
- *Interoperability* has been encountered already in the context of system architecture in Section 7.1. Interoperable *data* is encoded in standard and documented formats in order to enable system interoperability.
- *Reusable* data is associated with relevant provenance and licensing information.

While FAIR principles promote sharing of data, they do not on their own address the questions of responsible action introduced above. The FAIR principles are silent on issues of power, for example, even potentially exacerbating power imbalances in the case of *Indigenous data* (data created by or about First Nations people or lands). In response to this, the *CARE* (collective benefit, authority to control, responsibility, and ethics) principles were formulated specifically to provide guidance on protecting Indigenous peoples' data and rights (Carroll et al., 2020).

Indigenous data

CARE

- *Collective benefit* recognizes the imperative to share data in a way that promotes equity and inclusion.
- *Authority to control* is concerned with the rights and interests inherent in data, and the governance mechanisms that deliver those rights.
- *Responsibility* is connected with using data as the basis for building respectful and constructive relationships between Indigenous and non-Indigenous peoples.
- *Ethics* provide the framework to maximize the benefits of data sharing at the same time as minimizing potential harm.

Thus, CARE principles emphasize the purpose of data and the people who derive benefit from data, rather than simply the FAIR management of data. Some aspects of the CARE principles will already resonate with the critical GIS agenda, such as the emphasis on responsible and ethical action to empower those sectors of society who might otherwise be disadvantaged by the technology. One aspect of CARE in particular warrants further, more detailed discussion in the context of geospatial data, however: Indigenous data sovereignty.

Indigenous data sovereignty Data sovereignty concerns the management of data "consistent with the laws, practices, and customs of the nation-state in which it is located" (Snipp, 2016). Data sovereignty has direct parallels to issues of information privacy, discussed above. Where information privacy is connected with control of personal information by individuals, data sovereignty is concerned with control of data by nation-states. As distributed systems and the Internet today provide no geographical barriers to the movement of data, national governments around the world have moved to assert their control over any sovereign data that has legal, privacy, and national security implications.

data sovereignty

Indigenous data sovereignty refers to the right of Indigenous peoples to exercise control and ownership over Indigenous data including the collection, application, use, and sharing of data created by or about Indigenous people and their territories (Tauli-Corpuz, 2016). Although technology and data sharing is recognized widely as of enormous potential net benefit to society, those benefits (and costs) have not always fallen equally on different sectors of society. Indigenous data sovereignty has emerged as a priority in response to past inequitable impacts of data capture, sharing, and use on Indigenous

Indigenous data sovereignty

Box 10.5: Exercising Indigenous data sovereignty

GIS and spatial data have the potential to empower Indigenous communities, such as in support of land title claims and increased participation in decision-making processes (Briggs et al., 2020). Veland, Lynch, Bischoff-Mattson, Joachim, & Johnson (2014), for example, describe the co-design of a GIS application to support a Native Title Claim in Australia by the Yorta Yorta Nation. Indigenous Yorta Yorta people retained ownership and copyright of the spatial data generated through the claim, including the right to choose to keep secret certain knowledge, to be consulted on any intellectual property agreements, and to control approvals for future uses. Similarly, the ancient Budj Bim system of *kooyang* (short-finned eel, *Anguilla australis*) traps, part of the traditional lands of the Gunditjmara people of south-eastern Australia, was recognized as a UNESCO World Heritage site in 2019. In preparing the UNESCO application, precise LIDAR imagery uncovered more extensive evidence of past settlements dating back over 6,000 years. Importantly, the Indigenous Gunditjmara people retained ownership of the captured LIDAR imagery of the Gunditjmara lands—often not the case in other similar public projects—with perhaps more discoveries and uses of the data still to come in the future. A logical extension of Indigenous data sovereignty and CARE authority-to-control is the concept of "data repatriation," which involves the return of previously captured digital cultural objects, such as images, records, and of course geospatial data, to their Indigenous owners (Gardiner, McDonald, Byrne, & Thorpe, 2011).

people. The negative impacts have included the exploitation of Indigenous communities in connection with GIS and spatial data (see, for example, Eisner et al., 2012; Tripathi & Bhattarya, 2004).

Just as nation-states have asserted their rights to data sovereignty, so First Nations' aim to exercise the CARE principle of "authority to control" through Indigenous data sovereignty. For spatial data, effective authority to control requires lasting and meaningful collaboration between Indigenous communities and geographic information scientists (both Indigenous and non-Indigenous). Box 10.5 briefly explores some examples of how authority to control Indigenous spatial data is beginning to be exercised by Indigenous communities.

10.4.3 FATE and acting critically

FATE

The focus of FAIR, CARE, and Indigenous data sovereignty is on equitable capture, sharing, and use of data. The principles of *FATE* (fairness, accountability, transparency, and ethics) are a natural complement, concerned with equity in connection with algorithms, analyses, and automated decisions. FATE is emerging as an especially important framework for equity in GIS in the context of increasingly sophisticated "GeoAI" capabilities, already discussed in Chapter 9.

- *Fairness* attempts to combat implicit discrimination, bias, and inaccuracy in algorithms, safeguarding equity for people in machine-led decision processes.
- *Accountability* aims to ensure that the people involved in the development and operation of algorithms are responsible for their algorithm design decisions, akin to accountability in fair information practices (see Section 10.3.1).

- *Transparency* is concerned with openness about the design, limitations, data sources, and execution of algorithms, helping to increase people's trust in machines and their algorithmic outputs.
- *Ethics* in connection with algorithms is about adherence to accepted standards of conduct, such that the outputs of a machine-led process reflect the same expectations of behavior as a regular human-operated process.

In short, FATE aims to replicate the standards applied to human decision-making processes to decisions that are partially or even fully automated. Hence, FATE echoes the aspirations captured by human frameworks for ethical behavior, such as professional *codes of conduct*. Professional (human) codes of conduct generally embody six principles, namely that (GIS) professionals:

code of conduct

- act with *integrity* and *honesty*, never intentionally misleading;
- behave with *transparency*, being open about their actions and the motivations for those actions;
- are *accountable* for their actions, as already discussed in connection with FATE and fair information practices;
- respect the *confidentiality* of clients, colleagues, or data subjects where required;
- are *objective* in providing advice and arriving at decisions on the basis of evidence and expertise;
- remain *respectful* at all times to others, without prejudice, discrimination, or bias.

Figure 10.18 summarizes the different aspects of codes of conduct, as well as CARE, FATE, and fair information practices. Although each has its own focus and domain, Figure 10.18 provides indicative links to highlight where some of the main correspondences and relationships lie.

Combining all these elements with the critical GIS agenda at the beginning of the section, it is possible to suggest a framework for responsible action with GIS, based on work by Wilson (2016):

1. *Concept*: Preparing to use GIS critically should begin with an explicit understanding of the concepts behind any choice of action. In particular, this first stage clarifies what ways of knowing are being assumed or adopted, and what alternatives are being excluded as a consequence and why.

2. *Design*: With clarity around the concepts, the next step is to seek clarity about the technical framework chosen for the action. The second stage makes explicit the system constraints, data constraints, and analysis constraints imposed by the selected approach or technology, and why those specific options were chosen over the alternatives.

3. *Perform*: Performing the action (i.e., the GIS project, development, analysis, visualization) sits at the center of this framework. Most of this book is concerned primarily with this stage, and of understanding GIS and enacting GIS analyses.

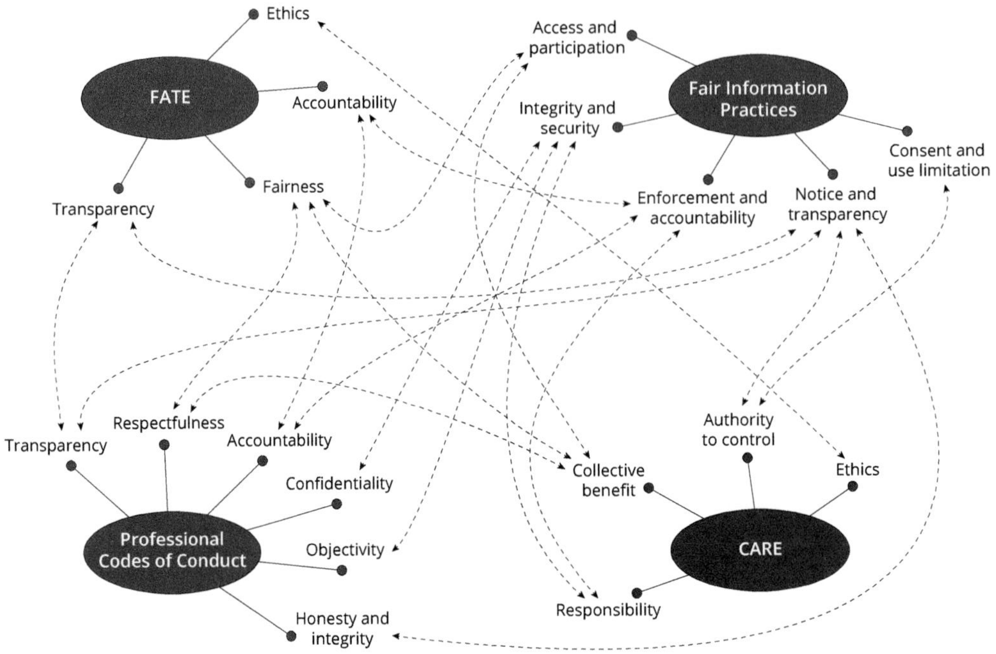

Figure 10.18: Summary and indicative relationships between fair information practices, CARE, FATE, and professional codes of conduct

4. *Validate*: Looking at the products of the action from stage 3, the validation stage attempts to place those outputs back into the context of the concept and design in stages 1 and 2. Beyond mere technical validity (e.g., whether the analysis was executed correctly), this stage aims to validate the action in its wider ethical context (see below for suggested ethical validity tests).

5. *Decide*: In the context of the previous four stages, the final step is to decide how to proceed with the action. In many cases—certainly if any ethical issues arise in connection with validation—the decision may be to learn from the critical process and reformulate with new concepts, explore alternative ways of knowing, or select different technical approaches to action. In all cases, actions should only proceed when it is clear whose perspective, power, and prestige will be enhanced by the choices taken in the previous stages.

In addressing stage 4—validating the chosen action—the answers to a number of questions posed by Davis (1997) can help illuminate the ethical implications of an action:

- Does my chosen action do less harm than the alternatives? (*Harm test*)
- Would I want my choices in deciding on this action published in tomorrow's newspaper? (*Publicity test*)
- Could I defend my choices to a scientific committee or a government inquiry? (*Defensibility test*)
- Would I still think this was a good choice if I were adversely affected by it? (*Reversibility test*)

- What might my peers or professional governing body say about my chosen action? (*Professional test*)
- What do my colleagues, staff, and managers say when I describe my chosen action and the rationale for taking it? (*Organizational test*)

Being critical can be a source of significant innovation, providing the space to appreciate different ways of knowing and alternative potential solutions, rather than only following well-trodden paths. Adopting a critical approach also provides a salient reminder that GIS is not a neutral tool, dispensing facts about the world; rather, it is an assemblage of powerful concepts, assumptions, and methodologies that together can not only be harnessed for the wider benefit of society, but also to silence, marginalize, and disadvantage certain members of society.

10.4.4 Revisiting John Snow

GIS courses and textbooks frequently begin with the time-honored story of John Snow, and his famous map of the 1854 Soho cholera outbreak in London (Figure 10.19). The map is often used as an exemplar of visual thinking, and the power of maps and mapping (and now GIS) to analyze spatial data and inform evidence-based decision-making. As the story is sometimes told, John Snow used his map to prove the new theory that cholera was caused by contaminated water, as opposed to the explanation of airborne "miasma" that was more widely accepted at the time. The clustering of deaths (shown by black bars) in the vicinity of one pump in particular—the Broad Street pump (see Figure 10.19 detail)—provides compelling evidence of the waterborne cause. Such evidence helped to convince the local authorities to remove the handle from the offending Broad Street pump, so ending the outbreak.

The story is, of course, essentially correct in its key aspects. Drinking contaminated water is indeed the mechanism for cholera transmission; the Broad Street pump was indeed the source of the outbreak; and John Snow's map does indeed illustrate the power of visual thinking, helping the viewer gain insight into an important truth through an improved understanding of the data.

Nevertheless, a critical rereading of the story also reveals a much richer, more complex, and more nuanced version of reality. For example, John Snow was convinced of the "germ theory" of cholera transmission long before the 1854 cholera outbreak. The first edition of John Snow's treatise "On the mode of communication of cholera" was published in 1849, several years before the Broad Street outbreak. The famous map, appearing in the second edition published in 1855, is therefore unlikely to have played any part in John Snow's reasoning and analytical processes. Rather, the map's power has been in subsequently convincing others of Snow's chain of reasoning.

It seems likely that John Snow was strongly influenced by earlier work in designing his map, and Thomas Shapter's published map of an earlier 1832 cholera outbreak in Exeter, UK, in particular (Figure 10.20). John Snow was

Figure 10.19: John Snow's 1855 map of the 1854 cholera outbreak in Soho, London, with close-up detail near the Broad Street pump (right) (Source: Wellcome Images, copyrighted work available under Creative Commons Attribution only license CC BY 4.0)

certainly an influential pioneer, but by no means the first to present data in the way he did.

Like many maps, part of the power of John Snow's map lies in what it does not show. John Snow lived in the vicinity of Soho, and he used his local network and connections in developing his ideas. In particular, John Snow gathered important knowledge from speaking with locals of the experiences of people who *lived* in the vicinity and did *not* catch cholera (such as the proprietor of the brewery on Broad Street, whose workers it transpired drank from the brewery's own, untainted well) and who did *not* live in the vicinity but *did* catch cholera (such by talking to the son of a deceased woman, who explained that his mother had water routinely brought to her from the Broad Street pump because she preferred the taste to that of a more proximal water source). Those personal experiences related to John Snow were undoubtedly important to the development of his thesis, but they are nowhere represented on the map. As a direct result of these omissions, the map is able to communicate a cleaner, crisper, more clinical-looking, and arguably more convincing image (see, for example, the discussion of cartographic abstraction in Section 8.2.1).

One other important detail suppressed in John Snow's map and many similar maps is *time*. The map aggregates data over the entire outbreak lasting several weeks. At the peak of the outbreak on 2 September more than 120 residents in the area died of cholera. By the time that John Snow had convinced the local authority to remove the pump handle a week later, the number of deaths had already subsided to fewer than 25 on 9 September. While shutting off the pump was unquestionably a reasoned and prudent course of action to prevent future outbreaks, it is likely that particular September outbreak may have, in the main, already run its course.

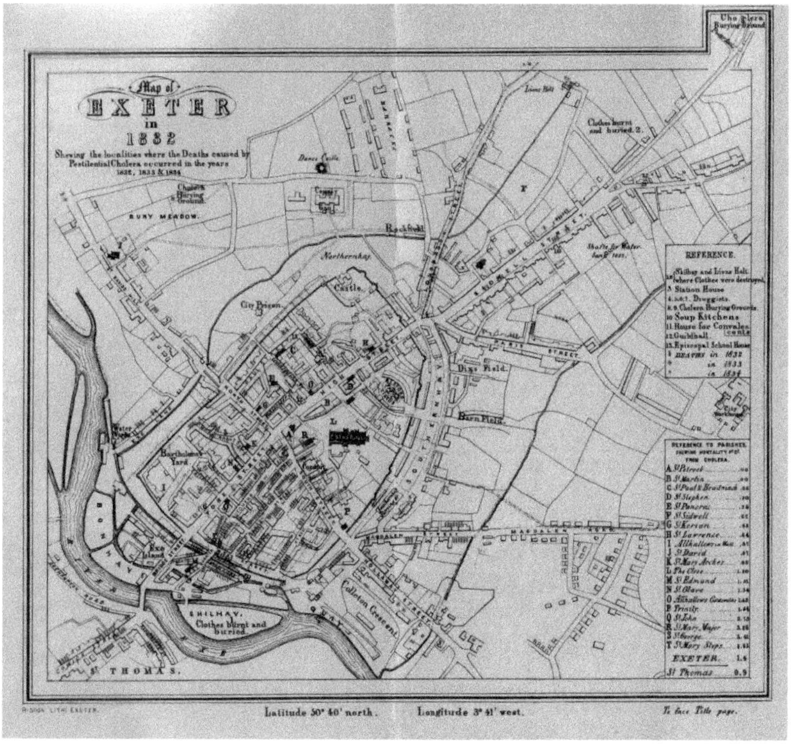

Figure 10.20: Thomas Shapter's 1849 map of "The History of the Cholera in Exeter in 1832" (Courtesy of the National Library of Medicine)

The value of GIS, maps, and mapping ultimately lies in the stories that we can tell with these tools. These stories should be founded on objective data, help us understand the world with new clarity, and told for the benefit of society, just like the story of John Snow's map. But we should not confuse those stories with reality, nor forget that the tools themselves are part of the story.

Reflections

There is always a risk in placing important topics in the final chapter of a book that they take on the semblance of an afterthought or addendum. Uncertainty, for example, is more frequently covered in earlier chapters in most GIS textbooks because it is an endemic and fundamental issue for GIS. However, like all the topics covered in this chapter, uncertainty is also among the more advanced topics in this book, both technically and conceptually. Hence, the sophistication and nuance required to handle uncertainty, together with its pervasive importance to GIS, elevated uncertainty to a position as a "capstone" topic connecting and integrating many of the topics covered previously.

Worboys & Clementini (2001) introduced and explored the typology of imperfection in Figure 10.2, the central idea at the foundation of our treatment of uncertainty. In addition to being simple (in essence, "inaccuracy = lack of correctness" and "imprecision = lack of detail"), these foundations support rigor and clarity when discussing uncertainty, often lacking in much of the literature on uncertainty. For those looking to contrast with alternative approaches, Couclelis (2003) begins with several of the themes here, such as imperfection and reasoning under uncertainty, but goes on to explore issues of knowledge and ignorance. Smithson (2012) has proposed an alternative taxonomy of ignorance, widely used in psychology, but the jumble of associated terms and contestable distinctions makes the taxonomy difficult to apply in practice. In contrast, Price & Shanks (2005) does approach the topic from a clear conceptual and theoretical basis in semiotics (the study of signs and meaning) and develops a simple taxonomy of data quality widely used in the field of information systems.

Imprecision and inaccuracy place an important role in location privacy, as we have seen. They also resonate strongly with the themes of critical GIS, setting GIS development and use in a wider context of responsible action. Uncertainty and imperfection are reminders that our information "can never fully or correctly reflect all aspects of reality" (p. 403). Most work on uncertainty in GIS inherently admits multiple possible different versions of reality, a short step to embracing different ways of knowing in GIS Duckham & Sharp (2005). One of the most succinct and engaging introductions to key issues surrounding critical GIS, and how that has evolved over the past 20 years, is Nadine Schuurman's two interviews with GIS pioneer Mike Goodchild (Schuurman, 1999, 2009). In the broader context of science in society, Latour (1993) and Hacking (1999) are two classic books that explore in detail the fascinating different ways in which scientific ideas can become and can create reality.

Appendix: Nutty Nuggets relational database

A

Nutty Nuggets relational database example

NUGGETS database schema

ADDRESS (ADDRESS ID, STREET_AD, ZIP, CITY, STATE)

CONTENT (OID, SID, WEIGHTG)

CUSTOMER (TEL, HID, TITLE, GNAME, FNAME)

ORDERS (ORDER NO, CUS_TEL, AID, MEAL, SIZE)

STOCK (INGREDIENT, QUANT, UNITS)

(See page 57.)

ADDRESS relation

ADDRESS_ID	STREET_AD	ZIP	CITY	STATE
10012-BS01	177a Bleecker St	10012	New York	NY
10560-GL01	1407 Graymalkin Lane	10560	North Salem	NY
11050-PA01	98 Pleasant Avenue	11050	Fort Washington	NY
11232-4201	738 42nd Street	11232	Brooklyn	NY
19104-CS01	85 Chestnut Street	19104	Philadelphia	PA
20002-VRD2	3501 Valley Road Drive	20002	Washington	DC
74012-RA01	554 Railroad Avenue	74012	Broxton	OK
94110-MS01	1128 Mission St	94110	San Francisco	CA

CUSTOMER relation

TEL	GNAME	FNAME	HID	TITLE
202-555-0125	Joanna	Cargill	20002-VRD2	Ms
213-555-0506	Roberto	Da Costa	94110-MS01	Mr
610-555-0195	Lorna	Dane	10560-GL01	Ms
757-555-0112	Bobby	Drake	11050-PA01	Mr
785-555-0189	Loki	Laufeyson	10012-BS01	Mx
939-555-0177	Jane	Foster	74012-RA01	Dr

ORDERS relation

ORDER_NO	CUS_TEL	AID	MEAL	SIZE
M066-22-06	610-555-0195	19104-CS01	Regular	2
M066-22-07	610-555-0195	19104-CS01	Regular	1
M066-22-08	610-555-0195	19104-CS01	Regular	2
M113-22-09	939-555-0177	74012-RA01	Vegan	1
M315-22-06	213-555-0506	94110-MS01	Lo carb	4
M315-22-07	213-555-0506	94110-MS01	Lo carb	4

CONTENT relation

OID	SID	WEIGHTG
M066-22-06	Chickpea	400
M066-22-06	Chili	10
M066-22-06	Rice	400
M066-22-06	Tomato	400
M066-22-07	Cheese	150
M066-22-07	Garlic	20
M066-22-07	Mushroom	400
M066-22-08	Carrot	300
M066-22-08	Corn	200
M066-22-08	Potato	300
M066-22-08	Vegetable stock	500
M113-22-09	Chickpea	400
M113-22-09	Garlic	20
M113-22-09	Lemon	200
M113-22-09	Mushroom	600
M315-22-06	Chickpea	800
M315-22-06	Chili	20
M315-22-06	Corn	600
M315-22-06	Tomato	600
M315-22-07	Cheese	350
M315-22-07	Garlic	30
M315-22-07	Mushroom	900

STOCK relation

INGREDIENT	QUANT	UNITS
Carrot	61	kg
Cheese	17	kg
Chickpeas	18	cans
Chili	750	g
Corn	15	kg
Garlic	8	kg
Lemon	8	kg
Mushroom	30	kg
Oil	42	l
Onion	25	kg
Potato	70	kg
Rice	64	kg
Tomato	104	cans
Vegetable stock	48	l

APPENDIX: DISCRETE MATHEMATICS PRIMER

B

THIS appendix provides a brief primer on the discrete mathematics structures and syntax used in the book, developed from an earlier primer that first appeared in Duckham (2013).

Sets

Sets are used throughout the book but introduced in some depth in Section 3.2. In summary, a *set* is a collection of "things," each called a *member* or an *element* of a set.[1] Some common sets are also identified with their own symbol, including:

- $\varnothing$, the *empty set* containing no elements, also written $\{\}$;
- $\mathbb{B}$, the Boolean set of two values (e.g., with members true/false, 1/0, on/off);
- $\mathbb{Z}$, the set of positive and negative integer numbers, including zero;
- $\mathbb{Q}$, the set of rationals—ratios of two integers; and
- $\mathbb{R}$, the set of reals—measurements on the number line.

The order of elements in a set is not significant.[2] A set that contains all the elements of another set, and perhaps some additional elements, is termed a *superset*. We write $X \subseteq Y$ to indicate Y is a superset of X, or equivalently that X is a *subset* of Y.[3] The empty set is a subset of every set, e.g., $\varnothing \subseteq X$. A set X is said to be a *proper* subset of Y, written $X \subset Y$, if X is a subset of Y and $X \neq Y$.

A set can be specified explicitly by itemizing each of its members. Alternatively, it is often easier to specify sets by *intension*, essential for sets with a large or infinite number of elements such as $\mathbb{Z}$ and $\mathbb{R}$. For example, another special set $\mathbb{N}$ (the set of natural numbers—positive integers from 1 to infinity) can be defined as $\mathbb{N} = \{x \in \mathbb{Z} | x > 0\}$ (read "the set containing numbers x in $\mathbb{Z}$ *such that* x is greater than 0").[4]

We often specify by intension subsets of the real numbers as *intervals*:

- $[a, b]$ is the set of reals between a and b including a and b, termed a *closed* interval;
- $]a, b[$ is the set of reals between a and b excluding a and b, termed an *open* interval; and

[1] For example, $A = \{a, b, c\}$ defines the set A with three members a, b, and c. If we wish to assert that a is an element of set A, we may write $a \in A$.

[2] So, set $A = \{a, b, c\} = \{b, c, a\} = \{c, a, b\} = \{c, b, a\} = \ldots$

[3] If $A = \{a, b, c\}$ and $B = \{a, b, c, d\}$ then $A \subseteq B$. It is also true that $A \subseteq A$, but A is not a proper subset of itself.

[4] The set $C = \{2, 3, 4, 5, 6\}$ can be specified as $C = \{x \in \mathbb{N} | 1 < x < 7\}$.

- $[a, b[$ is the set of reals between a and b including a but excluding b, termed a *semi-open* interval.

The *intersection* of two sets X and Y, written $X \cap Y$ is the set of elements in X and in Y. The *union* of X and Y, written $X \cup Y$ is the set of elements in X or in Y.[5] The set *difference*, $X \backslash Y$, is the set of elements in X but not in Y.[6]

[5] With reference to the sets above, $A \cap \{a, b, d\} = \{a, b\}$, $A \cup \{b, d\} = B$, and $A \cap C = \emptyset$.

[6] For example, $A \backslash B = \{b\}$.

The *cardinality* of a set X is the number of elements in that set, written $|X|$.[7] The *complement* of a set A is the set of elements *not* in that set, written A'. Set complement makes implicit reference to a *universal* set of all elements under consideration. The *power set* of a set X is the set of all subsets of that set (including the empty set), written $\mathcal{P}(X)$ or sometimes 2^X. Note that the power set of a set is still a set: the elements of a set may themselves be sets.[8] A *partition* of a set is a set of subsets such that each and every member of the original set appears in exactly one subset in the partition.[9] Therefore, a partition of a set is a subset of the powerset of that set.

[7] For the sets in the margins above, $|A| = 3$, $|B| = 4$, $|C| = 5$.

[8] For example, $\{a, c\} \in \mathcal{P}(A)$.

[9] The set $\{\{a, c\}, \{b\}, \{d\}\}$ is a partition of B.

Products and relations

The *product* of two sets X and Y, written $X \times Y$, is the set of all distinct pairs with the first element $x \in X$ and the second element $y \in Y$. A pair is usually written with enclosing parentheses, for example, (x, y).[10] Unlike the elements of a set, a pair is ordered, so $(x, y) \neq (y, x)$. Note that the product of two sets is, again, a set—albeit a set containing pairs as its elements.[11] The product of two sets can be extended to an n-ary product (product of n sets) in a straightforward way.[12]

[10] Pairs are sometimes written without parentheses or commas, e.g., xy.

[11] Thus, for example, $(b, d) \in A \times B$.

[12] For example, $(b, c, 3) \in A \times B \times C$).

A *relation* between two sets is simply a subset of the product of those sets.[13] Many important relations are from a set to itself, termed a *binary relation*. The set $\mathbb{R} \times \mathbb{R}$, often written $\mathbb{R}^2$, is the familiar points in the Cartesian plane (i.e., planar coordinates).[14]

[13] For example, $L = \{(d, a), (b, a), (c, b), (a, b)\}$ is a relation from B to A.

[14] Thus, for example, $(15.2, -1.9) \in \mathbb{R}^2$.

A binary relation R on X is said to be *reflexive* if for every $x \in X$, $(x, x) \in R$. A relation R is *irreflexive* if for every $x \in X$, $(x, x) \notin R$. A relation R on X is said to be *symmetric* if whenever $(x, y) \in R$ it is also the case that $(y, x) \in R$. A relation is *antisymmetric* relation if whenever $(x, y) \in R$, then $(y, x) \notin R$. Finally, a relation R on X is said to be *transitive* if whenever (x, y) and $(y, z) \in R$, $(x, z) \in R$.[15]

[15] For example, $N = \{(1, 2), (2, 1), (2, 5), (5, 2)\}$, a (binary) relation on $\mathbb{Z}$, is not transitive; is symmetric (and hence not antisymmetric); and is irreflexive (and hence not reflexive).

Functions

A *function* is a special case of a relation between two sets, X and Y, where every element in X is related to a unique element in Y. The first set, X, is the set of all permitted inputs to the function, termed the *domain*. The second set, Y, is the set of all possible outputs of the function, termed the *codomain*.

Although functions are a special case of relations, they have their own special formal syntax to highlight their particular constraints. A function f with domain X and codomain Y is specified as $f : X \to Y$ (read "f is a function from X to Y"). We can indicate that the function f relates a particular input $x \in X$ to a particular result $y \in Y$ by writing $y = f(x)$ or equivalently

$f(x) \mapsto y$ (read "f of x *maps to* y"). To highlight the correspondence with relations, if f were defined as a relation (rather than as a function) we might instead choose to write $(x, y) \in f$. Both are correct, the difference is a matter of style and convention.

A function where every element in the domain maps to a distinct (different) element in the codomain is termed an *injection*. A function where every element in the codomain has some element from the domain that maps to it is termed a *surjection*. A function that is both an injection and a surjection is termed a *bijection*.[16]

[16] For instance, function $g : A \rightarrow B$ where $g(a) = a, g(b) = b, g(c) = c$ is an injection, but not a surjection (and so not a bijection).

The *image* of a function f, denoted by f_*, is the set of elements that a function maps to.[17] Thus, the image of a function is necessarily a subset of the codomain for that function. Any injection has the property that there exists another function that maps back to elements in the domain from elements in the image of that function, termed an *inverse function*. A bijection has an inverse that maps back to elements in the domain from elements in the codomain of that function (since the image of a bijection is its codomain).

[17] For example, for the function $f : X \rightarrow Y$, the image of the function $f_* = \{y \in Y | y = f(x) \text{ for some } x \in X\}$.

Graphs

A *graph* is a pair $G = (V, E)$, where:

1. V is some set of elements, called *vertices* or *nodes*; and
2. E is a binary relation on V, $E \subseteq V \times V$, each element of which is called an *edge*.

If E is symmetric, then the graph G is termed *undirected* or *bidirected*; otherwise, G is a *directed* graph. For an undirected graph $G = (V, E)$, we may sometimes use the shorthand $\{x, y\} \in E$ instead of writing both $(x, y) \in E$ and $(y, x) \in E$ (since in an undirected graph E is symmetric, so if $(x, y) \in E$, then necessarily $(y, x) \in E$, and the ordering does not matter).

Graphs (and relations) have a natural graphical representation, with vertices depicted as dots and edges depicted as lines connecting two vertices. Directed graphs may be similarly depicted, with the addition of arrows to indicate the direction(s) of edges (as in Figure B.1).

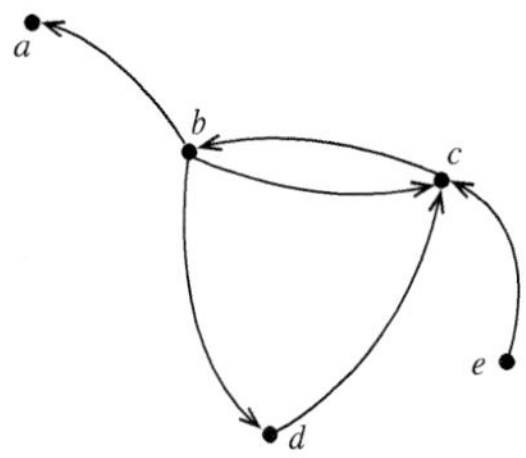

Figure B.1: Example directed graph, from Figure 2.14

Two nodes that are connected by an edge are said to be *adjacent*. An edge is said to be *incident on* the nodes it connects; likewise, a node is said to be *incident* with an edge that connects to it. The *degree* of a node is its number of incident edges. In a directed graph, we may distinguish between the number of incoming edges (a node's *in-degree*) and outgoing edges (a node's *out-degree*). The *order* of a graph is the number of nodes it contains.

A *path* is a sequence of adjacent nodes.[18] Two nodes are said to be *connected* if there exists a path between them. A graph is said to be connected if every pair of nodes in the graph is connected.

[18] For example, *bdc* is a path in *G* Figure B.1; *bcd* is not.

A *tree* is a (connected) graph where there exists exactly one path between any pair of nodes. A *rooted* tree is a tree with one node designated as the *root*.

A *planar* graph is a graph that can be drawn in the plane with edges only intersecting at nodes (i.e., no edges crossing). A graph that *can* be drawn in

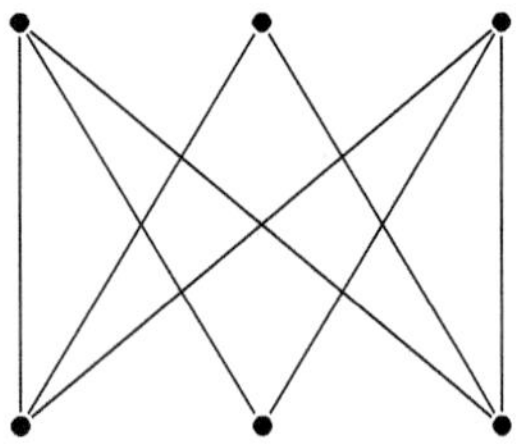

Figure B.2: Example non-planar graph, from Figure 3.42

the plane with no crossing edges is called planar; a graph that *is* drawn in the plane with no crossing edges is called *plane*. Thus all plane graphs are also planar; and all non-planar graphs (such as Figure B.2) can never be depicted as a plane graph.

Logic

Logical expressions in mathematics and computer science can be assigned a *truth value*. In the most basic (two-valued) logic, expressions have a truth value of either true or false, but not both. For example, the truth value of the expression "$5 < 10$" is true; the truth value of the expression "$1 = 2$" is false. We often use the formal symbols $\top$ (true) and $\bot$ (false).

The *conjunction* of two logical expressions $p \wedge q$ is true if *both* p *and* q are true; it is false otherwise. The *disjunction* of two logical expressions $p \vee q$ is true if *either* p *or* q are true; it is false if both p and q are false.[19] Other connectives include implication $\Longrightarrow$, equivalence $\Longleftrightarrow$ (if and only if), and negation $\neg$. The logical structure of connectives can be captured in truth tables, such as those in Figure B.3.

[19] For example, the truth value of "$a \in A \wedge a \in C$" is false; the truth value of "$5 < 10 \vee 1 = 2$" is true.

$\wedge$	$\top$	$\bot$
$\top$	$\top$	$\bot$
$\bot$	$\bot$	$\bot$

$\vee$	$\top$	$\bot$
$\top$	$\top$	$\top$
$\bot$	$\top$	$\bot$

$\neg$	$\top$
$\top$	$\bot$
$\bot$	$\top$

$\Longrightarrow$	$\top$	$\bot$
$\top$	$\top$	$\bot$
$\bot$	$\top$	$\top$

$\Longleftrightarrow$	$\top$	$\bot$
$\top$	$\top$	$\bot$
$\bot$	$\bot$	$\top$

Figure B.3: Truth tables for predicate logic connectives

Bibliography

Allen, J. F. (1983). Maintaining knowledge about temporal intervals. *Communications of ACM, 26*, 832–843.

Allen, J. F. (1984). Towards a general theory of action and time. *Artificial Intelligence, 23*, 123–154.

Alt, H., Efrat, A., Rote, G., & Wenk, C. (2003). Matching planar maps. *Journal of Algorithms, 49*(2), 262–283.

Amores, D., Tanin, E., & Vasardani, M. (2021). A proactive route planning approach to navigation errors. *International Journal of Geographical Information Science, 35*(6), 1094–1130.

Andrienko, G., Andrienko, N., Demsar, U., Dransch, D., Dykes, J., Fabrikant, S. I., Jern, M., Kraak, M.-J., Schumann, H., & Tominski, C. (2010). Space, time, and visual analytics. *International Journal of Geographical Information Science, 24*(10), 1577–1600.

Andrienko, N., & Andrienko, G. (2006). *Exploratory analysis of spatial and temporal data: A systematic approach.* Springer.

Ardagna, C. A., Cremonini, M., Damiani, E., Capitani di Vimercati, S. D., & Samarati, P. (2007). Location privacy protection through obfuscation-based techniques. In *Proc. IFIP Annual Conference on Data and Applications Security and Privacy* (pp. 47–60). Springer.

Ballas, D., Clarke, G., Franklin, R. S., & Newing, A. (2017). *GIS and the social sciences: Theory and applications.* Taylor & Francis.

Baralis, E., Dalla Valle, A., Garza, P., Rossi, C., & Scullino, F. (2017). SQL versus NoSQL databases for geospatial applications. In *Proc. IEEE International Conference on Big Data* (pp. 3388–3397).

Basiri, A., Lohan, E. S., de Silva, P. F., Peltola, P., Hill, C., & Moore, T. (2014). Overview of positioning technologies from fitness-to-purpose point of view. In *Proc. International Conference on Localization and GNSS (ICL-GNSS)* (pp. 1–7). IEEE.

Bast, H., Delling, D., Goldberg, A., Müller-Hannemann, M., Pajor, T., Sanders, P., Wagner, D., & Werneck, R. F. (2016). Route planning in transportation networks. In *Algorithm Engineering* (pp. 19–80). Springer.

Baumgart, B. (1975). A polyhedron representation for computer vision. In *Proc. AFIPS National Conference 44* (pp. 589–596).

Bearman, N., & Fisher, P. F. (2012). Using sound to represent spatial data in ArcGIS. *Computers & Geosciences, 46*, 157–163.

Beckmann, N., Kriegel, H.-P., Schneider, R., & Seeger, B. (1990). The R*-tree: An efficient and robust access method for points and rectangles. In *Proc. ACM SIGMOD International Conference on Management of Data* (pp. 322–331).

Belussi, A., Bertino, E., & Catania, B. (2002). Using spatial data access structures for filtering nearest neighbor queries. *Data & Knowledge Engineering*, *40*(1), 1–31.

Bender, E. M., Gebru, T., McMillan-Major, A., & Shmitchell, S. (2021). On the dangers of stochastic parrots: Can language models be too big? In *Proc. acm conference on fairness, accountability, and transparency (facct)* (pp. 610–623). doi:10.1145/3442188.3445922

Bertino, E., & Ooi, B. C. (1999). The indispensability of dispensable indexes. *IEEE Transactions on Knowledge and Data Engineering*, *11*(1), 17–27.

Bertino, E., Ooi, B. C., Sacks-Davis, R., Tan, K.-L., Zobel, J., Shidlovsky, B., & Catania, B. (1997). Temporal databases. In *Indexing Techniques for Advanced Database Systems* (pp. 113–149). Springer.

Bertolotto, M., De Floriani, L., & Marzano, P. (1994). An efficient representation for pyramidal terrain models. In N. Pissinou & K. Makki (Eds.), *Proc. Second ACM Workshop on Advances in Geographic Information Systems* (pp. 129–136). National Institute for Standards and Technology.

Bittner, T., Donnelly, M., & Smith, B. (2004). Endurants and perdurants in directly depicting ontologies. *AI Communications*, *17*(4), 247–258.

Bleisch, S., & Hollenstein, D. (2018). Exploratory geovisualizations for supporting the qualitative analysis and synthesis of place-related emotion data. *Cartographic Perspectives*, (91), 30–46.

Brewer, C. A. (2016). *Designing better maps: A guide for GIS users* (2nd Edition). ESRI press.

Briggs, C., Burfurd, I., Duckham, M., Guntarik, O., Kerr, D., McMillan, M., & San Martin Saldias, D. (2020). Bridging the geospatial gap: Data about space and indigenous knowledge of place. *Geography Compass*, *14*(11), e12542.

Brown, T., Mann, B., Ryder, N., Subbiah, M., Kaplan, J. D., Dhariwal, P., Neelakantan, A., Shyam, P., Sastry, G., Askell, A., Henighan, T., ... McCandlish, S. (2020). Language models are few-shot learners. *Advances in Neural Information Processing Systems*, *33*, 1877–1901.

Buchin, K., Speckmann, B., & Verbeek, K. (2011). Flow map layout via spiral trees. *IEEE Transactions on Visualization and Computer Graphics*, *17*(12), 2536–2544.

Buchin, M., Driemel, A., Kreveld, M. v., & Sacristán Adinolfi, V. (2011). Segmenting trajectories: A framework and algorithms using spatiotemporal criteria. *Journal of Spatial Information Science*, *3*, 33–63.

Burrough, P. A. (1981). Fractal dimension of landscapes and other environmental data. *Nature*, *294*, 240–242.

Burrough, P. A., McDonnell, R. A., & Lloyd, C. D. (2015). *Principles of geographical information systems* (3rd Edition). Oxford University Press.

Cao, H., & Wachowicz, M. (2019). The design of an IoT-GIS platform for performing automated analytical tasks. *Computers, Environment and Urban Systems*, *74*, 23–40.

Carroll, S. R., Garba, I., Figueroa-Rodríguez, O. L., Holbrook, J., Lovett, R., Materechera, S., Parsons, M., Raseroka, K., Rodriguez-Lonebear, D., Rowe, R., et al. (2020). The CARE principles for Indigenous data governance. *Data Science Journal*, *3*(1), 1–9.

Cedeno Jimenez, J. R., Zhao, P., Mansourian, A., & Brovelli, M. A. (2022). Geospatial blockchain: Review of decentralized geospatial data sharing systems. In E. Parseliunas, A. Mansourian, P. Partsinevelos, & J. Suziedelyte-Visockiene (Eds.), *Proc. 25th AGILE Conference on Geographic Information Science*.

Chazelle, B. (1991). Triangulating a simple polygon in linear time. *Discrete Computational Geometry*, *6*, 485–524.

Chrisman, N. R. (1978). Concepts of space as a guide to cartographic data structures. In G. Dutton (Ed.), *Proc. First International Advanced Study Symposium on Topological Data Structures for Geographic Information Systems* (pp. 1–19). Harvard Laboratory for Computer Graphics and Spatial Analysis.

Chrisman, N. R. (2006). *Exploring geographic information systems* (2nd Edition). John Wiley.

Claramunt, C., Parent, C., & Thériault, M. (1998). Design patterns for spatio-temporal processes. In *Data Mining and Reverse Engineering* (pp. 455–475). Springer.

Codd, E. F. (1970). A relational model for large shared data banks. *Communications of the ACM*, *13*(6), 377–387.

Çöltekin, A., Bleisch, S., Andrienko, G., & Dykes, J. (2017). Persistent challenges in geovisualization—A community perspective. *International Journal of Cartography*, *3*(sup1), 115–139.

Comber, L., & Brunsdon, C. (Eds.). (2020). *Geographical data science and spatial data analysis: An introduction in R*. SAGE.

Connolly, T., & Begg, C. (2014). *Database systems: A practical approach to design, implementation, and management* (6th Edition). Addison-Wesley.

Couclelis, H. (1992). People manipulate objects (but cultivate fields: Beyond the raster-vector debate in GIS. In A. U. Frank, I. Campari, & U. Formentini (Eds.), *Theories of Spatio-Temporal Reasoning in Geographic Space* (Vol. 639, pp. 65–77). Springer.

Couclelis, H. (2003). The certainty of uncertainty: GIS and the limits of geographic knowledge. *Transactions in GIS*, *7*(2), 165–175.

Couclelis, H. (2010). Ontologies of geographic information. *International Journal of Geographical Information Science*, *24*(12), 1785–1809.

Coxeter, H. S. M. (1961). *Introduction to geometry*. Wiley.

Crooks, A., Malleson, N., Manley, E., & Heppenstall, A. (Eds.). (2021). *Agent-based modeling and geographical information systems: A practical primer*. SAGE.

Damiani, M. L. (2014). Location privacy models in mobile applications: Conceptual view and research directions. *Geoinformatica*, *18*(4), 819–842.

Date, C. J. (2015). *SQL and relational theory: How to write accurate SQL code* (3rd Edition). O'Reilly.

Davis, M. (1997). Developing and using cases to teach practical ethics. *Teaching Philosophy*, *20*(4), 353–385.

De Floriani, L. (1989). A pyramidal data structure for triangle-based surface description. *IEEE Computer Graphics and Applications*, *9*(2), 67–78.

De Floriani, L., Magillo, P., & Puppo, E. (2000). Applications of computational geometry to geographic information systems. *Handbook of Computational Geometry*, *7*, 333–388.

De Floriani, L., Marzano, P., & Puppo, E. (1994a). Hierarchical terrain models: Survey and formalization. In *Proc. SAC'94* (pp. 323–327).

De Floriani, L., Marzano, P., & Puppo, E. (1994b). Line-of-sight communication on terrain models. *International Journal of Geographical Information Systems*, *8*(4), 329–342.

De Floriani, L., & Puppo, E. (1992). A hierarchical triangle-based model for terrain description. In A. U. Frank, I. Campari, & U. Formentini (Eds.), *Theories of Spatio-Temporal Reasoning in Geographic Space* (Vol. 639, pp. 236–251). Springer.

de Berg, M., Cheong, O., van Kreveld, M., & Overmars, M. (2008). *Computational geometry: Algorithms and applications* (3rd Edition). Springer.

Delling, D., & Wagner, D. (2007). Landmark-based routing in dynamic graphs. In *International Workshop on Experimental and Efficient Algorithms* (pp. 52–65). Springer.

Di Biase, D., MacEachren, A. M., Krygier, J., & Reeves, C. (1992). Animation and the role of map design in scientific visualization. *Cartography and Geographic Information Systems*, *23*(4), 345–370.

Dijkstra, E. W. (1959). A note on two problems in connexion with graphs. *Numerische Mathematik*, (1), 269–271.

Dodge, M., & Kitchin, R. (2002). *Atlas of cyberspace*. Addison-Wesley.

Dodge, S., Weibel, R., Ahearn, S. C., Buchin, M., & Miller, J. A. (2016). Analysis of movement data. *International Journal of Geographical Information Science*, *30*(5), 825–834.

Dodge, S., Weibel, R., & Lautenschütz, A.-K. (2008). Towards a taxonomy of movement patterns. *Information visualization*, *7*(3–4), 240–252.

Douglas, D. (1990). It makes me so CROSS. In D. J. Peuquet & D. F. Marble (Eds.), *Introductory Readings in Geographic Information Systems*. Taylor & Francis.

Doyle, J., Bertolotto, M., & Wilson, D. (2008). A survey of multimodal interfaces for mobile mapping applications. In *Map-based Mobile Services* (pp. 146–167). Springer.

Driemel, A. (2013). *Realistic analysis for algorithmic problems on geographical data* (Doctoral dissertation, Utrecht University).

Du, J., Wang, S., Ye, X., Sinton, D. S., & Kemp, K. (2022). GIS-KG: Building a large-scale hierarchical knowledge graph for geographic information science. *International Journal of Geographical Information Science, 36*(5), 873–897.

Duckham, M. (2013). *Decentralized spatial computing: Foundations of geosensor networks*. Springer.

Duckham, M. (2015). GI expertise. *Transactions in GIS, 19*(4), 499–515.

Duckham, M., Drummond, J. E., & Forrest, D. (2000). Assessment of error in digital vector data using fractal geometry. *International Journal of Geographical Information Science, 14*(1), 67–84.

Duckham, M., Gabela, J., Kealy, A., Khan, M., Legg, J., Moran, B., Rumi, S. K., Salim, F. D., Sharmeen, S., Tao, Y., Trentelman, K., & Vasardani, M. (2022). Explainable spatiotemporal reasoning for geospatial intelligence applications. *Transactions in GIS, 26*(6), 2455–2479.

Duckham, M., & Kulik, L. (2003). "Simplest" paths: Automated route selection for navigation. In *International Conference on Spatial Information Theory* (pp. 169–185). Springer.

Duckham, M., & Kulik, L. (2005). A formal model of obfuscation and negotiation for location privacy. In *Proc. International Conference on Pervasive Computing* (pp. 152–170). Springer.

Duckham, M., & Sharp, J. (2005). Uncertainty and geographic information: Computational and critical convergence. In P. F. Fisher & D. Unwin (Eds.), *Representing GIS* (Chap. 8, pp. 113–124). Wiley.

Duckham, M., Sun, Q., & Worboys, M. F. (2023). *GIS: A computing perspective* (3rd Edition). CRC Press.

Dwork, C. (2006). Differential privacy. In M. Bugliesi, B. Preneel, V. Sassone, & I. Wegener (Eds.), *Automata, Languages and Programming* (pp. 1–12). Springer.

Dwyer, R. A. (1987). A fast divide-and-conquer algorithm for constructing Delaunay triangulations. *Algorithmica, 2*, 137–151.

Edmonds, J. R. (1960). *A combinatorial representation for oriented polyhedral surfaces* (Master's thesis, University of Maryland).

Eisner, W. R., Jelacic, J., Cuomo, C. J., Kim, C., Hinkel, K. M., & Del Alba, D. (2012). Producing an indigenous knowledge Web GIS for Arctic Alaska communities: Challenges, successes, and lessons learned. *Transactions in GIS, 16*(1), 17–37.

Elmasri, R. A., & Navathe, S. B. N. (2016). *Fundamentals of database systems* (7th Edition). Pearson.

Elwood, S. (2010). Thinking outside the box: Engaging critical geographic information systems theory, practice and politics in human geography. *Geography Compass, 4*(1), 45–60.

Epp, S. S. (2018). *Discrete mathematics with applications* (4th Edition). Brooks/-Cole Cengage.

Fabrikant, S. I., & Buttenfield, B. P. (2001). Formalizing spaces for information access. *Annals of the Association of American Geographers, 91*, 263–280.

Fellows, M. R., Fomin, F. V., Lokshtanov, D., Rosamond, F., Saurabh, S., Szeider, S., & Thomassen, C. (2011). On the complexity of some colorful problems parameterized by treewidth. *Information and Computation, 209*(2), 143–153.

Fish, C., Goldsberry, K. P., & Battersby, S. (2011). Change blindness in animated choropleth maps: An empirical study. *Cartography and Geographic Information Science, 38*(4), 350–362.

Fox, J., Suryanata, K., Hershock, P., & Pramono, A. H. (2006). Mapping power: Ironic effects of spatial information technology. *Participatory learning and action, 54*(1), 98–105.

Frank, A. U. (1992). Qualitative spatial reasoning about distances and directions in geographic space. *Journal of Visual Languages & Computing, 3*(4), 343–371.

Freksa, C. (1992a). Temporal reasoning based on semi-intervals. *Artificial Intelligence, 54*(1–2), 199–227.

Freksa, C. (1992b). Using orientation information for qualitative spatial reasoning. In *Proc. International Conference GIS* (Vol. 639, pp. 162–178).

Freundschuh, S., & Egenhofer, M. J. (1997). Human conceptions of spaces: Implications for GIS. *Transactions in GIS, 2*(4), 361–375.

Gahegan, M. N. (1989). An efficient use of quadtrees in a geographical information system. *International Journal of Geographical Information System, 3*(3), 201–214.

Galton, A. (2012). States, processes and events, and the ontology of causal relations. In *Formal Ontology in Information Systems* (pp. 279–292). IOS Press.

Ganesan, D., Cerpa, A., Ye, W., Yu, Y., Zhao, J., & Estrin, D. (2004). Networking issues in wireless sensor networks. *Journal of Parallel and Distributed Computing, 64*(7), 799–814.

Garcia-Molina, H., Ullman, J., & Widom, J. (2013). *Database systems: The complete book* (New International Edition). Pearson.

Gärdenfors, P., & Rott, H. (1995). Belief revision. In D. M. Gabbay, C. J. Hogger, & J. A. Robinson (Eds.), *Handbook of Logic in Artificial Intelligence and Logic Programming* (Vol. 4, pp. 35–132). Oxford University Press.

Gardiner, G., McDonald, J., Byrne, A., & Thorpe, K. (2011). Respect, trust and engagement: Creating an Australian indigenous data archive. *Collection Building, 30*(4), 148–152.

Garey, M. R., Johnson, D. S., Preparata, F. P., & Tarjan, R. E. (1978). Triangulating a simple polygon. *Information Processing Letters, 7*, 175–179.

Gatrell, A. C. (1991). Concepts of space and geographical data. In D. J. Maguire, M. F. Goodchild, & D. W. Rhind (Eds.), *Geographical Information Systems* (Vol. 1, pp. 119–134). Longman.

Gaver, W. W. (1986). Auditory icons: Using sound in computer interfaces. *Human-Computer Interaction, 2*, 167–177.

Giblin, P. J. (1977). *Graphs, surfaces, and homology: An introduction to algebraic topology*. Chapman & Hall.

Golledge, R. G., Rice, M. T., & Jacobson, R. D. (2006). Multimodal interfaces for representing and accessing geospatial information. In *Frontiers of geographic information technology* (pp. 181–208). Springer.

Goodchild, M. F. (1988). Lakes on fractal surfaces: A null hypothesis for lake rich landscapes. *Mathematical Geology*, *20*(6), 615–629.

Goodchild, M. F. (2004). The validity and usefulness of laws in geographic information science and geography. *Annals of the Association of American Geographers*, *94*(2), 300–303.

Goodwin, S., Dykes, J., Slingsby, A., & Turkay, C. (2015). Visualizing multiple variables across scale and geography. *IEEE Transactions on Visualization and Computer Graphics*, *22*(1), 599–608.

Goyal, R., & Egenhofer, M. J. (2001). Similarity of cardinal directions. In C. Jensen, M. Schneider, B. Seeger, & V. Tsotras (Eds.), *Proc. International Symposium on Advances in Spatial and Temporal Databases (SSTD)* (pp. 36–58). Springer.

Grenon, P., & Smith, B. (2004). SNAP and SPAN: Towards dynamic spatial ontology. *Spatial Cognition and Computation*, *4*(1), 69–104.

Griffin, A. L. (2017). Cartography, visual perception, and cognitive psychology. In *The Routledge Handbook of Mapping and Cartography* (pp. 44–54). Routledge.

Guibas, L., & Stolfi, J. (1985). Primitives for the manipulation of general subdivisions and the computation of Voronoi diagrams. *ACM Transactions on Graphics*, *4*(2), 74–123.

Hacking, I. (1999). *The social construction of what?* Harvard University Press.

Hage, W. R. v., & Ceolin, D. (2013). The simple event model. In *Situation awareness with systems of systems* (pp. 149–169). Springer.

Haque, S., Kulik, L., & Klippel, A. (2006). Algorithms for reliable navigation and wayfinding. In *International Conference on Spatial Cognition* (pp. 308–326). Springer.

Harley, J. B. (1988). Maps, knowledge, and power. In D. Cosgrove & S. Daniels (Eds.), *The iconography of landscape* (pp. 277–312). Cambridge University Press.

Harrower, M., & Brewer, C. A. (2003). ColorBrewer.org: An online tool for selecting color schemes for maps. *The Cartographic Journal*, *40*(1), 27–37.

Hartsfield, N., & Ringel, G. (2003). *Pearls in graph theory: A comprehensive introduction*. Dover.

Harvey, F., Kwan, M.-P., & Pavlovskaya, M. (2005). Introduction: Critical GIS. *Cartographica*, *40*(4), 1–4.

Henle, M. (1979). *A combinatorial introduction to topology*. Freeman.

Heppenstall, A. J., Crooks, A. T., See, L. M., & Batty, M. (Eds.). (2012). *Agent-based models of geographical systems*. Springer.

Heywood, I., Cornelius, S., & Carver, S. (2011). *An introduction to geographical information systems* (4th Edition). Pearson.

Hightower, J., & Boriello, G. (2001). Location systems for ubiquitous computing. *IEEE Computer*, *34*(8), 57–66.

Holroyd, F., & Bell, S. B. (1992). Raster GIS: Models of raster encoding. *Computers & Geosciences, 18*(4), 419–426.

Hornsby, K., & Egenhofer, M. J. (2000). Identity-based change: A foundation for spatio-temporal knowledge representation. *International Journal of Geographical Information Science, 14*(3), 207–224.

Hornsby, K., & Egenhofer, M. J. (2002). Modeling moving objects over multiple granularities. *Annals of Mathematics and Artificial Intelligence, 36*(1), 177–194.

Kendall, E., & McGuinness, D. (2019). *Ontology engineering*. Morgan & Claypool.

Krygier, J. (1994). Sound and geographic visualization. In A. M. MacEachren & D. R. F. Taylor (Eds.), *Visualization in Modern Cartography* (pp. 1–12). Pergamon.

Kuipers, B. (1978). Modeling spatial knowledge. *Cognitive Science, 2*, 129–153.

Kwan, M.-P. (2001). Cyberspatial cognition and individual access to information: The behavioral foundation of cybergeography. *Environment and Planning B: Planning and Design, 28*(1), 21–37.

Lam, N. S.-N., & De Cola, L. (1993). *Fractals in geography*. Prentice-Hall.

Langran, G. (1992). *Time in geographic information systems*. Taylor & Francis.

Latour, B. (1993). *We have never been modern*. Harvard University Press.

Lee, D.-T., & Schachter, B. J. (1980). Two algorithms for constructing the Delaunay triangulation. *International Journal of Computer and Information Sciences, 9*(3), 219–242.

Lennes, N. J. (1911). Theorems on the simple finite polygon and polyhedron. *American Journal of Mathematics, 33*, 37–62.

Li, S., & Liu, W. (2010). Topological relations between convex regions. In *Proc. 24th AAAI Conference on Artificial Intelligence*.

Li, S., Dragicevic, S., Castro, F. A., Sester, M., Winter, S., Coltekin, A., Pettit, C., Jiang, B., Haworth, J., Stein, A., & Cheng, T. (2016). Geospatial big data handling theory and methods: A review and research challenges. *ISPRS Journal of Photogrammetry and Remote Sensing, 115*, 119–133.

Li, W. (2020). GeoAI: Where machine learning and big data converge in GIScience. *Journal of Spatial Information Science*, (20), 71–77.

Li, W., & Arundel, S. T. (2022). GeoAI and the future of spatial analytics. In B. Li, X. Shi, A.-X. Zhu, C. Wang, & H. Lin (Eds.), *New thinking in GIScience* (pp. 151–158). Springer.

Li, Y., Ouyang, S., & Zhang, Y. (2022). Combining deep learning and ontology reasoning for remote sensing image semantic segmentation. *Knowledge-Based Systems, 243*, 108469.

Li, Y., Zhao, Q., & Zhong, C. (2022). Gis and urban data science. *Annals of GIS, 28*(2), 89–92.

Lipschutz, S., & Lipson, M. (2021). *Schaum's outline of discrete mathematics* (4th Edition). McGraw-Hill.

Longley, P. A., Goodchild, M. F., Maguire, D. J., & Rhind, D. W. (Eds.). (2005). *Geographical information systems* (2nd Edition). John Wiley & Sons.

Longley, P. A., Goodchild, M. F., Maguire, D. J., & Rhind, D. W. (2015). *Geographic information systems and science* (4th Edition). John Wiley & Sons.

MacEachren, A. M. (1994a). Time as a cartographic variable. In H. M. Hearnshaw & D. J. Unwin (Eds.), *Visualization in Geographical Information Systems* (Chap. 13, pp. 115–130). John Wiley.

MacEachren, A. M. (1994b). Visualization in modern cartography: Setting the agenda. In A. M. MacEachren & D. Taylor (Eds.), *Visualization in Modern Cartography* (pp. 1–12). Pergamon.

MacEachren, A. M. (1995). *How maps work*. Guilford Press.

Maguire, D. J., Goodchild, M. F., & Rhind, D. W. (Eds.). (1991). *Geographical information systems* (1st Edition). Longman.

Mai, G., Janowicz, K., Yan, B., & Scheider, S. (2019). Deeply integrating linked data with geographic information systems. *Transactions in GIS*, *23*(3), 579–600.

Mai, G., Janowicz, K., Zhu, R., Cai, L., & Lao, N. (2021). Geographic question answering: Challenges, uniqueness, classification, and future directions. In P. Partsinevelos, P. Kyriakidis, & M. Kavouras (Eds.), *Proc. 24th AGILE Conference on Geographic Information Science*.

Mandelbrot, B. B. (1982). *The fractal geometry of nature*. Freeman.

Mao, H., Hu, Y., Kar, B., Gao, S., & McKenzie, G. (2017). GeoAI 2017 Workshop Report: The 1st ACM SIGSPATIAL International Workshop on GeoAI. *SIGSPATIAL Special*, *9*(3), 25.

Mark, D. M., Smith, B., & Tversky, B. (1999). Ontology and geographic objects: An empirical study of cognitive categorization. In *International Conference on Spatial Information Theory (COSIT)* (pp. 283–298). Springer.

Miller, C. L., & Laflamme, R. A. (1958). The digital terrain model—theory and application. *Photogrammetric Engineering*, *24*(3), 433–442.

Mitasova, H., Mitas, L., & Harmon, R. S. (2005). Simultaneous spline approximation and topographic analysis for LIDAR elevation data in open-source GIS. *IEEE Geoscience and Remote Sensing Letters*, *2*(4), 375–379.

Monmonier, M. (1996). *How to lie with maps* (2nd Edition). University of Chicago Press.

Montello, D. R. (1993). Scale and multiple psychologies of space. In A. U. Frank & I. Campari (Eds.), *Spatial Information Theory: A Theoretical Basis for GIS* (Vol. 716, pp. 312–321). Springer.

Mourelatos, A. P. D. (1978). Events, processes, and states. *Linguistics and Philosophy*, *2*, 415–434.

Munzner, T. (2014). *Visualization analysis and design*. CRC Press.

Nittel, S. (2009). A survey of geosensor networks: Advances in dynamic environmental monitoring. *Sensors*, *9*(7), 5664–5678.

Nittel, S., Stefanidis, A., Cruz, I., Egenhofer, M. J., Goldin, D., Howard, A., Labrinidis, A., Madden, S., Voisard, A., & Worboys, M. (2004). Report from the first workshop on geo sensor networks. *SIGMOD Record*, *33*(1), 141–144.

Norman, D. A. (1988). *The design of everyday things*. Doubleday.

Noy, N. F., & McGuinness, D. L. (2001). *Ontology development 101: A guide to creating your first ontology* (tech. rep. No. KSL-01-05). Stanford Knowledge Systems Laboratory.

Nuhn, E., & Timpf, S. (2022). Landmark weights—an alternative to spatial distances in shortest route algorithms. *Spatial Cognition & Computation*, 1–27.

Null, L., & Lobur, J. (2018). *The essentials of computer organization and architecture* (5th Edition). Jones & Bartlett Learning.

Nyerges, T. L., Couclelis, H., & McMaster, R. (2011). *The SAGE handbook of GIS and society*. SAGE Publications.

O'Rourke, J. (1997). *Computational geometry in C* (2nd Edition). Cambridge University Press.

O'Sullivan, D., & Unwin, D. (2010). *Geographic information analysis* (2nd Edition). John Wiley & Sons.

Ohori, K. A., Ledoux, H., & Stoter, J. (2015). An evaluation and classification of nD topological data structures for the representation of objects in a higher-dimensional GIS. *International Journal of Geographical Information Science*, *29*(5), 825–849.

Ohya, T., Iri, M., & Murota, K. (1984). A fast Voronoi-diagram algorithm with quaternary tree bucketing. *Information Processing Letters*, *18*(4), 227–231.

Orellana, D., & Renso, C. (2010). Developing an interactions ontology for characterising pedestrian movement behaviour. In M. Wachowicz (Ed.), *Movement-aware applications for sustainable mobility: Technologies and approaches* (pp. 62–86). IGI Global.

Oviatt, S. (1999). Mutual disambiguation of recognition errors in a multimodel architecture. In *Proc. SIGCHI conference on Human Factors in Computing Systems* (pp. 576–583).

Oviatt, S., Coulston, R., & Lunsford, R. (2004). When do we interact multimodally? cognitive load and multimodal communication patterns. In *Proc. 6th international conference on Multimodal interfaces* (pp. 129–136).

Parent, C., Spaccapietra, S., Renso, C., Andrienko, G., Andrienko, N., Bogorny, V., Damiani, M. L., Gkoulalas-Divanis, A., Macedo, J., Pelekis, N., et al. (2013). Semantic trajectories modeling and analysis. *ACM Computing Surveys*, *45*(4), 1–32.

Peterson, G. (2020). *GIS cartography: A guide to effective map design*. CRC Press.

Peuquet, D. J. (1984). A conceptual framework and comparison of spatial data models. *Cartographica*, *21*(4), 66–113.

Peuquet, D. J. (1994). It's about time: A conceptual framework for the representation of temporal dynamics in geographic information systems. *Annals of the Association of American Geographers*, *84*(3), 441–461.

Preece, J., Rogers, Y., & Sharp, H. (2015). *Interaction design: Beyond human-computer interaction* (4th Edition). John Wiley & Sons.

Preparata, F. P., & Shamos, M. I. (1985). *Computational geometry: An introduction*. Springer.

Price, R., & Shanks, G. (2005). A semiotic information quality framework: Development and comparative analysis. *Journal of Information Technology, 20*(2), 88–102.

Quinlan, J. (1986). Induction of decision trees. *Machine Learning, 1*, 81–106.

Reid, G., & Sieber, R. E. (2020). Learning from critiques of GIS for assessing the geoweb and Indigenous knowledges. *GeoJournal*, 1–19.

Reid, G., Sieber, R. E., & Blackned, S. (2020). Visions of time in geospatial ontologies from Indigenous peoples: A case study with the Eastern Cree in Northern Quebec. *International Journal of Geographical Information Science, 34*(12), 2335–2360.

Renolen, A. (2000). Modelling the real world: Conceptual modelling in spatiotemporal information system design. *Transactions in GIS, 4*(1), 23–42.

Rey, S. J., & Franklin, R. S. (Eds.). (2022). *Handbook of spatial analysis in the social sciences*. Edward Elgar.

Rigaux, P., Scholl, M. O., & Voisard, A. (2002). *Spatial databases: With application to GIS*. Morgan Kaufmann.

Roberts, J. C., Headleand, C., & Ritsos, P. D. (2015). Sketching designs using the five design-sheet methodology. *IEEE Transactions on Visualization and Computer Graphics, 22*(1), 419–428.

Romero, A., Gatta, C., & Camps-Valls, G. (2016). Unsupervised deep feature extraction for remote sensing image classification. *IEEE Transactions on Geoscience and Remote Sensing, 54*(3), 1349–1362. doi:10.1109/TGRS.2015.2478379

Samet, H. (1990). *Applications of spatial data structures: Computer graphics, image processing, and GIS*. Addison-Wesley.

Samet, H. (2006). *Foundations of multidimensional and metric data structures*. Morgan Kaufmann.

Saoub, K. R. (2021). *Graph theory: An introduction to proofs, algorithms, and applications*. CRC Press.

Schlieder, C. (1995). Reasoning about ordering. In *Proc. International Conference on Spatial Information Theory (COSIT)* (Vol. 988, pp. 341–349).

Schuurman, N. (1999). An interview with Michael Goodchild, January 6, 1998, Santa Barbara, California. *Environment and Planning D: Society and Space, 17*(1), 3–15.

Schuurman, N. (2000). *Critical GIS: Theorizing an emerging science* (Doctoral dissertation, University of British Columbia).

Schuurman, N. (2003). *GIS: A short introduction*. Blackwell.

Schuurman, N. (2009). An interview with Michael Goodchild: GIScience and social reordering in the new millennium. *The Information Society, 25*(5), 360–363.

Schuurman, N., & Pratt, G. (2002). Care of the subject: Feminism and critiques of GIS. *Gender, Place, and Culture, 9*(3), 291–299.

Siabato, W., Claramunt, C., Ilarri, S., & Manso-Callejo, M. Á. (2018). A survey of modeling trends in temporal GIS. *ACM Computing Surveys, 51*(2), 1–41.

Sieber, R. E., Robinson, P. J., Johnson, P. A., & Corbett, J. M. (2016). Doing public participation on the geospatial web. *Annals of the American Association of Geographers, 106*(5), 1030–1046.

Skupin, A., & Fabrikant, S. I. (2003). Spatialization methods: A cartographic research agenda for non-geographic information visualization. *Cartography and Geographic Information Science, 30*(2), 99–119.

Smith, B. (1995). On drawing lines on a map. In A. U. Frank & W. Kuhn (Eds.), *Spatial Information Theory: A Theoretical Basis for GIS* (Vol. 988), Springer.

Smith, B., & Varzi, A. C. (2000). Fiat and bona fide boundaries. *Philosophy and Phenomenological Research, 60*(2), 401–420.

Smith, T., Peuquet, D. J., Menon, S., & Agarwal, P. (1987). KBGIS-II: A knowledge-based geographical information system. *International Journal of Geographical Information Systems, 1*(2), 149–172.

Smithson, M. (2012). *Ignorance and uncertainty: Emerging paradigms.* Springer New York.

Snipp, C. M. (2016). What does data sovereignty imply: What does it look like? In T. Kukutai & J. Taylor (Eds.), *Indigenous data sovereignty: Toward an agenda* (pp. 39–56). ANU Press.

Sobel, D. (1995). *Longitude: The true story of a lone genius who solved the greatest scientific problem of his time.* Bloomsbury.

Sperber, D., & Wilson, D. (1995). *Relevance: Communication and cognition* (2nd Edition). Blackwell.

Stevens, S. S. (1946). On the theory of scales and measurement. *Science, 103*, 677–680.

Sullivan, D. (2015). *NoSQL for mere mortals.* Addison-Wesley Professional.

Sun, Q., Macleod, T., Both, A., Hurley, J., Butt, A., & Amati, M. (2021). A human-centred assessment framework to prioritise heat mitigation efforts for active travel at city scale. *Science of The Total Environment, 763*, 143033.

Szewczyk, R., Osterweil, E., Polastre, J., Hamilton, M., Mainwaring, A., & Estrin, D. (2004). Habitat monitoring with sensor networks. *Communications of the ACM, 47*(6), 34–40.

Tanenbaum, A. S. (2013). *Structured computer organization* (6th Edition). Pearson.

Tarjan, R. E., & Wyk, C. J. V. (1988). An $O(n \log \log n)$-time algorithm for triangulating a simple polygon. *SIAM Journal of Computing, 17*, 143–178.

Tauli-Corpuz, V. (2016). Preface. In T. Kukutai & J. Taylor (Eds.), *Indigenous data sovereignty: Toward an agenda* (pp. xxi–xxiii). ANU Press.

Timpf, S. (2002). Ontologies of wayfinding: A traveler's perspective. *Networks and Spatial Economics, 2*(1), 9–33.

Tobler, W. (1970). A computer movie simulating urban growth in the Detroit region. *Economic Geography*, *46*(2), 234–240.

Tobler, W. (1993). *Non-isotropic geographic modeling* (tech. rep. No. 93-1). National Center for Geographic Information and Analysis. In Three Presentations on Geographical Analysis and Modeling.

Tomlin, C. D. (1983). A map algebra. In *Proc. Harvard Computer Graphics Conference*.

Tripathi, N., & Bhattarya, S. (2004). Integrating indigenous knowledge and GIS for participatory natural resource management: State-of-the-practice. *The Electronic Journal of Information Systems in Developing Countries*, *17*(1), 1–13.

Tufte, E. R. (1990). *Envisioning information*. Graphics Press.

Tufte, E. R. (2001). *The visual display of quantitative information* (2nd Edition). Graphics Press.

Turing, A. M. (1950). Computing Machinery and Intelligence. *Mind*, *59*(236), 433–460.

Tversky, B., Morrison, J. B., & Betrancourt, M. (2002). Animation: Can it facilitate? *International Journal of Human-Computer Studies*, *57*(4), 247–262.

Van de Weghe, N., Cohn, A., Tré, G., & De Maeyer, P. (2006). A qualitative trajectory calculus as a basis for representing moving objects in geographical information systems. *Control and Cybernetics*, *35*, 97–119.

Van Kreveld, M., & Speckmann, B. (2007). On rectangular cartograms. *Computational Geometry*, *37*(3), 175–187.

Veland, S., Lynch, A., Bischoff-Mattson, Z., Joachim, L., & Johnson, N. (2014). All strings attached: Negotiating relationships of geographic information science. *Geographical Research*, *52*(3), 296–308.

Vilain, M. B., & Kautz, H. A. (1986). Constraint propagation algorithms for temporal reasoning. In *Proc. AAAI* (pp. 377–382).

Wang, J., Duckham, M., & Worboys, M. (2016). A framework for models of movement in geographic space. *International Journal of Geographical Information Science*, *30*(5), 970–992.

Want, R., Hopper, A., Falcao, V., & Gibbons, J. (1992). The Active Badge location system. *ACM Transactions on Information Systems*, *10*(1), 91–102.

Ward, M. O., Grinstein, G., & Keim, D. (2015). *Interactive data visualization: Foundations, techniques, and applications* (2nd Edition). John Wiley & Sons.

Westin, A. F. (1967). *Privacy and freedom*. Atheneum.

Wilensky, U., & Rand, W. (2015). *An introduction to agent-based modeling: Modeling natural, social, and engineered complex systems with NetLogo*. MIT Press.

Wilkinson, M. D., Dumontier, M., Aalbersberg, I. J., Appleton, G., Axton, M., Baak, A., Blomberg, N., Boiten, J.-W., da Silva Santos, L. B., Bourne, P. E., et al. (2016). The FAIR guiding principles for scientific data management and stewardship. *Scientific Data*, *3*(1), 1–9.

Wilson, M. (2016). *Critical GIS* (N. Clifford, M. Cope, T. Gillespie, & S. French, Eds.). SAGE.

Wolter, D., & Lee, J. H. (2010). Qualitative reasoning with directional relations. *Artificial Intelligence, 174*(18), 1498–1507.

Wood, J. D. (1996). *The geomorphological characterisation of digital elevation models* (Doctoral dissertation, University of Leicester).

Wood, Z., & Galton, A. (2009). A taxonomy of collective phenomena. *Applied Ontology, 4*(3–4), 267–292.

Worboys, M. F., & Clementini, E. (2001). Integration of imperfect spatial information. *Journal of Visual Languages & Computing, 12*(1), 61–80.

Yuan, M. (1999). Use of a three-domain representation to enhance GIS support for complex spatiotemporal queries. *Transactions in GIS, 3*(2), 137–159.

Zhang, T. Y., & Suen, C. Y. (1984). A fast parallel algorithm for thinning digital patterns. *Communications of the ACM, 27*(3), 236–239.

Zhang, X., Song, W., & Liu, L. (2014). An implementation approach to store GIS spatial data on NoSQL database. In *Proc. 22nd International Conference on Geoinformatics* (pp. 1–5).

Zhao, F., & Guibas, L. (2004). *Wireless sensor networks: An information processing approach.* Elsevier/Morgan-Kaufmann.

Zimmermann, K., & Freksa, C. (1996). Qualitative spatial reasoning using orientation, distance, and path knowledge. *Applied Intelligence, 6*(1), 49–58.

Zlatanova, S., & Gruber, M. (1998). 3D urban GIS on the Web: Data structuring and visualization. In *ISPRS Archives* (Vol. 32, pp. 691–699).

Zlatanova, S., Rahman, A. A., & Shi, W. (2004). Topological models and frameworks for 3D spatial objects. *Computers & Geosciences, 30*(4), 419–428.

Zubin, D. (1989). Natural language understanding and reference frames. In D. Mark, A. Frank, M. J. Egenhofer, S. Freundschuh, M. McGranaghan, & R. M. White (Eds.), *Languages of Spatial Relations: Initiative 2 Specialist Meeting Report* (pp. 13–16). National Center for Geographic Information and Analysis.

Index